Direct and Inverse Problems of Mathematical Physics

International Society for Analysis, Applications and Computation

Volume 5

The titles published in this series are listed at the end of this volume.

Direct and Inverse Problems of Mathematical Physics

Edited by

Robert P. Gilbert
University of Delaware

Joji Kajiwara
Kyushu University

and

Yongzhi S. Xu
University of Tennessee at Chattanooga

KLUWER ACADEMIC PUBLISHERS
DORDRECHT / BOSTON / LONDON

A C.I.P. Catalogue record for this book is available from the Library of Congress.

ISBN 978-1-4419-4818-2

Published by Kluwer Academic Publishers,
P.O. Box 17, 3300 AA Dordrecht, The Netherlands.

Sold and distributed in North, Central and South America
by Kluwer Academic Publishers,
101 Philip Drive, Norwell, MA 02061, U.S.A.

In all other countries, sold and distributed
by Kluwer Academic Publishers,
P.O. Box 322, 3300 AH Dordrecht, The Netherlands.

Printed on acid-free paper

CONTENTS

PREFACE

This volume consists of papers presented in the special sessions on "Wave Phenomena and Related Topics", and "Asymptotics and Homogenization" of the ISAAC'97 Congress held at the University of Delaware, during June 2-7, 1997. The ISAAC Congress coincided with a U.S.-Japan Seminar also held at the University of Delaware. The latter was supported by the National Science Foundation through Grant INT-9603029 and the Japan Society for the Promotion of Science through Grant MTCS-134.

It was natural that the participants of both meetings should interact and consequently several persons attending the Congress also presented papers in the Seminar. The success of the ISAAC Congress and the U.S.-Japan Seminar has led to the ISAAC'99 Congress being held in Fukuoka, Japan during August 1999. Many of the same participants will return to this Seminar. Indeed, it appears that the spirit of the U.S.-Japan Seminar will be continued every second year as part of the ISAAC Congresses. We decided to include with the papers presented in the ISAAC Congress and the U.S.-Japan Seminar several very good papers by colleagues from the former Soviet Union. These participants in the ISAAC Congress attended at their own expense.

This volume has the title **Direct and Inverse Problems of Mathematical Physics** which consists of the papers on scattering theory, coefficient identification, uniqueness and existence theorems, boundary controllability, wave propagation in stratified media, viscous flows, nonlinear acoustics, Sobolev spaces, singularity theory, pseudo differential operators, and semigroup theory.

We would like to thank the National Science Foundation and the Japanese Science Foundation who so generously supported our seminar. We would like to thank Ms. Pamela Irwin and Ginger Moore who helped in the organization of the Conference. Professors Wenbo Li, Rakesh, and Shangyou Zhang served on the organization committee. The following graduate students Min Fang, Zhongshan Lin, Nilima Nigam, Yvonne Ou, Alexander Panchenko and his wife Elena who helped at the registration desk. Finally, most of all we want to thank Pamela Irwin for her tireless effort with the preparation and formatting of the manuscripts. Without this help these proceedings would not have made it to publication.

ALGORITHMS OF THE ASYMPTOTIC NONLINEAR ANALYSIS

Alexander D. Bruno*

Department of Mathematics
Keldysh Institute of Applied Mathematics
Miusskaja sq. 4
Moscow 125047, Russia
e-mail: bruno@applmat.msk.su

Abstract:
All local and asymptotic first approximations of a polynomial, of a differential polynomial and of a system of such polynomials may be selected algorithmically. Here the first approximation of a solution of the system of equations is a solution of the corresponding first approximation of the system of equations. The power transformations induce linear transformations of vector exponents and commute with the operation of selecting first approximations. In a first approximation of a system of equations they allow to reduce number of parameters and to reduce the presence of some variables to the form of derivatives of their logarithms. If the first approximation is the linear system, then in many cases the system of equations can be transformed into the normal form by means of the formal change of coordinates. The normal form is reduced to the problem of smaller dimension by means of the power transformation. Combining these algorithms, in many problems we can resolve a singularity, find parameters determining properties of solutions and obtain the asymptotic ex-

*The work was partly supported by RFBR, Grant 96-01-01411.

R.P. Gilbert et al.(eds.), Direct and Inverse Problems of Mathematical Physics, 1–20.

pansions of solutions. Some applications from Mechanics, Celestial Mechanics and Hydrodynamics are indicated.

1.1 INTRODUCTION

1. Here we propose 4 algorithms and a list of their applications. Below they are enumerated with numbers of their Sections.

2. All local (or asymptotic) *first approximations* of a polynomial, of a differential polynomial and of a system of such polynomials may be selected algorithmically [1-6,10]. It allows to find all first approximations of a system of equations (algebraic and ordinary differential and partial differential). Here the first approximation of a solution of the system of equations is a solution of the corresponding first approximation of the system of equations. The algorithm consists of computations for each equation: of the set D of the vector exponents $Q_j \in \mathbb{R}^n$, of faces $\Gamma_k^{(d)}$ of the polyhedron M spanning D, of boundary subsets $D_k^{(d)} = \Gamma_k^{(d)} \cap D$ in $\mathbb{R}^n$ and of the normal cones $U_k^{(d)}$ of these faces $\Gamma_k^{(d)}$ in the dual space $\mathbb{R}_*^n$. The corresponding computer program see in [7].

3. The *power transformations* [1,2,10,13] induce linear transformations of vector exponents Q and commute with the operation of selecting first approximations. In a first approximation of a system of equations they allow to reduce number of parameters and to reduce the presence of some variables to the form of derivatives of their logarithms. Here the transformed full system is a regular perturbation of the transformed first approximation [14]. The algorithm consists of computation of a constant matrix of that linear transformation with prescribed properties [15].

4. If in a first approximation of a differential equation a variable presents in the form of its logarithm only, we consider the logarithm as a new variable instead of the initial one. Such *logarithmic transformation* allows to find asymptotics of solutions including logarithms.

5. If the first approximation is the linear system, then in many cases the system of equations can be transformed into the *normal form* by means of the formal change of coordinates. The normal form is reduced to the problem of less dimension by means of the power transformation. The algorithms of computation of the normal form are especially developed for ODE systems resolved with respect to derivatives [1,16,17,19].

6. Combining these algorithms, in many problems we can resolve a singularity, find parameters determining properties of solutions and obtain the asymptotic expansions of solutions. A *list of applications* is considered here.

1.2 FIRST APPROXIMATIONS

1. Polyhedra. Let in the space $\mathbb{R}^n$ with Cartesian coordinates $q_1, \ldots, q_n$ we have a finite set D of points $Q_j = (q_{1j}, \ldots, q_{nj})$, $j = 1, \ldots, m$. Let $\mathbb{R}^n_*$ denote the dual space such that for $P = (p_1, \ldots, p_n) \in \mathbb{R}^n_*$ and $Q = (q_1, \ldots, q_n) \in \mathbb{R}^n$ there is the scalar product $\langle P, Q\rangle = p_1q_1 + \ldots + p_nq_n$. For a fixed vector $P \neq 0$ let D_P denote such subset of the set D on which the scalar product $\langle P, Q\rangle$ has the maximal value c, that is

$$\langle P, Q_j\rangle = c \text{ for } Q_j \in D_P \text{ and } \langle P, Q_j\rangle < c \text{ for } Q_j \in D \setminus D_P. \qquad (2.1)$$

For the given set D and for each $P \neq 0$, we can find the corresponding subset D_P as follows. Let M denote the convex hull of the set D. The boundary ∂M of the *polyhedron* M consists of *faces* $\Gamma_k^{(d)}$ of different dimension d $(0 \leq d < n)$. Zerodimensional faces $\Gamma_k^{(0)}$ are vertices of the polyhedron M, onedimensional faces $\Gamma_k^{(1)}$ are its edges etc. According to (2.1) to each vector $P \in \mathbb{R}^n_*$ there corresponds in $\mathbb{R}^n$ a *supporting* to the set D *hyperplane* $L_P = \{Q : \langle P, Q\rangle = c\}$. It intersects the polyhedron M along some face $\Gamma_k^{(d)}$. The set $U_k^{(d)}$ of all vectors $P \in \mathbb{R}^n_*$ such, that $L_P \cap M = \Gamma_k^{(d)}$, is called the *normal cone* of the face $\Gamma_k^{(d)}$. If $\dim M = n$, the normal cone $U_k^{(n-1)}$ of the hyperface $\Gamma_k^{(n-1)}$ is a ray orthogonal to the face and directed out of the polyhedron M. The normal cone $U_k^{(n-2)}$ is a sector bounded by rays $U_l^{(n-1)}$ and $U_m^{(n-1)}$ if $\Gamma_k^{(n-2)} = \Gamma_l^{(n-1)} \cap \Gamma_m^{(n-1)}$, etc. The union of normal cones of all faces is $\mathbb{R}^n_* \setminus \{0\}$. Let us consider the *boundary subsets* $D_k^{(d)} = \Gamma_k^{(d)} \cap D$ of the set D.

Theorem 1.2.1 *If* $P \in U_k^{(d)}$, *then* $D_P = D_k^{(d)}$.

EXAMPLE 1. For $n = 2$ and

$$D = \{(3,0), (0,3), (1,1)\} \qquad (2.2)$$

the polygon M, faces $\Gamma_k^{(d)}$ and their normal cones are shown in Figure 1.

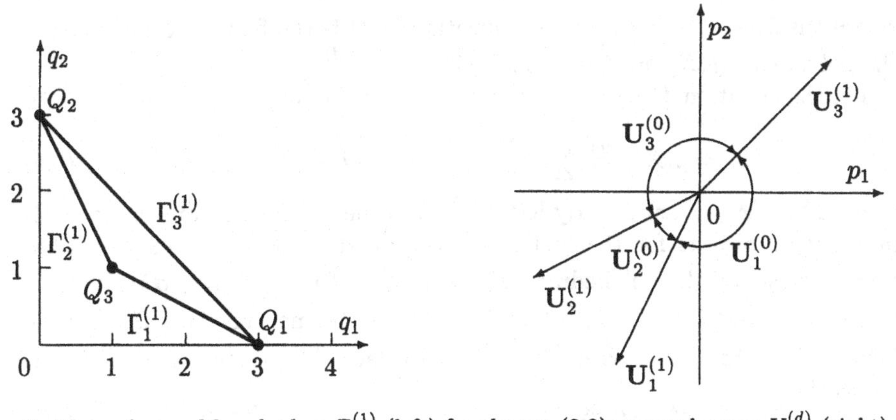

Fig. 1. The polygon M and edges $\Gamma_k^{(1)}$ (left) for the set (2.2); normal cones $U_k^{(d)}$ (right).

$$\begin{array}{llll} \text{Here } D_1^{(0)} = Q_1, & D_2^{(0)} = Q_3, & D_3^{(0)} = Q_2; \\ D_1^{(1)} = \{Q_1, Q_3\}, & D_2^{(1)} = \{Q_2, Q_3\}, & D_3^{(1)} = \{Q_1, Q_2\}; \\ U_1^{(1)} = -\lambda(1,2), & U_2^{(1)} = -\lambda(2,1), & U_3^{(1)} = -\lambda(1,1), \ \lambda > 0. \end{array}$$

If the set D contains infinitely many points Q_j, then M is the closure of the convex hull of the set D. The boundary ∂M consists of faces $\Gamma_k^{(d)}$. Each face $\Gamma_k^{(d)}$ has its boundary subset $D_k^{(d)} = \Gamma_k^{(d)} \cap D \subset \mathbb{R}^n$ and its normal cone $U_k^{(d)} \subset \mathbb{R}_*^n$ etc.

2. Algebraic equations. Let us consider a polynomial

$$f(X) \stackrel{\text{def}}{=} \sum f_Q X^Q \quad \text{for} \quad Q \in D, \tag{2.3}$$

where $X^Q = x_1^{q_1} \ldots x_n^{q_n}$, coefficients $f_Q \in \mathbb{C} \setminus 0$ and D is a set in $\mathbb{R}^n$. The set D is called the *support* of the polynomial. The convex hull M of the set D is called the *Newton polyhedron* of the polynomial f, and we can construct all accompanying objects as described above. Let $D_k^{(d)}$ be a boundary subset of the set D. The sum

$$\hat{f}_k^{(d)}(X) \stackrel{\text{def}}{=} \sum f_Q X^Q \quad \text{for} \quad Q \in D_k^{(d)} \tag{2.4}$$

is called the *truncation* of the polynomial $f(X)$ [1, 2]. It is a quasihomogeneous function with respect to the vector P. Now we consider a curve of the form

$$x_i = b_i \tau^{p_i}, \ \ b_i \neq 0, \ \ i = 1, \ldots, n, \tag{2.5}$$

where $\tau \to \infty$. On such a curve a monomial

$$X^Q = B^Q \tau^{\langle P, Q \rangle},$$

where $B = (b_1, \ldots, b_n)$ and $P = (p_1, \ldots, p_n)$. If $P \in U_k^{(d)}$, then on the curve (2.5)

$$f(X) = \hat{f}_k^{(d)}(B) \tau^c + \tau^c o(1),$$

where c is from (2.1). Thus, the truncation (2.4) is the first approximation of the polynomial $f(X)$ on curves (2.5) with $P \in U_k^{(d)}$.

Let the equation $f(X) = 0$ have a solution of the form

$$x_n = g(X') \stackrel{\text{def}}{=} \sum g_{Q'} X'^{Q'} \quad \text{for} \quad Q' \in D' \subset \mathbb{R}^{n-1}, \tag{2.6}$$

where $X' = (x_1, \ldots, x_{n-1})$. By letters with prime we shall denote the $(n-1)$-dimensional objects for the first $n-1$ coordinates. To the expansion (2.6) there correspond the polyhedron M', faces $\Gamma_l'^{(\delta)}$ of its boundary $\partial M'$ and the boundary subsets $D_l'^{(\delta)} = \Gamma_l'^{(\delta)} \cap D'$ in $\mathbb{R}^{n-1}$ and the normal cones $U_l'^{(\delta)}$ of the faces $\Gamma_l'^{(\delta)}$ in the dual space $\mathbb{R}_*^{n-1}$. To each face $\Gamma_l'^{(\delta)}$ there corresponds the truncation

$$\hat{g}_l^{(\delta)}(X') \stackrel{\text{def}}{=} \sum g_{Q'} X'^{Q'} \quad \text{for} \quad Q' \in D_l'^{(\delta)}. \tag{2.7}$$

It is the first approximation of the function $g(X')$ on curves

$$x_i = b_i \tau^{p_i}, \quad b_i \neq 0, \quad i = 1, \dots, n-1, \tag{2.8}$$

where $P' = (p_1, \dots, p_{n-1}) \in U'^{(\delta)}_l$. Moreover, on these curves according to (2.6) and (2.7)

$$x_n = \hat{g}_l^{(\delta)}(B')\tau^r + o(\tau^{r-\varepsilon}),$$

where $\varepsilon > 0$ and

$$r = \langle P', Q' \rangle \quad \text{for} \quad Q' \in \Gamma_l^{(\delta)}.$$

Hence, on curves (2.8) the solution (2.6) has the form $x_n = b_n \tau^r + o(\tau^{r-\varepsilon})$. Now let us select from the sum (2.3) its truncation (2.4) corresponding to the vector $P = (P', r)$, i.e. $p_n = r$ and $P \in U_k^{(d)}$.

Theorem 1.2.2 (3) . *Let $x_n = g(X')$ be a solution of the equation $f(X) = 0$, i.e. $f(X', g(X')) \equiv 0$. Let $\hat{g}_{P'}(X')$ be the truncation of the function $g(X')$ corresponding to the vector P' and $\hat{f}_P(X)$ be the truncation of the sum (2.4) corresponding to the vector $P = (P', r)$. Then $\hat{f}_P(X', \hat{g}_{P'}(X')) \equiv 0$, i.e. the first approximation of a solution is a solution of the corresponding first approximation of the equation.*

If we have found the first approximation (2.7) of the solution (2.6), then we can make the substitution $x_n = \hat{g}_l^{(\delta)}(X') + \tilde{x}_n$ into the equation $f(X) = 0$ and look for the second approximation for x_n as the first approximation for $\tilde{x}_n$ and so on.

EXAMPLE 2 (continuation of Example 1). Let $f = x_1^3 + x_2^3 - 3x_1x_2$, then the set D is (2.2) and truncations $\hat{f}_k^{(d)}$ are $\hat{f}_1^{(0)} = x_1^3$, $\hat{f}_2^{(0)} = -3x_1x_2$, $\hat{f}_3^{(0)} = x_2^3$, $\hat{f}_1^{(1)} = x_1^3 - 3x_1x_2$, $\hat{f}_2^{(1)} = x_2^3 - 3x_1x_2$ and $\hat{f}_3^{(1)} = x_1^3 + x_2^3$. We want to study solutions of the equation $f = 0$ for $x_1, x_2 \to 0$. Here $n = 2$ and (2.6) is an expansion of x_2 in powers of x_1. Its first approximation (2.7) is $x_2 = g_q x_1^q$ with $g_q \neq 0$ and $P = \text{const}\,(1, q)$. If (2.6) is a solution of the equation $f = 0$ with $P \neq 0$, then P cannot lie in $U_k^{(0)}$. Indeed if $P \in U_k^{(0)}$ then $\hat{f}_P = f_Q X^Q$ and $\hat{f}_P(x_1, g_q x_1^q) = f_Q g_q^{q_2} x_1^{q_1 + q q_2} \equiv 0$; but it is impossible if $g_q \neq 0$. If on the solution (2.6) $x_1, x_2 \to 0$ then $P < 0$, so $P \in U_1^{(1)}$ or $U_2^{(1)}$. If $P \in U_1^{(1)}$, i.e. $2p_1 = p_2 < 0$, then the first approximation $x_2 = b_2 x_1^2$ satisfies the truncated equation $x_1^3 - 3x_1x_2 = 0$, i.e. $x_2 = x_1^2/3$. Similarly, if $P \in U_2^{(1)}$, then the first approximation of the solution (2.6) is $x_1 = x_2^2/3$. Thus, two branches of the curve $f = 0$ pass through the origin $x_1 = x_2 = 0$ (see Figure 2). We can compute their asymptotic expansions of the form

$$x_2 = \sum_{m=1}^{\infty} g_m x_1^{r_m}, \quad \text{where} \quad r_{m-1} < r_m.$$

The algebraic curve $f = 0$ is called *folium of Descartes* and for $|x_1| + |x_2| \to \infty$ it has the asymptote $x_1 + x_2 = -1$, the first approximation of which $x_1 + x_2 = 0$ can be found from the truncated equation $\hat{f}_3^{(1)} = 0$.

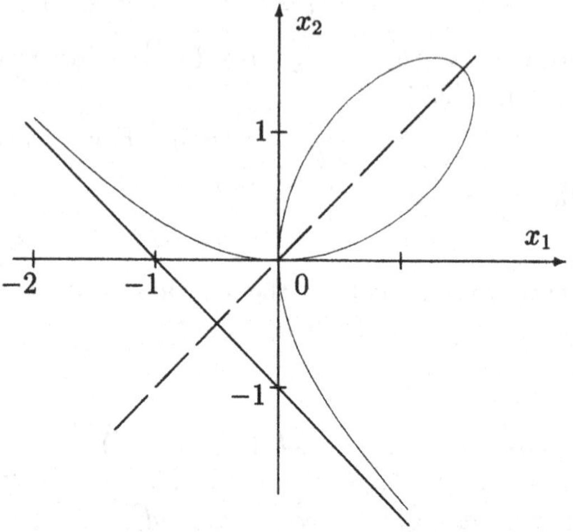

Fig. 2. Folium of Descartes.

3. Differential equations. The considered construction can be transferred to ordinary differential equations in the following manner. We define a *differential monomial* $a(X)$ as a product of powers of coordinates X and derivatives $d^m x_n / dx_{n-1}^m$, where integer $m > 0$. To the monomial $a(X)$ we put in correspondence the point $Q \in \mathbb{R}^n$: to the product const X^Q there corresponds the point Q, to the derivative $d^m x_n / dx_{n-1}^m$ there corresponds $Q = (0, \ldots, 0, -m, 1)$, and to the product of monomials there corresponds the sum of their vectors Q. To a *differential polynomial* $f(X)$, that is a sum of differential monomials, we put in correspondence in $\mathbb{R}^n$ the set $D(f)$ of points Q of its monomials. We call the set $D(f)$ as the *support* of the differential polynomial f. Here one point Q can correspond to several different monomials. For the set D, we construct as above the polyhedron M, its faces $\Gamma_k^{(d)}$, boundary subsets $D_k^{(d)}$ in $\mathbb{R}^n$ and their normal cones $U_k^{(d)}$ in the dual space $\mathbb{R}_*^n$. Theorem 2 holds for a solution (2.6) of the equation $f(X) = 0$ [5].

EXAMPLE 3. For the Emden–Fowler equation

$$f(x,y) \stackrel{\text{def}}{=} (x^2 y')' + x^\alpha y^\beta = 0, \quad \alpha, \beta \in \mathbb{R}, \quad \alpha \neq 0,\ \beta \neq 1, \ (\ ' \stackrel{\text{def}}{=} d/dx) \qquad (2.9)$$

we will study asymptotics of its solutions when $x \to 0$ or ∞. Its support D consists of two points $Q_1 = (0,1)$ and $Q_2 = (\alpha, \beta)$. So the polyhedron M is the interval $[Q_1, Q_2]$ and there are 3 boundary subsets: $D_1^{(0)} = \{Q_1\}$, $D_2^{(0)} = \{Q_2\}$ and $D_1^{(1)} = \{Q_1, Q_2\}$. Denote $R \stackrel{\text{def}}{=} Q_2 - Q_1 = (\alpha, \beta - 1)$, then $U_1^{(0)} = \{P : \langle P, R \rangle < 0\} = \{p_1\alpha + p_2(\beta - 1) < 0\}$, $U_2^{(0)} = \{P : \langle P, R \rangle > 0\}$ and $U_1^{(1)} = \{P : \langle P, R \rangle = 0\}$. The first truncated equation $\hat{f}_1^{(0)} \stackrel{\text{def}}{=} (x^2 y')' = 0$ has solutions

$$y = c_1 x^{-1} + c_0, \ c_i = \text{ const} . \qquad (2.10)$$

Their vector $P = \sigma(1,\nu)$, where $\nu = -1, 0$ and $\sigma = -1$, if $x \to 0$, and $\sigma = 1$, if $x \to \infty$. Formula (2.10) gives asymptotics of solutions of Equation (2.9) only if the corresponding $P = \sigma(1,\nu) \in U_1^{(0)}$. The second truncated equation $\hat{f}_2^{(0)} \stackrel{\text{def}}{=} x^\alpha y^\beta = 0$ has only trivial solutions $x = 0$ and $y = 0$ and is not interesting. The third truncated equation $\hat{f}_1^{(1)} = 0$ is Equation (2.9) itself. Its solutions with $P \in U_1^{(1)}$ have the form

$$y = bx^{-\alpha/(\beta-1)}, \tag{2.11}$$

where constant $b \neq 0$ satisfies the equation $\kappa b + b^\beta = 0$ with $\kappa \stackrel{\text{def}}{=} \alpha(\alpha - \beta + 1)(\beta - 1)^{-2}$. The equation for b has non zero solutions $b \in \mathbb{C}$ if $\kappa \neq 0$ only, but if $\kappa = 0$, i.e.

$$\alpha - \beta + 1 = 0, \tag{2.12}$$

the equation is trivial. Here (2.11) is an exact solution of Equation (2.9) and may be an asymptotics for some other solutions.

Analogously this approach works for a partial differential equation. Now there is a set of independent variables $(x_{l+1}, \dots, x_{n-1}) = X_2$ and derivatives have the form $\partial^{K_2} x_n / \partial X_2^{K_2}$, where integer $K_2 = (k_{l+1}, \dots, k_{n-1}) \geq 0$. To that derivative there corresponds the vector $Q = (0, \dots, 0, -K_2, 1) = (0, \dots, 0, -k_{l+1}, \dots, -k_{n-1}, 1)$. The remaining part of the construction is the same as above and Theorem 2 holds true again [5,6]. See an example in [4, 49].

4. Generalization. Now we want to generalize Theorem 2 for solutions $x_n = g(X')$ of more general nature than (2.6). A function $h(X')$ is called *pseudohomogeneous with respect to the vector* P', if along the curves (2.8) we have $h(X') = \phi(\tau, B')\tau^r$, where ϕ is a Laurent polynomial of $\log \tau$, whose coefficients are functions of B', and $r = \text{const}$. The function $\hat{g}_{P'}(X')$ is a truncation of the function $g(X')$ with respect to the vector P', if $\hat{g}_{P'}$ is pseudohomogeneous with respect to P' and along the curves (2.8) $g = \hat{g}_{P'} + o(\tau^{r-\varepsilon})$ with $\varepsilon > 0$. For such a definition of the asymptotical first approximation of a solution, Theorem 2 holds for both algebraic and differential equations. Moreover, Theorem 2 holds if coefficients of the algebraic or differential polynomial $f(X)$ are functions of X' which are pseudohomogeneous with respect to the vector P'. For a fixed vector P' it allows us to expand a solution $x_n = g(X')$ into an asymptotic series

$$g = \hat{g}_{P'} + \sum_{k=1}^{\infty} h_k$$

where h_k are pseudohomogeneous functions and along curves (2.8) $r(h_k) > r(h_{k+1})$.

EXAMPLE 4. Let the polynomial $f(x,y,z) \stackrel{\text{def}}{=} z^2 - x^2 - y^2 + \tilde{f}$, where $\tilde{f}$ contains terms with order greater than 2. We consider the equation $f = 0$ near the origin $x = y = z = 0$. The truncation $\hat{f}_P$ for $P = -(1,1,1)$ is $\hat{f}_P = z^2 - x^2 - y^2$. The truncated equation $\hat{f}_P = 0$ has two solutions $z = \omega \stackrel{\text{def}}{=} \pm\sqrt{x^2 + y^2}$. After the substitution $z = \omega + z_1$ we obtain $f(x, y, \omega + z_1) = 2\omega z_1 + h_3(x, y, \omega) +$

... $\stackrel{\text{def}}{=} f_1(x, y, z_1)$, where h_3 is a homogeneous polynomial on x, y, ω of order 3. Now the truncation $\hat{f}_1 = 0$ of the equation $f_1 = 0$, corresponding to the vector $P = -(1, 1, 3/2)$, is $\hat{f}_1 \stackrel{\text{def}}{=} 2\omega z_1 + h_3 = 0$. Its solution $z_1 = -h_3/2\omega$ gives the second term of the asymptotics. By these computations we can obtain the expansion

$$z = \omega + \sum_{i=2}^{\infty} g_{k_i}(x, y, \omega)/\omega^{l_i},$$

of the solution $z = g(x, y)$ of the equation $f = 0$. Here $g_m(x, y, \omega)$ are homogeneous polynomials of order m and $k_i - l_i \geq i$, $k_i - l_i \geq k_{i-1} - l_{i-1}$.

5. Algorithm. Thus to apply this approach we must have an algorithm giving all truncations of a polynomial. According to the beginning of Section 2 the algorithm must give the boundary subsets $D_k^{(d)}$ and their normal cones $U_k^{(d)}$ for any set $D \in \mathbb{R}^n$. Only for $n \leq 3$ it can be done by drawing some pictures. For $n > 3$ it requires some computations. Corresponding algorithm was developed in [2] and was written as a computer program in [7].

The Newton polygon, i.e. the polyhedron M for $n = 2$, was introduced by Puiseux [8]; Newton himself found one its edge only [9]. The polyhedron M (for any n) was introduced in [10] for a system of ordinary differential equations. Name "Newton polyhedron" was given by Gindikin [11]. See its history in [1,2]. Its other applications see in [1] and [12].

3. POWER TRANSFORMATIONS

Let n-vector X be divided into three parts: parameters X_1, independent variables X_2 and dependent variables X_3 with dimensions n_1, n_2 and n_3 respectively: $X = (X_1, X_2, X_3)$, $n = n_1 + n_2 + n_3$, $n_i \geq 0$. We consider the *power transformation*

$$\begin{aligned} \log X_1 &= W_{11} \log Y_1, \\ \log X_2 &= W_{21} \log Y_1 + W_{22} \log Y_2, \\ \log X_3 &= W_{31} \log Y_1 + W_{32} \log Y_2 + W_{33} \log Y_3. \end{aligned} \tag{3.1}$$

Here $\log X_1 = (\log x_1, \ldots, \log x_{n_1})^*$, W_{ii} are nondegenerate square matrices, W_{ij} are rectangle real matrices, $W = (W_{ij})$ is a square block matrix, the star denotes transposition.

Let the coordinate change (3.1) transform a differential polynomial $f(X)$ into $g(Y) = f(X)$. We want to study the relationship of their supports $D(f) = \{Q_j\}$ and $D(g) = \{S_j\}$.

Theorem 1.2.3 (13) . *Under the power transformation (3.1), supports $D(g) = \{S_j\}$ and $D(f) = \{Q_j\}$ are connected by the linear transformation*

$$S = W^* Q \tag{3.2}$$

and vectors of the dual space $\mathbb{R}_*^n$ are transformed as

$$R = W^{-1} P. \tag{3.3}$$

Here all sets in $\mathbb{R}^n$ and in $\mathbb{R}^n_*$ are also changed by the corresponding linear transformation (3.2) and (3.3). Hence the selection of truncations commutes with any power transformation. Moreover, the power transformation can be used to simplify the truncated equation by making its support parallel to a coordinate subspace.

Theorem 1.2.4 (13) . *Let $d = \dim M(f) < n$. There exists such a matrix W that, after the transformation (3.1), values of $n-d$ coordinates s_j are constants for all $S = (s_1, \ldots, s_n) \in D(g)$, where $g(Y) = f(X)$. Let $s_j =$ const. In $y_j^{-s_j} g(Y)$ the coordinate y_j is absent, if $j \le n_1$, and is present only in the form $\log y_j$, if $j > n_1$.*

Note that the multiplication of the polynomial $g(Y)$ by the factor Y^T induces the parallel translation of the set $D(g)$ by the vector T: $D(gY^T) = D(g) + T$. Thus, if $\dim M(f) < n$ then by the power transformation (3.1) and a multiplication by Y^T we can put the support of the polynomial f into d-dimensional coordinate subspace. It allows to reduce the dimension of the truncated problem. The initial equation is the regular perturbation of its truncation in the corresponding domain of the X-space, where the truncation is the first approximation. This property is preserved after any power transformation. So the domain can be made a vicinity of a coordinate subspace in the Y-space.

Theorem 1.2.5 (13,14) . *Let $\hat{f}_k^{(d)}(X)$ be a truncation of a differential polynomial $f(X)$. There exist such a matrix W and a vector T that, after the transformation* (3.1), *in $Y^T \hat{g}_k^{(d)}(Y)$ some $n-d$ coordinates y_j are either absent or in the form $\partial \log y_j$, where $\hat{g}_k^{(d)}(Y) = \hat{f}_k^{(d)}(X)$. Moreover, $D(Y^T g) \subset \mathbb{R}^n_+$, where $g(Y) = f(X)$ and $\mathbb{R}^n_+ = \{S \ge 0\}$.*

EXAMPLE 5 (continuation of Example 2). We make the edge $\Gamma_1^{(1)}$ parallel to the axis s_1 by means of the power transformation

$$\begin{cases} y_1 = x_1^2 x_2^{-1}, \\ y_2 = x_1^{-1} x_2, \end{cases} \qquad W^{-1} = \begin{pmatrix} 2 & -1 \\ -1 & 1 \end{pmatrix}$$

i.e.

$$\begin{cases} x_1 = y_1 y_2, \\ x_2 = y_1 y_2^2, \end{cases} \qquad W = \begin{pmatrix} 1 & 1 \\ 1 & 2 \end{pmatrix}. \tag{3.4}$$

Here

$$\begin{aligned} \hat{f}_1^{(1)} &= x_1^3 - 3x_1x_2 = y_1^3 y_2^3 - 3y_1^2 y_2^3 = y_1^2 y_2^3 (y_1 - 3), \\ f &= x_1^3 + x_2^3 - 3x_1x_2 = y_1^2 y_2^3 (y_1 + y_1 y_2^3 - 3). \end{aligned} \tag{3.5}$$

Reducing by $y_1^2 y_2^3$ we obtain the full equation

$$y_1 + y_1 y_2^3 - 3 = 0, \tag{3.6}$$

and the truncated equation $y_1 - 3 = 0$. Its root $y_1 = y_1^0 \stackrel{\text{def}}{=} 3$ is a simple one. Applying Implicit Function Theorem to the full equation (3.6) we can obtain $y_1 - 3$ as a power series in y_2. Here it can be found in the explicit form $y_1 = 3/(1+y_2^3)$. Substituting that expression into (3.4) we obtain the parametric representation of the branch $\mathcal{F}_1$:

$$x_1 = 3y_2/(1+y_2^3), \quad x_2 = 3y_2^2/(1+y_2^3).$$

Now the author recommend to the reader to draw sets D, M etc. for polynomials (3.5) and (3.6), and to compare them with Figure 1.

EXAMPLE 6 (continuation of Example 3). The power transformation $z = x^{\alpha/(\beta-1)}y$ transforms the equation (2.9) to the form $x^{-\alpha/(\beta-1)}[x^2z''+\kappa z+z^\beta] = 0$. But

$$z' = \frac{dz}{x d\log x}, \quad z'' = \frac{d^2z}{x^2(d\log x)^2} - \frac{dz}{x^2 d\log x}.$$

So the equation takes the form

$$d^2z/(d\log x)^2 - dz/d\log x + \kappa z + z^\beta = 0. \tag{3.7}$$

Generally speaking, if we have a system of equations then for each equation, we must consider its own Newton polyhedrons. That is, we must consider several Newton polyhedrons simultaneously [1,3,5]. But the system of ordinary differential equations

$$d\log X/dt \stackrel{\text{def}}{=} (\dot{\log X}) = F(X) \stackrel{\text{def}}{=} \sum F_Q X^Q \quad \text{for} \quad Q \in D \tag{3.8}$$

has one support D, one Newton polyhedron M and one set of corresponding objects. In particular, to each $P \in \mathbb{R}_*^n \setminus \{0\}$ there corresponds the truncated system

$$(\dot{\log X}) = \hat{F}_P(X). \tag{3.9}$$

The power transformation $\log X = W \log Y$ transforms System (3.8) into the system

$$(\dot{\log Y}) = W^{-1} \sum F_Q Y^S \quad \text{for} \quad Q \in D,$$

where $S = W^*Q$. If $\dim M(\hat{F}_P) = d < n$, then there exists such a matrix W that System (3.9) has the form

$$(\dot{\log Y}) = G(y_1, \ldots, y_d)Y^T.$$

After the substitution $dt_1 = Y^T dt$, we obtain the system

$$d\log Y/dt_1 = G(y_1, \ldots, y_d).$$

To solve it, we must solve only the d-dimensional subsystem

$$dy_i/dt_1 = y_i g_i(y_1, \ldots, y_d), \quad i = 1, \ldots, d.$$

There are two kinds of the power transformation (3.1): with arbitrary real matrix W and with unimodular matrix W (i.e. $\det W = \pm 1$) having integer elements. A power transformation of the second kind gives one-to-one correspondence between X and Y outside coordinate subspaces. To find such a matrix W one can use the continued fraction algorithm [1], if $n = 2$, and its generalizations [15], if $n > 2$. Particular cases of the power transformation were used long ago, but in the general form it was introduced in [10], see also [1].

4. LOGARITHMIC TRANSFORMATION

Let a differential polynomial $g(Y)$ be such that for some j coordinate $s_j = 0$ for all $S \in D(g)$. If y_j is an algebraic coordinate (i.e. a parameter, $j \leq n_1$), then g_j does not depend on y_j. If y is a differential coordinate (i.e. a variable, $j > n_1$), then g depends on $\log y_j$ only. Let J is the set of such indices $j > n_1$ that $s_j = 0$ for any $S \in D(g)$. We introduce new variables

$$v_j = \log y_j \quad \text{for} \quad j \in J,$$
$$v_k = y_k \quad \text{for} \quad k \notin J.$$

Denote $h(V) \stackrel{\text{def}}{=} g(Y)$. If g indeed depends of an y_j, $j \in J$, then the support $D(h)$ has points Q with $q_j \neq 0$. So we can find first approximations of $h(V)$ using its Newton polyhedron. Solutions of corresponding truncated equations can give the logarithmic asymptotics of solutions of the initial problem. Note that $v_j \to \infty$ when $y_j \to 0$ or ∞ and $j \in J$.

EXAMPLE 7 (continuation of Example 6). After the transformation $t = \log x$, Equation (3.7) has the form

$$\ddot{z} - \dot{z} + \kappa z + z^\beta = 0. \quad (\,^{\cdot} \stackrel{\text{def}}{=} d/dt) \tag{4.1}$$

If $\kappa \neq 0$, its support consists of 4 points:

$$Q_1 = (-2, 1),\ Q_2 = (-1, 1),\ Q_3 = (0, 1) \text{ and } Q_4 = (0, \beta),$$

where $q_1 = \operatorname{ord} t$, $q_2 = \operatorname{ord} z$. Since $t \to \infty$ then for studied solutions $z(t)$ with $t = \tau$, the vector $P = (p_1, p_2)$ has $p_1 = 1$. The Newton polygon of Equation (4.1) has only one edge $\Gamma_1^{(1)}$ having the normal vector P with $p_1 > 0$ (see Figure 3, b). The corresponding boundary subset $D_1^{(1)}$ consists of two points Q_3, Q_4 and the truncated equation is

$$\kappa z + z^\beta = 0.$$

Its solutions are values b indicated in Example 3.

If $\kappa = 0$, i.e. (2.12) holds, then support of Equation (4.1) consists of 3 points: Q_1, Q_2, Q_4. Again the Newton polygon has only one edge having the normal vector P with $p_1 > 0$ (see Figure 3, c). But now the corresponding boundary subset consists of two points Q_2, Q_4 and the truncated equation is

$$-\dot{z} + z^\beta = 0.$$

Its solutions are

$$z = \left(\frac{-\alpha}{t + c}\right)^{1/\alpha}, \quad c = \text{ const}.$$

Thus for $\alpha = \beta - 1$, Equation (2.9) has solutions with asymptotics

$$y \sim x^{-1}\left(\frac{-\alpha}{\log x}\right)^{1/\alpha}.$$

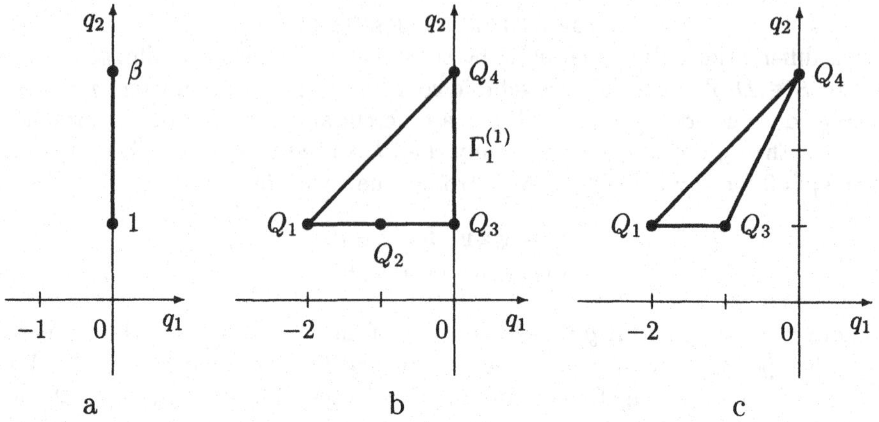

Fig. 3. Supports and Newton polygons for equations (3.7) (a), (4.1) with $\kappa \neq 0$ (b) and (4.1) with $\kappa = 0$ (c).

5. NORMAL FORM

Let $X \in \mathbb{R}^n$ or $\mathbb{C}^n$. In a neighborhood of the stationary point $X = 0$ we consider the ODE system

$$dX/dt \stackrel{\text{def}}{=} \dot{X} = AX + \Phi(X) \tag{5.1}$$

where Φ is a vector power series, $\Phi = O(\|X\|^2)$. We want to simplify System (5.1) by means of a formal transformation of coordinates

$$X = BY + \Xi(Y), \quad \det B \neq 0. \tag{5.2}$$

Let it bring System (5.1) into a form

$$\dot{Y} = CY + \Psi(Y). \tag{5.3}$$

Let the matrix $C = (c_{ij})$ be in the Jordan normal form. So its diagonal consists of its eigenvalues $(\lambda_1, \ldots, \lambda_n) = \Lambda$, and $c_{ij} = 0$ if $\lambda_i \neq \lambda_j$. Let us write components of the vector Ψ in (5.3) in the form

$$\psi_i(Y) \stackrel{\text{def}}{=} y_i g_i(Y) \stackrel{\text{def}}{=} y_i \sum g_{iQ} Y^Q, \quad i = 1, \ldots, n, \tag{5.4}$$

where $Q = (q_1, \ldots, q_n) \in \mathbb{Z}^n$ and $Q + E_i \geq 0$ and E_i is i-th unit vector. System (5.3), (5.4) is called *the resonant normal form,* if expansions (5.4) contain only such terms $g_{iQ}Y^Q$ for which the scalar product

$$\langle Q, \Lambda \rangle \stackrel{\text{def}}{=} q_1\lambda_1 + \ldots + q_n\lambda_n = 0. \tag{5.5}$$

THEOREM 6 [16, 17, 1]. *For each System* (5.1) *there exists a formal transformation* (5.2) *which transforms the* (5.1) *into its resonant normal form* (5.3).

Let Equation (5.5) have exactly k integer solutions $Q \in \mathbb{Z}^n$ which are linearly independent over $\mathbb{R}$. Then System (5.1) has the *k-fold resonance.*

THEOREM 7 [16, 17, 1]. *If System* (5.1) *has the k-fold resonance, then there exists such a power transformation $Y \to Z$ which reduces the resonant normal form* (5.3) *to the system*

$$(\dot{\log z_i}) = f_i(z_1, \ldots, z_k), \quad i = 1, \ldots, n. \tag{5.6}$$

Thus, to solve System (5.6) one must solve its subsystem of order k:

$$\dot{z}_i = z_i f_i(z_1, \ldots, z_k), \quad i = 1, \ldots, k. \tag{5.7}$$

System (5.7) has not a linear part. So to simplify System (5.7) we must find its truncated systems

$$\dot{z}_i = z_i \hat{f}_{ij}^{(d)}(z_1, \ldots, z_k), \quad i = 1, \ldots, k, \tag{5.8}$$

using the Newton polyhedron of System (5.7). System (5.8) is quasihomogeneous, so we can again use a power transformation to simplify it. To find asymptotics of solutions of System (5.6) we can use the logarithmic transformation (see Section 4)

$$v_i = \log z_i, \quad i = k+1, \ldots, n.$$

The question of analyticity of the normalizing transformation (5.2) for the analytic system (5.1) was solved in [17] in the following form. There are two conditions on the normal form (5.3): Condition ω on eigenvalues Λ (the small divisor condition) and Condition A on its nonlinear part Ψ. If both conditions are satisfied then (5.2) is analytic. If Condition A is violated then (5.2) can diverge. If $\operatorname{Re}\Lambda = 0$ then Condition A is very restrictive and usually (5.2) diverges in a whole neighborhood of the origin. But there are sets of dimension less then n on which the system (5.1) is transformed into the normal form by an analytical or smooth change of coordinates. A description of such sets is given below for the case $\operatorname{Re}\Lambda = 0$. By the normal form (5.3)-(5.5) we define the set $\mathcal{A} = \{Y : \psi_i = \lambda_i y_i \alpha, \ i = 1, \ldots, n\}$ where α is a free parameter. It is the set where Condition A is satisfied. Let $\mathbb{T}^l$ be an invariant l-dimensional torus of the system (5.3) and $\mu_{l+1}, \ldots, \mu_n$ be eigenvalues of its variational system. In the set $\mathcal{A}$ we isolate two subsets $\tilde{\mathcal{A}}$, $\mathcal{B}$: $\tilde{\mathcal{A}}$ consists of all stationary and periodic solutions, $\mathcal{B}$ consists of all invariant tori with all $\mu_j = 0$.

THEOREM 8 [1]. *On the set $\tilde{\mathcal{A}}$ the system* (5.1) *is analytically reduced to the normal form and the set $\tilde{\mathcal{A}}$ is analytical in X.*

THEOREM 9 [1]. *If Λ satisfies Condition ω then on the set $\mathcal{B}$ the system* (5.1) *is analytically reduced to the normal form and the set $\mathcal{B}$ is analytical in X.*

Bifurcations of periodic and conditionally periodic solutions are branching sets of these solutions over the subspace of parameters. To analyze the structure of the sets $\tilde{\mathcal{A}}$ and $\mathcal{B}$ one can use the method of a resolution of a singularity by means of Newton polyhedra and power transformations (see Sections 2 and 3). Analogously we can study a neighborhood of a periodic solution, of an invariant torus and of some families of such tori [1]. Note that $\tilde{\mathcal{A}} = \mathcal{A}$, if all λ_j are pairwise commensurable.

EXAMPLE 8. Let a system (5.1) have a small parameter $x_1 \stackrel{\text{def}}{=} \varepsilon$ (i.e. $\dot{\varepsilon} = 0$) and two variables x_2 and x_3. Then $\varepsilon = x_2 = x_3 = 0$ is its stationary point and $\lambda_1 = 0$. Let $\lambda_2 = -\lambda_3 = i = \sqrt{-1}$. Then Equation (5.5) is $q_2\lambda_2 + q_3\lambda_3 \stackrel{\text{def}}{=} i(q_2 - q_3) = 0$. Its integer solutions $Q = (q_1, q_2, q_3)$ have arbitrary q_1 and $q_2 = q_3$. So the normal form (5.3)-(5.5) of our System (5.1) is

$$\dot{\varepsilon} = 0, \quad \dot{y}_j = y_j(\lambda_j + g_j(\varepsilon, y_2 y_3)), \quad j = 2, 3.$$

The power transformation $\rho = y_2 y_3$, $y_3 = y_3$ transforms it to

$$\dot{\varepsilon} = 0, \quad \dot{\rho} = \rho(g_2(\varepsilon, \rho) + g_3(\varepsilon, \rho)), \quad \dot{y}_3 = y_3(\lambda_3 + g_3(\varepsilon, \rho)).$$

The set

$$\mathcal{A} = \{\varepsilon, y_2, y_3 : \ y_j g_j = \lambda_j y_j \alpha, \quad j = 2, 3\} = \mathcal{A}_1 \cup \mathcal{A}_2 \cup \mathcal{A}_3,$$

where $\mathcal{A}_1 = \{\varepsilon, y_2, y_3 : g_2 + g_3 = 0\}$ and $\mathcal{A}_j = \{y_j = 0\}$, $j = 2, 3$ are coordinate axis. Here $\tilde{\mathcal{A}} = \mathcal{A}$. By Theorem 8 the set $\mathcal{A}$ is analytic for analytical our initial system (5.1). If $g_j = b_j\varepsilon + c_j\rho + \ldots$, $j = 2, 3$, then the component set $\mathcal{A}_1$ is defined by the equation

$$g_2 + g_3 \equiv (b_2 + b_3)\varepsilon + (c_2 + c_3)\rho + \ldots = 0. \tag{5.9}$$

If $b_2 + b_3 \neq 0$ then by Implicit Function Theorem this equation has unique solution $\varepsilon = \varepsilon(\rho)$, i.e. $\mathcal{A}_1$ is a manifold. For a real system (5.1) with real x_j, we have that $y_2 = \overline{y}_3$, and $\rho = y_2 y_3$ is the square of the radius vector in real coordinates Rey_2, Imy_2, and Re$\mathcal{A}_1$ is a family of periodic solutions bifurcating from the origin. This is the Hopf bifurcation [18]. If $b_2 + b_3 = c_2 + c_3 = 0$, then Equation (5.9) has several branches of solutions. To find all branches one can use the Newton polygon for (5.9). Here Condition A is $g_2 + g_3 \equiv 0$.

If not all $\operatorname{Re}\lambda_j = 0$ then the system (5.1) has the invariant center manifold, corresponding to λ_j with $\operatorname{Re}\lambda_j = 0$. After a reduction of the system on the center manifold, we can apply the described approach.

Some partial differential equations can be written as an infinite system of ordinary differential equations, so called *evolution equation*. If the spectrum of its linear part has only finite number of points in the imaginary axis, then after reduction on the center manifold we obtain an ODE system, that can be transformed to the normal form and so on. If the spectrum has infinitely many points in the imaginary axis then after reduction on the center manifold we obtain an evolution equation again. It can be transformed to the resonant normal form. If the multiplicity of the resonance is finite then the normal form can be reduced to an ODE system by means of a power transformation. On the contrary, the power transformation can give new PDE which is more simple than initial one. Thus in any case the normalizing transformations followed by an appropriate power transformation simplifies a problem.

In a neighborhood of a stationary point of System (5.1) the algorithm for computation of the normal form and the normalizing transformation uses only arithmetic operations over coefficients of the right side series of System (5.1) [1, Ch. III, Section 1.7;19]. In a neighborhood of a periodic solution or an invariant torus the algorithm uses computation of some integrals as well as for PDE. There are several such algorithms [19] and their computer realizations, especially for Hamiltonian systems.

6. APPLICATIONS

In a concrete local (or asymptotic) nonlinear problem we must combine truncations, power transformations, logarithmic transformations and normalizations. Each of them diminish the dimension of the problem. So after several such steps, we obtain a problem which is simple enough to be solved. It can serve as an unperturbed problem for the transformed initial one. The approach allows to resolve a singularity, to study asymptotics of solutions, their branching, bifurcations, stability and so on. Below we list recent applications of the algorithms to problems from Celestial Mechanics and Hydrodynamics. Some examples see also in [1-6].

The restricted 3-body problem describes motion of a massless body under the Newtonian attraction of two bodies with masses 1 and μ. The planar circular problem assumes that orbits of two big bodies are circumferences and the massless body moves in the plane of these orbits. For small μ the problem has singular perturbations for orbits with close approach to the μ-mass body. In particular, the existence of stable periodic orbits with close approach to Jupiter and to Earth was shown in [22]. See [20–30].

The equation of the satellite motion around its mass center moving along an elliptical orbit with eccentricity e has a singularity when $e = 1$. Families of periodic solutions were computed for $e < 0.98$ and $e = 0.99$ and $e = 0.999$ [31]. Here for $e \to 1$ we study the limiting families of periodic solutions. Some of them have a fractal structure. See [31–38, 55, 56].

The problem of surface waves on water after reduction on the center manifold gives a 4-dimensional ODE system with 2 small parameters having a degenerate linear part. We have found its new periodic solutions and conditionally periodic solutions [48]. Sometimes computations demand 3 steps of reductions: a truncation, a normalizing transformation and another truncation. See [39–48].

Asymptotics of the stationary viscous fluid flow around a flat plate are found directly through the Newton polyhedron of the Helmholtz form of the Navier-Stokes equations [49]. It gives the mathematical background of the boundary layer theory. See [49, 50].

Other applications of these algorithms are following. Isolation of all branches of any algebraic or analytic space curve, and their computation with any accuracy [2]. A complicated bifurcation up to 10 periodic solutions from the stationary one in a 4-dimensional system (5.1) with one small parameter [51]. A way for estimation of number of limit cycles of such a polynomial system on the plane which is close to integrable one [52]. Some asymptotic properties of solutions of a finite-dimensional approximation of the Schrödinger equation were studied in [53, 54].

References

[1] Bruno, A. D. (1979). *Local Method of Nonlinear Analysis of Differential Equations*, Nauka, Moscow, (Russian) = Local Methods in Nonlinear Differential Equations, Springer- Verlag, Berlin, 1989 (English).

[2] Bruno A. D. and Soleev, A. (1991). *Local uniformization of branches of a space curve, and Newton polyhedra*, Algebra i Analis, Vol. 3(1), (pages 67–101), (R) = St. Petersburg Math. J., Vol. 3(1), (1992), (pages 53–82) (E).

[3] Bruno, A. D. and A. Soleev. (1994). *First approximations of algebraic equations*, Doklady Akademii Nauk, Vol. 335(3), (pages 277–278) (R) = Russian Ac. Sci. Doklady. Mathem., Vol. 49(2), (pages 291–293) (E).

[4] Bruno, A. D. (1995). *The Newton polyhedron in the Nonlinear Analysis*, Vestnik Moskovskogo Universiteta, Vol. 1(6), (pages 45–51), (R) = Mathem. Bulletin, Vol. 50, (1995) (E).

[5] Bruno, A. D. (1994). *First approximations of differential equations*, Doklady Akademii Nauk, Vol. 335(4), (pages 413–416), (R) = Russian Acad. Sci. Doklady. Mathem., Vol. 49(2), (pages 334–339) (E).

[6] Bruno, A. D. (1995). *General approach to the asymptotic analysis of singular perturbations*, In: Dynamical Systems and Chaos, edited, N. Aoki, K. Shiraiwa and Y. Takahashi, World Scientific, Singapore, Vol. 1, (pages 11-17).

[7] Soleev, A. and A. B. Aranson. (1994). *Computation of a polyhedron and of normal cones of its faces*, Preprint, Vol. 36, Inst. Appl. Math., Moscow, (R).

[8] Puiseux, V. (1850). *Recherches sur les fonctions algebriques*, J. de Math. pures et appl., Vol. 15, (pages 365–480).

[9] Newton, I. (1964). *A treatise of the method of fluxions and infinite series, with its application to the geometry of curve lines*, in The Mathematical Works of Isaac Newton, (Edited by Harry Woolf), Vol. 1, (pages 27–137), Johnson Reprint Corp., N.Y. and London (1964).

[10] Bruno, A. D. (1962). *The asymptotic behavior of solutions of nonlinear systems of differential equations*, Dokl. Akad. Nauk SSSR, Vol. 143(4), (pages 763–766), (R) = Soviet Math. Dokl., Vol. 3, (pages 464–467) (E).

[11] Gindikin, S. G. (1974). *Energy estimates and Newton polyhedra*, Trudy Mosk. Mat. Obsc., Vol. 31, (pages 189–236), (R) = Trans. Moscow Math. Soc., Vol. 31, (pages 193–246), (E).

[12] Khovanskii, A. G. (1983). *Newton polyhedra (resolution of singularities)*, In: Itogi Nauki i Tekhniki: Sovremennye Problemy Mat., Vol. 22, (pages 207–239), VINITI, Moscow, (R) = J. Sov. Math., Vol. 27, (1984), (pages 2811–2830), (E).

[13] Bruno, A. D. (1996). *Algorithms of Nonlinear Analysis*, Uspehi Matem. Nauk, Vol. 51(5), (R) = Russian Math. Surveys, Vol. 51(5), (E).

[14] Bruno, A. D. and Soleev, A. (1990). *Local uniformization of branches of an algebraic curve*, Preprint, Vol. 34, I.H.E.S., Paris.

[15] Bruno, A. D. and Parusnikov, V. I. (1997). *Comparison of different generalizations of continued fractions*, Matem. Zametki, Vol. 61(3), (pages 339–348), (R) = Math. Notes, Vol. 61(3), (1997) (E).

[16] Bruno, A. D. (1964). *Normal form of differential equations*, Dokl. Akad. Nauk SSSR, Vol. 157(6), (pages 1276–1279), (R) = Soviet Math. Dokl., Vol. 5, (1964), (pages 1105–1108), (E).

[17] Bruno, A. D. (1971). *Analytical form of differential equations*, Trudy Mosk. Mat. Obsc., Vol. 25, (pages 119–262); Vol. 26, (1972), (pages 199–239), (R) = Trans. Moscow Math. Soc., Vol.25, (1971), (pages 131–288), Vol. 26, (1972), (pages 199–239), (E).

[18] Marsden, J. E. and McCracken, M. (1976). *The Hopf Bifurcation and Its Applications*, Springer-Verlag, New York, 1976 (E) = Mir, Moscow, 1980, (R).

[19] Bruno, A. D. (1995). *Ways of computing a normal form*, Dokl. Akad. Nauk, Vol. 344(3), (pages 298–300), (R) = Russian Ac. Sci. Doklady Math., Vol. 52, (1995), (pages 200–202), (E).

[20] Bruno, A. D. (1990). *The Restricted 3-Body Problem*, Nauka, Moscow, (R) = Walter de Gruyter, Berlin, 1994 (E).

[21] Henon, M. (1968). *Sur les orbites interplanetaires qui rencontrent deux fois la Terre*, Bull. Astron., Vol. 3(3), (pages 377–402).

[22] Bruno, A. D. (1978). *On periodic flights round moon*, Preprint, Vol. 91, Inst. Appl. Math., Moscow, (R) = On periodic flybys of the moon, Celestial Mechanics, Vol. 24(3), (pages 255–268) (E).

[23] Bruno, A. D. (1992). *A general approach to the study of complex bifurcations*, Prikladnaja Mekhanika, Vol. 28(12), (pages 83–86) (R) = International Applied Mechanics, Vol. 28(12), (pages 849–853) (E).

[24] Bruno, A. D. (1994). *Singular perturbations in Hamiltonian Mechanics*, In: Hamiltonian Mechanics, (Edited by J. Seimenis), Plenum Press, New York, (pages 43–49).

[25] Henon, M. and M. Guyot. (1970). *Stability of periodic orbits in the restricted problem*, In: Periodic Orbits, Stability and Resonances, (G. E. O. Giacaglia, ed.), Reidel, Dordrecht, (pages 349–374).

[26] Bruno, A. D. and A. Soleev. (1995). *The Hamiltonian truncations of a Hamiltonian system*, Preprint, Vol. 55, Inst. Appl. Math., Moscow, (R).

[27] Bruno, A. D. and A. Soleev. (1995). *Newton polyhedra and Hamiltonian systems*, Vestnik Moskovskogo Universiteta, Vol. 1(6), (pages 84–86) (R) = Mathem. Bulletin, Vol. 50, (E).

[28] Bruno, A. D. and A. Soleev. (1996). *Hamiltonian truncated systems of a Hamiltonian system*, Doklady Akademii Nauk, Vol. 349(2), (pages 153–155) (R) = Russian Acad. Sci. Doklady. Mathematics, Vol. 54(1), (pages 512–514) (E).

[29] Bruno, A. D. (1993). *Simple (double, multiple) periodic solutions of the restricted three-body problem in the Sun-Jupiter case*, Preprints, Vols. 66, 67, 68, Inst. Appl. Math., Moscow, (R).

[30] Bruno, A. D. (1996). *Zero-multiple and retrograde periodic solutions of the restricted three-body problem*, Preprint, Vol. 93, Inst. Applied Math., Moscow, (R).

[31] Bruno, A. D. and V. Yu. Petrovich. (1994). *Computation of periodic oscillations of a satellite. Singular case*, Preprint, Vol. 44, Inst. Appl. Math., Moscow, (R).

[32] Bruno, A. D. and V. P. Varin. (1995). *First (second) limit problem for the equation of oscillations of a satellite*, Preprints, Vol. 124 & 128, Inst. of Applied Math., Moscow, (R).

[33] Bruno, A. D. and V. P. Varin. (1997). *Limit problems for the equation of oscillations of a satellite*, Celestial Mechanics, Vol. 67(1), (pages 1-40).

[34] Bruno, A. D. and V. Yu. Petrovich. (1994). *Regularization of oscillations of a satellite on a very stretched orbit.* Preprint, Vol. 4, Inst. Appl. Math., Moscow, (R).

[35] V. Varin, V. (1996). *The critical families of periodic solutions of the equation of oscillations of a satellite*, Preprint, Vol. 101, Inst. Appl. Math., Moscow, (R).

[36] Varin, V. (1997). *The critical subfamilies of the family K_0 of periodic solutions of the equation of oscillations of a satellite*, Preprint, Vol. 20, Inst. Appl. Math., Moscow, (R).

[37] Sadov, S. Yu. (1995). *Normal form of the equation of oscillations of a satellite in a singular case*, Matem. Zametki, Vol. 58(5), (pages 785–789) (R) = Mathematical Notes, Vol. 58(5), (pages 1234–1237) (E).

[38] Sadov, S. Yu. (1996). *Higher approximations of the method of averaging for the equation of plane oscillations of a satellite*, Preprint, Vol. 48, Inst. Appl. Math., Moscow, (R).

[39] Iooss, G. and K. Kirchgaessner, (1992). *Water waves for small surface tension: an approach via normal form*, Proceedings of the Royal Society of Edinburg, Vol. 122A, (pages 267–299).

[40] Arnold, V. I. (1971). *On matrices depending on parameters*, Uspehi Mat. Nauk, Vol. 26(2), (pages 101–114) (R) = Russ. Math. Surveys, Vol. 26(2), (pages 29–44) (E).

[41] Soleev, A. and A. B. Aranson. (1995). *The first approximation of a reversible system of ordinary differential equations*, Preprint, Vol. 28, Inst. of Applied Mathematics, Moscow, (R).

[42] Bruno, A. D. and Soleev, A. (1995). *Local analysis of singularity of a reversible system of ordinary differential equations: simple cases*, Preprint, Vol. 40, Inst. of Applied Mathematics, Moscow, (R).

[43] Bruno, A. D. and Soleev, A. (1995). *Local analysis of singularity of a reversible system of ordinary differential equations: complex cases*, Preprint, Vol. 47, Inst. of Applied Mathematics, Moscow, (R).

[44] Bruno, A. D. and Soleev, A. (1995). *Homoclinic solutions of a reversible system of ordinary differential equations*, Preprint, Vol. 54. Inst. of Applied Mathematics, Moscow, (R).

[45] Belitskii, G. R. (1975). *The normal form of local mappings*, Uspekhi Mat. Nauk, Vol. 30(1), (page 223), (R).

[46] Hammersley, J. M. and G. Mazarino. (1989). *Computational aspect of some autonomous differential equations*, Proc. Roy. Soc. London, Ser. A, Vol. 424, (pages 19–37).

[47] Bruno, A. D. and Soleev, A. (1995). Local analysis of singularities of a reversible ODE system. *Uspehi Mat. Nauk* **50**:6 (1995), 169–170 (R) = *Russian Math. Surveys* **50**:6 (1995), 1258–1259 (E).

[48] Bruno, A. D. and Soleev, A. (1995). *Bifurcations of solutions in a reversible ODE system*, Doklady Akademii Nauk, Vol. 345(5), (pages 590–592), (R) = Russian Acad. Sci. Doklady. Mathematics, Vol. 52(3), (pages 419–421), (E).

[49] Bruno, A. D. and Vasil'ev, M. M. (1995). *Newton polyhedra and the asymptotic analysis of the viscous fluid flow around a flat plate*, Preprint, Vol. 44, Inst. of Applied Math., Moscow, (R).

[50] Van Dyke, M. (1967). *Perturbation Methods in Fluid Mechanics.* Academic Press, New York, London, (E) = Mir, Moscow, (R).

[51] Bruno, A. D. (1988). *Bifurcation of the periodic solutions in the symmetric case of a multiple pair of imaginary eigenvlues*, in Numerical Solution of Ordinary Differential Equations, (Edited by S.S. Filippov), (pages 161–176), Inst. Appl. Math, Moscow, (R) = Selecta Math. formerly Sovietica, Vol. 12(1), (1993), (pages 1–12), (E).

[52] Bruno, A. D. (1990). *The normal form of a system, close to a Hamiltonian system*, Matem. Zametki, Vol. 48(5), (pages 35–46), (R) = Matem Notes, Vol. 48(5/6), (1991), (pages 1100–1108), (E).

[53] Sadov, S. Yu. (1994). *On a dynamic system arising from a finite-dimensional approximation of the Schrödinger equation*, Matem. Zametki, Vol. 56(3), (pages 118–133), (R) = Mathematical Notes, Vol. 56(3), (1994), (pages 960–971), (E).

[54] Bruno, A. D. and Sadov, S. Yu. (1995). *Formal integral of a divergence free system*, Matem. Zametki, Vol. 57(6), (pages 803–813), (R) = Math. Notes, Vol. 57(3), (1995), (pages 565–572), (E).

[55] Bruno, A. D. and Petrovich, V. Yu. (1997). *Computation of periodic oscillations of a satellite*, Matematicheskoe Modelirovanie, Mathematical Modelling, Vol. 9(6), (pages 82–94), (R).

[56] Sadov, S. Yu. (1997). *Plane motions of a near symmetric satellite about its mass center with rational rotation numbers*, Preprint, Vol. 31, Inst. Appl. Math., Moscow, (R).

Transmission loss in a depth-varying ocean over a poroelastic seabed*

James L. Buchanan†
Mathematics Department
United States Naval Academy
Annapolis, Maryland 21402 USA

Robert P. Gilbert
Department of Mathematical Sciences
University of Delaware
Newark, Delaware 19716 USA

Abstract

A Green's function representation for acoustic pressure in an ocean with depth-varying sound speed over a poroelastic seabed is derived via a residue computation. Transmission loss curves for upward and downward refracting sound speed profiles in the ocean are computed for three sediments of different coarseness.

1 Introduction

We consider an ocean of depth h over a semi-infinite poroelastic seabed which is assumed to obey the equations of the Biot model [2],[3]. A Green's function representation for acoustic pressure in a depth-varying ocean arising from a time-harmonic point source will be derived which consists of a summation over modal terms and three line integrals around branch cuts. This extends the representation for the case of an isovelocity ocean which was derived in [6].

2 Acoustic model for an ocean over a poroelastic seabed

We work in cylindrical coordinates. Suppressing the angular displacement components we use $\mathbf{u} := (u_r, u_z)$ and $\mathbf{U} := (U_r, U_z)$ to denote the frame and pore fluid displacement vectors respectively. The frame and fluid *dilatations* are defined to be

$$e_s := \nabla \cdot \mathbf{u} = \partial_r u_r + \frac{u_r}{r} + \partial_z u_z, \; \epsilon_s := \nabla \cdot \mathbf{U} = \partial_r U_r + \frac{U_r}{r} + \partial_z U_z.$$

In the Biot model the constitutive equations for an isotropic poroelastic material are

$$\sigma_{rr} = \lambda_s e_s + 2\mu_s e_{rr} + Q\epsilon_s, \; \sigma_{\phi\phi} \;=\; \lambda_s e_s + 2\mu_s e_{\phi\phi} + Q\epsilon_s$$

*This research was supported in part by the National Science Foundation through grant BES-9402539.
†Supported in part by a U.S. Naval Academy Recognition Grant.

R.P. Gilbert et al.(eds.), Direct and Inverse Problems of Mathematical Physics, 21–37.

$$\sigma_{zz} = \lambda_s e_s + 2\mu_s e_{zz} + Q\epsilon_s, \; \sigma_{rz} = 2\mu_s e_{rz}, \; \sigma = Qe_s + R\epsilon_s$$

where λ_s and μ_s are the Lamé coefficients of the porous skeleton, R is a Biot parameter measuring the pressure on the fluid required to force a certain volume of fluid into the sediment at constant volume, and Q measures the coupling of changes in the volume of the solid and fluid.

The equations of motion for the Biot model are

$$\partial_r \sigma_{rr} + \frac{1}{r}(\sigma_{rr} - \sigma_{\phi\phi}) + \partial_z \sigma_{rz} = \partial_{tt}(\rho_{11} u_r + \rho_{12} U_r) + b\partial_t(u_r - U_r)$$

$$\partial_r \sigma = \partial_{tt}(\rho_{12} u_r + \rho_{22} U_r) - b\partial_t(u_r - U_r)$$

$$\partial_r \sigma_{rz} + \frac{1}{r}\sigma_{rz} + \partial_z \sigma_{zz} = \partial_{tt}(\rho_{11} u_z + \rho_{12} U_z) + b\partial_t(u_z - U_z)$$

$$\partial_z \sigma = \partial_{tt}(\rho_{12} u_z + \rho_{22} U_z) - b\partial_t(u_z - U_z)$$

where ρ_{11} and ρ_{22} are density parameters for the solid and fluid, ρ_{12} is a density coupling parameter, and b is a dissipation parameter.

In [5],[4] Buchanan and Gilbert derived a system of differential equations for the Biot model assuming time harmonic oscillations $u_{zs}(r,z,t) = u_{zs}(r,z)e^{i\omega t}, \ldots$. We distinguish quantities in the ocean from the corresponding quantities in the seabed with subscripts o and s, e.g. λ_o and λ_s. In the case when the variation in the density of the ocean water is neglected and all seabed parameters are constant, the motions of the ocean and seabed are governed by the following set of differential equations and interface conditions for the acoustic pressure $p_o(r,z)$ and vertical displacement u_{zo} in the ocean and the dilatations and vertical displacements in the seabed.

Ocean surface:

$$p_o(r, 0) = 0$$

Ocean:

$$\begin{aligned} \nabla^2 p_o + k_0^2 n^2(z) p_o &= 0 \\ \partial_z p_o + \rho_o \omega^2 u_{zo} &= 0 \end{aligned}$$

Ocean-sediment interface:

$$u_{zo}(r,h-) = u_{zs}(r,h+) = U_{zs}(r,h+) \tag{1}$$

$$p_o(r,h-) = (\lambda_s + Q)e_s(r,h+) + 2\mu_s \partial_z u_{zs}(r,h+) + (Q+R)\epsilon_s(r,h+) \tag{2}$$

$$\partial_z u_{rs}(r,h+) + \partial_r u_{zs}(r,h+) = 0 \tag{3}$$

Sediment:

$$\Delta \begin{pmatrix} \tau \\ \sigma \\ u_{zs} \end{pmatrix} + A\partial_z \begin{pmatrix} \tau \\ \sigma \\ u_{zs} \end{pmatrix} + B \begin{pmatrix} \tau \\ \sigma \\ u_{zs} \end{pmatrix} = 0 \tag{4}$$

$$U_{zs} = -\frac{1}{p_{22}}(p_{12} u_{zs} + \partial_z \sigma)$$

$$\partial_r u_{rs} + \frac{u_{rs}}{r} - e_s + \partial_z u_{zs} = 0$$

$$\partial_r U_{rs} + \frac{U_{rs}}{r} - \epsilon_s + \partial_z U_{zs} = 0$$

Asymptotic condition at infinity:

$$e_s, \epsilon_s, u_{rs}, u_{zs}, U_{rs}, U_{zs} \to 0 \text{ as } z \to \infty.$$

In the above equations $k_0 = \omega/c_0$ where c_0 is some representative sound speed in the ocean and $n(z) = c_0/c(z)$ where $c(z)$ is the sound speed in the ocean at depth z.

$$\Delta := \partial_{rr} + \frac{1}{r}\partial_r + \partial_{zz}$$

is the azimuthally independent Laplacian and

$$p_{11} := \omega^2 \rho_{11} - i\omega b, \quad p_{12} := \omega^2 \rho_{12} + i\omega b, \quad p_{22} := \omega^2 \rho_{22} - i\omega b.$$

The interface conditions (1),(2) and (3) follow from the continuity of vertical displacement and normal stress at the ocean-sediment boundary and the vanishing of tangential stress on the sediment side of the interface. In the sediment τ and σ are new dependent variables defined by

$$\tau := (\lambda_s + 2\mu_s)e_s + Q\epsilon_s, \quad \sigma := Qe_s + R\epsilon_s.$$

The inverse transformation is

$$e_s = a_{11}\tau - a_{12}\sigma, \quad \epsilon_s = -a_{12}\tau + a_{22}\sigma \tag{5}$$

where

$$a_{11} = R/d, \quad a_{12} = Q/d, \quad a_{22} = (\lambda_s + 2\mu_s)/d, \quad d = R(\lambda_s + 2\mu_s) - Q^2.$$

In the case of constant seabed parameters the non-zero entries of the matrices A and B are

$$\begin{aligned} A_{31} &= a_{11} + \frac{a_{11}\lambda_s}{\mu_s} - \frac{a_{12}Q}{\mu_s}, \; A_{32} = -a_{12} - \frac{a_{12}\lambda_s}{\mu_s} + \frac{a_{22}Q}{\mu_s} - \frac{p_{12}}{p_{22}\mu_s} \\ B_{12} &= p_{12}a_{22} - p_{11}a_{12}, \; B_{21} = p_{12}a_{11} - p_{22}a_{12}, \; B_{33} = \frac{p_{11}p_{22} - p_{12}^2}{\mu_s}. \end{aligned}$$

In the sediment we seek separable solutions of the form

$$\begin{aligned} e_s(r,z) &= e_s(z)H_0^{(1)}(k_0\sqrt{\kappa}r), \; \epsilon_s(r,z) = \epsilon_s(z)H_0^{(1)}(k_0\sqrt{\kappa}r) \\ u_{zs}(r,z) &= u_{zs}(z)H_0^{(1)}(k_0\sqrt{\kappa}r), \; U_{zs}(r,z) = U_{zs}(z)H_0^{(1)}(k_0\sqrt{\kappa}r) \\ u_{rs}(r,z) &= u_{rs}(z)H_1^{(1)}(k_0\sqrt{\kappa}r), \; U_{rs}(r,z) = U_{rs}(z)H_1^{(1)}(k_0\sqrt{\kappa}r) \end{aligned} \tag{6}$$

where $H_j^{(1)}, j = 0,1$ are outgoing Hankel functions and $\mathrm{Im}(\kappa) \geq 0$ is required for solutions to approach zero as $r \to \infty$.

Substituting (6) into (4) gives the following system for the functions $\tau(z), \sigma(z)$ and $u_{zs}(z)$

$$\begin{aligned} \tau''(z) + a_\tau^2 \tau(z) + B_{12}\sigma(z) &= 0 & (7) \\ \sigma''(z) + B_{21}\tau(z) + a_\sigma^2 \sigma(z) &= 0 & (8) \\ u''_{zs}(z) + A_{31}\tau'(z) + A_{32}\sigma'(z) + a_u^2 u_{zs}(z) &= 0 & (9) \end{aligned}$$

where

$$\begin{aligned} a_\tau^2 &:= & B_{11} - k_0^2\kappa = p_{11}a_{11} - p_{12}a_{12} - k_0^2\kappa \\ a_\sigma^2 &:= & B_{22} - k_0^2\kappa = p_{22}a_{22} - p_{12}a_{12} - k_0^2\kappa \\ a_u^2 &:= & B_{33} - k_0^2\kappa = \frac{p_{11}}{\mu_s} - \frac{p_{12}^2}{p_{22}\mu_s} - k_0^2\kappa. \end{aligned}$$

Taking into account the asymptotic condition, the solution to the system (7) and (8) is

$$\begin{aligned} \tau(z) &= C_2 e^{im_+(z-h)} + C_3 e^{im_-(z-h)} \\ \sigma(z) &= -\frac{a_\tau{}^2 e^{im_+(z-h)} C_2}{B_{12}} + \frac{e^{im_+(z-h)} m_+{}^2 C_2}{B_{12}} - \frac{a_\tau{}^2 e^{im_-(z-h)} C_3}{B_{12}} + \frac{e^{im_-(z-h)} m_-{}^2 C_3}{B_{12}} \end{aligned}$$

where

$$m_\pm := \sqrt{\frac{a_\tau^2 + a_\sigma^2 \pm \sqrt{(a_\tau^2 - a_\sigma^2)^2 + 4B_{12}B_{21}}}{2}}$$

and the branch cut for the square root function is chosen so that $\mathrm{Im}(m_\pm) \geq 0$. The dilatations e_s and ϵ_s are obtained from (5). The solution to (9) is then

$$\begin{aligned} u_{zs} = {} & \frac{\frac{-i}{2}\, A_{31}\, e^{im_+(z-h)}\, m_+ C_2}{a_u\,(a_u - m_+)} + \frac{\frac{i}{2}\, a_\tau{}^2\, A_{32}\, e^{im_+(z-h)}\, m_+ C_2}{a_u\, B_{12}\,(a_u - m_+)} \\ & - \frac{\frac{i}{2}\, A_{32}\, e^{im_+(z-h)}\, m_+{}^3 C_2}{a_u\, B_{12}\,(a_u - m_+)} - \frac{\frac{i}{2}\, A_{31}\, e^{im_+(z-h)}\, m_+ C_2}{a_u\,(a_u + m_+)} + \frac{\frac{i}{2}\, a_\tau{}^2\, A_{32}\, e^{im_+(z-h)}\, m_+ C_2}{a_u\, B_{12}\,(a_u + m_+)} \\ & - \frac{\frac{i}{2}\, A_{32}\, e^{im_+(z-h)}\, m_+{}^3 C_2}{a_u\, B_{12}\,(a_u + m_+)} - \frac{\frac{i}{2}\, A_{31}\, e^{im_-(z-h)}\, m_- C_3}{a_u\,(a_u - m_-)} + \frac{\frac{i}{2}\, a_\tau{}^2\, A_{32}\, e^{im_-(z-h)}\, m_- C_3}{a_u\, B_{12}\,(a_u - m_-)} \\ & - \frac{\frac{i}{2}\, A_{32}\, e^{im_-(z-h)}\, m_-{}^3 C_3}{a_u\, B_{12}\,(a_u - m_-)} - \frac{\frac{i}{2}\, A_{31}\, e^{im_-(z-h)}\, m_- C_3}{a_u\,(a_u + m_-)} + \frac{\frac{i}{2}\, a_\tau{}^2\, A_{32}\, e^{im_-(z-h)}\, m_- C_3}{a_u\, B_{12}\,(a_u + m_-)} \\ & - \frac{\frac{i}{2}\, A_{32}\, e^{im_-(z-h)}\, m_-{}^3 C_3}{a_u\, B_{12}\,(a_u + m_-)} + e^{ia_u(z-h)}\, C_4. \end{aligned}$$

3 Green's function for pressure in the ocean

If the acoustic pressure field in the ocean derives from a point source located at depth $z = z_0$ then since $\lambda_o(z) = \rho_o c^2(z)$

$$\begin{aligned} \nabla^2 p_o + k_0^2 n^2(z) p_o &= -\frac{1}{2\pi r}\delta(r)\delta(z - z_0) \\ \partial_z p_o + \rho_o \omega^2 u_{zo} &= 0. \end{aligned}$$

We seek a Green's function representation

$$p_o(r,z,z_0) = \frac{k_0^2}{8\pi}\oint_{C_0} H_0^{(1)}(k_0\sqrt{\kappa r})G_2(z,z_0,\kappa)d\kappa \tag{10}$$

for the pressure in the ocean. The contour C_0 must enclose all singularities of G_2 and exclude those of $H_0^{(1)}(k_0\sqrt{\kappa r})$. We choose it to be the slit cut enclosing the positive real axis, oriented counter clockwise.

The depth Green's function G_2 satisfies the differential equation

$$\frac{d^2G_2}{dz^2} + a_o^2(z,\kappa)G_2 = -\delta(z-z_0)$$

where $a_o(z,\kappa) = k_0\sqrt{n^2(z)-\kappa}$ with the interface conditions

---

Surface: $G_2(0,z_0,\kappa) = 0$
Source depth:

$$G_2(z_0-,z_0,\kappa) = G_2(z_0+,z_0,\kappa)$$
$$\frac{dG_2}{dz}(z_0+,z_0,\kappa) - \frac{dG_2}{dz}(z_0-,z_0,\kappa) = -1$$

Ocean-sediment interface:

$$\frac{-1}{k_0^2\lambda_o(h-)}\frac{dG_2}{dz}(h-,z_0,\kappa) = u_{zs}(h+) = U_{zs}(h+)$$
$$G_2(h-,z_0,\kappa) = (\lambda_s + Q)e_s(h+) + 2\mu_s\frac{du_{zs}}{dz}(h+) + (Q+R)\epsilon_s(h+)$$
$$e_s'(h+) = u_{zs}''(h+) + k_0^2\kappa u_{zs}(h+).$$

Asymptotic condition:

$$e_s(z), \epsilon_s(z), u_{zs}(z) \to 0 \text{ as } z \to \infty.$$

Let φ_1 and φ_2 be solutions to

$$\frac{d^2\varphi}{dz^2} + a_o^2(z,\kappa)\varphi = 0$$

satisfying the conditions

$$\begin{aligned} \varphi_1(0) &= 0, \quad \varphi_1'(0) = 1, \\ \varphi_2(0) &= 1, \quad \varphi_2'(0) = 0. \end{aligned}$$

In the case of an isovelocity ocean

$$\varphi_1(z,\kappa) = \frac{\sin a_o(\kappa)z}{a_o(\kappa)}, \quad \varphi_2(z,\kappa) = \cos a_o(\kappa)z. \tag{11}$$

The functions φ_1 and φ_2 are entire functions of κ.

The general solution which satisfies the surface and source interface conditions is

$$G_2(z, z_0, \kappa) = C_1\varphi_1(z) - U(z - z_0)\frac{\varphi_1(z)\varphi_2(z_0) - \varphi_1(z_0)\varphi_2(z)}{W(z_0)}, \; 0 \le z \le h$$

where U is the unit step function and

$$W(z) = \varphi_1(z)\varphi_2'(z) - \varphi_1'(z)\varphi_2(z)$$

is non-vanishing and entire in κ. Substituting for $G_2, e_s, \epsilon_s, u_{zs}$ and U_{zs} into the ocean-sediment interface conditions gives a matrix equation

$$\mathcal{M}\begin{pmatrix} \frac{C_1}{\lambda_o} \\ C_2 \\ C_3 \\ C_4 \end{pmatrix} = \begin{pmatrix} \frac{1}{\lambda_o k_0^2}\frac{\varphi_1'(h)\varphi_2(z_0)-\varphi_1(z_0)\varphi_2'(h)}{W(z_0)} \\ \frac{1}{\lambda_o k_0^2}\frac{\varphi_1'(h)\varphi_2(z_0)-\varphi_1(z_0)\varphi_2'(h)}{W(z_0)} \\ -\frac{\varphi_1(h)\varphi_2(z_0)-\varphi_1(z_0)\varphi_2(h)}{W(z_0)} \\ 0 \end{pmatrix} \tag{12}$$

for the determination of the constants C_1, C_2, C_3 and C_4. The entries of the matrix $\mathcal{M}$ can be found in [6]. It is shown there that the singularities of G_2 are the zeroes of $\det(\mathcal{M})$ (the eigenvalues of the problem) and the $m_\pm$ and a_u branch cuts. The $m_\pm$ branch cuts go to the left horizontally from the points

$$m_{0\pm} = \frac{B_{11} + B_{22} \pm \sqrt{(B_{11} - B_{22})^2 + 4B_{12}B_{21}}}{2k_0^2}$$

while the a_u branch goes to the left horizontally from the point

$$a_{u0} = \frac{B_{33}}{k_0^2}.$$

From (12) we have

$$C_1 = \lambda_0\frac{\Delta_1}{\Delta_0}$$

where $\Delta_0 = \det(\mathcal{M})$ and $\Delta_1 = \det(\mathcal{M}_1)$ with

$$\mathcal{M}_1 = \begin{pmatrix} \frac{1}{\lambda_o k_0^2}\frac{\varphi_1'(h)\varphi_2(z_0)-\varphi_1(z_0)\varphi_2'(h)}{W(z_0)} & M_{12} & M_{13} & M_{14} \\ \frac{1}{\lambda_o k_0^2}\frac{\varphi_1'(h)\varphi_2(z_0)-\varphi_1(z_0)\varphi_2'(h)}{W(z_0)} & M_{22} & M_{23} & M_{24} \\ -\frac{\varphi_1(h)\varphi_2(z_0)-\varphi_1(z_0)\varphi_2(h)}{W(z_0)} & M_{32} & M_{33} & M_{34} \\ 0 & M_{42} & M_{43} & M_{44} \end{pmatrix}.$$

To compute the contour integral in (10) we introduce counter clockwise slit cuts $\mathcal{C}_{m_\pm}$ and $\mathcal{C}_{a_u}$ about the three branch cuts. The depth Green's function is now analytic outside of the contour $\mathcal{C}_{m_+} + \mathcal{C}_{m_-} + \mathcal{C}_{a_u}$ except at the eigenvalues $\{\kappa_n\}$. Computing the residues at the eigenvalues gives the following representation for pressure

Symbol	Parameter
ρ_f	Density of the pore fluid
ρ_r	Density of sediment grains
K_b	Complex frame bulk modulus
μ	Complex frame shear modulus
K_f	Fluid bulk modulus
K_r	Grain bulk modulus
β	Porosity
η	Viscosity of pore fluid
k	Permeability
α	Structure factor
a	Pore size parameter

Table 1: Parameters in the Biot-Stoll model

$$\begin{aligned} p_o(r,z,z_0) &= \sum_n \frac{k_0^2\lambda_0 i}{4} \frac{\Delta_1(\kappa_n)}{\frac{d\Delta_0}{d\kappa}(\kappa_n)} \varphi_1(z,\kappa) H_0^{(1)}(k_0\sqrt{\kappa_n} r) \\ &\quad - \frac{k_0^2}{8\pi} \oint_{C_{m_+}+C_{m_-}+C_{a_u}} G_2(z,z_0,\kappa) H_0^{(1)}(k_0\sqrt{\kappa} r) d\kappa. \end{aligned} \tag{13}$$

4 Numerical simulation of the effect of sound speed variability

The Biot-Stoll model for a poroelastic seabed depends upon the eleven parameters shown in Table 1. We compare the predictions of the Biot model for the upward and downward refracting profiles shown in Figure 1 over three seabeds whose Biot parameters $\lambda, \mu, R, Q, \rho_{11}, \rho_{12}, \rho_{22}, b$ are calculated from estimates of Biot-Stoll data taken from the literature. Two of the seabeds, a coarse sand and fine gravel sediment and a fine sand sediment were taken from Holland and Brunson [7], while the third, a coarse sand sediment was taken from Beebe, McDaniel and Rubano [1]. The formulas used to calculate the Biot parameters from the data in these articles are given in [5]. Table 2 shows the parameter estimates for these three seabeds. As can be seen from the estimates of porosity and permeability these sediments vary considerably in their degree of coarseness.

In the representation (13) the eigenvalues $\{\kappa_n\}$are found numerically by minimizing $|\Delta_0(\kappa)|$. This has been found more reliable than solving $\Delta_0(\kappa) = 0$ numerically. The derivatives $\frac{d\Delta_0}{d\kappa}(\kappa_n)$ are also computed numerically. The values of the modal functions φ_1 and φ_2 were calculated using a fourth order Runge-Kutta scheme. It was found that a step size of one meter produced very good agreement with the values of the modal functions given by (11) in the isovelocity case. In this article we shall be concerned with transmission loss in the far field. The contribution of the branch-cut integrals in (13) is expected to be slight at great distances from the transmitter and will be neglected. An assessment of the contribution of these integrals in the near field for an isovelocity ocean is given in [6].

Transmission loss is defined to be $TL = -20\log(N_0\,|p_0(r,z,z_0)|)$ and is measured in decibels.

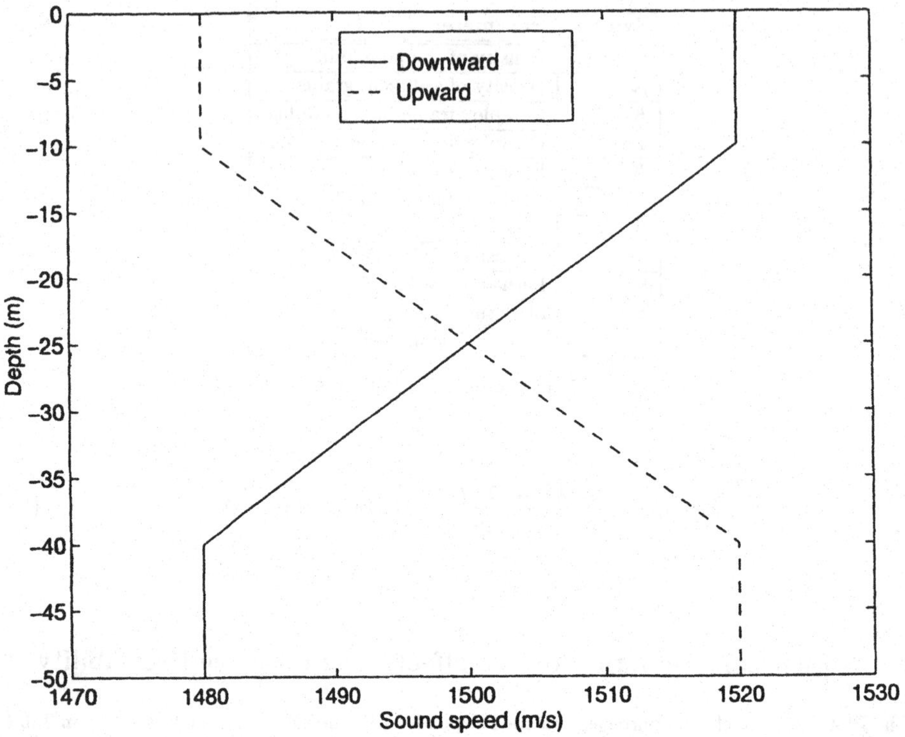

Figure 1: Upward and downward refracting sound speed profiles in an ocean of depth 50m.

Symbol	Fine sand	Coarse sand	Coarse sand, fine gravel
ρ_r	2670	2710	2680
K_b	$4.8 \times 10^7 + 6.7 \times 10^5 i$	$5.2 \times 10^7 + 7.4 \times 10^5 i$	$2.71 \times 10^7 + 9.0 \times 10^5 i$
μ	$6.7 \times 10^7 + 4.3 \times 10^6 i$	$7.4 \times 10^7 + 4.7 \times 10^6 i$	$1.25 \times 10^7 + 4.0 \times 10^5 i$
K_r	4.0×10^{10}	5.6×10^{10}	4.0×10^{10}
β	0.43	0.38	0.30
k	3.12×10^{-14}	7.5×10^{-11}	2.58×10^{-10}
α	1.25	1.25	1.25
a	1.19×10^{-6}	3.3×10^{-5}	1.31×10^{-4}
ρ_f	1000	1000	1000
K_f	2.39×10^9	2.4×10^9	2.4×10^9
η	1.01×10^{-3}	1.01×10^{-3}	1.01×10^{-3}

Table 2: Biot-Stoll parameters for three sand sediments. All dimensioned parameters are in MKS units.

For a point source the normalization factor is taken to be $N_0 = 4\pi$. This has the effect of making $TL = 0$ one meter from the source (hence dB re 1m).

The eigenvalues in this problem exhibit the "leaky modes" behavior common in underwater acoustics problems, that is, eigenvalues emerge from the branch cuts at certain frequencies. Those eigenvalues emerging from the m_+ branch cut are observed to migrate toward $\kappa = 1$ as frequency increases. Because $H_0^{(1)}(k_0\sqrt{\kappa}r)$ decays rapidly with increasing $\text{Im}\sqrt{\kappa}$, those eigenvalues lying near the positive real κ-axis will make the greatest contribution at long ranges. Figure 2 shows the eigenvalue maps for the upward and downward refracting profiles of Figure 1 at 100 Hz when the underlying seabed is the coarse sand and fine gravel sediment of Table 2.

Observe that for the eigenvalue nearest $\kappa = 1$ for the upward refracting profile is nearer to the real axis than the corresponding one for the downward refracting profile. Thus at long ranges where these eigenvalues dominate, the upward refracting profile will produce less transmission loss. For the eigenvalues of the secondary modes the situation is reversed. These modes propagate more strongly for the downward refracting profile.

The sound speed profile in the ocean will affect transmission loss in two ways. First it will channel acoustic energy toward or away from the receiver and second it will deflect energy toward or away from the bottom where it may absorbed into the sediment. To assess the relative influence of these two factors we consider transmission loss in an ocean of depth 50m with a source placed at depth 25m and a receiver which is either mid-depth (25m) or bottom mounted (49m). Figures 3 and 4 show transmission loss predictions for mid-depth and bottom mounted receivers in an ocean 50m in depth when the transmission frequency is 50Hz. Figures 5 and 6 show the predictions for the same bathymetry when the source frequency is 100Hz. At the high frequencies transmission loss is a complicated function of range, hence incoherent transmission loss, obtained by summing the moduli of the terms in the representation (13), is more enlightening. Incoherent transmission loss as a function of range for a source frequency of 100Hz is shown for mid-depth and bottom mounted receivers in Figures 7 and 8 respectively. In these figures transmission loss for an isovelocity ocean with a sound speed of 1500m/sec is also shown.

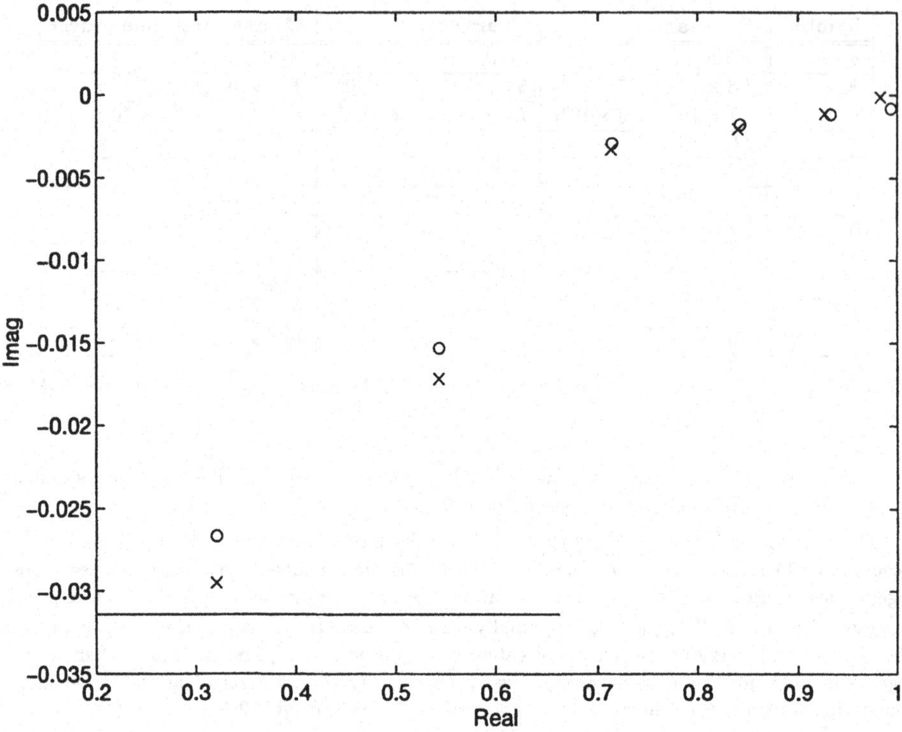

Figure 2: Eigenvalue maps for downward (o) and upward (x) refracting sound speed profiles for coarse sand, fine gravel at 100 Hz. The ocean depth is 50 m. The m_+-branch cut is shown at the lower left.

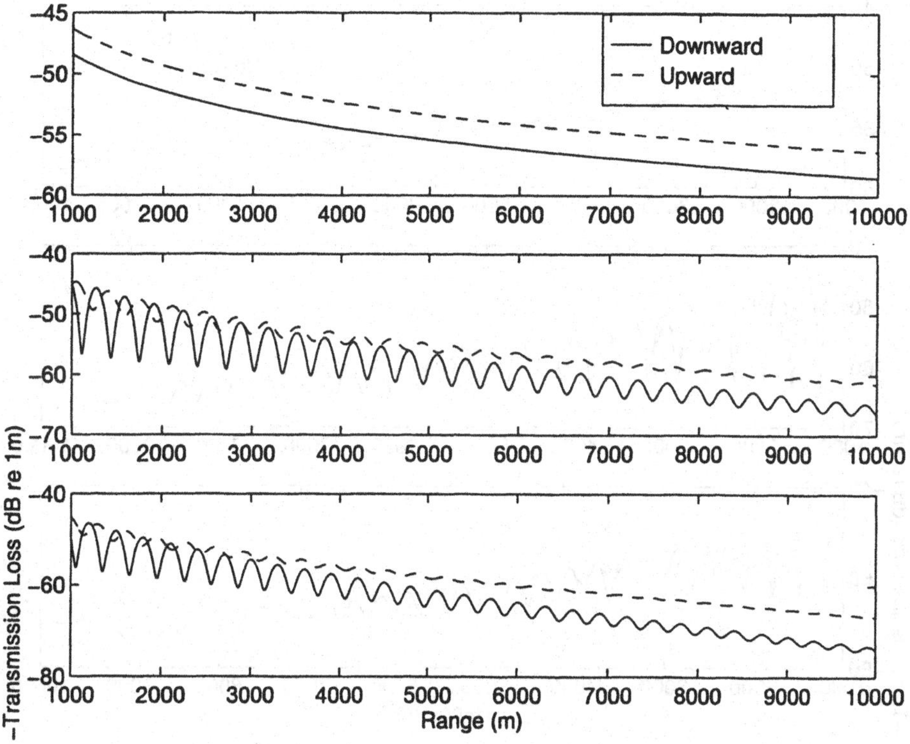

Figure 3: Transmission loss for a mid-depth receiver at 50 Hz. The ocean depth is 50m and both the source and receiver are at 25m. Top: fine sand; Middle: Coarse sand; Bottom: Coarse sand and fine gravel.

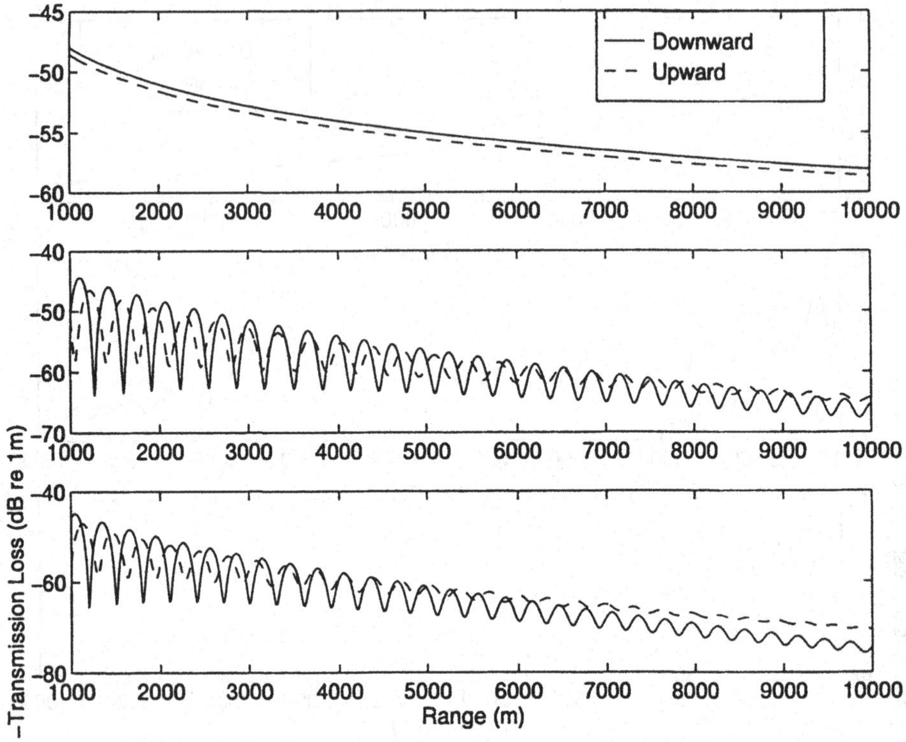

Figure 4: Transmission loss for a bottom-mounted receiver at 50 Hz. The ocean depth is 50m, the source is at 25m and the receiver is at 49m. Top: fine sand; Middle: Coarse sand; Bottom: Coarse sand and fine gravel.

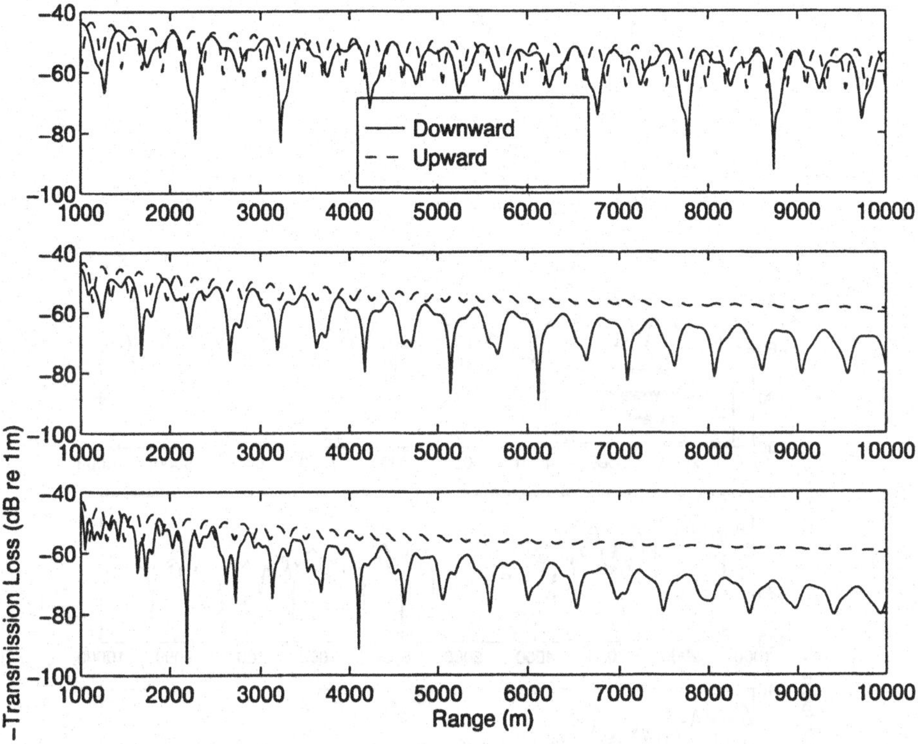

Figure 5: Transmission loss for a mid-depth receiver at 100 Hz. The ocean depth is 50m, the source is at 25m and the receiver is at 25m. Top: fine sand; Middle: Coarse sand; Bottom: Coarse sand and fine gravel.

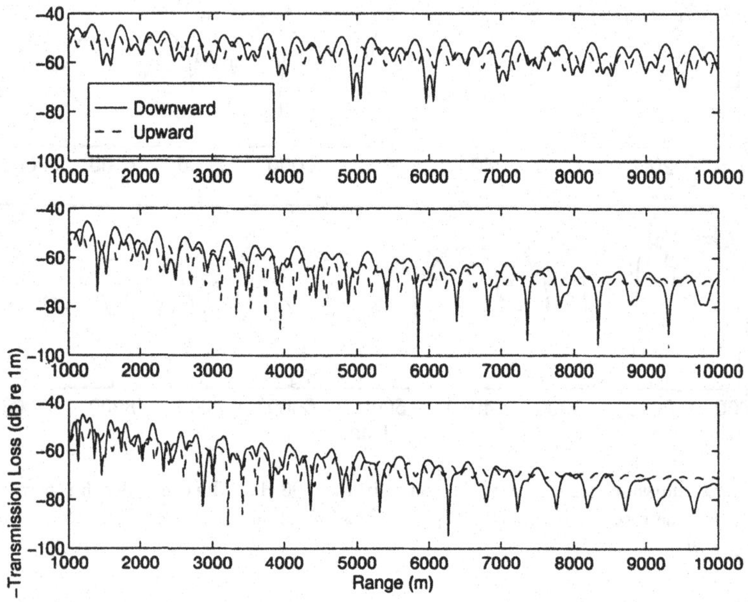

Figure 6: Transmission loss for a bottom-mounted receiver at 100 Hz. The ocean depth is 50m, the source is at 25m and the receiver is at 49m. Top: fine sand; Middle: Coarse sand; Bottom: Coarse sand and fine gravel.

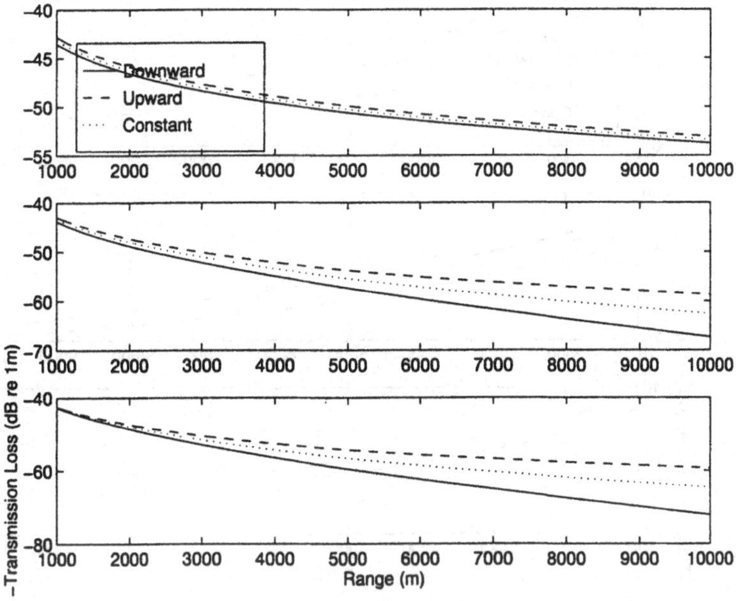

Figure 7: Incoherent transmission loss for a mid-depth receiver at 100 Hz. The ocean depth is 50m, the source is at 25m and the receiver is at 25m. Top: fine sand; Middle: Coarse sand; Bottom: Coarse sand and fine gravel. The speed in the isovelocity ocean is 1500 m/sec.

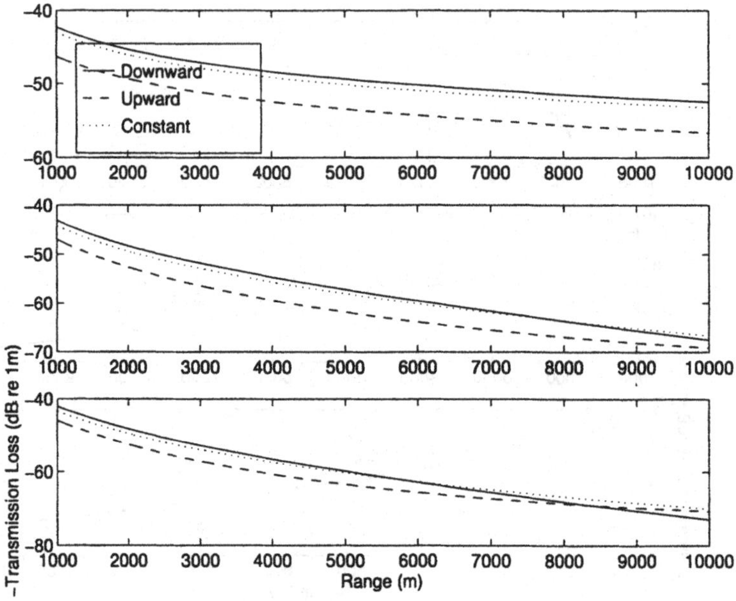

Figure 8: Incoherent transmission loss for a bottom-mounted receiver at 100 Hz. The ocean depth is 50m, the source is at 25m and the receiver is at 49m. Top: fine sand; Middle: Coarse sand; Bottom: Coarse sand and fine gravel. The speed in the isovelocity ocean is 1500 m/sec.

From the figures indicated above it can be seen that for a mid-depth receiver loss is greater for the downward refracting profile. Thus losses due to sediment absorption predominate. The coarseness of the sediment is consequential; the coarse sand and coarse sand and fine gravel sediments in Table 2 show a much greater discrepancy in loss between the two profiles than does the fine sand sediment. For a bottom mounted receiver deflection of energy away from the reciever by the upward refracting profile is more consequential than bottom loss at short ranges. Only in the coarsest of the three sediments does the effect of loss due to repeated interactions with the sediment become predominant in the last two kilometers of the 1-10 km range shown.

5 Conclusion

Based upon the results of these simulations, the coarseness of the sediment has a significant effect on transmission loss in an ocean with a depth-varying sound speed profile. Coarser sediments absorb more energy and thus a downward refracting profile which increases interaction with the sediment will cause greater loss. However at short ranges even for coarse sediments the increased loss due to absorption for a downward refracting profile may be less consequential than the deflection of energy away from a bottom mounted receiver by an upward refracting profile.

References

[1] J.H. Beebe, S.T. McDaniel, and L.A. Rubano. Shallow-water transmission loss prediction using the biot sediment model. *J. Acoust. Soc. Am.*, 71(6):1417–1426, 1982.

[2] M.A. Biot. Theory of propogation of elastic waves in a fluid-saturated porous solid. i. lower frequency range. *J. Acoust. Soc Am.*, 28(2):168–178, 1956.

[3] M.A. Biot. Theory of propogation of elastic waves in a fluid-saturated porous solid. ii. higher frequency range. *J. Acoust. Soc. Am.*, 28(2):179–191, 1956.

[4] J.L. Buchanan and R.P. Gilbert. Transmission loss in the far field over a seabed with rigid substrate assuming the biot sediment model. *J. Computational Acoustics*, 4(1):29–54, 1996.

[5] J.L. Buchanan and R.P. Gilbert. Transmission loss in the far field over a one-layer seabed assuming the biot sediment model. *ZAMM*, 77(2):121–135, 1997.

[6] J.L. Buchanan, R.P. Gilbert, and Y. Xu. Green's function representation for acoustic pressure over a poroelastic seabed. *Applicable Analysis*, 65:57–68, 1997.

[7] C.W. Holland and B.A. Brunson. The biot-stoll sediment model: An experimental assessment. *J. Acoust. Soc. Am.*, 84(4):1437–1443, 1988.

VARIATIONS ON (X,ψ) DUALITY

Robert Carroll

Mathematics Department
University of Illinois
Urbana, IL 61801
email: rcarroll@math.uiuc.edu

Abstract: We sketch some possible extensions of the (X,ψ) duality ideas of Faraggi-Matone [15, 16] in two directions. First one looks at a formulation in terms of Baker-Akhiezer (BA) functions of integrable systems and secondly we sketch some of the Olavo theory of [25] in order to indicate possible connections.

1.1 INTRODUCTION

We give in Section 2 some preliminary ideas in a general KP context toward a construction of formulas resembling those in the (X,ψ) duality of [15, 16], as expanded in [2, 3, 11]. In Section 3 we extract some ideas from the Olavo theory of [25] showing how the Schrödinger equation is related to classical statistical mechanics and then indicate possible connections to (X,ψ) duality.

1.2 DEVELOPMENT

Let us try to extend the (X,ψ) duality ideas of [2, 3, 10, 15, 16] to a more general context. The connection to quantum mechanics (QM) is removed but WKB ideas abound. We will sketch some themes and refer to [12, 13] for additional development. Thus first consider KP and recall from [4, 5, 6, 7, 8, 10, 21, 24] ($\bullet$) $L = \partial + \sum_1^\infty u_{n+1}\partial^{-n}$; $W = 1 + \sum_1^\infty w_n\partial^{-n}$; $L = W\partial W^{-1}$ (where $\partial \sim \partial/\partial x$) with equations ($\bullet\bullet$) $\partial_n\psi = L_+^n\psi$; $\psi = e^\xi(\tau_-/\tau)$; $\psi^* = e^{-\xi}(\tau_+/\tau)$ and

R.P. Gilbert et al.(eds.), Direct and Inverse Problems of Mathematical Physics, 39–52.

(• • •) $\tau_{\pm} = \tau(t_n \pm [1/n\lambda^n])$; $L\psi = \lambda\psi$; $L^*\psi = \lambda\psi^*$; $\xi = \sum_1^\infty t_n\lambda^n$. The transition to dispersionless form involves (cf. [2, 3, 4, 5, 6, 10, 11, 18, 20, 28] $T_n = \epsilon t_n$ ($x \sim t_1$, $X \sim T_1$) with

$$L_\epsilon = \epsilon\partial + \sum_1^\infty u_{n+1}(\epsilon, T)(\epsilon\partial)^{-n};\ \psi = e^{\frac{1}{\epsilon}S + O(1)};\ \tau = exp\left[\frac{1}{\epsilon^2}F + \left(\frac{1}{\epsilon}\right)\right] \quad (2.1)$$

where $u_{n+1}(\epsilon, T) = U_{n+1}(T) + O(\epsilon)$. Then passing $\epsilon \to 0$ yields for $P = S_X$

$$\lambda = P + \sum_1^\infty U_{n+1}P^{-n};\ B_n = L_+^n = \sum_0^n b_{nm}\partial^m \to \mathcal{B}_n = \sum_1^n b_{nm}P^m = \lambda_+^n; (2.2)$$

$$\partial_n S = \mathcal{B}_n(P);\ S = \sum_1^\infty T_n\lambda^n + \sum_1^\infty S_{j+1}\lambda^{-j};\ S_{j+1} = -\frac{\partial_j F}{j}$$

($\partial_n \sim \partial/\partial T_n$ in a dKP context) along with a Hamilton-Jacobi (HJ) theory (♣) $\dot{P}_n = dP/dT_n = \partial\mathcal{B}_n$; $\dot{X}_n = dX/dT_n = -\partial_P\mathcal{B}_n$. In fact one can say that dKP is characterized by (♣♣) $\lambda = P + \sum_1^\infty U_{n+1}P^{-n}$ and $\partial_n S = \mathcal{B}_n(P)$. This was enlarged in [2, 3, 11] (mainly for KdV) to an enhanced dKP theory involving all the ϵ terms (i.e. $\epsilon \not\to 0$ where $\epsilon \sim \hbar/\sqrt{2m}$). In that situation we looked at ($' \sim \partial/\partial X$) (♠) $-(\hbar^2/2m)\psi'' + V(X)\psi = E\psi$ as a "predispersionless" form of dKdV and introduced $\tilde{S} = S^0 + \sum_1^\infty \epsilon^j S^j$ (with $S^0 \sim S$ above) arising from $\tilde{F} = \sum_0^\infty \epsilon^j F^j$ ($F^0 \sim F$ above) via the vertex operator equation (VOE) $\psi = exp(\xi)\tau_-/\tau$. This led to a full WKB type theory to all orders of ϵ and for KdV this could be connected to the (X, ψ) duality of [15, 16] (as well as to Riemann surfaces and Seiberg-Witten = SW theory).

A natural generalization now can be developed as follows. One asks for a "prepotential" $\mathcal{F}$ such that $\psi^* = \partial\mathcal{F}/\partial\psi$ with $\mathcal{F}' = \psi'\psi^*$ and $\psi = \psi(X)$, $X = X(\psi)$ is to be a basic dependence (independent of other "time" variables). One can motivate this via HJ theory for example where X and P are distinguished variables. Then one expects $\mathcal{F} = (1/2)\psi\psi^* + G$ where necessarily $\mathcal{F}' = (1/2)\psi'\psi^* + (1/2)\psi(\psi^*)' + G' = \psi'\psi^*$ implies (♠♠) $\psi'\psi^* - \psi(\psi^*)' = 2G'$. In the KdV situation with $\psi^* \sim \bar{\psi}$ (♠♠) is a Wronskian and $G' = 1/i\epsilon$ with $G = X/i\epsilon$. Here we consider $\epsilon\partial_2\psi = (\epsilon^2\partial^2 + 2U)\psi$ and $-\epsilon\partial_2\psi^* = (\epsilon^2\partial^2 + 2U)\psi^*$ where $u = u_2$ ($\partial \sim \partial/\partial X$ and $\partial_n \sim \partial/\partial T_n$ in a dKP context). Multiplying by ψ^* and ψ respectively and subtracting we have $\epsilon\partial_2(\psi\psi^*) = \epsilon\partial(\psi^*\epsilon\psi' - \epsilon(\psi^*)'\psi)$ so to mimic KdV we could ask for situations where $\partial_2(\psi\psi^*) = 0$. Then $\psi^*\epsilon\psi' - \epsilon(\psi^*)'\psi = c \sim 2\epsilon G'$ (c independent of X) leading to $G' = c/2\epsilon$ and $G = cX/2\epsilon = \alpha X/\epsilon$ so that $\mathcal{F} = (1/2)\psi\psi^* + (\alpha X/\epsilon)$. Note that $\partial_2(\psi\psi^*) = 0$ holds for KdV where the even variables are absent of course but a priori it could also hold in more general situations. We recall here that $\psi\psi^*$ is an extremely important creature in studying symmetries and Hamiltonian theory. Thus in particular we recall from [7, 8, 10, 21, 24] ($u \sim u_2$) that $\partial_n u = \partial s_{n+1} = K_{n+1}(u)$ and

$$\partial_3 S = \frac{1}{4}\partial^3 S + 3\partial(uS) + \frac{3}{4}\partial^{-1}\partial_2^2 S = K'[S]\ (K \sim K_4) \quad (2.3)$$

for symmetries $\mathcal{S} \sim K_n$ (this is the linearized KP equation) while for conserved densities $\gamma \sim s_n$

$$\partial_3\gamma = \frac{1}{4}\partial^3\gamma + 3u\partial\gamma + \frac{3}{4}\partial^{-1}\partial_2^2\gamma \tag{2.4}$$

(adjoint linearized KP equation); note $\mathcal{S}_n = \partial s_n$ with $s_0 = 1$, $s_1 = 0$, and we recall also $\psi\psi^* = \sum_0^\infty (s_n/\lambda^n)$. Since $s_n = \partial^{-1}\partial_{n-1}u$ from [7, 21], $\partial_2 s_n = 0 \sim \partial_2 u = 0$ so that is sufficient for $\partial_2(\psi\psi^*) = 0$ (recall also $u_2 = \partial^2 log\tau$). This does not necessarily correspond to KdV (or a 2-reduction of KP) where no t_{2n} are present. Let us note that in general a formula $\mathcal{F} = (1/2)\psi\psi^* + G$ could involve a complicated G. Thus e.g. $\partial_2(\psi\psi^*) \neq 0$ specifies $G'' = (1/2\epsilon)\partial_2(\psi\psi^*)$ where one is thinking of variables T_n.

Thus consider $G'' = (1/2\epsilon)\partial_2(\psi\psi^*)$ in T_n variables. One of the features of (X, ψ) duality in [15, 16] was the Legendre forms (KdV situation now)

$$-\frac{X}{i\epsilon} = \psi^2\frac{\partial\mathcal{F}}{\partial(\psi^2)} - \mathcal{F};\ -\mathcal{F} = \phi\left(\frac{X_\phi}{i\epsilon}\right) - \frac{X}{i\epsilon} \tag{2.5}$$

where $\phi = \partial\mathcal{F}/\partial(\psi^2) = \bar{\psi}/2\psi$. Recall in classical mechanics $\dot{Q} = \partial_p H$ and $\dot{p} = -\partial_q H$ with $p\dot{q} - L = p\partial_p H - L = H$. Thus the Legendre duality in (2.5) pairs $-X/i\epsilon$ and ψ^2 with $\mathcal{F} \sim H$ or $-\mathcal{F}$ and ϕ with $X/i\epsilon \sim H$. The connection to QM here required no extra variables in the fundamental relations $\psi = \psi(X)$, $X = X(\psi)$. On the other hand from $\mathcal{F} = (1/2)\psi\bar{\psi} + (X/i\epsilon)$ in the KdV situation one introduced "variables" or better parameters T_{2n+1} (and λ) in $\mathcal{F}$ via their presence in $|\psi|^2$. This does not remove the fundamental X dependence in $\psi = \psi(X)$ or $\mathcal{F} = \mathcal{F}(X)$; it simply marks how the KdV framework infuses (λ, T_n) parameters into $\mathcal{F}$ and these parameters will eventually be fixed in any event. However X or better $X/i\epsilon$ stands alone here. Is there any reason why G cannot contain other parameters (T_n, λ)? Consider the Legendre aspects here. Thus $\psi'\psi^* = (1/2)\psi'\psi^* + (1/2)\psi\psi'(\partial\psi^*/\partial\psi) + G'$ implies $(\partial\psi^*/\partial\psi) = (1/\psi)\left[\psi^* - (2G'/\psi')\right]$. Here for $X_\psi = 1/\psi'$ we have $\partial_\psi = 2\psi\partial/\partial(\psi^2)$ and $\phi = \partial\mathcal{F}/\partial(\psi^2) = \psi^*/2\psi$ in analogy to KdV. Also $X_\phi = X_\psi\psi_\phi$ with $\psi_\phi \sim 1/\phi_\psi$ where $\phi_\psi = -(\psi^*/2\psi^2) + (1/2\psi)(\partial\psi^*/\partial\psi) = -G'/\psi^2\psi'$ so $X_\phi = -\psi^2/G'$ or $\psi^2 = -G'X_\phi$. Hence first $\psi^2(\partial\mathcal{F}/\partial(\psi^2) = (1/2)\psi\psi^*$ leading to $-G = \psi^2(\partial\mathcal{F}/\partial(\psi^2)) - \mathcal{F}$ and $\psi^2 = -G'/X_\phi$ with $\phi = \psi^*/2\psi$ implies $\phi(-G'X_\phi) = (1/2)\psi\psi^*$ leading to $-\mathcal{F} = \phi(G'X_\phi) - G$. This suggests Legendre duality for $-G$ and ψ^2 with $\mathcal{F} \sim H$ but duality for $-\mathcal{F}$ and ϕ seems to require $G' = c$ and $G = cX \sim H$. We note also that at the S or $\tilde{S}$ level $\mathcal{F} = (1/2)\psi\psi^* + G$ leads to $\tilde{\mathcal{F}}' = (1/2\epsilon)(\tilde{P} + \tilde{P}^*)(2\tilde{\mathcal{F}} - 2G) + G'$ so that $\tilde{\mathcal{F}}' - (1/\epsilon)(\tilde{P} + \tilde{P}^*)\tilde{\mathcal{F}} = G' - (1/\epsilon)(\tilde{P} + \tilde{P}^*)G$ but the meaning eludes us.

Now following [2, 3, 11], in order to balance ϵ terms, we will be using a full ϵ expansion in S and F so e.g. from $(\bullet)$ – $(\bullet\bullet\bullet)$ we have first for simple $F = F^0$, $\psi = exp(\xi)(\tau_-/\tau)$ and $\psi^* = exp(-\xi)(\tau_+/\tau)$

$$\frac{1}{\epsilon}S = \frac{1}{\epsilon}\xi + \frac{1}{\epsilon^2}\left[F^0\left(T_n - \frac{\epsilon}{n\lambda^n}\right) - F^0(T_n)\right]; \tag{2.6}$$

$$\frac{1}{\epsilon}S^* = -\frac{1}{\epsilon}\xi + \frac{1}{\epsilon^2}\left[F^0\left(T_n + \frac{\epsilon}{n\lambda^n}\right) - F^0(T_n)\right]$$

Consequently

$$log(\psi\psi^*) = \frac{1}{\epsilon}(S + S^*) = 2\sum_1^\infty \lambda^{-m-n}\frac{F^0_{mn}}{mn} + O(\epsilon^2) \tag{2.7}$$

and we note also $\partial_2(\psi\psi^*) = 0 \sim F_2 = \partial F/\partial T_2 = 0$. Extending further for $F \sim \tilde{F} = \sum_0^\infty \epsilon^n F^n$ we get

$$F\left(T_n - \frac{\epsilon}{n\lambda^n}\right) - F(T_n) = \sum_{k=0}^\infty \epsilon^k\left[-\epsilon\sum_1^\infty\left(\frac{F^k_n}{n\lambda^n}\right) + \frac{\epsilon^2}{2}\sum\frac{F^k_{mn}}{mn}\lambda^{-m-n} + O(\epsilon^3)\right] \tag{2.8}$$

leading to $S \to \tilde{S} = \sum_1^\infty T_n\lambda^n + \sum_0^\infty \epsilon^k \sum_1^\infty S^k_{j+1}\lambda^{-j} = S^0 + \sum_1^\infty \epsilon^k \sum_1^\infty S^k_{j+1}\lambda^{-j}$ and one has

$$\frac{1}{\epsilon}\sum_0^\infty \epsilon^k\left[-\sum_1^\infty\left(\frac{F^k_n}{n}\right)\lambda^{-n} + \frac{\epsilon}{2}\sum\left(\frac{F^k_{mn}}{nm}\right)\lambda^{-m-n} + O(\epsilon^2)\right] =$$

$$= \frac{1}{\epsilon}\sum_0^\infty \epsilon^k \sum_1^\infty S^k_{j+1}\lambda^{-j} \tag{2.9}$$

(note here that lower indices correspond to derivatives and upper indices are position markers except for S^k_{j+1} where $j+1$ is a position marker). Hence in particular

$$\sum_1^\infty S^0_{j+1}\lambda^{-j} = -\sum_1^\infty\left(\frac{F^0_n}{n}\right)\lambda^{-n}; \tag{2.10}$$

$$\sum_1^\infty S^1_{j+1}\lambda^{-j} = -\sum_1^\infty\left(\frac{F^1_n}{n}\right)\lambda^{-n} + \frac{1}{2}\sum_1^\infty\left(\frac{F^0_{nm}}{mn}\right)\lambda^{-m-n}$$

leading to $S^0_{j+1} = -F^0_j/j$ and $S^k_2 = -F^k_1$ with

$$S^k_{j+1} = -\frac{F^k_j}{j} + \frac{1}{2}\sum_1^{j-1}\left(\frac{F^{k-1}_{m,(j-m)}}{m(j-m)}\right) + \cdots + (-1)^{k+1}\frac{1}{(k+1)!} \sum_{\sum m_i = j}\frac{F^0_{m_1,\cdots,m_{k+1}}}{\prod_1^{k+1} m_i} \tag{2.11}$$

for $k \geq 1$, together with

$$\tilde{P} = \tilde{S}_X = P + \sum_1^\infty \epsilon^k \sum_1^\infty \partial_X S^k_{j+1}\lambda^{-j} = P + \sum_1^\infty \epsilon^k S^k_X;\ S^k = \sum_1^\infty S^k_{j+1}\lambda^{-j} \tag{2.12}$$

where S^k_{j+1} is given in (2.11) and $P^k = P_k = \sum_1^\infty \partial_X S^k_{j+1}\lambda^{-j}$.

Let us note that since $\gamma \sim \psi\psi^*$ satisfies (2.4) this serves in a way as a replacement for the Gelfand-Dickey (GD) equation in the KdV situation to provide a differential equation for $\mathcal{F}$. Indeed for $\partial_2\gamma = 0$ we have $\epsilon\partial_3\gamma = (1/4)\epsilon^3\partial^3\gamma + 3U\epsilon\partial\gamma$. Then write $\psi\psi^* = \gamma = 2[\mathcal{F} - (\alpha X/\epsilon)]$ to obtain

$$\partial_3\mathcal{F} = \frac{1}{4}\epsilon^2\mathcal{F}''' + 3U\left(\mathcal{F}' - \frac{\alpha}{\epsilon}\right) \tag{2.13}$$

Since $u = \partial^2 \log\tau$ implies that $U = \epsilon^2\partial^2(\tilde{F}/\epsilon^2) = \tilde{F}_{11}$ we have a linkage between $\tilde{F}$ and $\mathcal{F}$ (as in [2, 3, 11]) and this led to a connection between SW variables a_i and $\mathcal{F}$ in the Riemann surface situation. A similar connection therefore exists here as well in a finite zone KP situation.

Next in order to find relations between the $F^j_{ab\cdots q}$ we can proceed as in [11] (indicated also in [2]) and write the differential Fay identity in the form

$$\frac{1}{\epsilon^2}\sum_1^\infty \lambda^{-m-n}\chi_n(-\epsilon\tilde{\partial})\chi_m(-\epsilon\tilde{\partial})\tilde{F} = log\,\dot{\Xi}(\lambda,\epsilon) \tag{2.14}$$

where

$$\lambda - \Xi(\lambda,\epsilon) = -\frac{1}{\epsilon}\sum_1^\infty \chi_s(-\epsilon\tilde{\partial})\lambda^{-s}\tilde{F}_1;\ \dot{\Xi}(\lambda,\epsilon) = 1 + \frac{1}{\epsilon}\sum_1^\infty(-s)\lambda^{-s-1} \chi_s(-\epsilon\tilde{\partial})\tilde{F}_1 \tag{2.15}$$

This leads to $\Xi = \tilde{P} = \tilde{S}_X$ and formulas

$$\dot{\Xi} = \sum_0^\infty \chi_\ell(\tilde{Z}_1 = 0, \tilde{Z}_2, \cdots, \tilde{Z}_\ell)\lambda^{-\ell};\ \tilde{Z}_j = \sum_{m+n=j}\chi_n(-\epsilon\tilde{\partial})\chi_m(-\epsilon\tilde{\partial})\frac{\tilde{F}}{\epsilon^2} \tag{2.16}$$

$$\sum_0^\infty \chi_\ell(\tilde{Z}_1 = 0, \tilde{Z}_2, \cdots, \tilde{Z}_\ell)\lambda^{-\ell} = \dot{\Xi} = 1 - \frac{1}{\epsilon}\sum_2^\infty(\ell-1)\lambda^{-\ell}\chi_{\ell-1}\tilde{F}_1 \Rightarrow$$

$$\Rightarrow -\frac{1}{\epsilon}(\ell-1)\chi_{\ell-1}(-\epsilon\tilde{\partial})\tilde{F}_1 = \chi_\ell(\tilde{Z}_1 = 0, \tilde{Z}_2, \cdots, \tilde{Z}_\ell)\ \ (\ell \geq 2) \tag{2.17}$$

Next we observe that with $\mathcal{F} = (1/2)\psi\psi^* + (\alpha X/\epsilon)$ one has for $\psi = exp(\tilde{S}/\epsilon)$ and $\psi^* = exp(\tilde{S}^*/\epsilon)$ the formulas $\psi' = (\tilde{P}/\epsilon)\psi$ and $(\psi^*)' = (\tilde{P}^*/\epsilon)\psi^*$ so $\epsilon\psi'\psi^* - \psi\epsilon(\psi^*)' = 2\alpha$ implies

$$\psi\psi^*(\tilde{P} - \tilde{P}^*) = 2\alpha \Rightarrow \frac{1}{\epsilon}(\tilde{S} + \tilde{S}^*) = \tag{2.18}$$

$$= log\left(\frac{2\alpha}{\tilde{P} - \tilde{P}^*}\right) \Rightarrow \frac{1}{\epsilon}(\tilde{P} + \tilde{P}^*) = -\frac{\tilde{P}' - (\tilde{P}^*)'}{\tilde{P} - \tilde{P}^*}$$

This appears to involve restrictions on some of the compatibility conditions from (2.14) (as in [2, 11]) and in the KdV situation it essentially forces $F^{2j+1} = 0$. However one should not regard this as a constraint since (2.18) is a natural consequence of the WKB format (cf. [11] and Remark 3.2). Writing now $\tilde{P}^* = -\lambda + (1/\epsilon)\sum_1^\infty \lambda^{-j}\chi_j(\epsilon\tilde{\partial})\tilde{F}_1$ to go with $\tilde{P} = \Xi$ of (2.15) we have

$$\tilde{P} + \tilde{P}^* = \frac{1}{\epsilon}\sum_1^\infty \lambda^{-j}\left[\chi_j(\epsilon\tilde{\partial}) + \chi_j(-\epsilon\tilde{\partial})\right]\tilde{F}_1; \tag{2.19}$$

$$\tilde{P} - \tilde{P}^* = -2\lambda - \frac{1}{\epsilon}\sum_1^\infty \lambda^{-j}\left[\chi_j(\epsilon\tilde{\partial}) - \chi_j(-\epsilon\tilde{\partial})\right]\tilde{F}_1$$

In terms of (2.6) - (2.7) we have

$$\tilde{P} + \tilde{P}^* = \epsilon\sum_1^\infty \frac{F^0_{mn}}{mn}\lambda^{-m-n} + O(\epsilon^3);\ \tilde{P} - \tilde{P}^* = 2\lambda - 2\sum_1^\infty \frac{F_{1m}}{m}\lambda^{-m} + O(\epsilon^2) \tag{2.20}$$

Hence (2.18) can be written as

$$\left[\sum_1^\infty \lambda^{-m-n}\frac{F^0_{mn}}{mn} + O(\epsilon^2)\right] \times \left[\lambda - \sum_1^\infty \frac{F^0_{1m}}{m}\lambda^{-m} + O(\epsilon^2)\right] + $$

$$+\sum_1^\infty \frac{F^0_{11m}}{m}\lambda^{-m} + O(\epsilon^2) = 0 \tag{2.21}$$

The leading term is then

$$\left(\lambda - \sum_1^\infty \frac{F^0_{1m}}{m}\lambda^{-m}\right)\sum_1^\infty \frac{F^0_{mn}}{mn}\lambda^{-m-n} + \sum_1^\infty \frac{F^0_{11m}}{m}\lambda^{-m-n} = 0 \tag{2.22}$$

which should be checked as arising also from (2.17) under suitable circumstances. The pattern in (2.20) suggests further that $\tilde{P} = P_0 + \epsilon P_1 + \epsilon^2 P_2 + \cdots$ and $\tilde{P}^* = -P_0 + \epsilon P_1 - \epsilon^2 P_2 + \cdots$ so

$$\tilde{P} + \tilde{P}^* = 2\sum_0^\infty \epsilon^{2j+1}P_{2j+1};\ \tilde{P} - \tilde{P}^* = 2\sum_0^\infty \epsilon^{2j}P_{2j} \tag{2.23}$$

This is actually the same pattern as in [2, 11] leading to

$$2\left(\sum_0^\infty \epsilon^{2j}P_{2j+1}\right) \times \left(\sum_0^\infty \epsilon^{2j}P_{2j}\right) + \sum_0^\infty \epsilon^{2j}P'_{2j} = 0 \tag{2.24}$$

and there it was found to be consistent with compatibility relations determined by (2.17) (with $F^{2j+1} = 0$ as restriction).

It would be nice to have an attractive formula such as $L^2\psi = \lambda^2\psi$ leading to $\tilde{P}^2 + \epsilon\tilde{P}' + 2\tilde{F}_{11} = \lambda^2$ as for KdV in [2, 11] but here we will have to resort to the general formula inverting (2.2). Thus we can use (2.15) for $\tilde{P} = \Xi$. Note that an equation $\epsilon\partial_2\psi = 0 = \epsilon^2\partial^2\psi + 2u\psi$ is of no use here; ψ is also independent of T_2 in KdV. Thus via (2.15) we have for $\tilde{F} = \sum_0^\infty \epsilon^j F^j$

$$\tilde{P} = \lambda + \frac{1}{\epsilon}\sum_1^\infty \lambda^{-j}\chi_j(-\epsilon\tilde{\partial})\tilde{F}_1 \Rightarrow \tag{2.25}$$

$$\Rightarrow P_0 = \lambda - \sum_1^\infty \lambda^{-j}\frac{F^0_{1j}}{j};\ P_1 = \sum_1^\infty \lambda^{-j}\left(-\frac{F^1_{1j}}{j} + \frac{1}{2}\sum_{m+n=j}\frac{F^0_{1mn}}{mn}\right);\ \cdots$$

A priori there is no requirement on reality; for example in KP1 one can have $T_{2n} = i\hat{T}_{2n}$ with real potentials u_i and in this spirit F^j real is natural. Also $\psi\psi^* = \sum_0^\infty s_n/\lambda^n$ shows that $\psi\psi^*$ will be real for real u_j (and λ real) since the s_n are differential polynomials in the u_j. Recall that all of the u_j can be expressed in terms of u via operations involving ∂ and ∂^{-1}. However a formula of the form $s_n = (2\tau^2)^{-2}\partial\partial_{n-1}\tau\cdot\tau$ (in bilinear notation), corresponding to $s_n = \partial^{-1}\partial_{n-1}u$, shows that with complex times s_n could be imaginary.

Note that the expansion of Hamilton-Jacobi (HJ) theory to the full ϵ context is not trivial. Thus e.g. in the KdV situation one has $L_+^2\psi = L^2\psi = \lambda^2\psi$ generating $\tilde{P}^2 + \epsilon\tilde{P}' + 2U = \lambda^2$ which is not $\tilde{P}^2 + 2U = \lambda^2$; thus one cannot simply replace P by $\tilde{P}$ (cf. [3] for more on this). We should rather define $\tilde{\mathcal{B}}_n = \partial_n\tilde{S}$ ($\tilde{S}$ being well determined via the $F^j_{ab\cdots q}$). Then $\partial_n\tilde{P} = \hat{\partial}\tilde{\mathcal{B}}_n$ and we can probably think of $\tilde{\mathcal{B}}_n = \tilde{\mathcal{B}}_n(X, \tilde{P}, \tilde{P}', \cdots)$ so that $\hat{\partial}\tilde{\mathcal{B}}_n = \partial\tilde{\mathcal{B}}_n + \sum_0^N(\partial\tilde{\mathcal{B}}_n/\partial\tilde{P}^{(k)})\partial^{k+1}\tilde{P}$ (N suitably large, depending on n). The latter terms can be associated to the form $< (\delta\tilde{\mathcal{B}}_n/\delta\tilde{P}), \delta\tilde{P} > \sim < \sum_0^N(-\partial)^k(\partial\mathcal{B}_n/\partial\tilde{P}^{(k)}), \partial\tilde{P} >$ but this doesn't seem to be enough to mimic HJ theory. Note the standard procedure here from [5, 20] would involve $(d\tilde{P}/dT_n) = \partial_n\tilde{P} + \partial_X\tilde{P}(dX/dT_n) = \hat{\partial}\tilde{\mathcal{B}}_n + \partial_X\tilde{P}(dX/dT_n)$ leading weakly to $(d\tilde{P}/dT_n) = \partial\tilde{\mathcal{B}}_n$; $(dX/dT_n) = -\sum_0^N(-\partial)^k(\partial\tilde{\mathcal{B}}_n/\partial\tilde{P}^{(k)})$. Even when the P_i in $\tilde{P}$ can be written in terms of $P_0 \sim P$ (as in KdV - cf. [11]) one still must take into account $\partial\tilde{\mathcal{B}}_n/\partial P_0^{(k)}$.

1.3 REMARKS ON THE OLAVO THEORY

As indicated in [2, 3] there is an apparent connection between (X, ψ) duality and the theory of Olavo [25] which derives quantum mechanics (QM) via the density matrix and statistical mechanics using an infinitesimal Wigner-Moyal (WM) transformation ((3.1) below). We sketch here some ideas leading to the Schrödinger equation **(SE)** following [25] and indicate some possible directions for development in connection with (X, ψ) duality. We do not feel qualified to make ultimate judgements about the physics but it certainly appears to be plausible enough and in any case represents a refreshing attempt to deal with epistemology (cf. [14, 19, 26]); the mathematics seems perfectly correct in the sense normally acceptable in physics. In any case, given an apparent historical reluctance to deal with interpretive questions in QM (cf. [29]) we feel the present sketch is appropriate and hopefully it is faithfully rendered. The first derivation of **(SE)** (Article 1 in [25]) goes as follows. There are three axioms: **(A)** Newtonian mechanics is valid for all particles which constitute the systems in the "ensemble". **(B)** For an isolated system the joint probability density function is conserved, i.e. $(d/dt)F(x,p,t) = 0$ (we relace X by x in this section). **(C)** The Wigner-Moyal infinitesimal transformation defined by (3.1) below is adequate for the description of any non-relativistic quantum system (note the emphasis here on the infinitesimal aspect which seems to be a new idea).

$$\rho\left(x + \frac{\delta x}{2}, x - \frac{\delta x}{2}, t\right) = \int F(x,p,t) exp\left(i\frac{p\delta x}{\hbar}\right) dp \tag{3.1}$$

Then from **(B)** one has $dF/dt = \partial F/\partial t + (dx/dt)(\partial F/\partial x) + (dp/dt)(\partial F/\partial p)$ and using $dx/dt = p/m$ with $dp/dt = f = -\partial V/\partial x$, upon multiplying the resulting equation by $exp(ip\delta x/\hbar)$ and integrating there results (•♠♠) $-\partial\rho/\partial t + (i\hbar/m)\partial^2\rho/\partial x\partial(\delta x) - (i/\hbar)\delta V\rho = 0$ where one writes $(\partial V/\partial x)\delta x = \delta V(x) = V(x+(\delta x/2)) - V(x-(\delta x/2))$ and uses the fact that $[F(x,p,t)exp(ip\delta x/\hbar)]_{-\infty}^{\infty} = 0$. Changing variables via $y = x + (\delta x/2)$ and $y' = x - (\delta x/2)$ (•♠♠) can be rewritten as

$$\left\{\frac{\hbar^2}{2m}\left[\frac{\partial^2}{\partial y^2} - \frac{\partial^2}{\partial (y')^2}\right] - [V(y) - V(y')]\right\}\rho(y,y',t) = -i\hbar\frac{\partial}{\partial t}\rho(y,y',t) \tag{3.2}$$

which is called Schrödinger's first equation for the density function. Now one assumes **(D)** $\rho(y,y',t) = \psi^*(y',t)\psi(y,t)$ and **(E)** $\psi(y,t) = A(y,t)exp\,(is(y,t)/\hbar)$ and expands ρ in **(D)** in terms of δx to get two equations **(F)** $\partial\chi/\partial t + (\partial/\partial x)[\chi(s'/m)] = 0$ and then **(G)** $(\partial/\partial x)[(1/2m)(s')^2 + V + \partial s/\partial t - (\hbar^2/2mA)(\partial^2 A/\partial x^2)] = 0$ where we have changed notation a bit in using $\chi = A^2 = lim\,\rho[x+(\delta x/2), x-(\delta x/2)]$ as $\delta x \to 0$. Here χ is the standard probability density in configuration space. Now **(G)** can be rewritten as (♠♠♠) $(1/2m)\,(\partial s/\partial x)^2 + V + (\partial s/\partial t) - (\hbar^2/2mA)(\partial^2 A/\partial x^2) = c = 0$ which is **(SE)** $(\hbar^2/2m)\partial^2\psi/\partial x^2 - V(x)\psi = -i\hbar\partial\psi/\partial t$ (if $\psi = Aexp(is/\hbar)$ is substituted in **(SE)** one obtains **(G)**). In summary: If one can write **(D)** then ψ given by **(E)** satisfies **(SE)** along with the equation of continuity **(F)**. With suitable definitions of operators etc. **(SE)** has the operator form $\hat{H}\psi = i\hbar(\partial/\partial t)\psi$. Note that one has now $[\hat{x},\hat{p}] = i\hbar$ and it can be shown that in fact (•♡•) $\Delta x\Delta p \geq \hbar/2$ (Heisenberg uncertainty) as a consequence of writing the density function in the product form **(D)**. In this spirit, instead of representing a fundamental property of nature the uncertainty principle simply represents a limitation of the description based on **(SE)** (i.e. quantum mechanics as developed above is only applicable to problems where the density function can be decomposed as in **(D)**). Further one remarks that the dispersion relations do not impose any constraint upon the behavior of nature but only upon our capacity to describe nature by means of quantum theory so defined. If e.g. $\Delta q\Delta p < \hbar/2$ in some situation then this quantum theory does not apply or will not give good results.

Since the hypotheses **(D)** leads to possibly negative values of F (cf. (3.3) below) a different construction was developed in Article 16 of [25] to produce a **(SE)** with **(D)** replaced by (♣♣♣) $F(x,p,t) = \phi^*(x,p,t)\phi(x,p,t)$ where ϕ is a probability amplitude on phase space. Some problems indicated in [14] are dismissed via the distinction that operators $\hat{x}' \sim x$ and $\hat{p}' \sim -i\hbar\partial/\partial(\delta x)$ apply when working on the density function. The uncertainty principle (•♡•) is not developed in Article 16 but the same kind of reasoning would assert that the QM determined by the **(SE)** derived there (see below) holds when $F = \phi^*\phi$ and not necessarily in other situations. Returning to (♠♠♠) one can consider this as a Hamilton Jacobi (HJ) equation with an effective potential $V_{eff} = V(x) - (\hbar^2/2mA)\partial^2 A/\partial x^2$ (the last term being D. Bohm's quantum potential - cf. [19] and see (3.9) below). Thus formally with initial condition $p = s_X$ one can write $\dot{p} = -\partial_x V_{eff}(x) = -\partial_x[V - (\hbar^2/2mA)A'']$. The integration of this with $p = s_x$ will give a series of trajectories equivalent to the force lines associated to V_{eff}. The resolution method goes as follows: First **(SE)** must be solved in order to obtain the "probability amplitudes" referring to the ensemble. Then the effective potential, which will act as a statistical field for the ensemble is constructed. This potential should not be considered as a real potential but a ficticious one which acts as a field in reproducing through trajectories the statistical results of the original equation (♠♠♠).

Now following Article 16 of [25] one can resolve the problem of producing positive definite phase space densities. The possibly negative values arising from a formal inversion

$$F(x,p,t) = \int \rho\left(x + \frac{\delta x}{2}, x - \frac{\delta x}{2}\right) exp\left(-i\frac{p\delta x}{\hbar}\right) d(\delta x) \tag{3.3}$$

can be attributed to the stipulation **(D)** and the theory is modified as follows. One assumes (♣♣♣) above so F is positive definite, and defines $\rho = Z_Q$ as before by (3.1). Also one defines also a characteristic amplitude $\xi(x',x,t) = \int e^{ipx'/\hbar}\phi(x,p,t)dp$ so that (•♠•) $Z_Q(x,\delta x,t) = \int \xi^*(x',x,t)\xi(x'+\delta x,x,t)dx'$ which is a convolution. The characteristic amplitude can be negative without violating any physical principles and we note that (•♠•) can also be written as $Z_Q(x,\delta x,t) = \int \xi^*(x',x,t)e^{i\delta x\hat{p}'/\hbar}\xi(x',x,t)dx'$ with $\hat{p}' = -i\hbar\partial/\partial x'$. Putting this Z_Q into (•♠♠) one obtains an equation for $\xi(x',x,t)$ and writing $\xi(x',x,t) = \psi(x',t)\psi^*(x,t)$ leads to $Z_Q(x,\delta x,t) = \psi(x,t)\psi^*(x,t)\int \psi^*(x',t)$ $exp\,(\delta x[\partial/\partial x'])\,\psi(x',t)dx'$. Using these formulas one obtains the Schrödinger equation $(\hat{p}^2/2m)\psi(x,t)+V(x)\psi(x,t) = E\psi(x,t)$ with $\hat{p} = -i\hbar\partial/\partial x$ where $E \sim \bar{E} = \int\int[(p^2/2m) + V(x)]F(x,p,t)dxdp$ (cf. (♠)). Further $\psi^*(x,t)\psi(x',t) = \int e^{ipx'/\hbar}\phi(x,p,t)dp$ which implies (note factors of 2π are occasionally missing) $\phi(x,p,t) = \psi^*(x,t)\phi(p,t)$;
$\phi(p,t) = \int e^{-ipx'/\hbar}\psi(x',t)dx'$ leading to $F(x,p,t) = |\psi(x,t)|^2|\phi(p,t)|^2$ which is a natural probability density here (see [14, 25] for more discussion). Note that the operator constructions here are somewhat different from those of Article 1 for example.

A thermodynamic derivation of **(SE)** is given in Article 15 of [25] (cf. also

Article 11). The main idea is to derive the classical **(SE)** of quantum mechanics (QM) from classical ideas and in particular to show the thermodynamic character of the quantum formalism. Thus one begins with the Liouville equation in the form $\partial_t F(x,p,t)+(p/m)\partial_x F(x,p,t)-\partial_x V\partial_p F(x,p,t) = (dF/dt)(x,p,t) = 0$ with $\int_{-\infty}^{\infty} F dp = \rho(x,t)$; $\int_{-\infty}^{\infty} pF dp = p(x,t)\rho(x,t)$ where $p(x,t)$ is the so called macroscopic momentum. This second equation is a definition which can also be phrased as $\int pF dp = \rho < p >$ with $< p >= p(x,t)$. Then $\partial_t \rho(x,t) + (1/m)\partial_x[p(x,t)\rho(x,t)] = 0$. Note x, p, t are independent variables and $\int_{-\infty}^{\infty} F_p dp = F|_{-\infty}^{\infty} = 0$ for "reasonable" distributions. Also $\int_{-\infty}^{\infty} pF_x dp = \partial_x \int_{-\infty}^{\infty} pF dp = \partial_x[p(x,t)\rho(x,t)]$. Next multiply the Liouville equation by p and integrate using $\int_{-\infty}^{\infty} p^2 F(x,p,t)dp = M_2(x,t)$ to get

$$\frac{1}{m}\partial_x[M_2 - p^2(x,t)\rho] + \rho\left[\partial_t p(x,t) + \partial_x\left(\frac{p^2(x,t)}{2m}\right) + \partial_x V\right] = 0 \tag{3.4}$$

The first term involves $M_2 - p^2(x,t)\rho = \int[p^2 - p^2(x,t)]F dp = \int[p-p(x,t)]^2 F dp$ since $\int p(x,t)Fpdp = p^2(x,t)$. Thus one writes $\overline{(\delta p)^2}\rho(x,t) = \int[p-p(x,t)]^2 F dp = M_2 - p^2(x,t)\rho$ and looks for a functional expression for $\overline{(\delta p)^2}$. In this direction, given $\Omega(x,t) \sim$ system accessible states for x in the range $[x, x+\delta x]$ and $S = k\,log\Omega(x,t)$ the entropy, the equal a priori probability postulate implies $\rho(x,t) \propto \Omega(x,t) = exp[(1/k)S(x,t)]$ (cf. [26]). Let $S_{eq}(x,t)$ correspond to the thermodynamic equilibrium configuration entropy ($\delta x = 0$) where S is a maximum with $(\partial S_{eq}/\partial x)|_{\delta x=0} = 0$ and let ρ_{eq} be the corresponding density. Some argument leads to $\rho(x,\delta x,t) = \rho_{eq}\,exp[-\gamma(\delta x)^2]$ (where $\gamma = (1/2k)|\partial^2 S_{eq}/\partial x^2|$) and then the mean quadratic dispersion $\overline{(\delta x)^2}$ associated with such fluctuations $\delta\rho$ is given by

$$\overline{(\delta x)^2} = \frac{\int(\delta x)^2 exp[-\gamma(\delta x)^2]d(\delta x)}{\int exp[-\gamma(\delta x)^2]d(\delta x)} = \frac{1}{2\gamma} \tag{3.5}$$

A priori there is no relation between such fluctuations and those related to momenta but one can impose the restriction that in the thermodynamic equilibrium situation one should have (•♣•) $\overline{(\delta p)^2(\delta x)^2} = \hbar^2/4$. This is not clear but it implies (for $\rho = exp(S/k)$) (♣♢♡♠) $\rho\overline{(\delta p)^2} = -(\hbar^2\rho/4)\partial^2 log\rho$. Writing now $\rho = A^2$; $p(x,t) = \partial s/\partial x = s'$ one puts (♣♢♡♠) into (3.4) to get

$$A^2(x,t)\partial_x\left[\frac{\partial s}{\partial x} + \frac{1}{2m}\left(\frac{\partial s}{\partial x}\right)^2 + V(x) - \frac{\hbar^2}{2mA}\partial_x^2 A\right] = 0 \tag{3.6}$$

Then as indicated above, this equation plus $\rho_t + (1/m)\partial_x[p\rho] = 0$ is equivalent to the **(SE)** $-(\hbar^2/2m)\psi_{xx} + V(x)\psi = i\hbar\psi_t$; $\psi = Aexp(is(x,t)/\hbar)$. One notes that the condition $p(x,t) = \partial_x s$ is in fact some sort of restriction - i.e. p is a gradient of some function s. Thus three axioms are involved, namely: **(H)** Newtonian particle mechanics is valid for every individual system composing an ensemble. **(I)** The Liouville equation is valid for the description of the ensemble behavior. **(J)** In a thermodynamic equilibrium situation one should have the restriction (•♣•). In addition, the restriction represented by $\rho = A^2$, $p = s'$ must be applicable at this equilibrium situation. One can also make a direct comparison of formulas here with those of Article 1 of [25] and it is shown that (♣◇♡♠) can be derived from first principles without using **(D)** or (•♣•) (perhaps lending credence to (•♣•)). Finally the first two postulates **(H)** and **(I)** above are the same as the postulates **(A)** and **(B)** in the Article 1 of [25] so, regarding the emergence of the **(SE)**, the hypothesis **(J)** above, or simply (♣◇♡♠), is formally equivalent to assuming the validity of the WM transformation together with the hypothesis that $\rho \sim Z_Q$ in (3.1) can be written as $Z_Q(x, \delta x, t) = \psi^*(x-(\delta x/2), t)\psi(x+(\delta x/2), t)$ (which is **(D)**). Alternatively we can also say that **(J)**, or (♣◇♡♠), is equivalent to $F = \phi\phi^*$ as in (♣♣♣) since one arrives at the **(SE)** in both cases.

In [3], based on Articles 1-15 of [25] we suggested a formula

$$\hat{\mathcal{F}} = \frac{1}{2}\psi\left(X + \frac{\delta X}{2}\right)\bar{\psi}\left(X - \frac{\delta X}{2}\right) + \frac{X}{i\epsilon} = \frac{1}{2}\rho\left(X + \frac{\delta X}{2}, X - \frac{\delta X}{2}\right) + \frac{X}{i\epsilon} \quad (3.7)$$

which in [2] was expressed following Article 16 of [25] as $\hat{\mathcal{F}} = (1/2)Z_Q(x, \delta x, t) + (X/i\epsilon)$ with $Z_Q \sim \int F(x,p,t)exp(ip\delta x/\hbar)dp$. We recall that in [15] the equation $|\psi|^2 = 2\mathcal{F} - (2X/i\epsilon)$ of (X, ψ) duality is interperted as describing the space variable as a macroscopic thermodynamic quantity with the microscopic information encoded in the prepotential. Then QM can be reformulated in terms of (X, ψ) duality with the **(SE)** replaced by the third order equation (3.10) for example. Here $\hbar$ can be considered as the scale of the statistical system (cf. [1, 15]). These comments from [15] seem completely adaptable to a connection such as (3.7) with the theory of [25]. Thus suppose we define $\hat{\mathcal{F}}$ as in (3.7) and use **(D)** - **(E)** so that $\chi = A^2 = lim\, \rho[X+(\delta X/2), X-(\delta X/2)]$ as $\delta X \to 0$, leading to **(SE)**. Then $\hat{\mathcal{F}} \to \mathcal{F} = (1/2)\chi + (X/i\epsilon)$ with the mixing equation (•♠♠) in the background. Further if the t dependence is restricted to $t = t_2$ of the form $exp(-iEt/\hbar)$ for suitable t then the **(SE)** has the form (♠). If we modify (3.7) now in the light of Article 16 of [25], one can set $\hat{\mathcal{F}} = (1/2)Z_Q + (X/i\epsilon)$ and again $Z_Q \to |\psi|^2(x,t)$ as $\delta x \to 0$.

Regarding thermodynamic behavior now, the idea is to let the systems (SY) composing an ensemble interact with a neighborhood (O) called the heat bath. The interaction is considered sufficiently feeble so as to allow one to write a Hamiltonian $H(q,p)$ for SY not depending on the degrees of freedom of (O). The system O is necessary only as a means of imposing its temperature T upon SY. Now in a state of equilibrium there is a canonical probability distribution $F(q,p) = Cexp(-2\beta H(q,p))$ where $2\beta = 1/K_BT$ with K_B being the Boltzmann constant, T the absolute temperature, and C some normalization constant. The Hamiltonian may be written $H(q,p) = \sum_1^N (p_n^2/2m_n) + V(q_1, \cdots, q_N)$ where NK_BT represents the energy of the reservoir (O). Using the Wigner-Moyal transformation we get a formula which yields $\rho = C_1 exp[-2\beta V(q)] \cdot exp[-\sum(m_n/4\beta\hbar^2)(\delta q_n)^2]$. Some analysis (using **(D)**) leads to $\psi(q,t) = \sqrt{C_3}exp[-\beta V(q)] \cdot exp[-iEt/\hbar]$ with

$\rho = C_3 exp\{-2\beta[V(q) + (1/8)\sum(\delta q_n)^2(\partial^2 V/\partial q_n^2)]\}$ and taking $C_1 = C_3 = 0$ one has $\rho_{eq} = exp(-2\beta V(q))$ with $\rho = \rho_{eq}(q + (\delta q/2)) = \rho_{eq}(q - (\delta q/2))$ (Taylor expansion of V). Then $-\sum_1^N(\hbar^2/2m_n)(\partial^2\psi/\partial q_n^2) + V\psi = E\psi$ with $E = V(q^0) + NK_BT$ where q^0 represents the mechanical equilibrium point. One can now establish (motivated by **(D)**) a connection between the microscopic entities of the quantum formalism and the macroscopic description given by thermodynamics (we remark that this connection should also prevail locally in various situations without recourse to **(D)**). Thus define the free energy $F_G(q) = V(q)$ such that $F_G = -K_BTlog(\psi^*(q)\psi(q))$. Writing entropy as $\mathcal{S} = K_Blog(\psi^*(q)\psi(q))$ we have $F_G = -T\mathcal{S}$ and it must be emphasized here that one is dealing with a "local" entropy of $\mathcal{S}(q)$ as spelled out earlier via $\Omega(x,t)$, $(x \sim q)$; it is this locality which suggests an adaption to (X,ψ) duality. Thus if we now express ψ as $exp(\tilde{S}/\epsilon)$ as in [2, 3, 11] one has $\chi = A^2 = |\psi|^2 = exp(2\Re\tilde{S}/\epsilon)$ which implies $log\,A^2 = 2\Re\tilde{S}/\epsilon \sim \mathcal{S}/K_B$ or $\mathcal{S} \sim 2K_B\Re\tilde{S}/\epsilon$. A priori there seems to be no reason not to think of $K_B\Re\tilde{S}/\epsilon$ as an entropy term, given that $|\psi|^2$ refers to a statistical system, and consequently (recall from [2, 3, 11, 15, 16] that $-(X/i\epsilon) = \psi^2(\partial\mathcal{F}/\partial(\psi^2) - \mathcal{F}$ or equivalently $-\mathcal{F} = \phi(X_\phi/i\epsilon) - (X/i\epsilon)$ for $\phi = \partial\mathcal{F}/\partial(\psi^2))$

$$\mathcal{F} \sim \frac{1}{2}exp\left[\frac{\mathcal{S}}{K_B}\right] + \frac{X}{i\epsilon};\ \frac{1}{2}exp\left[\frac{\mathcal{S}}{K_B}\right] \sim \psi^2\frac{\partial\mathcal{F}}{\partial(\psi^2)} \sim -\phi\frac{X_\phi}{i\epsilon} \tag{3.8}$$

The first equation of (3.8) then would represent a macroscopic formula for $\mathcal{F}$, relative to the statistical system involving ψ but at this point we suggest the analogies here only in a most heuristic spirit.

Next we recall that the (X,ψ) duality theme in [2, 3, 11, 15, 16] was based on the equation (♠) $-(\hbar^2/2m)\psi'' + V(X)\psi = E\psi$ with $\mathcal{F} = (1/2)\psi\bar{\psi} + (X/i\epsilon)$ and a WKB formulation via $\psi = exp[\tilde{S}/\epsilon]$ was employed ($\epsilon = \hbar/\sqrt{2m}$) with $\tilde{S} = \sum_0^\infty \epsilon^j S^j$ (S^{2j+1} real and S^{2j} imaginary for k real, $\lambda = -ik$). It was shown in [11] how this is related to the formulation of [17, 22] for $\psi = Aexp[is/\hbar]$ where (in an expanded dKdV theory) $s \sim \sqrt{2m}\Im\tilde{S}$ and $log\,A = (1/\epsilon)\tilde{S}_{odd} = (1/\epsilon)\Re\tilde{S}$. One knows that A satisfies

$$(s')^2 = 2m(E - V) + \frac{\hbar^2}{2}\left[\frac{3}{2}\left(\frac{s''}{s'}\right)^2 - \frac{s'''}{s'}\right] \tag{3.9}$$

and $A = c(s')^{-1/2}$ so for $\tilde{P} = \tilde{S}'$ the condition $|\psi|^2\Im\tilde{P} = -1$ is automatically satisfied for $c = i$ (this arises since $\Im\tilde{P} < 0$). On the other hand the (expanded) $\tilde{\mathcal{F}} = \mathcal{F}_\epsilon$ satisfies a Gelfand-Dickey (GD) type equation

$$\epsilon^2\mathcal{F}''' + \left(\mathcal{F}' - \frac{1}{i\epsilon}\right)(8F'' + 4E) + 4F'''\left(\mathcal{F} - \frac{X}{i\epsilon}\right) = 0 \tag{3.10}$$

where $V = -2F^0_{XX}$ and $\tilde{V} = -2\tilde{F}_{XX}$ for $\tilde{F} = \sum_0^\infty \epsilon^{2j}F^{2j}$ ($E = -\lambda^2$). Since ψ and $\bar{\psi}$ satisfy (♠) we know that the square eigenfunctions $\psi\bar{\psi}$, ψ^2, and $(\bar{\psi})^2$ satisfy the GD equation. Now it turns out that our enhanced dKdV corresponds to a standard WKB expansion for s in (even) powers of $\hbar$. Further from [17] a general solution of (3.9) is given by $s' = \pm\sqrt{2m}(a\psi^2 + b\bar{\psi}^2 + \psi\bar{\psi})^{-1}$ where ψ, $\bar{\psi}$ are normalized solutions of (♠) so we introduce a constraint $\int |\psi|^2 dX = 1$. Here one recalls that $|\psi|^2 = -1/\Im\tilde{P} = exp[(1/\epsilon)\Re\tilde{S}] = exp[\sum_0^\infty \epsilon^{2j} S^{2j+1}] = -i/\sum_0^\infty \epsilon^{2j} S_X^{2j}$ so $\int |\psi|^2 = \sum_0^\infty \epsilon^{2\ell} H_\ell = h(\epsilon)$. Since there is no reason to assume $h(\epsilon) = 1$ for any ϵ one will have to scale $\psi \to c\psi$ in order to insure this constraint. This amounts to a normalization of a BA function and should introduce no problems. Now the GD equation (3.10) for $\mathcal{F}$ can be written in terms of $\Xi = \psi\bar{\psi}$, or ψ^2 , or $\bar{\psi}^2$ as **(GD)** $\epsilon^2\Xi''' - 4V\Xi' - 2V'\Xi + 4E\Xi' = 0$. so given s' as above one wonders about connections between (3.9) and **(GD)**. The connection is of course obvious since in particular $s' = c/|\psi|^2$ satisfies (3.9) and $|\psi|^2 = c/s'$ satisfies **(GD)**. Thus in fact (3.9) is a differential equation whose solutions are reciprocals of square eigenfunctions and in this sense it is similar, or even equivalent to, the GD equation.

References

[1] Bonelli, G. and M. Matone, (1996). Phys. Rev. Lett., Vol. 76, (pages 4107-4110).

[2] Carroll, R. (1997). hep-th 9705229, Nucl. Phys. B, Vol. 502, to appear.

[3] Carroll, R. hep-th 9607219, 9610216, and 9702138.

[4] Carroll, R. and Y. Kodama, (1995). Jour. Phys. A, Vol. 28, (pages 6373-6387).

[5] Carroll, R. (1994). Jour. Nonlin. Sci., Vol. 4, (pages 519-544); Teor. Mat. Fizika, Vol. 99, (1994), (pages 220-225).

[6] Carroll, R. (1995). Proc. NEEDS'94, World Scientific, (pages 24-33).

[7] Carroll, R. (1993/1995). Applicable Anal., Vol. 49, (1993), (pages 1-31); Vol. 56, (1995), (pages 147-164).

[8] Carroll, R. (1991). Topics in soliton theory, North-Holland.

[9] Carroll, R. and J.H. Chang, (1997). Applicable Anal., Vol. 64, (pages 343-378).

[10] Carroll, R. (1996). Repts. Math. Phys., Vol. 37, (pages 1-21).

[11] Carroll, R. Proc. Conf. Supersymmetry and integrable models, UIC, June, 1997, Springer, to appear.

[12] Carroll, R. In preparation.

[13] Chang, J.H. Thesis, Univ. of Illinois, In preparation.

[14] Cohen, L. (1966). Jour. Math. Phys., Vol. 7, (pages 781-786).

[15] Faraggi, A. and M. Matone, (1997). Phys. Rev. Lett., Vol. 78, (pages 163-166).

[16] Faraggi, A. and M. Matone. hep-th 9705108.

[17] Floyd, E. quant-ph 9707051.

[18] Guha, P. and K. Takasaki, solv-int 9705013.

[19] Holland, P. (1993). The quantum theory of motion, Cambridge Univ. Press.

[20] Kodama, Y. and J. Gibbons, (1990). Workshop on nonlinear and turbulent processes in physics, World Scientific, (pages 166-180).

[21] Matsukidaira, J., J. Satsuma, and W. Strampp, (1990). Jour. Math. PHys., Vol. 31, (pages 1426-1434).

[22] Messiah, A. (1995). Quantum mechanics, Vol. 1, North-Holland.

[23] Nakatsu, T. and K. Takasaki, (1996). Mod. Phys. Lett. A, Vol. 11, (pages 157-168).

[24] Ohta, Y., J. Satsuma, D. Takahashi, and T. Tokihiro, (1988). Prog. Theor. Phys., Supp. 94, (pages 210-244).

[25] Olavo, L. quant-ph 9503020, 9503021, 9503022, 9503024, 9503025, 9509012, 9509013, 9511028, 9511039, 9601002, 9607002, 9607003, 9609003, 9609023, 9703006, 9704004.

[26] Reif, F. (1965). Fundamentals of statistical and thermal physics, McGraw-Hill.

[27] Seiberg, N. and E. Witten, (1994). Nucl. Phys. B, Vol. 426, (pages 19-52); Vol. 431, (pages 484-550).

[28] Takasaki, K. and T. Takebe, (1992/1995). Inter. Jour. Mod. Phys. A, Supp. 1992, (pages 889-922); Rev. Math. Phys., Vol. 7, (1995), (pages 743-808).

[29] Tegmark, M. quant-ph 9709032.

UNIQUENESS OF CONTINUATION THEOREMS

Matthias M. Eller

Abstract: We prove a sharp uniqueness of continuation result for two classical systems of partial differential equations. The proof is based on a special Carleman type estimate developed by Tataru.

1.1 INTRODUCTION

Questions concerning the uniqueness of solutions are of great importance in almost all areas of partial differential equations. The study of these questions has a long history, however, in recent years applied problems such as inverse problems and control problems have posed new ones. Unique continuation for solutions of p artial differential equations is one of them and can be described as follows. Consider a function or distribution which satisfies a homogeneous partial differential equation in an open set Ω. This set is divided into two parts by a surface which passes through a fixed point $x_0 \in \Omega$ and $u \equiv 0$ on one side of the hypersurface. Uniqueness of continuation means that these conditions imply that u vanishes in a full neighborhood of x_0. It is clear that uniqueness of continuation depends on various conditions:

- the smoothness of coefficients of the differential operator
- the regularity assumptions on the solution
- the geometry of the hypersurface

We like to point out that uniqueness of continuation is a local property. Compactness arguments often lead to conclusions of a global nature. We will not pursue that direction here.

R.P. Gilbert et al.(eds.), Direct and Inverse Problems of Mathematical Physics, 53–105.

The uniqueness of continuation problem can be also formulated as the Cauchy problem. The uniq ueness question is then the following: Do zero Cauchy data on the boundary of a domain Ω (or on part of it) for a solution to a partial differential equation imply that the solution vanishes close to the boundary. For applications it is usually the Cauchy problem which is of interest. From the mathematical point of view the formulation as the uniqueness of continuation problem has some advantages. In particular we do not need to be concerned with the Cauchy data and the regularity assump tions on the Cauchy data for a specific differential operator.

Up to the beginning of the 1990s two classical results were well-known. The first one is Holmgren's theorem which seems optimal with respect to the geometry of the hypersurface, however, it requires a differential operator with analytic coefficients. On the other hand there is Hörmander's theorem which admits differential operators with just continuously differentiable coefficients, but requires the hypersurface to be strongly pse udo-convex. And this pseudoconvexity condition is somehow rigid. However, it provided optimal results with respect to second-order elliptic operators.

Many evolution processes in time are governed by parabolic or hyperbolic equations. For many applications, the coefficients are time independent, they are functions of the space variable only. The problem was whether the conclusion of Holmgren's theorem would hold for these operators even when the coefficients are just continuously differentia ble. Robbiano [9] and Hörmander [5] made the first attempts, but it was Tataru [10] who proved the conjecture in 1995.

The proof of Hörmander's theorem is based on apriori estimates for the differential operator which contain a weight function and a large parameter. These estimates were introduced first by Carleman in the 1930s. Tataru also uses these kind of estimates in connection with a pseudo-differential operator. This allows him to require a less restrictive versi on of the strong pseudoconvexity condition which for a second order hyperbolic operator is satisfied for non-characteristic surfaces.

We are mainly interested in proving uniqueness of continuation for some classical systems of partial differential equations. This is a wide open field since there are no estimates of Carleman's type for systems of partial differential equations. There are several problems which occur when one tries to prove these estimates for systems. The crucial point is the

characteristics of the system.

Nevertheless, we found a way to look at two classical systems. We will consider Maxwell's system and the elasticity system, both in the isotropic case with the coefficients independent in time. Our main tool is a reduction of these systems to equations which are weakly coupled, i.e. the coupling exists through lower order terms only.

In section 2 we introduce function spaces and differential operators. Our approach is based on Sobolev spaces, however, we will need some special spaces of distributions as well. Some basic facts, like regularization and Friedrich's lemma are briefly mentioned. The subsection about differential operators discusses how these operators commute with exponential functions and with a special pseudo-differential operator. We like to point out that we consider only operators with real valued coefficients. The last subsection of this section discusses differential quadratic forms. It provides a special integration by parts techniqu e which is crucial for proving Carleman type estimates.

Section 3 is dedicated to the geometric conditions on surfaces in connection with differential operators. After discussing non-characteristic surfaces briefly we try to give a detailed description of strongly pseudo-convex surfaces.

The next section contains the classical results about uniqueness of continuation. We mention Holmgren's theorem and give a complete proof of Hörmander's theorem. This proof is based on differential quadratic forms. The last subsection in this chapter is dedicated especially to Tataru's result. The use of techniques developed in the previous sections and in the proof of Hörmander's theorem enables us to give a new proof of this theorem.

In the fifth section we prove uniqueness of continuation results for Maxwell's system and the equations of linear elasticity under minimal regularity assumptions.

Sections 3 through 5 represent a joined work together with Victor Isakov.

1.2 PRELIMINARIES

1.2.1 Function spaces and norms

Here we will introduce the function spaces and their norms which we will use later on. Let Ω be an open, non-empty simply connected subset of the Euclidean space $\mathbf{R}^n$. Usually x_0 will denote a fixed point in Ω. The ball with center at x_0 and radius δ is denoted by $B_\delta(x_0)$. We will always assume $B_\delta(x_0) \subset \Omega$.

Sometimes we will decompose the n dimensional vector x into two vectors. Let $\mathbf{R}^n = \mathbf{R}^{n''} \times \mathbf{R}^{n'}$ and $x = (x'', x')$. Everytime we use this decomposition we assume $\Omega = \Omega'' \times \Omega'$ where Ω'' is an open non-empty simply connected subset of $\mathbf{R}^{n''}$ and Ω' is an open simply connected subset of $\mathbf{R}^{n'}$. By $\mathbf{R}_+$ we denote the positive real numbers and by $\mathbf{R}_-$ the negative real numbers.

By $C_0^\infty(\Omega)$ we denote the space of test functions, i.e. the functions with compact support which have derivatives of arbitrary order. The space of all continuous linear forms on $C_0^\infty(\Omega)$ will be denoted $\mathcal{D}'(\Omega)$, the space of distributions. Furthermore $C^\infty(\Omega)$ denotes the space of infinitely often differentiable functions and $\mathcal{E}'(\Omega)$ is the its dual space, the space of compactly supported distributions. By $\mathcal{S}(\mathbf{R}^n)$ we denote the Schwartz space and by $\mathcal{S}'(\mathbf{R}^n)$ the space of tempered distributions.

Let $D_j = -i\partial_j$ and $D = (D_1, D_2, ..., D_n)$. By $C^k(\bar{\Omega})$ we denote the space of k-times continuously differentiable functions in Ω which can be extended to functions in $C^m(\mathbf{R}^n)$. A norm for this space is given by

$$||u||_{C^k(\Omega)} = \sum_{|\alpha|\leq k} \sup_{x\in\Omega} |D^\alpha u| \, .$$

Here $\alpha = (\alpha_1, \alpha_2, ..., \alpha_n)$ denotes a multiindex, i.e. α is a n-dimensional vector consisting of non-negative integers and $|\alpha| = |\alpha_1| + ... + |\alpha_n|$. The space $L^2(\Omega)$

is the space of all measurable square integrable functions in Ω with finite

$$\|u\|^2_{L^2(\Omega)} = \int_\Omega |u|^2 .$$

For a non-negative integer s $H^s(\Omega)$ denotes the L^2-based Sobolev space with the norm

$$\|u\|^2_{H^s(\Omega)} = \sum_{|\alpha|\leq s} \|D^\alpha u\|^2_{L^2(\Omega)} .$$

When we consider the space $H^s(\mathbf{R}^n)$ this norm can be equivalently defined by means of

$$\|u\|^2_{H^s(\mathbf{R}^n)} = (2\pi)^{-n} \int (1+|\xi|^2)^s |\hat{u}(\xi)|^2 d\xi$$

where $\hat{u}$ is the Fourier transform of u. This definition extends to $s \in \mathbf{R}$. We will need some weighted norms in these spaces as well. For $\tau > 0$, $H^s_\tau(\mathbf{R}^n)$ is the same space as $H^s(\mathbf{R}^n)$ but equipped with the norm

$$\|u\|^2_{H^s_\tau(\mathbf{R}^n)} = (2\pi)^{-n} \int (\tau^2+|\xi|^2)^s |\hat{u}(\xi)|^2 d\xi .$$

An equivalent norm is given by

$$\|u\|^2_{H^s_\tau(\mathbf{R}^n)} = \sum_{|\alpha|\leq s} \tau^{2(s-|\alpha|)} \|D^\alpha u\|^2_{L^2(\mathbf{R}^n)}$$

provided s is a non-negative integer. That allows us to consider $H^s_\tau(\Omega)$ as well. Finally we need to consider anisotropic Sobolev spaces. For $r, s \in \mathbf{R}$ the space $H^{s,r}_\tau(\mathbf{R}^n)$ is defined as the space of distributions such that

$$\|u\|^2_{H^{s,r}_\tau(\mathbf{R}^n)} = (2\pi)^{-n} \int (\tau^2+|\xi|^2)^s (\tau^2+|\xi''|^2)^r |\hat{u}(\xi)|^2 d\xi$$

is finite. If s and r are non-negative integers an equivalent norm can be introduced by

$$\|u\|^2_{H^{s,r}_\tau(\mathbf{R}^n)} = \sum_{\substack{|\alpha|\leq s\\ |\beta''|\leq r}} \tau^{2(s+r-|\alpha|-|\beta''|)} \|D^\alpha D^{\beta''} u\|_{L^2(\mathbf{R}^n)} \tag{1.1}$$

where $\beta'' = (\beta_1, .., \beta_{n''}, 0, .., 0)$.

The space $H^{s,r}_{0,\tau}(\Omega)$ is defined as the completion of $C_0^\infty(\Omega)$ with respect to the norm (1.1). We like to point out that the dual space of $H^{s,r}_{0,\tau}(\Omega)$ is $H^{-s,-r}_\tau(\Omega)$.

More anisotropic function sp aces will be needed later on. For that sake we introduce Banach space valued functions and distributions.

Let $\mathcal{S}(\mathbf{R}^{n''}, H^s(\Omega'))$ denote the set of rapidly decreasing infinitely often differentiable functions on $\mathbf{R}^{n''}$ with values in $H^s(\Omega')$. By, $\mathcal{D}'(\Omega'', H^s(\Omega'))$ we denote the space of distributions on Ω'' with values in $H^s(\Omega')$. We remark that $u \in \mathcal{E}'(\Omega'', H^s_0(\Omega'))$ implies that there exist a positive integer r such that $u \in H^{s,-r}_{0,\tau}(\Omega)$.

On $C^k(\bar{\Omega})$ we introduce the weighted norm

$$\|u\|_{C^k_\tau(\Omega)} = \sum_{|\alpha|\le k} \tau^{-|\alpha|}\|D^\alpha u\|_{C(\Omega)} \,.$$

For later reference we state the following result.

Lemma 1 *Let r be an integer and let $u \in H^{0,r}_\tau(\Omega)$ and $v \in C^{|r|}(\bar{\Omega})$. Then there exists a constant C independent of τ such that*

$$\|uv\|_{H^{0,r}_\tau(\Omega)} \le C\|u\|_{H^{0,r}_\tau(\Omega)}\|v\|_{C^{|r|}_\tau(\Omega)} \,. \tag{1.2}$$

Proof: Let $r \ge 0$. Then using Leibniz' rule we obtain

$$\begin{aligned}
\|uv\|^2_{H^{0,r}_\tau(\Omega)} &= \sum_{|\alpha''|\le r} \tau^{2(r-|\alpha|)}\|D^{\alpha''}(uv)\|^2_{L^2(\Omega)} \\
&\le C\sum_{|\alpha''|\le r} \tau^{2(r-|\alpha''|)} \sum_{\beta''\le\alpha''} \sup_{x\in\Omega}|D^{\alpha''-\beta''}v|^2\|D^{\beta''}u\|^2_{L^2(\Omega)} \\
&\le C\sum_{|\alpha''|\le r}\sum_{\beta''\le\alpha''} \tau^{-2(|\alpha''|-|\beta''|)} \sup_{x\in\Omega}|D^{\alpha''-\beta''}v|^2\tau^{2(r-|\beta''|)}\|D^{\beta''}u\|^2_{L^2(\Omega)} \\
&\le C\sum_{|\gamma''|\le r} \tau^{-2|\gamma''|}\|D^{\gamma''}v\|^2_{C(\Omega)} \sum_{|\beta''|\le r} \tau^{2(r-|\beta''|)}\|D^{\beta''}u\|^2_{L^2(\Omega)} \\
&\le C\|u\|^2_{H^{0,r}_\tau(\Omega)}\|v\|^2_{C^{|r|}_\tau(\Omega)} \,.
\end{aligned}$$

which proves the lemma in that case. For $r < 0$ the proof follows by duality. Then

$$\begin{aligned}
\|uv\|_{H^{0,r}_\tau(\Omega)} &= \sup_{\phi\in H^{0,-r}_{0,\tau}(\Omega)} \frac{|\langle uv,\phi\rangle|}{\|\phi\|_{H^{0,-r}_\tau(\Omega)}} \\
&= \sup_{\phi\in H^{0,-r}_{0,\tau}(\Omega)} \frac{|\langle u,v\phi\rangle|}{\|\phi\|_{H^{0,r}_\tau(\Omega)}\|v\|_{C^{-r}_\tau(\Omega)}}\|v\|_{C^{-r}_\tau(\Omega)} \\
&\le C \sup_{\phi\in H^{0,-r}_{0,\tau}(\Omega)} \frac{|\langle u,v\phi\rangle|}{\|v\phi\|_{H^{0,-r}_\tau(\Omega)}}\|v\|_{C^{-r}_\tau(\Omega)} = C\|u\|_{H^{0,r}_\tau(\Omega)}\|v\|_{C^{-r}_\tau(\Omega)}
\end{aligned}$$

which concludes the proof in that case. □

Now we mention briefly the concept of regularization in Sobolev spaces, for reference see theorem 2.2.10 in [3].

Lemma 2 *Let $\phi \in C^\infty_0(\mathbf{R}^n)$ and assume $\int \phi = 1$. If $u \in H^s(\mathbf{R}^n)$ where s is a positive intege r, then the regularizations*

$$u_\epsilon(x) = \frac{1}{\epsilon^n}\int u(y)\phi\left(\frac{x-y}{\epsilon}\right)dy \tag{1.3}$$

converge to u in $H^s(\mathbf{R}^n)$ for $\epsilon \to 0$.

Later we will use regularization in anisotropic spaces with respect to certain variables only. This can be done without any problems since we can consider the remaining variables as parameters.

In one particular case we will consider regularization of di stributions. For $T > 0$ consider $u \in \mathcal{D}'((0,T))$. For $\delta > 0$ there exist an ϵ_0 such that $\phi(\frac{x-\cdot}{\epsilon}) \in C_0^\infty((0,T))$ for $\epsilon \leq \epsilon_0$. Consequently

$$u_\epsilon(x) = \frac{1}{\epsilon^n}\left\langle u, \phi\left(\frac{x-\cdot}{\epsilon}\right)\right\rangle \in C^\infty((\delta, T-\delta))$$

and we have

$$u_\epsilon \longrightarrow u \quad \text{in } \mathcal{D}'((\delta, T-\delta))$$

by duality.

One important application of regular ization is Friedrich's lemma. For a proof we refer to [4], lemma 17.1.5.

Lemma 3 *Let $u \in L^2(\mathbf{R}^n)$ and $a \in C^1(\mathbf{R}^n)$. Then*

$$\|(aD_jv)_\epsilon - a(D_jv)_\epsilon\|_{L^2(\mathbf{R}^n)} \longrightarrow 0$$

for $\epsilon \to 0$.

1.2.2 *Differential operators*

Let Ω be an open simply connected subset in the n-dimensional Euclidean space $\mathbf{R}^n$. Let $P(x,D)$ denote a linear partial differential operator

$$P(x,D) = \sum_{|\alpha|\leq m} a_\alpha(x) D^\alpha \,.$$

The $a_\alpha(x)$ are functions from Ω into $\mathbf{C}$ and we will refer to them as coefficients. Moreover m is the order of the differential operator and $D_j = -i\partial/\partial x_j$. We will always assume $a_\alpha \in L^\infty(\Omega)$. The principal symbol is

$$p(x,\xi) = \sum_{|\alpha|=m} a_\alpha(x)\xi^\alpha \,.$$

Here ξ denotes the dual variable to x. By $\bar{p}$ we denote the complex conjugate of the symbol, i.e.

$$\bar{p}(x,\xi) = \sum_{|\alpha|=m} \bar{a}_\alpha(x)\xi^\alpha \,.$$

For two symbols p and q we define the Poisson bracket by

$$\{p,q\}(x,\xi) = \sum_{j=1}^{n}\left(\frac{\partial p}{\partial \xi_j}(x,\xi)\frac{\partial q}{\partial x_j}(x,\xi) - \frac{\partial p}{\partial x_j}(x,\xi)\frac{\partial q}{\partial \xi_j}(x,\xi)\right) \,. \tag{1.4}$$

In the following we will use the abbreviations

$$p_{(j)}(x,\xi) = \frac{\partial p(x,\xi)}{\partial x_j}$$
$$p^{(j)}(x,\xi) = \frac{\partial p(x,\xi)}{\partial \xi_j}$$

where we point out that $p_{(j)}$ will denote the derivative with respect to the jth component of the first variable only. This will be of importance when we consider ξ which depend on x as well.

Let $\varphi \in C^m(\bar{\Omega})$ and $\tau \in \mathbf{R}$. Later on we will need the following result about the commutator of a differential operator and the exponential function $e^{\tau\varphi}$.

Lemma 4 *Let $u \in H^m(\Omega)$ and let $v = e^{\tau\varphi}u$. Then*

$$e^{\tau\varphi}P(x,D)u = P(x, D + i\tau\nabla\varphi(x))v \ . \tag{1.5}$$

where

$$P(x, D + i\tau\nabla\varphi(x)) = \sum_{|\alpha|\leq m} a_\alpha(x)(D + i\tau\nabla\varphi(x))^\alpha \ . \tag{1.6}$$

Proof: We need to show that

$$e^{\tau\varphi}D^\alpha(e^{-\tau\varphi}v) = (D - \tau D\varphi)^\alpha v \ .labelexp* \tag{1.7}$$

That can be done by induction with respect to α. For $|\alpha| = 1$ we have

$$e^{\tau\varphi}D_j(e^{\tau\varphi}v) = -\tau D_j\varphi v + D_j v = (D_j - \tau D_j\varphi)v$$

which is exactly (??) in this case. Next we assume that (??) holds for all $|\alpha| \leq \nu$. Set $\alpha^* = \alpha_j + \alpha$ where $|\alpha| = \nu$ and $|\alpha_j| = 1$. We obtain

$$\begin{aligned} (D - \tau D\varphi)^{\alpha^*} v &= (D - \tau D\varphi)^{\alpha_j}(D - \tau D\varphi)^\alpha v \\ &= (D_j - \tau D_j\varphi)e^{\tau\varphi}D^\alpha(e^{-\tau\varphi}v) \\ &= e^{\tau\varphi}\Big(D^{\alpha^*}(e^{-\tau\varphi}v) + \tau D_j\varphi D^\alpha(e^{-\tau\varphi}v) - \tau D_j\varphi D^\alpha(e^{-\tau\varphi}v)\Big) \\ &= e^{\tau\varphi}D^{\alpha^*}(e^{-\tau\varphi}v) \end{aligned}$$

which shows that (??) is also true for $|\alpha| = \nu + 1$. □.

For later purpose we decompose the operator

$$P(x, D + i\tau\nabla\ varphi(x)) = P_\tau(x,D) + R(x,D) \tag{1.8}$$

where

$$P_\tau(x,D) = \sum_{|\alpha|=m} a_\alpha(x) \sum_{\beta\leq\alpha} \binom{\alpha}{\beta} (i\tau\nabla\varphi(x))^{\alpha-\beta} D^\beta$$

is the principal part, i.e. the part of the operator which is homogeneous in (τ, D) of order m. The operator $R(x, D)$ consists only of terms which are homogeneous of lower order in (τ, D). This provides

$$\|R(x,D)v\|^2_{L^2(\Omega)} \leq C\|v\|^2_{H^{m-1}_\tau(\Omega)} \tag{1.9}$$

where C depends only on the L^∞-norm of the coefficients of P and the C^m-norm of φ in Ω. The pseudo-differential symbol of the operator $P_\tau(x, D)$ is

$$p_\tau(x,\xi) = p(x, \xi + i\tau\nabla\varphi(x)) \,. \tag{1.10}$$

Recall that we split $x = (x'', x')$ and assume $\Omega = \Omega'' \times \Omega'$. For $\tau > 0$ we consider an integral operator acting on $e^{\tau\varphi}u$

$$Q^\varphi_{\lambda,\tau}u(x) = \left(\frac{\tau}{2\pi}\right)^{\frac{n''}{2}} \int_{\Omega''} e^{-\frac{\tau}{2\lambda}(x''-y'')^2} e^{\tau\varphi(y'',x')}u(y'',x')dy'' \tag{1.11}$$

which is defined for $u \in \mathcal{E}'(\Omega'', L^2(\Omega'))$ in the sense of distribution. The pseudo-differential symbol is

$$q^\varphi_{\lambda,\tau}(\xi) = e^{-\frac{\lambda}{2\tau}|\xi''|^2} \tag{1.12}$$

and we no tice that the kernel of the integral operator

$$K(x'', y'') = \left(\frac{\tau}{2\pi}\right)^{\frac{n''}{2}} e^{-\frac{\tau}{2\lambda}(x''-y'')^2} \tag{1.13}$$

resembles the heat kernel. In some sense λ/τ is the time variable. Sometimes we will also write

$$Q^\varphi_{\lambda,\tau}u = e^{-\frac{\lambda}{2\tau}|D''|^2}(e^{\tau\varphi}u) \,.$$

The operator is a smoothing one, i.e.

$$Q^\varphi_{\lambda,\tau}u \in \mathcal{S}(\mathbf{R}^{n''}, L^2(\Omega')) \,.$$

The next lemma discusses the commutator of Q and P in a special case.

Lemma 5 *Let φ be a second degree polynomial in x and let P be a partial differential operator with real coefficients independent of x''. Moreover, let $u \in \mathcal{E}'(\Omega'', H^m(\Omega'))$. Then*

$$Q^\varphi_{\lambda,\tau}P(x,D)u = P(x, D + i\tau\nabla\varphi(x) - \lambda\nabla^2\varphi D'')Q^\varphi_{\lambda,\tau}u \tag{1.14}$$

where $\nabla^2\varphi$ denotes the constant matrix of the second derivatives of φ and

$$(\nabla^2\varphi D'')_j = \sum_{k=1}^{n''} \partial_{jk}\varphi D_k \,.$$

Proof: We have to show that

$$Q^\varphi_{\lambda,\tau}D^\alpha u = (D + i\tau\nabla\varphi(x) - \lambda\nabla^2\varphi D'')^\alpha Q^\varphi_{\lambda,\tau}u \,.$$

Like before we will do this by induction with respect to α. Since φ is a second degree polynomial we have

$$\partial_j\varphi(x) = \sum_{k=1}^{n} a_{jk}x_k + b_j$$

and

$$\partial_{jk}\varphi(x) = a_{jk} .$$

Set $v = e^{\tau\varphi}u$ and use lemma 4. For $|\alpha| = 1$ we obtain

$$\begin{aligned}
Q^{\varphi}_{\lambda,\tau}D_ju(x) &= \int_{\Omega''} K(x'',y'')\Big(D_j + i\tau\sum_{k=1}^{n''} a_{jk}y_k \\
&\quad + i\tau\sum_{k=n''+1}^{n} a_{jk}x_k + i\tau b_j\Big)v(y'',x')dy'' \\
&= D_j\int_{\Omega''} K(x'',y'')v(y'',x')dy'' \\
&\quad + i\tau\Big(\sum_{k=n''+1}^{n} a_{jk}x_k + b_j\Big)\int_{\Omega''} K(x'',y'')v(y'',x')dy'' \\
&\quad - \lambda\sum_{k=1}^{n''} a_{jk}D_k\int_{\Omega''} K(x'',y'')v(y'',x')dy'' \\
&\quad + i\tau\sum_{k=1}^{n''} a_{jk}x_k\int_{\Omega''} K(x'',y'')v(y'',x')dy'' \\
&= D_jQ^{\varphi}_{\lambda,\tau}u(x) + i\tau\partial_j\varphi(x)Q^{\varphi}_{\lambda,\tau}u(x) - \lambda\sum_{k=1}^{n''}\partial_{jk}\varphi D_kQ^{\varphi}_{\lambda,\tau}u(x)
\end{aligned}$$

Next we assume the result is true for $|\alpha| \le \nu$. Set $\alpha^* = \alpha_j + \alpha$ where $|\alpha| = \nu$ and $|\alpha_j| = 1$. We get

$$\begin{aligned}
&(D + i\tau\nabla\varphi(x) - \lambda\nabla^2\varphi D'')^{\alpha^*}Q^{\varphi}_{\lambda,\tau}u(x) \\
&= (D + i\tau\nabla\varphi(x) - \lambda\nabla^2\varphi D'')^{\alpha_j}(D + i\tau\nabla\varphi - \lambda\nabla^2\varphi D'')^{\alpha}Q^{\varphi}_{\lambda,\tau}u \\
&= (D_j + i\tau\nabla\varphi(x) - \lambda\nabla^2\varphi D'')Q^{\varphi}_{\lambda,\tau}D^{\alpha}u(x) \\
&= Q^{\varphi}_{\lambda,\tau}D_jD^{\alpha}u(x)
\end{aligned}$$

and the lemma is proved. □

The statement is valid for some functions which require less regularity in x'.

Corollary 6 *The claim of lemma 5 holds for $u \in \mathcal{E}'(\Omega'', H_0^{m-1}(\Omega'))$ with $P(x,D)u \in \mathcal{E}'(\Omega'', L^2(\Omega'))$ provided P has a C^1 principal symbol.*

Proof: We will make use of regularization in x'. Then by lemma 2 the sequence $\{u_\epsilon\} \subset \mathcal{E}'(\Omega'', C_0^\infty(\Omega'))$ converges for $\epsilon \to 0$ to u in the topology of

$\mathcal{E}'(\Omega'', H^{m-1}(\Omega'))$ and (1.14) is valid for all u_ϵ. The left hand side of (1.14) is well defined since $P(x,D)u \in \mathcal{E}'(\Omega'', L^2(\Omega'))$ by hypothesis and repeated use of lemma 3 shows that

$$Q^{\varphi}_{\lambda,\tau} P(x,D)u_\epsilon \longrightarrow Q^{\varphi}_{\lambda,\tau} P(x,D)u \tag{1.15}$$

in $L^2(\Omega)$.

We claim the equivalent statement is true for the right hand side. The only thing we have to worry about are the differentiations of order m. Out of those the differentiations in x'' directio n do not cause a problem since the operator $Q^{\varphi}_{\lambda,\tau}$ is a smoothing one. We can expand

$$P(x, D + i\tau\nabla\varphi(x) - \lambda\nabla^2\varphi D'') = P(x,D) + S(x,D)$$

and observe that $S(x,D)$ is an operator which contains no differentiation of order m in x' only. That provides

$$S(x,D)Q^{\varphi}_{\lambda,\tau} u_\epsilon \longrightarrow S(x,D)Q^{\varphi}_{\lambda,\tau} u$$

in $L^2(\Omega)$. Next we observe that

$$\begin{aligned} P(x,D)Q^{\varphi}_{\lambda,\tau} u &= \int K(x'',y'')P(y'',x',D)(e^{\tau\varphi(y'',x')}u(y'',x'))dy'' \\ &= Q^{\varphi}_{\lambda,\tau} P(x,D)u + Q^{\varphi}_{\lambda,\tau} T(x,D)u \end{aligned}$$

where $T(x,D)$ is a differential operator of order $m-1$. The claim follows then from (1.15). □

Inspecting the proof of lemma 5 carfully, we can see that φ needs to be a second degree polynomial only on *supp* u.

Finally, we remar k that the operator $P(x, D - \lambda\nabla^2\varphi D'' + i\tau\nabla\varphi(x))$ allows a decomposition as in (1.8) with the same properties. We denote the pseudo-differential symbol of $P_\tau(x, D - \lambda\nabla^2\varphi D'')$ by $p_\tau(x, \xi - \lambda\nabla^2\varphi\xi'')$.

1.2.3 Differential quadratic forms

This subsection is purely technical. It will supply the tools which are needed to prove Carleman type estimates. Our exposition follows Hörmander [3], sect ion 8.2. We included it in order to give a self contained treatment of the theory.

The operator acting on u and $\bar{u}$

$$F(Du, \overline{Du}) = \sum_{\substack{|\alpha|,|\beta|\le m \\ |\alpha|+|\beta|\le\mu}} a_{\alpha\beta} D^\alpha u \overline{D^\beta u} \qquad u \in C^m(\mathbf{R}^n)$$

is called a differential quadratic form. Here $a_{\alpha\beta}$ are constants and the sum is finite. The form is of order (μ, m). With every differential quadratic form we associate a polynomial

$$F(\zeta, \bar{\zeta}) = \sum_{\substack{|\alpha|,|\beta|\le m \\ |\alpha|+|\beta|\le\mu}} a_{\alpha\beta} \zeta^\alpha \bar{\zeta}^\beta \qquad \zeta \in \mathbf{C}^n .$$

The correspondence between these complex valued polynomials and differential quadratic forms is one-to-one. Let $G^k(D,\bar{D})$, $k = 1,2,..,n$ be differential forms and set

$$F(Du,\overline{Du}) = \sum_{k=1}^{n} \frac{\partial}{\partial x_k} G^k(Du,\overline{Du}) . \tag{1.16}$$

The product rule of differentiation provides

$$\frac{\partial}{\partial x_k} D^\alpha u \overline{D^\beta u} = i\Big((D_k D^\alpha u)\overline{D^\beta u} - D^\alpha u \overline{D_k D^\beta u}\Big)$$

which leads to the identity

$$F(\zeta,\bar{\zeta}) = i\sum_{k=1}^{n} (\zeta_k - \bar{\zeta}_k) G^k(\zeta,\bar{\zeta}) \tag{1.17}$$

and implies $F(\xi,\xi) = 0$ for all $\xi \in \mathbf{R}^n$. The first lemma shows that this condition is sufficient for (1.16).

Lemma 7 *Let $F(D,\bar{D})$ be a differential quadratic form such that $F(\xi,\xi) = 0$ for all $\xi \in \mathbf{R}^n$. Then there exist differential quadratic forms $G^k(D,\bar{D})$ such that (1.16) holds. Furthermore,*

$$G^k(\xi,\xi) = -\frac{1}{2}\frac{\partial}{\partial \eta_k} F(\xi + i\eta, \xi - i\eta)\Big|_{\eta=0} \quad \textit{for all } \xi \in \mathbf{R}^n . \tag{1.18}$$

If F is of order (μ, m), G can be chosen of order $(\mu - 1, m')$ where

$$m' = \begin{cases} m & \textit{if} \quad \mu = 2m \\ m-1 & \textit{if} \quad \mu < 2m \end{cases} .$$

Proof: Since $F(\xi,\xi) = 0$ we can expand

$$F(\xi + i\eta, \xi - i\eta) = \sum_{k=1}^{n} \eta_k g^k(\xi,\eta) . \tag{1.19}$$

This is a representation of the form (1.16) with $\zeta = \xi + i\eta$ which implies $i\eta = (\zeta - \bar{\zeta})/2$. We set

$$G^k(\xi + i\eta, \xi - i\eta) = -\frac{1}{2} g^k(\xi,\eta)$$

and differentiation of (1.19) gives (1.18) .

Finally, we need to discuss the order of G. We introduce a new notion. Two polynomials F_1 and F_2 of order (μ, m) are congruent if their difference can be written as a sum in (1.17) for some G^k of order $(\mu - 1, m')$. We write then $F_1 \equiv F_2$. We claim that

$$\zeta^\alpha \bar{\zeta}^\beta \equiv \zeta^{\alpha'} \bar{\zeta}^{\beta'}$$

where $\alpha+\beta=\alpha'+\beta'$, and both sides are of order (μ, m). Consider at first the case $\mu<2m$. That implie s either $|\alpha|<m$ or $|\beta|<m$. If $|\alpha|<m$ use

$$\bar{\zeta}_j=\zeta_j-(\zeta_j-\bar{\zeta}_j)$$

and when $|\beta|<m$ use

$$\zeta_j=\bar{\zeta}_j+(\zeta_j-\bar{\zeta}_j)\,.$$

Both formulas show that one factor in a product $\zeta^\alpha\bar{\zeta}^\beta$ can be replaced by its complex conjugate without leaving the congruence class. This can be done repeatedly and proves the claim when $\mu<2m$. The case $\mu=2m$ can be dealt with the same argument except when $|lpha|+|\alpha'|=|\beta|+|\beta'|=2m$. Then we replace a product $\zeta_j\bar{\zeta}_k$ by its complex conjugate using

$$\zeta_j\bar{\zeta}_k-\zeta_k\bar{\zeta}_j=(\zeta_j-\bar{\zeta}_j)\bar{\zeta}_k-(\zeta_k-\bar{\zeta}_k)\bar{\zeta}_j\,.$$

The claim is proved.

That means that every differential quadratic form of order (μ, m) is congruent to a differential quadratic form

$$F_1(\zeta,\bar{\zeta})=\sum_{\substack{|\alpha|,|\beta|\le m\\|\alpha|+|\beta|\le\mu}}a_{\alpha\beta}\zeta^\alpha\bar{\zeta}^\beta$$

where there is at most one non-zero term for each multiindex sum $\alpha+\beta$. Since $F_1(\xi,\xi)=0$ it follows that $a_{\alpha\beta}=0$ for all multiindices. That shows $F\equiv 0$ and proves the lemma. □

Now we can consider differential quadratic forms with variable coefficients

$$F(x,Du,\overline{Du})=\sum_{\substack{|\alpha|,|\beta|\le m\\|\alpha|+|\beta|\le\mu}}a_{\alpha\beta}(x)D^\alpha u\overline{D^\beta u}\quad u\in C_0^m(\Omega)$$

where $a_{\alpha\beta}(x)\in C^r(\bar{\Omega})$. Similar to lemma 7 we have the following results.

Lemma 8 *Let $F(x,D,\bar{D})$ be a differential quadratic form of order (μ,m) and assume that $F(x,\xi,\xi)=0$ for all $x\in\Omega$ and $\xi\in\mathbf{R}^n$. Then there exist a differential quadratic form $G(x,D,\bar{D})$ of order $(\mu-1,m')$ with coefficients in $C^{r-1}(\bar{\Omega})$ such that*

$$\int F(x,Du,\overline{Du})=\int G(x,Du,\overline{Du})\quad\text{for all } u\in C_0^m(\bar{\Omega})\,. \tag{1.20}$$

Moreover,

$$G(x,\xi,\xi)=\frac{1}{2}\sum_{k=1}^n\frac{\partial^2}{\partial x_k\partial\eta_k}F(x,\xi+i\eta,\xi-i\eta)\Big|_{\eta=0}\,. \tag{1.21}$$

Proof: Consider the set of constant differential quadratic forms of order (μ,m) with $F(\xi,\xi)=0$. That are polynomials in ζ and $\bar{\zeta}$ over $\mathbf{C}^n$ which form a vector space. Let $F_1, ...F_N$ be a basis in this space. By lemma 7 there exist $G_j^k(D,\bar{D})$ such that

$$F_j(D,\bar{D}) = \sum_{k=1}^{n} \frac{\partial}{\partial x_k} G_j^k(D,\bar{D}) .$$

Since the differential quadratic form under consideration satisfies $F(x,\xi,\xi) = 0$ there are $a_j(x) \in C^r(\bar{\Omega})$ such that

$$F(x,D,\bar{D}) = \sum_{j=1}^{N} a_j(x) F_j(D,\bar{D}) = \sum_{j=1}^{N}\sum_{k=1}^{n} a_j(x) \frac{\partial}{\partial x_k} G_j^k(D,\bar{D}) .$$

Now we set

$$G(x,D,\bar{D}) = -\sum_{j=1}^{N}\sum_{k=1}^{n} \partial_k a_j(x) G_j^k(D,\bar{D})$$

which is of order $(\mu - 1, m')$ by lemma 7 and integration by parts proves (1.20).

Finally, we prove (1.21). By (1.18) we have that

$$G_j^k(\xi,\xi) = -\frac{1}{2}\frac{\partial}{\partial \eta_k} F_j(\xi + i\eta, \xi - i\eta) .$$

Together with the two formulas above this concludes the proof. □

1.3 GEOMETRY OF HYPERSURFACES

Here we will study the conditions on surfaces which will be needed to prove uniqueness of continuation. Usually we use surfaces which are given in the neighborhood of a point as level surfaces of continuous or differentiable functions.

Let $x_0 \in \Omega$ and $\varphi \in C^\nu(\bar{\Omega})$ be real valued and $\nabla\varphi(x_0) \neq 0$. Then the set

$$S = \{x \in \Omega \ : \ \varphi(x) = \varphi(x_0)\} \tag{1.22}$$

defines an oriented C^ν-hyp ersurface in a neighborhood of x_0. We call the set

$$\{x \in \Omega \ : \ \varphi(x) > \varphi(x_0)\}$$

the positive side of S.

1.3.1 Non-characteristic surfaces

Characteristics are of high importance in the theory of partial differential equations. We begin with the definition of the characteristic set.

Definition 9 *The characteristic set Char P of a linear partial differential operator is defined by*

$$Char\ P = \{(x,\xi) \in \Omega \times \mathbf{R}^n \setminus \{0\} \ : \ p(x,\xi) = 0\} . \tag{1.23}$$

Next we introduce the concept of a characteristic surface.

Definition 10 *The C^1-surface S defined by (1.22) is characteristic at x_0 with respect to P if $(x_0, \nabla\varphi(x_0)) \in Char\ P$.*

A surface is a characteristic surface if it is characteristic at each of its points. Consequently, we have to make the following distinction. A surface is non-characteristi c at x_0 with respect to P if $(x_0, \nabla\varphi(x_0)) \notin Char\ P$. However, a non-characteristic surface is a surface which is non-characteristic at every point.

The property of being non-characteristic is stable with respect to small perturbations.

Lemma 11 *Let P be a differential operator with continuous symbol. Suppose the level surface (1.22) of $\varphi \in C^1(\bar{\Omega})$ is non-characteristic at x_0. Then there exist $\epsilon > 0$ and $\delta > 0$ su ch that every $\psi \in C^1(\bar{\Omega})$ with*

$$|D^\alpha(\varphi - \psi)| < \epsilon \qquad \text{for } x \in B_\delta(x_0)\,,\, |\alpha| \leq 1$$

has non-characteristic level surfaces at every point in $B_\delta(x_0)$.

Proof: By hypothesis $p(x, \nabla\varphi(x))$ depends continuously on x and $\nabla\varphi(x)$. □

1.3.2 Pseudo-convex surfaces

This subsection is consist mainly of results which are given in [3], section 8.6 and [4], section 28.3. Throughout this subsection we consider C^2-surfaces given as a level surface of a real valued function $\varphi \in C^2(\bar{\Omega})$. Let p denote the principal symbol of a linear partial differential operator P. Moreover, we recall the symbol $p_\tau(x, \xi)$ introduced in (1.10). Let Γ be a subspace of $\mathbf{R}^n$.

Definition 12 *A C^2-surface S defined by (1.22) in Ω such that $\nabla\varphi(x_0) \neq 0$ is pseudo-co nvex with respect to P on Γ at $x_0 \in \Omega$ if*

$$\begin{aligned} Re\Big\{\bar{p}, \{p, \varphi\}\Big\}(x_0, \xi) &> 0 \quad \text{for all } \xi \in \Gamma \setminus \{0\} \\ \text{when} \quad p(x_0, \xi) &= \textstyle\sum_{j=1}^n p^{(j)}(x_0, \xi)\partial_j\varphi(x_0) = 0\,. \end{aligned} \tag{1.24}$$

The same surface is called strongly pseudo-convex with respect to P on Γ at x_0 when in addition to (1.24)

$$\begin{aligned} \tfrac{1}{2i\tau}\{\bar{p}_{-\tau}, p_\tau\}(x_0, \xi) &> 0 \quad \text{for all } \xi \in \Gamma, \tau > 0 \\ \text{when} \quad p_\tau(x_0, \xi) &= \textstyle\sum_{j=1}^n p_\tau^{(j)}(x_0, \xi)\partial_j\varphi(x_0) = 0\,. \end{aligned} \tag{1.25}$$

We like to point out that this definition is actually independent of the function φ which is used to represent the surface.

Before we proceed we want to give an example. Let $P = \Delta$, i.e. $p = |\xi|^2$ and $\Gamma = \mathbf{R}^n$. A short calculation shows that

$$\Big\{\bar{p}, \{p, \varphi\}\Big\}(x_0, \xi) = 4\xi^T\varphi''(x_0)\xi\,,$$

where φ'' denotes the matrix of second derivatives of φ. The right hand side is exactly the term which appears in a convexity condition for φ. However, the pseudo-convexity condition for the Laplacian is not relevant since *Char* P is empty.

Looking at the strong pseudo-convexity condition we discover

$$\frac{1}{2i\tau}\{p_{-\tau},p_\tau\} = \overline{\xi + i\tau\nabla\varphi(x_0)}^T \varphi''(x_0)(\xi + i\tau\nabla\varphi(x_0)) .$$

This is a convexity condition as well. This condition has to be verified on the set

$$\{(\xi,\tau)\ :\ p_\tau(x_0,\xi) = \sum_{j=1}^n p_\tau^{(j)}(x_0,\xi)\partial_j\varphi(x_0) = 0,\ \ \tau > 0\} .$$

However, this set is empty.

¿From that result it follows that every C^2-surface is strongly pseudo-convex with respect to any second-order elliptic operator at any point.

Restricting ourselves to operators with real coefficients we will be able to simplify this definition slightly. At first we need the following technical result.

Proposition 13 *Let P be a operator with real coefficients. Let $x_0 \in \Omega$ and $\varphi \in C^2(\bar\Omega)$ with $\nabla\varphi(x_0) \neq 0$. Then*

$$\lim_{\tau\to 0}\frac{1}{2i\tau}\{p_{-\tau},p_\tau\}(x_0,\xi) = \Big\{p,\{p,\varphi\}\Big\}(x_0,\xi)$$

for all $\xi \in \mathbf{R}^n$.

Proof: By hypothesis we have $\bar p = p$. For convenience we drop (x_0,ξ) throughout the proof. We calculate

$$\{p_{-\tau},p_\tau\} = 2i\tau\sum_{j,k=1}^n \partial^2_{jk}\varphi\, p_{-\tau}^{(j)}p_\tau^{(k)} + \sum_{j=1}^n\Big(p_{-\tau}^{(j)}p_{\tau(j)} - p_{-\tau(j)}p_\tau^{(j)}\Big) \tag{1.26}$$

and

$$\Big\{p,\{p,\varphi\}\Big\} = \sum_{j,k=1}^n \partial^2_{jk}\varphi\, p^{(j)}p^{(k)} + \sum_{j=1}^n\{p,p^{(j)}\}\partial_j\varphi .$$

This means we need to prove

$$\left|\frac{1}{2i\tau}\sum_{j=1}^n(p_{-\tau}^{(j)}p_{\tau(j)} - p_{-\tau(j)}p_\tau^{(j)}) \quad - \quad \sum_{k=1}^n\{p,p^{(k)}\}\,\partial_k\varphi\right| \leq C\tau|\xi + i\tau\nabla\varphi(x_0)|^{2m-3} . \tag{1.27}$$

Notice that

$$\begin{aligned}\sum_{k=1}^n\{p,p^{(k)}\}\,\partial_k\varphi &= \sum_{j,k=1}^n(p^{(j)}p^{(k)}_{(j)} - p_{(j)}p^{(jk)})\,\partial_k\varphi \\ &= \frac{1}{2i}\sum_{j=1}^n\frac{\partial}{\partial\sigma}\Big(p_{-\sigma}^{(j)}p_{\sigma(j)} - p_{-\sigma(j)}p_\sigma^{(j)}\Big)\Big|_{\sigma=0}\end{aligned}$$

and Taylor's formula gives

$$\begin{aligned}\sum_{j=1}^{n}(p^{(j)}_{-\tau}p_{\tau(j)} - p_{-\tau(j)}p^{(j)}_{\tau}) &= \{p,p\} + \tau\sum_{j=1}^{n}\frac{\partial}{\partial\sigma}\left(p^{(j)}_{-\sigma}p_{\sigma(j)} - p_{-\sigma(j)}p^{(j)}_{\sigma}\right)\Big|_{\sigma=0} \\ &\quad + \frac{\tau^2}{2}\sum_{j=1}^{n}\frac{\partial^2}{\partial\sigma^2}\left(p^{(j)}_{-\sigma}p_{\sigma(j)} - p_{-\sigma(j)}p^{(j)}_{\sigma}\right)\Big|_{\sigma=\eta}\end{aligned}$$

where $\eta \in (0,\tau)$. We use this to obtain the following estimate

$$\begin{aligned}&\left|\frac{1}{2i}\sum_{j=1}^{n}(p^{(j)}_{-\tau}p_{\tau(j)} - p_{-\tau(j)}p^{(j)}_{\tau}) - \tau\sum_{k=1}^{n}\{p,p^{(k)}\}\,\partial_k\varphi\right| \\ &\le |\{p,p\}| + C\tau^2|\xi + i\eta\nabla\varphi|^{2m-3}\,.\end{aligned}$$

Because of $\{p,p\} = 0$, (1.52) is proved.□

This lemma shows that the Poisson bracket in (1.25) has a meaning when $\tau = 0$. It allows us to find a simpler condition for strong pseudo-convex surfaces.

Lemma 14 *Let P be a differential operator with real coefficients and let S be strongly pseudo-convex with respect to P on Γ at x_0. Then there exist a $C > 0$ such that*

$$\begin{array}{rcll}\frac{1}{2i\tau}\{\bar{p}_{-\tau},p_\tau\}(x_0,\xi) & \ge & C(|\xi|^2+\tau^2)^{m-1} & \textit{for all } \xi\in\Gamma, \tau\ge 0 \\ \textit{when}\;\; p_\tau(x_0,\xi) & = & \sum_{j=1}^{n} p^{(j)}_\tau(x_0,\xi)\partial_j\varphi(x_0) = 0 & \end{array} \qquad (1.28)$$

where the surface S is a level surface of φ.

Proof: Let Σ denote the unit sphere in the (ξ,τ) space and define

$$A = \left\{(\xi,\tau)\in\Sigma\;:\tau\ge 0,\; p_\tau(x_0,\xi) = \sum_{j=1}^{n}p^{(j)}_\tau(x_0,\xi)\partial_j\varphi(x_0) = 0\right\}. \qquad (1.29)$$

The set A is closed in Σ. By hypothesis we have

$$\frac{1}{2i\tau}\{p_{-\tau},p_\tau\}(x_0,\xi) > 0 \quad \text{for } \tau > 0,\; (\xi,\tau)\in A$$

and by proposition 29 and by hypothesis

$$\lim_{\tau\to 0}\frac{1}{2i\tau}\{p_{-\tau},p_\tau\}(x_0,\xi) = \left\{p,\{p,\varphi\}\right\} > 0$$

on $p(x_0,\xi) = \sum_{j=1}^{n}p^{(j)}(x_0,\xi)\partial_j\varphi(x_0) = 0$. Consequently

$$\frac{1}{2i\tau}\{p_{-\tau},p_\tau\}(x_0,\xi) > 0 \quad \text{for } (\xi,\tau)\in A$$

and since A is closed in Σ we introduce

$$C = \inf_{(\xi,\tau)\in A}\frac{1}{2i\tau}\{p_{-\tau},p_\tau\}(x_0,\xi)\,.$$

The statement of the lemma follows now by homogeneity. □

The next lemma shows that we can represent a strongly pseudo-convex surface as a level surface of a function φ with an even simpler condition for strong pseudo-convexity.

Lemma 15 *Let P be a differential oper ator with real coefficients and let S be strongly pseudo-convex with respect to P on Γ at x_0. Then there exist a $\varphi \in C^2(\bar{\Omega})$ and $C > 0$ such that S is a level surface for φ and*

$$\begin{array}{rcll} \frac{1}{2i\tau}\{\bar{p}_{-\tau}, p_\tau\}(x_0,\xi) & \geq & C(|\xi|^2+\tau^2)^{m-1} & \textit{for all } \xi \in \Gamma, \tau \geq 0 \\ \textit{when} \quad p_\tau(x_0,\xi) & = & 0\,. & \end{array} \tag{1.30}$$

Proof: Since S is strongly pseudo-convex at x_0 there exist a $\psi \in C^2(\bar{\Omega})$ such that (1.53) is satisfied. Set $\varphi = e^{\lambda\psi}$ with $\lambda > 0$. We have

$$\begin{array}{rcl} \partial_j\varphi & = & \lambda\varphi\partial_j\psi \quad \text{and} \\ \partial^2_{jk}\varphi & = & \lambda^2\varphi\partial_j\psi\partial_k\psi + \lambda\varphi\partial^2_{jk}\psi\,. \end{array} \tag{1.31}$$

In order to do t he next step in the proof we need to refine our notation. Set

$$\begin{array}{rcl} p_{\tau,\varphi}(x_0,\xi) & = & p(x_0,\xi+i\tau\nabla\varphi(x_0)) \quad \text{and} \\ p_{\sigma,\psi}(x_0,\xi) & = & p(x_0,\xi+i\sigma\nabla\psi(x_0))\,. \end{array}$$

Then we choose $\sigma = \lambda\varphi(x_0)\tau$ which implies $p_{\tau,\varphi}(x_0,\xi) = p_{\sigma,\psi}(x_0,\xi)$. That gives together with (1.56) (we omit (x_0,ξ) for convenience)

$$\sum_{j,k=1}^n \partial^2_{jk}\varphi p^{(j)}_{-\tau,\varphi}p^{(k)}_{\tau,\varphi} = \lambda\varphi\left(\sum_{j,k=1}^n \partial^2_{jk}\psi p^{(j)}_{-\sigma,\psi}p^{(k)}_{\sigma,\psi} + \lambda\left|\sum_{j=1}^n p^{(j)}_{\sigma,\psi}\partial_j\psi\right|^2\right).$$

Hence,

$$\frac{1}{2i\tau}\{p_{-\tau,\varphi},p_{\tau,\varphi}\} = \lambda\varphi\left(\frac{1}{2i\sigma}\{p_{-\tau,\sigma},p_{\tau,\sigma}\} + \lambda\left|\sum_{j=1}^n p^{(j)}_{\sigma,\psi}\partial_j\psi\right|^2\right) \tag{1.32}$$

and the right hand side is greater or equal to $C\lambda\varphi(|\xi|^2+\sigma^2)^{m-1}$ on $p_{\tau,\psi} = \sum_{k=1}^n p^{(j)}_{\tau,\psi}\partial_j\psi = 0$. However, choosing λ large enough we can ensure the same estimate on the larger set $p_{\sigma,\psi} = 0$. This follows from the fact that the last term in (1.57) is homogeneous of order $2(m-1)$ in (ξ,σ) and positive provided $\sum_{k=1}^n p^{(j)}_{\tau,\psi}\partial_j\psi \neq 0$. Consequently

$$\frac{1}{2i\tau}\{p_{-\tau,\varphi},p_{\tau,\varphi}\}(x_0,\xi) \geq C\lambda\varphi(x_0)(|\xi|^2+\sigma^2)^{(m-1)}$$

and replacing σ by $\lambda\varphi(x_0)\tau$ proves (1.55). □

The last result makes the following definition useful.

Definition 16 *Let $\varphi \in C^2(\bar{\Omega})$ be real valued with $\nabla\varphi(x_0) \neq 0$. We say that φ is strongly pseudo-convex with respect to P on Γ at x_0 if it satisfies (1.55).*

The next result follows directly from lemma 30. It shows that the strong pseudo-convexity is stable with respect to small perturbations.

Corollary 17 *Let P be an operator with real coefficients and C^1 principal symbol. Suppose the level surface of $\varphi \in C^2(\bar{\Omega})$ is strongly pseudo-convex at with respect to P on Γ at x_0. Then there exist $\epsilon > 0$ and $\delta > 0$ such that every $\psi \in C^2(\bar{\Omega})$ with*

$$|D^\alpha(\varphi - \psi)| < \epsilon \qquad \text{for } x \in B_\delta(x_0)\,,\, |\alpha| \leq 2 \tag{1.33}$$

has strongly pseudo-convex level surfaces at every point in $B_\delta(x_0)$.

Proof: We define the following function

$$F(\psi, x, \xi, \tau) = \frac{1}{2i\tau}\{p_{-\tau,\psi}, p_{\tau,\psi}\}$$

where $\psi \in C^2(\Omega)$ and $x \in \Omega$. This function is continuous with respect to all variable s and with respect to the first and second derivatives of ψ. The pseudo-convexity conditions imply $F(\varphi, x_0, \xi, \tau) \geq C$ on A where A is defined in (1.54). Consequently, there exists a neighborhood A' of A in Σ and $F(\varphi, x_0, \xi, \tau) \geq C/2$ on A'. By continuity of p we can find $\epsilon > 0$ and $\delta > 0$ such that for ψ satisfying (1.58)

$$\{(\xi, \tau) \in \Sigma \,:\, p_{\tau,\psi}(x, \xi) = \sum_{j=1}^{n} p^{(j)}_{\tau,\psi}(x, \xi)\partial_j\psi(x) = 0\} \subset A'$$

for all $x \in B_\delta(x_0)$. Finally, by continuity of F and if necessary, by making ϵ and δ smaller, we can guarantee $F(\psi, x, \xi, \tau) \geq C/4$ on A' for all $x \in B_\delta(x_0)$. □

A similar result can be proved with respect to small perturbations with respect to the operator $P(x, D)$.

Corollary 18 *Let P be an operator with real coefficients and C^1 principal symbol. Suppose the level surface of $\varphi \in C^2(\bar{\Omega})$ is strongly pseudo-convex with respect to P on Γ at x_0. Then there exist a $\epsilon > 0$ such that the level surface of φ is strongly pseudo-convex with respect to $\tilde{P}$ on Γ at x_0 provided*

$$|D^\beta(a_\alpha(x_0) - \tilde{a}_\alpha(x_0))| \leq \epsilon \qquad \text{for } |\beta| \leq 1\,,\, |\alpha| = m$$

where $\tilde{a}_\alpha$ denotes the coefficients of the operator $\tilde{P}$.

Proof: The proof is analog to the proof of the previous corollary and will be omitted.

Finally, we state a result which relates strong pseudo-convex and non-characteristic surfaces.

Lemma 19 *Let S be non-characteristic at x_0. Then the surface S is strongly pseudo-convex at x_0 with respect to P on $\{0\}$.*

Proof: We have to check that (1.25) is satisfied since the pseudo-convexity condition (1.24) does not apply when $\Gamma = \{0\}$. We have

$$p_\tau(x_0, 0) = p(x_0, i\tau\nabla\varphi(x_0)) = (i\tau)^m p(x_0, \nabla\varphi(x_0)) \neq 0$$

for $\tau > 0$ since S is non-characteristic at x_0.

1.4 UNIQUENESS OF CONTINUATION THEOREMS FOR SOLUTIONS OF PARTIAL DIFFERENTIAL EQUATIONS

This section consist of three different uniqueness of continuation theorems. The first one is Holmgren's theorem. We state it here since it is necessary for a good understanding of the further results. Moreover, its conclu sion seems to be optimal with respect to the geometry of the hypersurface. Since it is a well known result we will omit the proof here. The second result is Hörmander's theorem which provides uniqueness of continuation for a large family of operators under a more restrictive condition on the geometry of the hypersurface. The hypersurface is assumed to be strongly pseudo-convex with respect to the operator. The third result is a recent one which shows that the conclusion of Holmgren 's theorem holds for a larger class of operators.

Before we proceed we give a formal definition of the uniqueness of continuation property where we make use of an oriented hypersurfaces defined by (1.22).

Definition 20 *Let $P(x, D)$ be a partial differential operator and S an oriented hypersurface in Ω. We say that the uniqueness of continuation property holds with respect to P across S at $x_0 \in S$ for a distribution u if there exists a neighbor hood $V(x_0) \subset \Omega$ such that $P(x, D)u = 0$ in Ω and $u \equiv 0$ on the positive side of S imply that u vanishes in $V(x_0)$.*

We mention that this definition is not entirely accurate since we need to impose conditions on the operator and on the distribution u for $P(x, D)u$ to be well defined. Here we will assume that these conditions are met.

1.4.1 *Holmgren's theorem*

Theorem 21 *Let $P(x, D)$ be a partial differential operator with analytic coefficients. Moreover, assume that the oriented C^1-hypersurface S is non-characteristic at x_0. Then the uniqueness of continuation property holds with respect to P across S at x_0 for every $u \in D'(\Omega)$.*

For a proof of this theorem we refer to [3] section 5.3. The following corollary is obvious when we observe corollary 35.

Corollary 22 *Let P be a differential operator with constant coefficients a nd S be an oriented C^2-hypersurface in Ω. If the surface is strongly pseudo-convex with respect to P on $\{0\}$ at x_0, then the uniqueness of continuation property holds with respect to P across S at x_0 for all $u \in D'(\Omega)$.*

1.4.2 Hörmander's theorem

This is the second classical result about uniqueness of continuation. We restrict ourselves to operators with real-valued coefficients.

Theorem 23 *Let P b e a differential operator with real coefficients and C^1 principal symbol. Assume that the oriented C^2-hypersurface S is strongly pseudo-convex with respect to P on $\mathbf{R}^n$ at x_0. Then the uniqueness of continuation property holds with respect to P across S at x_0 for every $u \in H^{m-1}(\Omega)$.*

The main tool for the proof of this theorem is a Carleman type estimate which will be established in the following theorem. Up to minor det ails it can be found in [3] as theorem 8.4.2. Our proof will follow the proof given there for the most part and relies on integration by parts using differential quadratic forms. We like to point out that a second proof is given in [4] chapter 28. That proof is based on pseudo-differential calculus and a version of the sharp Gårding inequality also known as Fefferman-Phong inequality. We felt that the proof in [3] is easier and applies to operators with C^1 p rincipal symbol whereas the other proof requires much more technique and is valid only for operators with C^∞ principal symbol.

Theorem 24 *Let P be a differential operator with real coefficients and C^1 principal symbol. Moreover, let $\varphi \in C^m(\bar{\Omega})$ be strongly pseudo-convex with respect to P on $\mathbf{R}^n$ at x_0. Then there exist a $\delta > 0$ and a constant $K > 0$ such that*

$$\sum_{|\alpha|<m} \tau^{2(m-|\alpha|)-1} \int |D^\alpha u|^2 e^{2\tau\varphi} \leq K \int |P(x,D)u|^2 e^{2\tau\varphi} \tag{1.34}$$

for all $u \in C_0^\infty(B_\delta(x_0))$ and τ large enough.

Proof: At first we notice that with $v = e^{\tau\varphi}u$ and (4) this estimate can be written as

$$\tau\|v\|^2_{H_\tau^{m-1}(\mathbf{R}^n)} \leq K\|P(x, D + i\tau\nabla\varphi(x))v\|^2_{L^2(\mathbf{R}^n)} \,. \tag{1.35}$$

We claim that this estimate is proved when we show

$$\tau\|v\|^2_{H_\tau^{m-1}(\mathbf{R}^n)} \leq \frac{K}{4}\|P_\tau(x,D)v\|^2_{L^2(\mathbf{R}^n)} \,. \tag{1.36}$$

When we use (1.8) we get

$$|P_\tau(x,D)v|^2 \leq 2|P(x, D + i\tau\nabla\varphi(x))v|^2 + 2|R(x,D)v|^2$$

which leads together with (1.9) to

$$\|P_\tau(x,D)v\|^2_{L^2(\mathbf{R}^n)} \leq 2\|P(x, D + i\tau\nabla\varphi(x))v\|^2_{L^2(\mathbf{R}^n)} + 2C\|v\|^2_{H_\tau^{m-1}(\mathbf{R}^n)} \,.$$

Assume now that (1.61) is true. Then our last estimate implies

$$(\tau - 2C)\|v\|^2_{H^{m-1}_\tau(\mathbf{R}^n)} \le \frac{K}{2}\|P(x, D + i\tau\nabla\varphi(x))v\|^2_{L^2(\mathbf{R}^n)}$$

and choosing $\tau > 4C$ proves (1.60).

The proof of (1.60) will be based on integration by parts in the integral

$$\int |P_\tau(x, D)v|^2$$

where we make use of differential quadratic forms. We start with a trivial estimate

$$\int |P_\tau(x, D)v|^2 \ge \int \left(|P_\tau(x, D)v|^2 - |P_{-\tau}(x, D)v|^2\right) .$$

The reason for this last estimate is that we will interpret the right hand side as a differential quadratic form $F(x, D, \bar{D})$ to which we can apply lemma 8. The associated polynomial is

$$F(x, \zeta, \bar{\zeta}) = |p_\tau(x, \zeta)|^2 - |p_\tau(x, \bar{\zeta})|^2$$

and we notice that $F(x, \xi, \xi) = 0$ for all $\xi \in \mathbf{R}^n$. We need to have a closer look at this polynomi al. It is a polynomial in τ of order $2m$ as well. We can expand it into powers of τ

$$F(x, \zeta, \bar{\zeta}) = \sum_{j=0}^{2m} \tau^j F^{(j)}(x, \zeta, \bar{\zeta})$$

and observe that $F^{(0)}(x, \zeta, \bar{\zeta}) = |p(x, \zeta)|^2 - |p(x, \zeta)|^2 = 0$ which shows that $\tau^{-1}F(x, \zeta, \bar{\zeta})$ is a polynomial associated to a differential quadratic form as well. Moreover $F^{(2m)}(x, \zeta, \bar{\zeta}) = |p(x, i\nabla\varphi(x))|^2 - |p(x, i\nabla\varphi(x))|^2 = 0$. We apply lemma 8 to each of the $F^{(j)}(x, \zeta, \bar{\zeta})$. Those are differential quadratic forms of order $(2m - j, m)$. Consequently, there exist differential quadratic forms $G^{(j)}(x, \zeta, \bar{\zeta})$ of order $(2m - j - 1, m - 1)$ such that with

$$G(x, \zeta, \bar{\zeta}) = \sum_{j=1}^{2m-1} \tau^{j-1} G^{(j)}(x, \zeta, \bar{\zeta}) = \sum_{j=1}^{2m-1} \tau^{j-1} \sum_{\substack{|\alpha|,|\beta| \le m-1 \\ |\alpha|+|\beta| \le 2m-j-1}} a^{(j)}_{\alpha\beta}(x)\zeta^\alpha \bar{\zeta}^\beta \tag{1.37}$$

we obtain

$$\tau \int G(x, Dv, \overline{Dv}) = \int F(x, Dv, \overline{Dv}) .$$

Moreover, (1.20) gives

$$G(x, \xi, \xi) \;=\; \frac{1}{2\tau} \sum_{k=1}^{n} \frac{\partial^2}{\partial x_k \partial \eta_k} F(x, \xi + i\eta, \xi - i\eta)\Big|_{\eta=0}$$

$$= \frac{1}{2\tau}\sum_{k=1}^{n}\frac{\partial}{\partial x_k \partial \eta_k}(p_{-\tau}(x,\xi-i\eta)p_\tau(x,\xi+i\eta)$$
$$-p_{-\tau}(x,\xi+i\eta)p_\tau(x,\xi-i\eta))\Big|_{\eta=0}$$
$$= \frac{1}{2\tau}\sum_{k=1}^{n}\frac{\partial}{\partial x_k}\Big(-ip_{-\tau}^{(k)}(x,\xi)p_\tau(x,\xi)+ip_{-\tau}(x,\xi)p_\tau^{(k)}(x,\xi)$$
$$-ip_{-\tau}^{(k)}(x,\xi)p_\tau(x,\xi)+ip_{-\tau}(x,\xi)p_\tau^{(k)}(x,\xi)\Big)$$
$$= \frac{1}{i\tau}\sum_{k=1}^{n}\frac{\partial}{\partial x_k}\Big(p_{-\tau}^{(k)}(x,\xi)p_\tau(x,\xi)-p_{-\tau}(x,\xi)p_\tau^{(k)}(x,\xi)\Big)$$
$$= \frac{1}{i\tau}\Big(\{p_{-\tau},p_\tau\}(x,\xi)$$
$$+\sum_{k=1}^{n}\Big(\frac{\partial}{\partial x_k}p_{-\tau}^{(k)}(x,\xi)p_\tau(x,\xi)-p_{-\tau}(x,\xi)\frac{\partial}{\partial x_k}p_\tau^{(k)}(x,\xi)\Big)\Big).$$

Hence,

$$G(x,\xi,\xi) = \frac{1}{i\tau}\{p_{-\tau},p_\tau\}(x,\xi)$$

for all $(x,\xi)\in\Omega\times\mathbf{R}^n$ that satisfy $p_\tau(x,\xi)=0$ and $\tau\geq 0$. By hypothesis

$$G(x_0,\xi,\xi)\geq C \quad \text{for all } (\xi,\tau)\in\Sigma \text{ with } p_\tau(x_0,\xi)=0\,,\tau\geq 0\,.$$

Hence, we can find a constant $\tilde{C}$ such that

$$G(x_0,\xi,\xi)+\tilde{C}|p_\tau(x_0,\xi)|^2\geq C \quad \text{for all } (\xi,\tau)\in\Sigma \tag{1.38}$$

which by homogeneity implies that

$$(|\xi|^2+\tau^2)^{m-1}\leq C_1G(x_0,\xi,\xi)+C_2|p_\tau(x_0,\xi)|^2(|\xi|^2+\tau^2)^{-1}\,.$$

Now we multiply this inequality by $|\hat{v}(\xi)|^2$, integrate an d use Parseval's identity and obtain

$$\|v\|_{H_\tau^{m-1}(\mathbf{R}^n)}\leq C_1\int G(x_0,Dv,\overline{Dv})+C_2\|P_\tau(x_0,D)v\|^2_{H_\tau^{-1}(\mathbf{R}^n)}\,. \tag{1.39}$$

For $\epsilon>0$ there exist a $\delta>0$ such that

$$|a_\alpha(x)-a_\alpha(x_0)| \leq \epsilon \tag{1.40}$$
$$|\partial_j\varphi(x)-\partial_j\varphi(x_0)| \leq \epsilon \tag{1.41}$$
$$|a_{\alpha\beta}^{(j)}(x)-a_{\alpha\beta}^{(j)}(x_0)| \leq \epsilon \tag{1.42}$$

for all $x\in B_\delta(x_0)$. The first and the second estimate are based on the continuity of the coefficients a_α and the continuity of the gradient of φ. For the third estimate we notice that the coefficients of the differential quadratic form G

depend on the coefficients a_α and its first derivatives and on the first and second derivatives of φ. Since $a_\alpha \in C^1(\bar{\Omega})$ and $\varphi \in C^m(\bar{\Omega})$ the coefficients of the differential quadratic form are continuous as well.

Since $supp\, v \subset B_\delta(x_0)$ we can estimate

$$\begin{aligned}
&\left| \int G(x, Dv, \overline{Dv}) - \int G(x_0, Dv, \overline{Dv}) \right| \\
&\leq \sum_{j=0}^{2m-2} \tau^j \int |G^{(j+1)}(x, Dv, \overline{Dv}) - G^{(j+1)}(x_0, Dv, \overline{Dv})| \\
&\leq \sum_{j=0}^{2m-2} \tau^j \sum_{\substack{|\alpha|,|\beta| \leq m-1 \\ |\alpha|+|\beta| \leq 2m-j-2}} \int |a_{\alpha\beta}^{(j+1)}(x) - a_{\alpha\beta}^{(j+1)}(x_0)||D^\alpha v||D^\beta v| \\
&\leq \epsilon \sum_{j=0}^{2m-2} \tau^j \sum_{\substack{|\alpha|,|\beta| \leq m-1 \\ |\alpha|+|\beta| \leq 2m-j-2}} \int |D^\alpha v||D^\beta v| \\
&\leq \epsilon \sum_{|\alpha|,|\beta| \leq m-1} \tau^{2(m-1)-|\alpha|-|\beta|} \int |D^\alpha v||D^\beta v| \\
&\leq \epsilon \sum_{|\alpha| \leq m-1} \tau^{2(m-1-|\alpha|)} \int |D^\alpha v|^2 \leq \epsilon \|v\|^2_{H_\tau^{m-1}(\mathbf{R}^n)} \,. \qquad (1.43)
\end{aligned}$$

We need to get a similar estimate for the last term in (1.64). We have

$$\begin{aligned}
&\|P_\tau(x_0, D)v - P_\tau(x, D)v\|_{H_\tau^{-1}(\mathbf{R}^n)} \\
&\leq \left\| P_\tau(x_0, D)v - \sum_{|\alpha|=m} a_\alpha(x) \sum_{\beta \leq \alpha} \binom{\alpha}{\beta} (i\tau\nabla\varphi(x_0))^{\alpha-\beta} D^\beta v \right\|_{H_\tau^{-1}(\mathbf{R}^n)} \\
&+ \left\| \sum_{|\alpha|=m} a_\alpha(x) \sum_{\beta \leq \alpha} \binom{\alpha}{\beta} (i\tau\nabla\varphi(x_0))^{\alpha-\beta} D^\beta v - P_\tau(x, D)v \right\|_{H_\tau^{-1}(\mathbf{R}^n)} \\
&\leq \epsilon \| \sum_{|\alpha| \leq m} (D + i\tau\varphi(x_0))^\alpha v\|_{H_\tau^{-1}(\mathbf{R}^n)} + \epsilon\|P(x, D + i\tau\mathbf{1})v\|_{H_\tau^{-1}(\mathbf{R}^n)} \\
&\leq \epsilon C_3 \|v\|_{H_\tau^{m-1}(\mathbf{R}^n)} \qquad (1.44)
\end{aligned}$$

where C_3 depends on ly on P and $\nabla\varphi(x_0)$. Hence, by the triangle inequality

$$\begin{aligned}
\|P_\tau(x_0, D)v\|^2_{H_\tau^{-1}(\mathbf{R}^n)} &\leq 2\|P_\tau(x, D)v\|^2_{H_\tau^{-1}(\mathbf{R}^n)} + 2C_3^2\epsilon^2\|v\|^2_{H_\tau^{m-1}(\mathbf{R}^n)} \\
&\leq 2\tau^{-2}\|P_\tau(x, D)v\|^2_{L^2(\mathbf{R}^n)} + 2C_3^2\epsilon^2\|v\|^2_{H_\tau^{m-1}(\mathbf{R}^n)}
\end{aligned}$$

Combining everything we obtain

$$(1 - C_1\epsilon - C_2C_3^2\epsilon^2)\|v\|_{H_\tau^{m-1}(fR^n)} \leq (C_1\tau^{-1} + C_2\tau^{-2})\|P_\tau(x, D)v\|^2_{L^2(\mathbf{R}^n)}$$

Finally we choose $\epsilon > 0$ as small that

$$1 - C_1\epsilon - C_2C_3\epsilon^2 > \frac{1}{2}$$

and multiply with 2τ and we obtain (1.61) for $\tau > C_2$ and $K/4 = 2C_1 + 2$. □

Next we show that the Carleman type estimate is valid for a larger class of functions.

Corollary 25 *Under the same hypothesis as in theorem 40 the estimate (??) is valid for $u \in H_0^{m-1}(B_\delta(x_0))$ such that $P(x,D)u \in L^2(B_\delta(x_0))$.*

Proof: By lemma 2 there exists a sequence of mollifiers $u_\epsilon \subset C_0^\infty(B_\delta(x_0))$ such that

$$u_\epsilon \longrightarrow u \quad \text{in } H^{m-1}(B_\delta(x_0)) \quad \text{for } \epsilon \to 0\,.$$

We need to show that this implies

$$P(x,D)u_\epsilon \longrightarrow P(x,D)u \quad \text{in } L^2(B_\delta(x_0)) \quad \text{for } \epsilon \to 0\,.$$

This can be shown as follows

$$\begin{aligned} &\|P(x,D)u_\epsilon - P(x,D)u\|_{L^2(B_\delta(x_0))} \\ &\leq \|P(x,D)u_\epsilon - (P(x,D)u)_\epsilon\|_{L^2(B_\delta(x_0))} \\ &+\|(P(x,D)u)_\epsilon - P(x,D)u\|_{L^2(B_\delta(x_0))} \to 0\,. \end{aligned}$$

Here the first term tends to zero because of lemma 3 and the second one since $P(x,D)u \in L^2(B_\delta(x_0))$. □

Now we can prove Hörmander's theorem.

Proof of theorem 39: Since S is a strongly pseudo-convex surface we can always find a strongly pseudo-convex function φ such that S is the level surface $\varphi(x) = \varphi(x_0)$. This is proved in lemma 31. Let

$$\psi(x) = \nabla\varphi(x_0)\cdot(x - x_0) + \frac{1}{2}(x - x_0)^T\nabla^2\varphi(x_0)(x - x_0) - 3\gamma|x - x_0|^2\,.$$

where $\gamma > 0$ is chosen to be as small that $\psi(x)$ is strongly pseudo-convex in $B_\delta(x_0)$. Without loss of generality assume that $\{|x - x_0| \leq 2\delta\} \subset \Omega$ and

$$\psi(x) \leq \varphi(x) - \varphi(x_0) - 2\gamma|x - x_0|^2$$

for $|x - x_0| \leq 2\delta$. Let χ be a smooth cutoff function

$$\chi(x) = \begin{cases} 1 & \text{if} \quad x \geq -\gamma\delta^2 \\ 0 & \text{if} \quad x \leq -2\gamma\delta^2 \end{cases}$$

and define

$$\tilde{u}(x) = \begin{cases} \chi(\psi(x))u(x) & \text{if } |x - x_0| \leq 2\delta \\ 0 & \text{otherwise} \end{cases}\,.$$

We will apply (1.59) to $\tilde{u}$. We can do so since

$$\begin{aligned}
supp\, \tilde{u} &\subset supp\, u \cap supp\, \chi(\psi) \cap B_{2\delta}(x_0) \\
&\subset \{\varphi(x) \le \varphi(x_0)\} \cap \{\psi(x) \ge -2\gamma\delta^2\} \cap B_{2\delta}(x_0) \\
&\subset \{\varphi(x) \le \varphi(x_0)\} \cap \{\varphi(x) - \varphi(x_0) - 2\gamma|x - x_0|^2 > -2\gamma\delta^2\} \\
&\subset B_\delta(x_0)\, .
\end{aligned}$$

Moreover, $\tilde{u} \in H_0^{m-1}(B_\delta(x_0))$ and $P(x,D)\tilde{u} \in L^2(B_\delta(x_0))$ with

$$supp\, P(x,D)\tilde{u} \subset G = \{\psi \le -\gamma\delta^2\} \cap B_\delta(x_0)\, . \tag{1.45}$$

By G^c we denote the complement of G in $B_\delta(x_0)$, i.e. $G^c = B_\delta(x_0) \setminus G$. Now we use the Carleman type estimate with weight function ψ and obtain

$$\begin{aligned}
e^{-2\tau\gamma\delta^2} \sum_{|\alpha|\le m-1} \tau^{2(m-|\alpha|)-1} \int_{G^c} |D^\alpha u|^2 &\le \sum_{|\alpha|\le m-1} \tau^{2(m-|\alpha|)-1} \int |D^\alpha \tilde{u}|^2 e^{2\tau\psi} \\
&\le K \int |P(x,D)\tilde{u}|^2 e^{2\tau\psi} \\
&\le e^{-2\tau\gamma\delta^2} K \int |P(x,D)\tilde{u}|^2
\end{aligned}$$

which gives

$$\sum_{|\alpha|\le m-1} \tau^{2(m-1-|\alpha|)} \int_{G^c} |D^\alpha u|^2 \le \tau^{-1} K \int |P(x,D)\tilde{u}|^2\, .$$

For $\tau \to \infty$ we obtain

$$u \equiv 0 \qquad \text{for all } x \in G^c\, .$$

This set is a neighborhood of x_0 and that proves the theorem. □

1.4.3 A recent result about uniqueness of continuation

In this subsection we like to present a result which is somehow between Holmgren's and Hörmander's theorem. In 1995 Tataru published his results in [10]. Here we w ill present a special case of his theorem which we will apply in the following section to two systems of partial differential equations.

Let us recall that corollary 38 gives the uniqueness of continuation property for operators with constant coefficients across a surface that is strongly pseudo-convex with respect to the operator on $\{0\}$. On the other hand Hörmander's theorem gives the same property for operators with C^1 principal symbol across a surface that is strongly pseudo-convex with respect to the operator on $\mathbf{R}^n$. Indeed, between those two results can be interpolated.

As in subsection 1.2.1 we decompose $\mathbf{R}^n = \mathbf{R}^{n''} \times \mathbf{R}^{n'}$ and $x = (x'', x')$ and $\xi = (\xi'', \xi')$. Throughout this subsection $\Omega = \Omega'' \times \Omega'$ where Ω'' is an open simply connected subset of $\mathbf{R}^{n''}$ and Ω' is an open simply connected subset of $\mathbf{R}^{n'}$. In the following

$$\Gamma = \{\xi \in \mathbf{R}^n\ :; \xi'' = 0\}\, .$$

Then we consider a partial differential operator with real coefficients that depend only on x'. For those operators we will prove the following result, which is a special case of theorem 2 in [10].

Theorem 26 *Let P be a linear partial differential operator with real coefficients independent of x'' and C^1 principal symbol. Moreover, let S be strongly pseudo-convex with respect to P on Γ at x_0. Then the uniqueness of co ntinuation property holds with respect to P across S at x_0 for $u \in \mathcal{D}'(\Omega'', H^{m-1}(\Omega'))$.*

Similar to Hörmander's theorem the proof is based on Carleman type estimates. In complete analogy with (39) we obtain the following result. Again, this lemma is a special case of lemma 2.4 in [10]. We give here an independent proof relying on differential quadratic forms.

Lemma 27 *Let P satisfy the same conditions as in theorem 42. Furthermore, let φ be a second degree polynomial in x which is strongly pseudo-convex at x_0 with respect to P on Γ. Then for $\lambda \leq \lambda_0$ there exist a $\delta > 0$ and $K > 0$ such that*

$$\begin{aligned} \tau\|v\|^2_{H^{m-1}_\tau(\mathbf{R}^n)} &\leq K\Big(\|P(x, D - \lambda\nabla^2\varphi D'' + i\tau\nabla\varphi(x))v\|^2_{L^2(\mathbf{R}^n)} \\ &\quad + \tau \sum_{|\alpha|\leq m-1} \|D''^\alpha v\|^2_{L^2(\mathbf{R}^n)}\Big) labelcarta \end{aligned} \tag{1.46}$$

for all $v \in C_0^\infty(B_\delta(x_0))$ and large τ.

Proof: We notice the obvious analogy to (1.60). This lemma is more or less a corollary of theorem 40. That means, the proof follows exactly the one of theorem 40 with a few modifications. Instead of the operator $P_\tau(x, D)$ and its symbol $p_\tau(x, \xi)$ we will work with $P_\tau(x, D - \lambda\nabla^2\varphi D'')$ and $p_\tau(x, \xi - \lambda\, nabla^2\varphi\xi'')$, respectively. We notice that the these two symbols coincide on Γ. Since the strong pseudo-convexity condition holds on Γ, (1.63) is valid only on $\Sigma \cap \Gamma$. However, we can replace the differential quadratric form G by the differential quadratic form G_λ generated by $p_\tau(x, \xi - \lambda\nabla^2\varphi\xi'')$ as long as λ is small enough. This is possible since the strong pseudo-convexity condition is stab le with respect to small perturbations, see also lemma 34. That provides

$$G_\lambda(x_0, \xi, \xi) + \tilde{C}|p_\tau(x_0, \xi - \lambda\nabla^2\varphi\xi'')|^2 + |\xi''|^2 \geq C \quad \text{for all } (\xi, \tau) \in \Sigma\,.$$

By homogeneity

$$(|\xi|^2 + \tau^2)^{m-1} \leq C_1 G_\lambda(x_0, \xi, \xi) + C_2 \frac{p_\tau(0, \xi - \lambda\nabla^2\varphi\xi'')|^2}{(|\xi|^2 + \tau^2)} + C_4|\xi''|^{2(m-1)}$$

and we proceed as before. The last term in this inequality turns into the last term of (??). □

In order to obtain a meaningful Carleman type estimate we want to set $v = Q^\varphi_{\lambda,\tau} u$ where $Q^\varphi_{\lambda,\tau}$ is the integral operator (1.11). However, this function is not compactly supported so lemma 43 is not directly applicable. Since *supp* $u \subset B_\delta(x_0)$ we know that v is supported in the cylindrical set

$$supp\, v = \mathbf{R}^{n''} \times B_\delta(x_0') \,. \tag{1.47}$$

In the proof of theorem 40 and lemma 43 the support of v in a δ neighborhood of 0 is essential only for the three estimates (1.65) - (1.67). The first one is still valid in the cylindrical set (1.72) since the coefficients of P do not depend on x''. In order to save (1.66) we have to modify φ outside of $B_\delta(x_0)$. We k now that φ is a second order polynomial, i.e.

$$\varphi(x) = \frac{1}{2}(x-x_0)^T\nabla^2\varphi(x_0)(x-x_0) + \nabla\varphi(x_0)\cdot(x-x_0) + \varphi(x_0) \,.$$

We introduce a new function $\tilde{\varphi}$ by

$$\tilde{\varphi}(x) = \chi(x)\frac{1}{2}(x-x_0)^T\nabla^2\varphi(x_0)(x-x_0) + \nabla\varphi(x_0)\cdot(x-x_0) \tag{1.48}$$

where $\chi \in C_0^\infty(B_{2\delta}(x_0))$ with $\chi \equiv 1$ in $B_\delta(x_0)$. Lemma 5 is still valid with φ replaced by $\tilde{\varphi}$ provided *supp* $u \subset B_\delta(x_0)$ as the remark after corollary 6 shows. In addition, we observe that

$$\nabla\tilde{\varphi}(x) - \nabla\tilde{\varphi}(x_0) = \chi(x)\nabla^2\varphi(x_0)(x-x_0) + \nabla\chi(x)\frac{1}{2}(x-x_0)^T\nabla^2\varphi(x_0)(x-x_0)$$

which provides

$$|\nabla\tilde{\varphi}(x) - \nabla\tilde{\varphi}(x_0)| \leq C\delta$$

where the constant C depends only on $\nabla^2\varphi$. Of course, that is only true when we choose a special cut off function χ, see also theorem 1.4.1 and formula (1.4.2) in [4].

This estimate replaces now (1.66) and that means (1.69) will be valid for v supported in (1.72).

The only estimate which we can not extend to the whole set (1.72) is (1.67) since the coefficients of the differential quadratic form G depend also on th e second derivatives of $\tilde{\varphi}$ and we can not guarantee that $|\partial_{jk}\tilde{\varphi}(x) - \partial_{jk}\tilde{\varphi}(x_0)|$ is uniformly small in (1.72). Hence we modify the estimate (1.68) like follows.

Lemma 28 *For $\epsilon > 0$ there exists a $\delta \in (0,1)$ such that*

$$\left|\int G_\lambda(x, Dv, \overline{Dv}) - \int G_\lambda(x, Dv, \overline{Dv})\right| \leq \epsilon\|v\|^2_{H_\tau^{m-1}(\mathbf{R}^n)} + e^{-\delta\tau}\|u\|^2_{H_\tau^{m-1,-r}(\mathbf{R}^n)} \tag{1.49}$$

for all $u \in C_0^\infty(B_\delta(x_0))$ provided τ is large enough.

Proof: Our goal is to split the integration in (1.74) into two parts. One will be an integration over a certain neighborhood of x_0 which will lead to the first term in (1.74) by virtue of (1.67). The integration over the complement of this neighborhood will mak e use of the exponential decay of v and lead to the last term in (1.74).

We assume that $x_0 = 0$ $\delta < 1$ a priori. Then there exist a $c_1 > 0$ such that $\varphi \le c_1\delta$ for $x \in B_\delta(x_0)$. Let $w = e^{\tau\varphi}u$ and let α' be a multiindex which refers to differentiations in x' only. Then lemma 1 gives

$$\begin{aligned}
\|D^{\alpha'}w\|^2_{H^{0,-r}_\tau(\mathbf{R}^n)} &= \|D^{\alpha'}(e^{\tau\varphi}u)\|^2_{H^{0,-r}_\tau(\mathbf{R}^n)} \\
&= \|\sum_{\beta'\le\alpha'}\binom{\alpha'}{\beta'}D^{\beta'}e^{\tau\varphi}D^{\alpha'-\beta'}u\|^2_{H^{0,-r}_\tau(\mathbf{R}^n)} \\
&\le C\sum_{\beta'\le\alpha'}\|D^{\beta'}e^{\tau\varphi}\|^2_{C^r_\tau(B_\delta(x_0))}\|D^{\alpha'-\beta'}u\|^2_{H^{0,-r}_\tau(\mathbf{R}^n)} \\
&\le C\sum_{\beta'\le\alpha'}\tau^{2|\beta'|} \qquad (1.50)
\end{aligned}$$

Restricting ourselves to operators with real coefficients we will be able to simplify this definition slightly. At first we need the following technical result.

Proposition 29 *Let P be a operator with real coefficients. Let $x_0 \in \Omega$ and $\varphi \in C^2(\bar\Omega)$ with $\nabla\varphi(x_0) \ne 0$. Then*

$$\lim_{\tau\to 0}\frac{1}{2i\tau}\{p_{-\tau}, p_\tau\}(x_0,\xi) = \Big\{p, \{p,\varphi\}\Big\}(x_0,\xi)$$

for all $\xi \in \mathbf{R}^n$.

Proof: By hypothesis we have $\bar p = p$. For convenience we drop (x_0,ξ) throughout the proof. We calculate

$$\{p_{-\tau}, p_\tau\} = 2i\tau\sum_{j,k=1}^n \partial^2_{jk}\varphi\, p^{(j)}_{-\tau}p^{(k)}_\tau + \sum_{j=1}^n\left(p^{(j)}_{-\tau}p_{\tau(j)} - p_{-\tau(j)}p^{(j)}_\tau\right) \qquad (1.51)$$

and

$$\Big\{p, \{p,\varphi\}\Big\} = \sum_{j,k=1}^n \partial^2_{jk}\varphi\, p^{(j)}p^{(k)} + \sum_{j=1}^n\{p, p^{(j)}\}\partial_j\varphi\,.$$

This means we need to prove

$$\begin{aligned}
\left|\frac{1}{2i\tau}\sum_{j=1}^n(p^{(j)}_{-\tau}p_{\tau(j)} - p_{-\tau(j)}p^{(j)}_\tau) \quad - \quad \sum_{k=1}^n\{p, p^{(k)}\}\,\partial_k\varphi\right| \\
\le\ C\tau|\xi + i\tau\nabla\varphi(x_0)|^{2m-3}\,. \qquad (1.52)
\end{aligned}$$

Notice that

$$\begin{aligned}
\sum_{k=1}^n\{p, p^{(k)}\}\,\partial_k\varphi &= \sum_{j,k=1}^n(p^{(j)}p^{(k)}_{(j)} - p_{(j)}p^{(jk)})\,\partial_k\varphi \\
&= \frac{1}{2i}\sum_{j=1}^n\frac{\partial}{\partial\sigma}\left(p^{(j)}_{-\sigma}p_{\sigma(j)} - p_{-\sigma(j)}p^{(j)}_\sigma\right)\Big|_{\sigma=0}
\end{aligned}$$

and Taylor's formula gives

$$\begin{aligned}\sum_{j=1}^{n}(p_{-\tau}^{(j)}p_{\tau(j)} - p_{-\tau(j)}p_{\tau}^{(j)}) &= \{p,p\} + \tau\sum_{j=1}^{n}\frac{\partial}{\partial\sigma}\left(p_{-\sigma}^{(j)}p_{\sigma(j)} - p_{-\sigma(j)}p_{\sigma}^{(j)}\right)\Big|_{\sigma=0} \\ &\quad + \frac{\tau^2}{2}\sum_{j=1}^{n}\frac{\partial^2}{\partial\sigma^2}\left(p_{-\sigma}^{(j)}p_{\sigma(j)} - p_{-\sigma(j)}p_{\sigma}^{(j)}\right)\Big|_{\sigma=\eta}\end{aligned}$$

where $\eta \in (0,\tau)$. We use this to obtain the following estimate

$$\begin{aligned}&\left|\frac{1}{2i}\sum_{j=1}^{n}(p_{-\tau}^{(j)}p_{\tau(j)} - p_{-\tau(j)}p_{\tau}^{(j)}) - \tau\sum_{k=1}^{n}\{p,p^{(k)}\}\,\partial_k\varphi\right| \\ &\leq |\{p,p\}| + C\tau^2|\xi + i\eta\nabla\varphi|^{2m-3}\,.\end{aligned}$$

Because of $\{p,p\} = 0$, (1.52) is proved.□

This lemma shows that the Poisson bracket in (1.25) has a meaning when $\tau = 0$. It allows us to find a simpler condition for strong pseudo-convex surfaces.

Lemma 30 *Let P be a differential operator with real coefficients and let S be strongly pseudo-convex with respect to P on Γ at x_0. Then there exist a $C > 0$ such that*

$$\begin{aligned}\tfrac{1}{2i\tau}\{\bar{p}_{-\tau},p_\tau\}(x_0,\xi) &\geq C(|\xi|^2+\tau^2)^{m-1} \quad \textit{for all}\ \xi\in\Gamma, \tau\geq 0 \\ \textit{when}\quad p_\tau(x_0,\xi) &= \textstyle\sum_{j=1}^{n} p_\tau^{(j)}(x_0,\xi)\partial_j\varphi(x_0) = 0\end{aligned} \tag{1.53}$$

where the surface S is a level surface of φ.

Proof: Let Σ denote the unit sphere in the (ξ,τ) space and define

$$A = \left\{(\xi,\tau)\in\Sigma\ : \tau\geq 0,\ p_\tau(x_0,\xi) = \sum_{j=1}^{n}p_\tau^{(j)}(x_0,\xi)\partial_j\varphi(x_0) = 0\right\}. \tag{1.54}$$

The set A is closed in Σ. By hypothesis we have

$$\frac{1}{2i\tau}\{p_{-\tau},p_\tau\}(x_0,\xi) > 0 \quad \text{for } \tau > 0,\ (\xi,\tau)\in A$$

and by proposition 29 and by hypothesis

$$\lim_{\tau\to 0}\frac{1}{2i\tau}\{p_{-\tau},p_\tau\}(x_0,\xi) = \Big\{p,\{p,\varphi\}\Big\} > 0$$

on $p(x_0,\xi) = \sum_{j=1}^{n}p^{(j)}(x_0,\xi)\partial_j\varphi(x_0) = 0$. Consequently

$$\frac{1}{2i\tau}\{p_{-\tau},p_\tau\}(x_0,\xi) > 0 \quad \text{for } (\xi,\tau)\in A$$

and since A is closed in Σ we introduce

$$C = \inf_{(\xi,\tau)\in A}\frac{1}{2i\tau}\{p_{-\tau},p_\tau\}(x_0,\xi)\,.$$

The statement of the lemma follows now by homogeneity. □

The next lemma shows that we can represent a strongly pseudo-convex surface as a level surface of a function φ with an even simpler condition for strong pseudo-convexity.

Lemma 31 *Let P be a differential oper ator with real coefficients and let S be strongly pseudo-convex with respect to P on Γ at x_0. Then there exist a $\varphi \in C^2(\bar\Omega)$ and $C > 0$ such that S is a level surface for φ and*

$$\begin{array}{rcll} \frac{1}{2i\tau}\{\bar p_{-\tau}, p_\tau\}(x_0,\xi) & \geq & C(|\xi|^2+\tau^2)^{m-1} & \text{for all } \xi\in\Gamma, \tau\geq 0 \\ \text{when} \quad p_\tau(x_0,\xi) & = & 0\,. & \end{array} \tag{1.55}$$

Proof: Since S is strongly pseudo-convex at x_0 there exist a $\psi \in C^2(\bar\Omega)$ such that (1.53) is satisfied. Set $\varphi = e^{\lambda\psi}$ with $\lambda > 0$. We have

$$\begin{array}{rcl} \partial_j\varphi & = & \lambda\varphi\partial_j\psi \quad \text{and} \\ \partial^2_{jk}\varphi & = & \lambda^2\varphi\partial_j\psi\partial_k\psi + \lambda\varphi\partial^2_{jk}\psi\,. \end{array} \tag{1.56}$$

In order to do t he next step in the proof we need to refine our notation. Set

$$\begin{array}{rcl} p_{\tau,\varphi}(x_0,\xi) & = & p(x_0,\xi+i\tau\nabla\varphi(x_0)) \quad \text{and} \\ p_{\sigma,\psi}(x_0,\xi) & = & p(x_0,\xi+i\sigma\nabla\psi(x_0))\,. \end{array}$$

Then we choose $\sigma = \lambda\varphi(x_0)\tau$ which implies $p_{\tau,\varphi}(x_0,\xi) = p_{\sigma,\psi}(x_0,\xi)$. That gives together with (1.56) (we omit (x_0,ξ) for convenience)

$$\sum_{j,k=1}^n \partial^2_{jk}\varphi p^{(j)}_{-\tau,\varphi}p^{(k)}_{\tau,\varphi} = \lambda\varphi\left(\sum_{j,k=1}^n \partial^2_{jk}\psi p^{(j)}_{-\sigma,\psi}p^{(k)}_{\sigma,\psi} + \lambda\left|\sum_{j=1}^n p^{(j)}_{\sigma,\psi}\partial_j\psi\right|^2\right)\,.$$

Hence,

$$\frac{1}{2i\tau}\{p_{-\tau,\varphi},p_{\tau,\varphi}\} = \lambda\varphi\left(\frac{1}{2i\sigma}\{p_{-\tau,\sigma},p_{\tau,\sigma}\} + \lambda\left|\sum_{j=1}^n p^{(j)}_{\sigma,\psi}\partial_j\psi\right|^2\right) \tag{1.57}$$

and the right hand side is greater or equal to $C\lambda\varphi(|\xi|^2+\sigma^2)^{m-1}$ on $p_{\tau,\psi} = \sum_{k=1}^n p^{(j)}_{\tau,\psi}\partial_j\psi = 0$. However, choosing λ large enough we can ensure the same estimate on the larger set $p_{\sigma,\psi} = 0$. This follows from the fact that the last term in (1.57) is homogeneous of order $2(m-1)$ in (ξ,σ) and positive provided $\sum_{k=1}^n p^{(j)}_{\tau,\psi}\partial_j\psi \neq 0$. Consequently

$$\frac{1}{2i\tau}\{p_{-\tau,\varphi},p_{\tau,\varphi}\}(x_0,\xi) \geq C\lambda\varphi(x_0)(|\xi|^2+\sigma^2)^{(m-1)}$$

and replacing σ by $\lambda\varphi(x_0)\tau$ proves (1.55). □

The last result makes the following definition useful.

Definition 32 *Let $\varphi \in C^2(\bar{\Omega})$ be real valued with $\nabla\varphi(x_0) \neq 0$. We say that φ is strongly pseudo-convex with respect to P on Γ at x_0 if it satisfies (1.55).*

The next result follows directly from lemma 30. It shows that the strong pseudo-convexity is stable with respect to small perturbations.

Corollary 33 *Let P be an operator with real coefficients and C^1 principal symbol. Suppose the level surface of $\varphi \in C^2(\bar{\Omega})$ is strongly pseudo-convex at with respect to P on Γ at x_0. Then there exist $\epsilon > 0$ and $\delta > 0$ such that every $\psi \in C^2(\bar{\Omega})$ with*

$$|D^\alpha(\varphi - \psi)| < \epsilon \qquad \textit{for } x \in B_\delta(x_0)\,, |\alpha| \leq 2 \tag{1.58}$$

has strongly pseudo-convex level surfaces at every point in $B_\delta(x_0)$.

Proof: We define the following function

$$F(\psi, x, \xi, \tau) = \frac{1}{2i\tau}\{p_{-\tau,\psi}, p_{\tau,\psi}\}$$

where $\psi \in C^2(\Omega)$ and $x \in \Omega$. This function is continuous with respect to all variable s and with respect to the first and second derivatives of ψ. The pseudo-convexity conditions imply $F(\varphi, x_0, \xi, \tau) \geq C$ on A where A is defined in (1.54). Consequently, there exists a neighborhood A' of A in Σ and $F(\varphi, x_0, \xi, \tau) \geq C/2$ on A'. By continuity of p we can find $\epsilon > 0$ and $\delta > 0$ such that for ψ satisfying (1.58)

$$\{(\xi, \tau) \in \Sigma \ : \ p_{\tau,\psi}(x, \xi) = \sum_{j=1}^{n} p_{\tau,\psi}^{(j)}(x, \xi)\partial_j\psi(x) = 0\} \subset A'$$

for all $x \in B_\delta(x_0)$. Finally, by continuity of F and if necessary, by making ϵ and δ smaller, we can guarantee $F(\psi, x, \xi, \tau) \geq C/4$ on A' for all $x \in B_\delta(x_0)$. □

A similar result can be proved with respect to small perturbations with respect to the operator $P(x, D)$.

Corollary 34 *Let P be an operator with real coefficients and C^1 principal symbol. Suppose the level surface of $\varphi \in C^2(\bar{\Omega})$ is strongly pseudo-convex with respect to P on Γ at x_0. Then there exist a $\epsilon > 0$ such that the level surface of φ is strongly pseudo-convex with respect to $\tilde{P}$ on Γ at x_0 provided*

$$|D^\beta(a_\alpha(x_0) - \tilde{a}_\alpha(x_0))| \leq \epsilon \qquad \textit{for } |\beta| \leq 1\,, |\alpha| = m$$

where $\tilde{a}_\alpha$ denotes the coefficients of the operator $\tilde{P}$.

Proof: The proof is analog to the proof of the previous corollary and will be omitted.

Finally, we state a result which relates strong pseudo-convex and non-characteristic surfaces.

Lemma 35 *Let S be non-characteristic at x_0. Then the surface S is strongly pseudo-convex at x_0 with respect to P on $\{0\}$.*

Proof: We have to check that (1.25) is satisfied since the pseudo-convexity condition (1.24) does not apply when $\Gamma = \{0\}$. We have

$$p_\tau(x_0, 0) = p(x_0, i\tau\nabla\varphi(x_0)) = (i\tau)^m p(x_0, \nabla\varphi(x_0)) \neq 0$$

for $\tau > 0$ since S is non-characteristic at x_0.

1.5 UNIQUENESS OF CONTINUATION THEOREMS FOR SOLUTIONS OF PARTIAL DIFFERENTIAL EQUATIONS

This section consist of three different uniqueness of continuation theorems. The first one is Holmgren's theorem. We state it here since it is necessary for a good understanding of the further results. Moreover, its conclu sion seems to be optimal with respect to the geometry of the hypersurface. Since it is a well known result we will omit the proof here. The second result is Hörmander's theorem which provides uniqueness of continuation for a large family of operators under a more restrictive condition on the geometry of the hypersurface. The hypersurface is assumed to be strongly pseudo-convex with respect to the operator. The third result is a recent one which shows that the conclusion of Holmgren 's theorem holds for a larger class of operators.

Before we proceed we give a formal definition of the uniqueness of continuation property where we make use of an oriented hypersurfaces defined by (1.22).

Definition 36 *Let $P(x, D)$ be a partial differential operator and S an oriented hypersurface in Ω. We say that the uniqueness of continuation property holds with respect to P across S at $x_0 \in S$ for a distribution u if there exists a neighbor hood $V(x_0) \subset \Omega$ such that $P(x, D)u = 0$ in Ω and $u \equiv 0$ on the positive side of S imply that u vanishes in $V(x_0)$.*

We mention that this definition is not entirely accurate since we need to impose conditions on the operator and on the distribution u for $P(x, D)u$ to be well defined. Here we will assume that these conditions are met.

1.5.1 Holmgren's theorem

Theorem 37 *Let $P(x, D)$ be a partial differential operator with analytic coefficients. Moreover, assume that the oriented C^1-hypersurface S is non-characteristic at x_0. Then the uniqueness of continuation property holds with respect to P across S at x_0 for every $u \in D'(\Omega)$.*

For a proof of this theorem we refer to [3] section 5.3. The following corollary is obvious when we observe corollary 35.

Corollary 38 *Let P be a differential operator with constant coefficients a nd S be an oriented C^2-hypersurface in Ω. If the surface is strongly pseudo-convex with respect to P on $\{0\}$ at x_0, then the uniqueness of continuation property holds with respect to P across S at x_0 for all $u \in D'(\Omega)$.*

1.5.2 Hörmander's theorem

This is the second classical result about uniqueness of continuation. We restrict ourselves to operators with real-valued coefficients.

Theorem 39 *Let P b e a differential operator with real coefficients and C^1 principal symbol. Assume that the oriented C^2-hypersurface S is strongly pseudo-convex with respect to P on $\mathbf{R}^n$ at x_0. Then the uniqueness of continuation property holds with respect to P across S at x_0 for every $u \in H^{m-1}(\Omega)$.*

The main tool for the proof of this theorem is a Carleman type estimate which will be established in the following theorem. Up to minor det ails it can be found in [3] as theorem 8.4.2. Our proof will follow the proof given there for the most part and relies on integration by parts using differential quadratic forms. We like to point out that a second proof is given in [4] chapter 28. That proof is based on pseudo-differential calculus and a version of the sharp Gårding inequality also known as Fefferman-Phong inequality. We felt that the proof in [3] is easier and applies to operators with C^1 p rincipal symbol whereas the other proof requires much more technique and is valid only for operators with C^∞ principal symbol.

Theorem 40 *Let P be a differential operator with real coefficients and C^1 principal symbol. Moreover, let $\varphi \in C^m(\bar{\Omega})$ be strongly pseudo-convex with respect to P on $\mathbf{R}^n$ at x_0. Then there exist a $\delta > 0$ and a constant $K > 0$ such that*

$$\sum_{|\alpha|<m} \tau^{2(m-|\alpha|)-1} \int |D^\alpha u|^2 e^{2\tau\varphi} \leq K \int |P(x,D)u|^2 e^{2\tau\varphi} \tag{1.59}$$

for all $u \in C_0^\infty(B_\delta(x_0))$ and τ large enough.

Proof: At first we notice that with $v = e^{\tau\varphi}u$ and (4) this estimate can be written as

$$\tau\|v\|^2_{H_\tau^{m-1}(\mathbf{R}^n)} \leq K\|P(x, D + i\tau\nabla\varphi(x))v\|^2_{L^2(\mathbf{R}^n)} \,. \tag{1.60}$$

We claim that this estimate is proved when we show

$$\tau\|v\|^2_{H_\tau^{m-1}(\mathbf{R}^n)} \leq \frac{K}{4}\|P_\tau(x,D)v\|^2_{L^2(\mathbf{R}^n)} \,. \tag{1.61}$$

When we use (1.8) we get

$$|P_\tau(x,D)v|^2 \leq 2|P(x, D + i\tau\nabla\varphi(x))v|^2 + 2|R(x,D)v|^2$$

which leads together with (1.9) to

$$\|P_\tau(x,D)v\|^2_{L^2(\mathbf{R}^n)} \leq 2\|P(x, D + i\tau\nabla\varphi(x))v\|^2_{L^2(\mathbf{R}^n)} + 2C\|v\|^2_{H_\tau^{m-1}(\mathbf{R}^n)} \,.$$

Assume now that (1.61) is true. Then our last estimate implies

$$(\tau - 2C)\|v\|^2_{H^{m-1}_\tau(\mathbf{R}^n)} \leq \frac{K}{2}\|P(x, D + i\tau\nabla\varphi(x))v\|^2_{L^2(\mathbf{R}^n)}$$

and choosing $\tau > 4C$ proves (1.60).

The proof of (1.60) will be based on integration by parts in the integral

$$\int |P_\tau(x, D)v|^2$$

where we make use of differential quadratic forms. We start with a trivial estimate

$$\int |P_\tau(x, D)v|^2 \geq \int \left(|P_\tau(x, D)v|^2 - |P_{-\tau}(x, D)v|^2\right).$$

The reason for this last estimate is that we will interpret the right hand side as a differential quadratic form $F(x, D, \bar{D})$ to which we can apply lemma 8. The associated polynomial is

$$F(x, \zeta, \bar{\zeta}) = |p_\tau(x, \zeta)|^2 - |p_\tau(x, \bar{\zeta})|^2$$

and we notice that $F(x, \xi, \xi) = 0$ for all $\xi \in \mathbf{R}^n$. We need to have a closer look at this polynomi al. It is a polynomial in τ of order $2m$ as well. We can expand it into powers of τ

$$F(x, \zeta, \bar{\zeta}) = \sum_{j=0}^{2m} \tau^j F^{(j)}(x, \zeta, \bar{\zeta})$$

and observe that $F^{(0)}(x, \zeta, \bar{\zeta}) = |p(x, \zeta)|^2 - |p(x, \zeta)|^2 = 0$ which shows that $\tau^{-1}F(x, \zeta, \bar{\zeta})$ is a polynomial associated to a differential quadratic form as well. Moreover $F^{(2m)}(x, \zeta, \bar{\zeta}) = |p(x, i\nabla\varphi(x))|^2 - |p(x, i\nabla\varphi(x))|^2 = 0$. We apply lemma 8 to each of the $F^{(j)}(x, \zeta, \bar{\zeta})$. Those are differential quadratic forms of order $(2m - j, m)$. Consequently, there exist differential quadratic forms $G^{(j)}(x, \zeta, \bar{\zeta})$ of order $(2m - j - 1, m - 1)$ such that with

$$G(x, \zeta, \bar{\zeta}) = \sum_{j=1}^{2m-1} \tau^{j-1} G^{(j)}(x, \zeta, \bar{\zeta}) = \sum_{j=1}^{2m-1} \tau^{j-1} \sum_{\substack{|\alpha|, |\beta| \leq m-1 \\ |\alpha|+|\beta| \leq 2m-j-1}} a^{(j)}_{\alpha\beta}(x)\zeta^\alpha \bar{\zeta}^\beta \tag{1.62}$$

we obtain

$$\tau \int G(x, Dv, \overline{Dv}) = \int F(x, Dv, \overline{Dv}).$$

Moreover, (1.20) gives

$$G(x, \xi, \xi) \quad = \quad \frac{1}{2\tau} \sum_{k=1}^{n} \frac{\partial^2}{\partial x_k \partial \eta_k} F(x, \xi + i\eta, \xi - i\eta)\Big|_{\eta=0}$$

$$\begin{aligned}
&= \frac{1}{2\tau}\sum_{k=1}^{n}\frac{\partial}{\partial x_k\partial\eta_k}(p_{-\tau}(x,\xi-i\eta)p_\tau(x,\xi+i\eta)\\
&\quad -p_{-\tau}(x,\xi+i\eta)p_\tau(x,\xi-i\eta))\Big|_{\eta=0}\\
&= \frac{1}{2\tau}\sum_{k=1}^{n}\frac{\partial}{\partial x_k}\Big(-ip_{-\tau}^{(k)}(x,\xi)p_\tau(x,\xi)+ip_{-\tau}(x,\xi)p_\tau^{(k)}(x,\xi)\\
&\quad -ip_{-\tau}^{(k)}(x,\xi)p_\tau(x,\xi)+ip_{-\tau}(x,\xi)p_\tau^{(k)}(x,\xi)\Big)\\
&= \frac{1}{i\tau}\sum_{k=1}^{n}\frac{\partial}{\partial x_k}\Big(p_{-\tau}^{(k)}(x,\xi)p_\tau(x,\xi)-p_{-\tau}(x,\xi)p_\tau^{(k)}(x,\xi)\Big)\\
&= \frac{1}{i\tau}\Big(\{p_{-\tau},p_\tau\}(x,\xi)\\
&\quad +\sum_{k=1}^{n}\Big(\frac{\partial}{\partial x_k}p_{-\tau}^{(k)}(x,\xi)p_\tau(x,\xi)-p_{-\tau}(x,\xi)\frac{\partial}{\partial x_k}p_\tau^{(k)}(x,\xi)\Big)\Big).
\end{aligned}$$

Hence,

$$G(x,\xi,\xi)=\frac{1}{i\tau}\{p_{-\tau},p_\tau\}(x,\xi)$$

for all $(x,\xi)\in\Omega\times\mathbf{R}^n$ that satisfy $p_\tau(x,\xi)=0$ and $\tau\geq 0$. By hypothesis

$$G(x_0,\xi,\xi)\geq C \quad \text{for all } (\xi,\tau)\in\Sigma \text{ with } p_\tau(x_0,\xi)=0\,,\tau\geq 0\,.$$

Hence, we can find a constant $\tilde{C}$ such that

$$G(x_0,\xi,\xi)+\tilde{C}|p_\tau(x_0,\xi)|^2\geq C \quad \text{for all } (\xi,\tau)\in\Sigma \tag{1.63}$$

which by homogeneity implies that

$$(|\xi|^2+\tau^2)^{m-1}\leq C_1G(x_0,\xi,\xi)+C_2|p_\tau(x_0,\xi)|^2(|\xi|^2+\tau^2)^{-1}\,.$$

Now we multiply this inequality by $|\hat{v}(\xi)|^2$, integrate an d use Parseval's identity and obtain

$$||v||_{H_\tau^{m-1}(\mathbf{R}^n)}\leq C_1\int G(x_0,Dv,\overline{Dv})+C_2||P_\tau(x_0,D)v||^2_{H_\tau^{-1}(\mathbf{R}^n)}\,. \tag{1.64}$$

For $\epsilon>0$ there exist a $\delta>0$ such that

$$|a_\alpha(x)-a_\alpha(x_0)| \;\leq\; \epsilon \tag{1.65}$$
$$|\partial_j\varphi(x)-\partial_j\varphi(x_0)| \;\leq\; \epsilon \tag{1.66}$$
$$|a_{\alpha\beta}^{(j)}(x)-a_{\alpha\beta}^{(j)}(x_0)| \;\leq\; \epsilon \tag{1.67}$$

for all $x\in B_\delta(x_0)$. The first and the second estimate are based on the continuity of the coefficients a_α and the continuity of the gradient of φ. For the third estimate we notice that the coefficients of the differential quadratic form G

depend on the coefficients a_α and its first derivatives and on the first and second derivatives of φ. Since $a_\alpha \in C^1(\bar{\Omega})$ and $\varphi \in C^m(\bar{\Omega})$ the coefficients of the differential quadratic form are continuous as well.

Since $supp\, v \subset B_\delta(x_0)$ we can estimate

$$\begin{aligned}
&\left|\int G(x, Dv, \overline{Dv}) - \int G(x_0, Dv, \overline{Dv})\right| \\
&\leq \sum_{j=0}^{2m-2} \tau^j \int |G^{(j+1)}(x, Dv, \overline{Dv}) - G^{(j+1)}(x_0, Dv, \overline{Dv})| \\
&\leq \sum_{j=0}^{2m-2} \tau^j \sum_{\substack{|\alpha|,|\beta|\leq m-1 \\ |\alpha|+|\beta|\leq 2m-j-2}} \int |a_{\alpha\beta}^{(j+1)}(x) - a_{\alpha\beta}^{(j+1)}(x_0)||D^\alpha v||D^\beta v| \\
&\leq \epsilon \sum_{j=0}^{2m-2} \tau^j \sum_{\substack{|\alpha|,|\beta|\leq m-1 \\ |\alpha|+|\beta|\leq 2m-j-2}} \int |D^\alpha v||D^\beta v| \\
&\leq \epsilon \sum_{|\alpha|,|\beta|\leq m-1} \tau^{2(m-1)-|\alpha|-|\beta|} \int |D^\alpha v||D^\beta v| \\
&\leq \epsilon \sum_{|\alpha|\leq m-1} \tau^{2(m-1-|\alpha|)} \int |D^\alpha v|^2 \leq \epsilon \|v\|^2_{H_\tau^{m-1}(\mathbf{R}^n)} \, . \qquad (1.68)
\end{aligned}$$

We need to get a similar estimate for the last term in (1.64). We have

$$\begin{aligned}
&\|P_\tau(x_0, D)v - P_\tau(x, D)v\|_{H_\tau^{-1}(\mathbf{R}^n)} \\
&\leq \left\| P_\tau(x_0, D)v - \sum_{|\alpha|=m} a_\alpha(x) \sum_{\beta\leq\alpha} \binom{\alpha}{\beta} (i\tau\nabla\varphi(x_0))^{\alpha-\beta} D^\beta v \right\|_{H_\tau^{-1}(\mathbf{R}^n)} \\
&+ \left\| \sum_{|\alpha|=m} a_\alpha(x) \sum_{\beta\leq\alpha} \binom{\alpha}{\beta} (i\tau\nabla\varphi(x_0))^{\alpha-\beta} D^\beta v - P_\tau(x, D)v \right\|_{H_\tau^{-1}(\mathbf{R}^n)} \\
&\leq \epsilon \| \sum_{|\alpha|\leq m} (D + i\tau\varphi(x_0))^\alpha v\|_{H_\tau^{-1}(\mathbf{R}^n)} + \epsilon\|P(x, D + i\tau\mathbf{1})v\|_{H_\tau^{-1}(\mathbf{R}^n)} \\
&\leq \epsilon C_3 \|v\|_{H_\tau^{m-1}(\mathbf{R}^n)} \qquad (1.69)
\end{aligned}$$

where C_3 depends on ly on P and $\nabla\varphi(x_0)$. Hence, by the triangle inequality

$$\begin{aligned}
\|P_\tau(x_0, D)v\|^2_{H_\tau^{-1}(\mathbf{R}^n)} &\leq 2\|P_\tau(x, D)v\|^2_{H_\tau^{-1}(\mathbf{R}^n)} + 2C_3^2\epsilon^2\|v\|^2_{H_\tau^{m-1}(\mathbf{R}^n)} \\
&\leq 2\tau^{-2}\|P_\tau(x, D)v\|^2_{L^2(\mathbf{R}^n)} + 2C_3^2\epsilon^2\|v\|^2_{H_\tau^{m-1}(\mathbf{R}^n)}
\end{aligned}$$

Combining everything we obtain

$$(1 - C_1\epsilon - C_2C_3^2\epsilon^2)\|v\|_{H_\tau^{m-1}(\underline{f}R^n)} \leq (C_1\tau^{-1} + C_2\tau^{-2})\|P_\tau(x, D)v\|^2_{L^2(\mathbf{R}^n)}$$

Finally we choose $\epsilon > 0$ as small that

$$1 - C_1\epsilon - C_2C_3\epsilon^2 > \frac{1}{2}$$

and multiply with 2τ and we obtain (1.61) for $\tau > C_2$ and $K/4 = 2C_1 + 2$. □

Next we show that the Carleman type estimate is valid for a larger class of functions.

Corollary 41 *Under the same hypothesis as in theorem 40 the estimate (??) is valid for $u \in H_0^{m-1}(B_\delta(x_0))$ such that $P(x, D)u \in L^2(B_\delta(x_0))$.*

Proof: By lemma 2 there exists a sequence of mollifiers $u_\epsilon \subset C_0^\infty(B_\delta(x_0))$ such that

$$u_\epsilon \longrightarrow u \quad \text{in } H^{m-1}(B_\delta(x_0)) \quad \text{for } \epsilon \to 0\,.$$

We need to show that this implies

$$P(x, D)u_\epsilon \longrightarrow P(x, D)u \quad \text{in } L^2(B_\delta(x_0)) \quad \text{for } \epsilon \to 0\,.$$

This can be shown as follows

$$\begin{aligned} &\|P(x,D)u_\epsilon - P(x,D)u\|_{L^2(B_\delta(x_0))} \\ &\leq \|P(x,D)u_\epsilon - (P(x,D)u)_\epsilon\|_{L^2(B_\delta(x_0))} \\ &+\|(P(x,D)u)_\epsilon - P(x,D)u\|_{L^2(B_\delta(x_0))} \to 0\,. \end{aligned}$$

Here the first term tends to zero because of lemma 3 and the second one since $P(x, D)u \in L^2(B_\delta(x_0))$. □

Now we can prove Hörmander's theorem.

Proof of theorem 39: Since S is a strongly pseudo-convex surface we can always find a strongly pseudo-convex function φ such that S is the level surface $\varphi(x) = \varphi(x_0)$. This is proved in lemma 31. Let

$$\psi(x) = \nabla\varphi(x_0) \cdot (x - x_0) + \frac{1}{2}(x - x_0)^T \nabla^2\varphi(x_0)(x - x_0) - 3\gamma|x - x_0|^2\,.$$

where $\gamma > 0$ is chosen to be as small that $\psi(x)$ is strongly pseudo-convex in $B_\delta(x_0)$. Without loss of generality assume that $\{|x - x_0| \leq 2\delta\} \subset \Omega$ and

$$\psi(x) \leq \varphi(x) - \varphi(x_0) - 2\gamma|x - x_0|^2$$

for $|x - x_0| \leq 2\delta$. Let χ be a smooth cutoff function

$$\chi(x) = \begin{cases} 1 & \text{if} \quad x \geq -\gamma\delta^2 \\ 0 & \text{if} \quad x \leq -2\gamma\delta^2 \end{cases}$$

and define

$$\tilde{u}(x) = \begin{cases} \chi(\psi(x))u(x) & \text{if } |x - x_0| \leq 2\delta \\ 0 & \text{otherwise} \end{cases}\,.$$

We will apply (1.59) to $\tilde{u}$. We can do so since

$$\begin{aligned} supp\,\tilde{u} &\subset supp\,u \cap supp\,\chi(\psi) \cap B_{2\delta}(x_0) \\ &\subset \{\varphi(x) \le \varphi(x_0)\} \cap \{\psi(x) \ge -2\gamma\delta^2\} \cap B_{2\delta}(x_0) \\ &\subset \{\varphi(x) \le \varphi(x_0)\} \cap \{\varphi(x) - \varphi(x_0) - 2\gamma|x-x_0|^2 > -2\gamma\delta^2\} \\ &\subset B_\delta(x_0)\,. \end{aligned}$$

Moreover, $\tilde{u} \in H_0^{m-1}(B_\delta(x_0))$ and $P(x,D)\tilde{u} \in L^2(B_\delta(x_0))$ with

$$supp\,P(x,D)\tilde{u} \subset G = \{\psi \le -\gamma\delta^2\} \cap B_\delta(x_0)\,. \tag{1.70}$$

By G^c we denote the complement of G in $B_\delta(x_0)$, i.e. $G^c = B_\delta(x_0) \setminus G$. Now we use the Carleman type estimate with weight function ψ and obtain

$$\begin{aligned} e^{-2\tau\gamma\delta^2} \sum_{|\alpha|\le m-1} \tau^{2(m-|\alpha|)-1} \int_{G^c} |D^\alpha u|^2 &\le \sum_{|\alpha|\le m-1} \tau^{2(m-|\alpha|)-1} \int |D^\alpha \tilde{u}|^2 e^{2\tau\psi} \\ &\le K \int |P(x,D)\tilde{u}|^2 e^{2\tau\psi} \\ &\le e^{-2\tau\gamma\delta^2} K \int |P(x,D)\tilde{u}|^2 \end{aligned}$$

which gives

$$\sum_{|\alpha|\le m-1} \tau^{2(m-1-|\alpha|)} \int_{G^c} |D^\alpha u|^2 \le \tau^{-1} K \int |P(x,D)\tilde{u}|^2\,.$$

For $\tau \to \infty$ we obtain

$$u \equiv 0 \qquad \text{for all } x \in G^c\,.$$

This set is a neighborhood of x_0 and that proves the theorem. □

1.5.3 A recent result about uniqueness of continuation

In this subsection we like to present a result which is somehow between Holmgren's and Hörmander's theorem. In 1995 Tataru published his results in [10]. Here we w ill present a special case of his theorem which we will apply in the following section to two systems of partial differential equations.

Let us recall that corollary 38 gives the uniqueness of continuation property for operators with constant coefficients across a surface that is strongly pseudo-convex with respect to the operator on $\{0\}$. On the other hand Hörmander's theorem gives the same property for operators with C^1 principal symbol across a surface that is strongly pseudo-convex with respect to the operator on $\mathbf{R}^n$. Indeed, between those two results can be interpolated.

As in subsection 1.2.1 we decompose $\mathbf{R}^n = \mathbf{R}^{n''} \times \mathbf{R}^{n'}$ and $x = (x'', x')$ and $\xi = (\xi'', \xi')$. Throughout this subsection $\Omega = \Omega'' \times \Omega'$ where Ω'' is an open simply connected subset of $\mathbf{R}^{n''}$ and Ω' is an open simply connected subset of $\mathbf{R}^{n'}$. In the following

$$\Gamma = \{\xi \in \mathbf{R}^n \;:\,;\xi'' = 0\}\,.$$

Then we consider a partial differential operator with real coefficients that depend only on x'. For those operators we will prove the following result, which is a special case of theorem 2 in [10].

Theorem 42 *Let P be a linear partial differential operator with real coefficients independent of x'' and C^1 principal symbol. Moreover, let S be strongly pseudo-convex with respect to P on Γ at x_0. Then the uniqueness of co ntinuation property holds with respect to P across S at x_0 for $u \in \mathcal{D}'(\Omega'', H^{m-1}(\Omega'))$.*

Similar to Hörmander's theorem the proof is based on Carleman type estimates. In complete analogy with (39) we obtain the following result. Again, this lemma is a special case of lemma 2.4 in [10]. We give here an independent proof relying on differential quadratic forms.

Lemma 43 *Let P satisfy the same conditions as in theorem 42. Furthermore, let φ be a second degree polynomial in x which is strongly pseudo-convex at x_0 with respect to P on Γ. Then for $\lambda \leq \lambda_0$ there exist a $\delta > 0$ and $K > 0$ such that*

$$\begin{aligned}\tau \|v\|^2_{H^{m-1}_\tau(\mathbf{R}^n)} \quad &\leq \quad K\Big(\|P(x, D - \lambda\nabla^2\varphi D'' + i\tau\nabla\varphi(x))v\|^2_{L^2(\mathbf{R}^n)} \\ &\quad +\tau \sum_{|\alpha|\leq m-1} \|D''^\alpha v\|^2_{L^2(\mathbf{R}^n)}\Big) labelcarta \qquad (1.71)\end{aligned}$$

for all $v \in C_0^\infty(B_\delta(x_0))$ and large τ.

Proof: We notice the obvious analogy to (1.60). This lemma is more or less a corollary of theorem 40. That means, the proof follows exactly the one of theorem 40 with a few modifications. Instead of the operator $P_\tau(x, D)$ and its symbol $p_\tau(x, \xi)$ we will work with $P_\tau(x, D - \lambda\nabla^2\varphi D'')$ and $p_\tau(x, \xi - \lambda\ nabla^2\varphi\xi'')$, respectively. We notice that the these two symbols coincide on Γ. Since the strong pseudo-convexity condition holds on Γ, (1.63) is valid only on $\Sigma \cap \Gamma$. However, we can replace the differential quadratric form G by the differential quadratic form G_λ generated by $p_\tau(x, \xi - \lambda\nabla^2\varphi\xi'')$ as long as λ is small enough. This is possible since the strong pseudo-convexity condition is stab le with respect to small perturbations, see also lemma 34. That provides

$$G_\lambda(x_0, \xi, \xi) + \tilde{C}|p_\tau(x_0, \xi - \lambda\nabla^2\varphi\xi'')|^2 + |\xi''|^2 \geq C \quad \text{for all } (\xi, \tau) \in \Sigma\,.$$

By homogeneity

$$(|\xi|^2 + \tau^2)^{m-1} \leq C_1 G_\lambda(x_0, \xi, \xi) + C_2 \frac{p_\tau(0, \xi - \lambda\nabla^2\varphi\xi'')|^2}{(|\xi|^2 + \tau^2)} + C_4|\xi''|^{2(m-1)}$$

and we proceed as before. The last term in this inequality turns into the last term of (??). □

In order to obtain a meaningful Carleman type estimate we want to set $v = Q^\varphi_{\lambda,\tau} u$ where $Q^\varphi_{\lambda,\tau}$ is the integral operator (1.11). However, this function is not compactly supported so lemma 43 is not directly applicable. Since *supp* $u \subset B_\delta(x_0)$ we know that v is supported in the cylindrical set

$$supp\, v = \mathbf{R}^{n''} \times B_\delta(x_0') \,. \tag{1.72}$$

In the proof of theorem 40 and lemma 43 the support of v in a δ neighborhood of 0 is essential only for the three estimates (1.65) - (1.67). The first one is still valid in the cylindrical set (1.72) since the coefficients of P do not depend on x''. In order to save (1.66) we have to modify φ outside of $B_\delta(x_0)$. We k now that φ is a second order polynomial, i.e.

$$\varphi(x) = \frac{1}{2}(x-x_0)^T\nabla^2\varphi(x_0)(x-x_0) + \nabla\varphi(x_0)\cdot(x-x_0) + \varphi(x_0)\,.$$

We introduce a new function $\tilde{\varphi}$ by

$$\tilde{\varphi}(x) = \chi(x)\frac{1}{2}(x-x_0)^T\nabla^2\varphi(x_0)(x-x_0) + \nabla\varphi(x_0)\cdot(x-x_0) \tag{1.73}$$

where $\chi \in C_0^\infty(B_{2\delta}(x_0))$ with $\chi \equiv 1$ in $B_\delta(x_0)$. Lemma 5 is still valid with φ replaced by $\tilde{\varphi}$ provided *supp* $u \subset B_\delta(x_0)$ as the remark after corollary 6 shows. In addition, we observe that

$$\nabla\tilde{\varphi}(x) - \nabla\tilde{\varphi}(x_0) = \chi(x)\nabla^2\varphi(x_0)(x-x_0) + \nabla\chi(x)\frac{1}{2}(x-x_0)^T\nabla^2\varphi(x_0)(x-x_0)$$

which provides

$$|\nabla\tilde{\varphi}(x) - \nabla\tilde{\varphi}(x_0)| \le C\delta$$

where the constant C depends only on $\nabla^2\varphi$. Of course, that is only true when we choose a special cut off function χ, see also theorem 1.4.1 and formula (1.4.2) in [4].

This estimate replaces now (1.66) and that means (1.69) will be valid for v supported in (1.72).

The only estimate which we can not extend to the whole set (1.72) is (1.67) since the coefficients of the differential quadratic form G depend also on the second derivatives of $\tilde{\varphi}$ and we can not guarantee that $|\partial_{jk}\tilde{\varphi}(x) - \partial_{jk}\tilde{\varphi}(x_0)|$ is uniformly small in (1.72). Hence we modify the estimate (1.68) like follows.

Lemma 44 *For $\epsilon > 0$ there exists a $\delta \in (0,1)$ such that*

$$\left|\int G_\lambda(x, Dv, \overline{Dv}) - \int G_\lambda(x, Dv, \overline{Dv})\right| \le \epsilon\|v\|^2_{H_\tau^{m-1}(\mathbf{R}^n)} + e^{-\delta\tau}\|u\|^2_{H_\tau^{m-1,-\tau}(\mathbf{R}^n)} \tag{1.74}$$

for all $u \in C_0^\infty(B_\delta(x_0))$ provided τ is large enough.

Proof: Our goal is to split the integration in (1.74) into two parts. One will be an integration over a certain neighborhood of x_0 which will lead to the first term in (1.74) by virtue of (1.67). The integration over the complement of this neighborhood will make use of the exponential decay of v and lead to the last term in (1.74).

We assume that $x_0 = 0$ $\delta < 1$ a priori. Then there exist a $c_1 > 0$ such that $\varphi \le c_1\delta$ for $x \in B_\delta(x_0)$. Let $w = e^{\tau\varphi}u$ and let α' be a multiindex which refers to differentiations in x' only. Then lemma 1 gives

$$\begin{aligned}
\|D^{\alpha'}w\|^2_{H^{0,-r}_\tau(\mathbf{R}^n)} &= \|D^{\alpha'}(e^{\tau\varphi}u)\|^2_{H^{0,-r}_\tau(\mathbf{R}^n)} \\
&= \|\sum_{\beta'\le\alpha'}\binom{\alpha'}{\beta'}D^{\beta'}e^{\tau\varphi}D^{\alpha'-\beta'}u\|^2_{H^{0,-r}_\tau(\mathbf{R}^n)} \\
&\le C\sum_{\beta'\le\alpha'}\|D^{\beta'}e^{\tau\varphi}\|^2_{C^r_\tau(B_\delta(x_0))}\|D^{\alpha'-\beta'}u\|^2_{H^{0,-r}_\tau(\mathbf{R}^n)} \\
&\le C\sum_{\beta'\le\alpha'}\tau^{2|\beta'|}e^{2c_1\delta\tau}\|D^{\alpha'-\beta'}u\|^2_{H^{0,-r}_\tau(\mathbf{R}^n)} \\
&\le Ce^{2c_1\delta\tau}\sum_{\gamma'\le\alpha'}\tau^{2(|\alpha'|-|\gamma'|)}\|D^{\gamma'}u\|^2_{H^{0,-r}_\tau(\mathbf{R}^n)} \\
&\le Ce^{2c_1\delta\tau}\|u\|^2_{H^{|\alpha'|,-r}_\tau(\mathbf{R}^n)}\,. \qquad (1.75)
\end{aligned}$$

By $a^{(j)}_{\alpha\beta}$ we denote the coefficients of the differential quadratic form G_λ. For $\epsilon > 0$ choose $\delta > 0$ such that

$$\begin{aligned}
&|a^{(j)}_{\alpha\beta}(x) - a_{\alpha\beta}e^{2c_1\delta\tau}\|D^{\alpha'-\beta'}u\|^2_{H^{0,-r}_\tau(\mathbf{R}^n)} \\
&\qquad\le Ce^{2c_1\delta\tau}\sum_{\gamma'\le\alpha'}\tau^{2(|\alpha'|-|\gamma'|)}\|D^{\gamma'}u\|^2_{H^{0,-r}_\tau(\mathbf{R}^n)} \\
&\qquad\le Ce^{2c_1\delta\tau}\|u\|^2_{H^{|\alpha'|,-r}_\tau(\mathbf{R}^n)}\,. \qquad (1.76)
\end{aligned}$$

By $a^{(j)}_{\alpha\beta}$ we denote the coefficients of the differential quadratic form G_λ. For $\epsilon > 0$ choose $\delta > 0$ such that

$$|a^{(j)}_{\alpha\beta}(x) - a^{(j)}_{\alpha\beta}(x_0)| \le \epsilon \qquad \text{for } x \in A\times B_\delta(x_0') \qquad (1.77)$$

where $A = \{x'' \in \mathbf{R}^{n''} \;:\; |x''| \le 2\delta + 4\sqrt{(c_1+1)\delta\lambda}\}$.

Next we give a pointwise estimate of v and its derivatives. We have

$$v(x) = Q^\varphi_{\lambda,\tau}u(x) = \langle K(x'',\cdot), w(\cdot,x')\rangle$$

where K is the kernel defined by (1.13). This gives

$$\begin{aligned}
&|D^\alpha v(x)| \\
&= |D^\alpha\langle K(x'',\cdot), w(\cdot,x')\rangle| = |\langle D^{\alpha''}K(x'',\cdot), D^{\alpha'}w(\cdot,x')\rangle| \\
&\le \|D^{\alpha''}K(x'',\cdot)\|_{H^r_\tau(B_\delta(x_0''))}\|D^{\alpha'}w(\cdot,x')\|_{H^{-r}_\tau(B_\delta(x_0''))} \\
&\le C\tau^r\|D^{\alpha''}K(x'',\cdot)\|_{C^r_\tau(B_\delta(x_0''))}\|D^{\alpha'}w(\cdot,x')\|_{H^{-r}_\tau(B_\delta(x_0''))} \\
&\le C\left(\frac{\tau}{\lambda}\right)^{\frac{n''}{2}+r+|\alpha''|}(1+|x''|^{r+|\alpha''|})e^{-\frac{\tau}{2\lambda}(|x''|-\delta)^2}\|D^{\alpha'}w(\cdot,x')\|_{H^{-r}_\tau(B_\delta(x_0''))}\,.
\end{aligned}$$

Here we used lemma 1 and the fact that

$$\sup_{y''\in B_\delta(x_0'')} e^{-\frac{\tau}{2\lambda}(x''-y'')^2} \le e^{-\frac{\tau}{2\lambda}(|x''|-\delta)^2} .$$

Then using the estimate above and (1.76) we obtain

$$\begin{aligned}
&\int_{A^c\times\mathbf{R}^{n'}} |D^\alpha v|^2 \\
\le& C\left(\frac{\tau}{\lambda}\right)^{n''+2r+2|\alpha''|}\int_{A^c}(1+|x''|^{2(r+|\alpha''|)})e^{-\frac{\tau}{4\lambda}|x''|^2}dx''e^{2c_1\delta\tau}\|u\|^2_{H_\tau^{|\alpha'|,-r}(\mathbf{R}^n)} \\
\le& Ce^{-2\delta\tau}\left(\frac{\tau}{\lambda}\right)^{n''+2r+2|\alpha''|}\int_{A^c}(1+|x''|^{2(r+|\alpha''|)})e^{-\frac{\tau}{8\lambda}|x''|^2}dx''\|u\|^2_{H_\tau^{|\alpha'|,-r}(\mathbf{R}^n)} \\
\le& Ce^{-2\delta\tau}\left(\frac{\tau}{\lambda}\right)^{n''+2r+2|\alpha''|}\|u\|^2_{H_\tau^{|\alpha'|,-r}(\mathbf{R}^n)} .
\end{aligned}$$

Here we used the fact that

$$e^{-\frac{\tau}{\lambda}(|x''|-\delta)^2} \le e^{-\frac{\tau}{4\lambda}|x''|^2} \qquad \text{for } x''\in A^c$$

and that

$$e^{2c_1\delta\tau-\frac{\tau}{8\lambda}|x''|^2} \le e^{-2\delta\tau} \qquad \text{for } x''\in A^c .$$

Moreover,

$$\int_{A^c}(1+|x''|^{2(r+|\alpha''|)})e^{-\frac{\tau}{8\lambda}|x''|^2}dx''$$

is bounded since the integrant is rapidly decreasing.

Now we can split the integration in (1.74) into two parts: one integration will be over $A\times\mathbf{R}^{n'}$ and that one can be estimated as in (1.68) when we recall that v is supported in (1.72).

The other integration over $A^c\times\mathbf{R}^{n'}$ will be based on the estimate above. For that part it will be also crucial to observe that

$$|a_{\alpha\beta}^{(j)}(x)-a_{\alpha\beta}^{(j)}(x_0)|\le C \qquad \text{for } x\in A^c\times B_\delta(x_0')$$

which follows from the fact that the coefficients of G_λ are continuous and that the second drivatives of $\tilde\varphi$ vanish outside of $B_{2\delta}(x_0)$.

Following (1.68) we get

$$\begin{aligned}
&\left|\int_{A^c\times\mathbf{R}^{n'}} G_\lambda(x,Dv,\overline{Dv}) - \int_{A^c\times\mathbf{R}^{n'}} G_\lambda(x,Dv,\overline{Dv})\right| \\
\le& C\sum_{|\alpha|\le m-1}\tau^{2(m-1-|\alpha|)}\int_{A^c\times\mathbf{R}^{n'}}|D^\alpha v|^2 \\
\le& C\sum_{|\alpha|\le m-1}\tau^{2(m-1-|\alpha|)}e^{-2\delta\tau}\left(\frac{\tau}{\lambda}\right)^{n''+2r+2|\alpha''|}\|u\|^2_{H_\tau^{|\alpha'|,-r}(\mathbf{R}^n)} \\
\le& e^{-\delta\tau}\|u\|^2_{H_\tau^{m-1,-r}(\mathbf{R}^n)}
\end{aligned}$$

provided τ is large enough to satisfy the estimate

$$C \sum_{|\alpha|\le m-1} \tau^{2(m-1-|\alpha|)} \left(\frac{\tau}{\lambda}\right)^{n''+2r+2|\alpha''|} \le e^{\delta\tau} .$$

□

The next lemma takes care of the last term in (??). This lemma is given in [10], lemma 3.2.

Lemma 45 *Let φ be a second degree polynomial in x such that $\varphi(x_0) = 0$. Furthermore, let k and r be a positive integers. Then there exists a constant C such that*

$$\sum_{|\alpha|\le k} \|D''^{\alpha} v\|^2_{L^2(\mathbf{R}^n)} \le C\Big(\tau^{2k}\|v\|^2_{L^2(\mathbf{R}^n)} + e^{-\delta\tau}\|u\|^2_{H^{0,-r}_\tau(\mathbf{R}^n)}\Big) \tag{1.78}$$

for τ large and $u \in H^{0,-r}_\tau(B_\delta(x_0))$ and $\delta \le \min\{1, \lambda\}$.

Proof: Like before we use $v = Q^{\varphi}_{\lambda,\tau} u$ and $w = e^{\tau\varphi}u$. Since φ is a second degree polynomial and $\varphi(x_0) = 0$ there exist a $c_1 > 0$ such that $\varphi(x) \le c_1\delta$ for $x \in B_\delta(x_0)$. By lemma 1 we have

$$\|w\|_{H^{0,-r}_\tau(\mathbf{R}^n)} \le C\|e^{\tau\varphi}\|_{C^r_\tau(B_\delta(x_0))}\|u\|_{H^{0,-r}_\tau(\mathbf{R}^n)} \le Ce^{c_1\delta\tau}\|u\|_{H^{0,-r}_\tau(\mathbf{R}^n)} . \tag{1.79}$$

Next we estimate the left hand side in (1.78) using Parseval's identity.

$$\begin{aligned}\sum_{|\alpha|\le k} \|D''^{\alpha} v\|^2_{L^2(\mathbf{R}^n)} &= (2\pi)^n \int_{\mathbf{R}^n} |\xi''|^{2k}|\hat v(\xi)|^2 d\xi \\ &= (2\pi)^n \int_A |\xi''|^{2k}|\hat v(\xi)|^2 d\xi + (2\pi)^n \int_{A^c} |\xi''|^{2k}|\hat v(\xi)|^2 d\xi\end{aligned}$$

where $A = \{\xi \in \mathbf{R}^n \; : \; |\xi''|^2 \le c_2\tau^2\delta/\lambda\}$ where c_2 is a constant and its size will be given later. The first integral can be estimated directly

$$(2\pi)^n \int_A |\xi''|^{2k}|\hat v(\xi)|^2 d\xi \le (c_2\delta/\ lambda)^k \tau^{2k}\|v\|^2_{L^2(\mathbf{R}^n)} .$$

For the second integral we recall the symbol (1.12) and get

$$(2\pi)^n \int_{A^c} |\xi''|^{2k}|\hat v(\xi)|^2 d\xi \le \sup_{\xi\in A^c} |\xi''|^{2k} e^{-\frac{\lambda}{\tau}|\xi''|^2}(|\xi''|^2 + \tau^2)^r \|w\|^2_{H^{0,-r}_\tau(\mathbf{R}^n)} .$$

For τ large enough the supremum is attained at $|\xi''|^2 = c_2\tau^2\delta/\lambda$ which gives

$$\begin{aligned}&\sum_{|\alpha|\le k} \|D''^{\alpha} v\|^2_{L^2(\mathbf{R}^n)} \\ &\le (c_2\delta/\lambda)^k\tau^{2k}\|v\|^2_{L^2(\mathbf{R}^n)} + (c_2\delta/\lambda)^k\tau^{2k}(1 + c_2\delta/\lambda)^r\tau^{2r}e^{-c_2\delta\tau}\|w\|^2_{H^{0,-r}_\tau(\mathbf{R}^n)} \\ &\le c_2^k\tau^{2k}\|v\|^2_{L^2(\mathbf{R}^n)} + c_2^k(1 + c_2)^r\tau^{2r+2k}e^{-c_2\delta\tau}\|w\|^2_{H^{0,-r}_\tau(\mathbf{R}^n)} .\end{aligned}$$

Next choose $c_2 > c_1 + 1$ and for τ large enough we have

$$\tau^{2r+2k} \leq e^{(c_2-c_1-1)\delta\tau}$$

which provides

$$\begin{aligned}\sum_{|\alpha|\leq k} \|D''^{\alpha} v\|^2_{L^2(\mathbf{R}^n)} &\leq c_2^k\left(\tau^{2k}\|v\|^2_{L^2(\mathbf{R}^n)} + (1+c_2)^r e^{-(c_1+1)\delta\tau}\|w\|^2_{H^{0,-r}_\tau(\mathbf{R}^n)}\right)\\ &\leq c_2^k\left(\tau^{2k}\|v\|^2_{L^2(\mathbf{R}^n)} + C(1+c_2)^r e^{-\delta\tau}\|u\|^2_{H^{0,-r}_\tau(\mathbf{R}^n)}\right)\end{aligned}$$

which proves the lemma after introducing a new constant. □

The Carleman type estimate for $v = Q^{\varphi}_{\lambda,\tau}u$ has now the following form (compare with theorem 1 in [10]). The proof follows from lemma 43, lemma 44, lemma 45 and lemma 5.

Theorem 46 *Let P satisfy the same conditions as in theorem 42. Furthermore, let φ be a second degree polynomial in with $\varphi(x_0) = 0$ which is strongly pseudo-convex with respect to P on Γ at x_0. Then for $\lambda \leq \lambda_0$ there exist a $\delta > 0$ and $K > 0$ such that*

$$\tau\|Q^{\tilde\varphi}_{\lambda,\tau}u\|^2_{H^{m-1}_\tau(\mathbf{R}^n)} \leq K\left(\|Q^{\tilde\varphi}_{\lambda,\tau}P(x,D)u\|^2_{L^2(\mathbf{R}^n)} + e^{-\delta\tau}\|u\|^2_{H^{m-1,-r}_\tau(\mathbf{R}^n)}\right) \tag{1.80}$$

for all $u \in C_0^\infty(B_\delta(x_0))$ provided τ is large enough. Here $\tilde\varphi$ is the function introduced in (1.73).

Again, we can extend the class of functions for which this estimate holds.

Corollary 47 *Under the hypotheses in theorem 46 the estimate (1.80) is valid for $u \in H^{m-1,-r}_{0,\tau}(B_\delta(x_0))$ with $P(x,D)u \in H^{0,-r}_\tau(B_\delta(x_0))$.*

Proo f: At first we notice that lemma 45 is valid for $u \in H^{m-1,-r}_{0,\tau}(B_\delta(x_0))$. We know that (??) is valid for $v \in \mathcal{S}(\mathbf{R}^{n''}, C_0^\infty(B_\delta(x_0')))$. So we use a regularization in x'-variables only and show as in corollary 41 that (??) holds for $v \in \mathcal{S}(\mathbf{R}^{n''}, H_0^{m-1}(B_\delta(x_0)))$ with $P(x,D)v \in \mathcal{S}(\mathbf{R}^{n''}, L^2(B_\delta(x_0)))$. Finally, the application of corollary refQP* concludes the proof. □

Before we can prove now Tataru's theorem we need the following proposition about the integral operator (1.11). This is proposition 4.1 in [10].

Proposition 48 *Let $u \in H^{0,-r}(\mathbf{R}^n)$ be a function with compact support. Furthermore, let $\varphi \in C^r(\mathbf{R}^n)$ and assume*

$$\|Q^{\varphi}_{\lambda,\tau}u\|_{L^2(\mathbf{R}^n)} \longrightarrow 0 \qquad \text{as } \tau \to \infty\,. \tag{1.81}$$

Then $u \equiv 0$ in $\{x \in \mathbf{R}^n \ : \ \varphi(x) > 0\}$.

Proof: Let $\nu > 0$ and $g \in C_0^\infty(\mathbf{R}^n)$ and let $f = e^{-\nu|D''|^2}g$. Then we introduce a distribution $h \in \mathcal{E}'(\mathbf{R})$ acting on $w \in C^\infty(\mathbf{R})$ by

$$h(w) = \langle fw(\varphi), u\rangle\,.$$

Its Fourier transform is an entire function and is given by

$$\hat{h}(\tau) = \langle f e^{i\tau\varphi}, u\rangle$$

and $\hat{h}$ has at most exponential growth at infinity. Moreover, for real τ we have

$$|\hat{h}(\tau)| \leq C\|e^{i\tau\varphi} f\|_{H^{0,r}_\tau(\mathbf{R}^n)} \|u\|_{H^{0,-r}_\tau(\mathbf{R}^n)} \leq C\tau^r$$

and

$$\hat{h}(i\tau) = \langle f, e^{\tau\varphi}u\rangle = \langle e^{(\frac{\lambda}{2\tau}-\nu)|D''|^2} g, Q^\varphi_\tau u\rangle \longrightarrow 0$$

for $\tau \to \infty$. That means the function $\hat{h}(\tau)(\tau+i)^{-r}$ is bounded on the real axis and the positive imaginary axis and has at most exponential growth at infinity. By the Phragmen-Lindelöf principle (4.2 Corollary p.139 in [1]) applied to $\hat{h}(\tau)(\tau+i)^{-r}$ we conclude that

$$|\hat{h}(\tau)| \leq C(1+|\tau|)^r \qquad \text{for } Im\,\tau \geq 0\,.$$

Using the Paley-Wiener-Schwartz theorem (theorem 7.3.1 in [4]) w e obtain that h is a distribution of order r and

$$\sup_{x\in supp\ h} \langle x, Im\,\tau\rangle = 0 \qquad \text{for } Im\,\tau \geq 0\,.$$

That implies that h is supported on the negative real axis, $supp\ h \subset \mathbf{R}_-$. Hence, $h(w) = 0$ for all $w \in C^\infty(\mathbf{R})$ with $supp\, w \subset \mathbf{R}_+$. For $\nu \to 0$ we get

$$\langle g w(\varphi), u\rangle = 0 \qquad \text{for all } g \in C_0^\infty(\mathbf{R}^n)$$

and w as above. Thus $\langle g, u\rangle = 0$ whenever $g \in C_0^\infty(\mathbf{R}^n)$ is supported in $\{x \in \mathbf{R}^n \,:\, \varphi(x) > 0\}$. □

Now we can prove the main theorem of this section.

Proof of theorem 42: The proof is identical to that of theorem 39 up to (1.70) except that $\tilde{u} \in H^{m-1,-r}_{0,\tau}(B_\delta(x_0))$ and $P(x,D)\tilde{u} \in H^{0,-r}_\tau(B_\delta(x_0))$ and ψ has to be modified as in (??).

Now we want to use the Carleman type estimate (1.80). We start with estimating the first term in the right hand side

$$\begin{aligned} \|Q^\psi_{\lambda,\tau} P(x,D)\tilde{u}\|^2_{L^2(\mathbf{R}^n)} &= \|e^{-\frac{\lambda}{2\tau}|\xi''|^2} \widehat{e^{\tau\psi}P(x,D)\tilde{u}}\|^2_{L^2(\mathbf{R}^n)} \\ &\leq \sup_{\xi\in\mathbf{R}^n} (\tau^2+|\xi''|^2)^r e^{-\frac{\lambda}{\tau}|\xi''|^2} \|e^{\tau\psi}P(x,D)\tilde{u}\|^2_{H^{0,-r}_\tau(\mathbf{R}^n)} \\ &\leq C\tau^{2r}\|e^{\tau\psi}\|^2_{C^r_\tau(G)} \|P(x,D)\tilde{u}\|^2_{H^{0,-r}_\tau(\mathbf{R}^n)} \\ &\leq C\tau^{2r} e^{-2\gamma\delta^2\tau} \|P(x,D)\tilde{u}\|^2_{H^{0,-r}_\tau(\mathbf{R}^n)} \end{aligned} \tag{1.82}$$

where we used Parseval's identity, lemma 1 and (1.70). Hence, for τ large enough we can use (1.80) and get

$$\tau\|Q^\psi_{\lambda,\tau}\tilde{u}\|^2_{H^{m-1}_\tau(\mathbf{R}^n)} \leq C(\tau^{2r}e^{-2\gamma\delta^2\tau} + e^{-\delta\tau}) \leq C\tau^{2r}e^{-2\gamma\delta^2\tau}\,.$$

That leads to

$$\|Q_{\lambda,\tau}^{\psi+\frac{\gamma\delta^2}{2}}\tilde{u}\|^2_{H_\tau^{m-1}(\mathbf{R}^n)} = \|Q_{\lambda,\tau}^{\psi}\tilde{u}\|^2_{H_\tau^{m-1}(\mathbf{R}^n)}e^{\gamma\delta^2\tau} \leq C\tau^{2r}e^{-\gamma\delta^2\tau} \to 0$$

as $\tau \to \infty$. Finally, proposition 48
applied to $\psi + \gamma\delta^2/2$ shows that

$$\tilde{u} \equiv 0 \quad \text{in } V = \left\{x \in B_\delta(x_0) \,:\, \psi(x) \geq -\frac{\gamma\delta^2}{2}\right\}.$$

However, $\tilde{u} = u$ in this set and V is a neighborhood of x_0. □

1.6 UNIQUENESS OF CONTINUATION THEOREMS FOR TWO CLASSICAL SYSTEMS OF PARTIAL DIFFERENTIAL EQUATIONS

In this section we will use in particular Tataru's result in order to prove a uniqueness of continuation theorem for Maxwell's system and for the elasticity system. Since Carleman type estimates cannot be directly applied to systems we will need to reduce these systems to equations whose principal parts is scalar. Both systems have the property that the variables decompose into time and into space and that moreover, the coefficients in the system can be assumed to be time independent.

Throughout this section we will need the usual operators of vector analysis. Let $\mathbf{e}_1, \mathbf{e}_2, \mathbf{e}_3$ be t he unit vectors in coordinate direction in $\mathbf{R}^3$. Let $\mathbf{f} = (f_1, f_2, f_3)$ be a vector valued function, i.e. a function from $\mathbf{R}^3$ into $\mathbf{R}^3$. Then the divergence of $\mathbf{f}$ is given by

$$div\,\mathbf{f} = \partial_1 f_1 + \partial_2 f_2 + \partial_3 f_3$$

and the rotation is given by

$$curl\,\mathbf{f} = \begin{vmatrix} \mathbf{e}_1 & \mathbf{e}_2 & \mathbf{e}_3 \\ \partial_1 & \partial_2 & \partial_3 \\ f_1 & f_2 & f_3 \end{vmatrix}.$$

We will con sider also function spaces of vector valued function. However, we will not change the notation, i.e. $\mathbf{f} \in L^2(\Omega)$ means $f_j \in L^2(\Omega)$ for $j = 1, 2, 3$.

1.6.1 Maxwell's system

We consider Maxwell's system in a three-dimensional domain Ω'. The three space variables are denoted by x' and the time variable by $t = x''$ and we introduce $x = (t, x') = (x'', x_1, x_2, x_3)$. Let $\mathbf{E}(x)$ and $\mathbf{H}(x)$ be vector-valued functions which denote the electric and magnetic field intensity, i.e.

$$\mathbf{E}(x) = (E_1(x), E_2(x), E_3(x)) \quad \text{and} \quad \mathbf{H}(x) = (H_1(x), H_2(x), H_3(x)).$$

Furthermore, $\varepsilon(x')$ and $\mu(x')$ denote the electric permittivity and the magnetic permeability. Let $\sigma(x')$ be the electric conductivity. We assume them to be positive functions of the space variable only. Moreover, let $\varrho(x)$ be the electrical charge density and $\mathbf{j}(x)$ be the electrical curren t density. The evolution of the electric and the magnetic field in time in Ω' is given by Maxwell's equations

$$\begin{aligned} \varepsilon\partial_t\mathbf{E} + \sigma\mathbf{E} - curl\ \mathbf{H} &= \mathbf{j} \\ \mu\partial_t\mathbf{H} + curl\ \mathbf{E} &= 0 \\ div(\varepsilon\mathbf{E}) &= 4\pi\varrho \\ div(\mu\mathbf{H}) &= 0 \end{aligned} \tag{1.83}$$

for $x \in (0,T) \times \Omega'$ where $T > 0$. Note that we consider Maxwell's system in the isotropic case. The electric permittivity and the magnetic permeability are scalar functions. For the anisotropic case we would have to replace them with 3×3 matrices. For an introduction into this case we refer to chapter 8 in [7].

When we discuss uniqueness of continuation we consider the homogeneous system, i.e. $\mathbf{j} = 0$ and $\varrho = 0$. We will transform (1.83) into a system of second order equations with scalar principal part. To do so we differentiate the first equation with respect to t and apply the *curl* to the second equation. We obtain

$$\begin{aligned} \varepsilon\partial_t^2\mathbf{E} + \sigma\partial_t\mathbf{E} - \partial_t curl\ \mathbf{H} &= 0 \\ curl(\mu\partial_t\mathbf{H}) + curl\ curl\ \mathbf{E} &= 0\,. \end{aligned}$$

Now we use the fact that $curl(\mu\mathbf{H}) = \mu\, curl\,\mathbf{H} + \nabla\mu \times \mathbf{H}$ and that $curl\ curl\ \mathbf{E} = -\Delta\mathbf{E} + \nabla\ div\ \mathbf{E}$ and combine the two equations into

$$\mu\varepsilon\partial_t^2\mathbf{E} - \Delta\mathbf{E} + \nabla\mu \times \partial_t\mathbf{H} + \nabla\ div\ \mathbf{E} + \mu\sigma\partial_t\mathbf{E} = 0\,.$$

Here we make use of the third equation in (1.83) which can be written as

$$div\ \mathbf{E} = -\frac{\nabla\varepsilon}{\varepsilon}\mathbf{E}$$

and replace the last term in the previous equation. We obtain

$$\mu\varepsilon\partial_t^2\mathbf{E} - \Delta\mathbf{E} + \nabla\mu \times \partial_t\mathbf{H} - \nabla\left(\frac{\nabla\varepsilon}{\varepsilon}\mathbf{E}\right) + \mu\sigma\partial_t\mathbf{E} = 0\,. \tag{1.84}$$

We can do th e same procedure for $\mathbf{H}$ and obtain

$$\mu\varepsilon\partial_t^2\mathbf{H} - \Delta\mathbf{H} - \nabla\varepsilon \times \partial_t\mathbf{E} - \nabla\left(\frac{\nabla\mu}{\mu}\mathbf{H}\right) + curl(\sigma\mathbf{E}) = 0\,. \tag{1.85}$$

The six equations (1.84) and (1.85) represent the whole Maxwell system and form a new system of six second order PDEs. This system is coupled only through the lower order terms. The principal part of each of the six equations is a second order hyperbolic operator and will be denoted by

$$\Box_{\varepsilon\mu} = \varepsilon\mu\partial_t^2 - \Delta\,. \tag{1.86}$$

We introduce a vector-valued function $\mathbf{u} = (\mathbf{E}, \mathbf{H})$ which allows as to write (1.84) and (1.85) as

$$\Box_{\varepsilon\mu} u_k = \sum_{j=1}^{6} \sum_{i=0}^{3} a_{ij}^k \partial_i u_j + \sum_{j=1}^{6} b_j^k u_j \,. \tag{1.87}$$

where $a_{ij}^k, b_j^k \in C(\bar{\Omega}')$ provided $\varepsilon, \mu \in C^2(\bar{\Omega}')$ and $\sigma \in C^1(\bar{\Omega}')$.

Our main result is the following

Theorem 49 *Let $\Omega = (0,T) \times \Omega'$ and let S be an oriented C^2-surface in Ω that is non-characteristic at x_0 with respect to the hyperbolic operator (1.86). Furthermore, assume $\varepsilon, \mu \in C^2(\bar{\Omega}')$ and $\sigma \in C^1(\bar{\Omega}')$. Then the uniqueness of continuation property holds with respect to Maxwell's system across S at x_0 for* $\mathbf{E}, \mathbf{H} \in \mathcal{D}'((0,T), H^1(\Omega'))$.

Proof: Let p be the symbol of the hyperbolic operator $\Box_{\varepsilon\mu}$ and let ξ'' be the dual variable to t. We claim that the surface S is the level surface of a function φ that is strongly pseudo-convex at x_0 with respect to p on $\Gamma = \{\xi'' = 0\}$. Since S is non-characteristic at x_0 there exists a function $\phi \in C^2(\bar{\Omega})$ such that S is a level surface of ϕ and $p(x_0, \nabla\phi(x_0)) \neq 0$. On the other hand, the strong pseudo-convexity condition (1.25) is imposed on

$$p(x_0, \xi + i\tau\nabla\phi(x_0)) = \partial_\tau p(x_0, \xi + i\tau\nabla\phi(x_0)) = 0 \,. \tag{1.88}$$

In our case, $p(x_0, \xi + i\tau\nabla\phi(x_0))$ is a second degree polynomial in $i\tau$ with real coefficients and its leading coefficient does not vanish. In order to satisfy (1.88) the polynomial needs to have a double zero. However, that is possi ble only if $i\tau$ is real, i.e. τ is purely imaginary. That shows that condition (1.25) does not apply. The condition (1.24) never applies since

$$p(x, \xi) = \xi_1^2 + \xi_2^2 + \xi_3^2 > 0 \qquad \text{on } \Gamma \,.$$

Hence, the function ϕ satisfies the conditions of definition 12 and using lemma 31 we find a function $\varphi \in C^2(\bar{\Omega})$ that is strongly pseudo-convex at x_0 with respect to p on Γ. Consequently, the hypotheses of the orem 46 are fulfilled with the operator (1.86) under consideration.

The proof proceeds now as the proof of theorem 39 and we modify ψ as in (1.73). As before, we set

$$\tilde{u_k}(x) = \begin{cases} \chi(\psi(x)) u_k(x) & \text{if } |x - x_0| \leq 2\delta \\ 0 & \text{otherwise} \end{cases} \qquad k = 1, 2, .., 6$$

and show $supp\ \tilde{u_k} \subset B_\delta(x_0))$. Since $u_k \in \mathcal{D}'((0,T), H^1(\Omega'))$ we know tha t $\tilde{u_k} \in H_{0,\tau}^{1,-r}(B_\delta(x_0))$ for some positive integer r. We compute

$$\Box_{\varepsilon\mu}\tilde{u_k} = \chi \Box_{\varepsilon\mu} u_k + u_k \Box_{\varepsilon\mu} \chi + 2\nabla\chi \cdot \nabla u_k + 2\varepsilon\mu \partial_t \chi \partial_t u_k \,. \tag{1.89}$$

Since u satisfies Maxwell's system we have $\Box_{\varepsilon\mu}\tilde{u} \in H_\tau^{0,-r}(B_\delta(x_0))$. Now we can use the Carleman type estimate (1.80) for each of the $\tilde{u_k}$. In order to get a m eaningful result we need to estimate the first term in the right hand side of (1.80). This term is the square of the L^2-norm of $Q^{\psi}_{\lambda,\tau}\Box_{\varepsilon\tau}\tilde{u_k}$. For that we will use the expansion (1.89) and we notice that the three last terms in this expansion can be estimated as in (1.82) since these terms are supported in G. That means they can be estimated by $C\tau^{2r}e^{-2\gamma\delta^2\tau}\|u_k\|^2_{H^{1,-r}_\tau(\mathbf{R}^n)}$. The only di fficulty occurs with the first term in (1.89). However using (1.87) and $\chi\partial_i u_j = \partial_i\tilde{u_j} - \partial_i\chi u_j$ we have

$$\begin{aligned}
\|Q^{\psi}_{\lambda,\tau}\chi\Box_{\varepsilon\mu}u_k\|^2_{L^2(\mathbf{R}^n)} &\leq \sum_{j=1}^{6}\left\|Q^{\psi}_{\lambda,\tau}\left(\chi\sum_{i=0}^{3}a^k_{ij}\partial_i u_j + b^k_j\tilde{u_j}\right)\right\|^2_{L^2(\mathbf{R}^n)}\\
&\leq \sum_{j=1}^{6}\left\|Q^{\psi}_{\lambda,\tau}\left(\sum_{i=0}^{3}a^k_{ij}\partial_i\tilde{u_j} - \sum_{i=0}^{3}a^k_{ij}\partial_i\chi u_j + b^k_j\tilde{u_j}\right)\right\|^2_{L^2(\mathbf{R}^n)}\\
&\leq C\sum_{j=1}^{6}\Big(\sum_{i=0}^{3}(\|Q^{\psi}_{\lambda,\tau}\partial_i\tilde{u_j}\|^2_{L^2(\mathbf{R}^n)} + \|Q^{\psi}_{\lambda,\tau}\partial_i\chi u_j\|^2_{L^2(\mathbf{R}^n)})\\
&\quad + \|Q^{\psi}_{\lambda,\tau}\tilde{u_j}\|^2_{L^2(\mathbf{R}^n)}\Big)\\
&\leq C\sum_{j=1}^{6}\Big(\sum_{i=0}^{3}(\|(D_i + i\tau\partial_i\psi - \lambda\nabla^2\ psiD_t)Q^{\psi}_{\lambda,\tau}\tilde{u_j}\|^2_{L^2(\mathbf{R}^n)}\\
&\quad + \tau^{2r}e^{-2\gamma\delta^2\tau}\|\partial_i\chi u_j\|^2_{H^{0,-r}_\tau(\mathbf{R}^n)}) + \|Q^{\psi}_{\lambda,\tau}\tilde{u_j}\|^2_{L^2(\mathbf{R}^n)}\Big)\\
&\leq C\sum_{j=1}^{6}\left(\|Q^{\psi}_{\lambda,\tau}\tilde{u_j}\|^2_{H^1_\tau(\mathbf{R}^n)} + \tau^{2r}e^{-2\gamma\delta^2\tau}\|u_j\|^2_{H^{0,-r}_\tau(\mathbf{R}^n)}\right)
\end{aligned}$$

where we used lemma 5 and the same technique as i n estimate (1.82).

For each of the $\tilde{u_k}$ we use now the Carleman type estimate (1.80) in connection with the estimate above and obtain

$$\begin{aligned}
\tau\|Q^{\psi}_{\lambda,\tau}\tilde{u_k}\|^2_{H^1_\tau(\mathbf{R}^n)} \quad &\leq \quad C\sum_{j=1}^{6}(\|Q^{\psi}_{\lambda,\tau}\tilde{u_j}\|^2_{H^1_\tau(\mathbf{R}^n)} + \tau^{2r}e^{-2\gamma\delta^2\tau}\|u_j\|_{H^{1,-r}_\tau(\mathbf{R}^n)})\\
&\quad + Ce^{-\delta\tau}\|\tilde{u_k}\|^2_{H^{1,-r}_\tau(\mathbf{R}^n)}
\end{aligned}$$

a nd after adding this estimate over k we get

$$\begin{aligned}
\tau\sum_{k=1}^{6}\|Q^{\psi}_{\lambda,\tau}\tilde{u_k}\|^2_{H^1_\tau(\mathbf{R}^n)} \quad &\leq \quad C\sum_{k=1}^{6}(\|Q^{\psi}_{\lambda,\tau}\tilde{u_k}\|^2_{H^1_\tau(\mathbf{R}^n)}\\
&\quad + \tau^{2r}e^{-2\gamma\delta^2\tau}\|u_j\|_{H^{1,-r}_\tau(\mathbf{R}^n)} + e^{-\delta\tau}\|\tilde{u_k}\|^2_{H^{1,-r}_\tau(\mathbf{R}^n)}).
\end{aligned}$$

Choosing $\tau > 2C$ we can subtract the first term on the right hand side to the left hand side and concl ude the proof as the proof of theorem 42. □

The regularity of the solution to Maxwell's system can be slightly lowered. Let $\mathbf{E}, \mathbf{H} \in \mathcal{D}'((0,T), L^2(\Omega'))$ satisfy Maxwell's system. This solution has to be understood as a distributional solution in t and a weak solution in x'. Then $\mathbf{E}$ and $\mathbf{H}$ satisfy the second order system (1.84) and (1.85) in the sense of distributions. Let $\delta > 0$ and regularize both functions in t dire ction. Then for $\epsilon \leq \epsilon_0$ we obtain sequences $\{\mathbf{E}_\epsilon(s, x')\} \subset C_0^\infty((\delta, T-\delta), L^2(\Omega'))$ and $\{\mathbf{H}_\epsilon(s, x')\} \subset C_0^\infty((\delta, T-\delta), L^2(\Omega'))$ such that

$$\begin{array}{lcl} \mathbf{E}_\epsilon & \longrightarrow & \mathbf{E} \\ \mathbf{H}_\epsilon & \longrightarrow & \mathbf{H} \end{array} \quad \text{in } \mathcal{D}'((\delta, T-\delta), L^2(\Omega')) \tag{1.90}$$

for $\epsilon \to 0$. Fo r a justification of this procedure we refer to lemma 2 and the following remarks.

Since differentiation and regularization can be interchanged (see for example lemma 7.3 in [2]) we know that (1.84) and (1.85) are valid for $\mathbf{E}_\epsilon$ and $\mathbf{H}_\epsilon$ as well. Now consider (1.84) at a fixed $s \in (\delta, T-\delta)$

$$\Delta \mathbf{E}_\epsilon = \mu\varepsilon\partial_s^2 \mathbf{E}_\epsilon + \nabla\mu \times \partial_s \mathbf{H}_\epsilon - \nabla\left(\frac{\nabla\varepsilon}{\varepsilon}\mathbf{E}_\epsilon\right) + \mu\sigma\partial_s \mathbf{E}_\epsilon \,.$$

We denote the right hand side in this equation by $\mathbf{F}$. This is an elliptic equation in all three components with a parameter s

$$\Delta E_{j\epsilon} = F_j \qquad j = 1, 2, 3\,.$$

Let $\psi \in C_0^\infty(\Omega')$. The product of $E_{j\epsilon}$ and ψ satisfies the following equation

$$\Delta(\psi E_{j\epsilon}) = \psi F_j - \nabla\psi \cdot \nabla E_{j\epsilon} - \Delta\psi E_{jesilon}$$

with the right hand side in $H^{-1}(\Omega')$. In addition to that $\psi E_{j\epsilon}$ has zero Dirichlet data. Elliptic regularity theory (theorem 8.19 in [8]) provides $\psi E_{j\epsilon} \in H_0^1(\Omega')$ and the estimate

$$\|\psi E_{j\epsilon}\|_{H^1(\Omega')} \leq C\|G_j\|_{H^{-1}(\Omega')}$$

where G_j denotes the right hand side in the equation above. This right hand side is continuous (actually C^∞) in s and that proves $\psi E_{j\epsilon}$ is continuous in s as well. Hence,

$$E_{j\epsilon} \in C((\delta, T-\delta), H^1_{loc}(\Omega'))\,.$$

The same procedure can be repeated for all derivatives with respect to s and we obtain

$$\partial_s^\nu \mathbf{E}_\epsilon \in C((\delta, T-\delta), H^1_{loc}(\Omega')) \quad \text{for all } \nu\,.$$

Hence, we have $\mathbf{E}_\epsilon \in C^\infty((\delta, T-\delta), H^1_{loc}(\Omega'))$. We obtain the same result for $\mathbf{H}_\epsilon$ when we make use of (1.85).

Now we can prove the following corollary.

Corollary 50 *Under the same assumptions the conclusion of theorem 49 holds for* $\mathbf{E}, \mathbf{H} \in \mathcal{D}'((0, T), L^2(\Omega'))$.

Proof: By assumption $\mathbf{E}$ and $\mathbf{H}$ vanish on the positive side of the surface S. Then we choose ϵ_1 as small that the regularizations $\mathbf{E}_\epsilon$ and $\mathbf{H}_\epsilon$ vanish on the positive side of S_{ϵ_1} for $\epsilon \leq \epsilon_1$ and the surface S_{ϵ_1} is close to the surface S and hence non-characteristic with respect to the operator (1.86) at

$$x_{\epsilon_1} = \inf_{x \in S_{\epsilon_1}} dist(x_0, x) .$$

We apply theorem 49 to $\mathbf{E}_\epsilon$ and $\mathbf{H}_\epsilon$ and the surface S_{ϵ_1} at x_{ϵ_1} and obtain a neighborhood V where $\mathbf{E}_\epsilon$ and $\mathbf{H}_\epsilon$ vanish. Consequently $\mathbf{E}$ and $\mathbf{H}$ will vanish there as well. Finally, when ϵ_1 is chosen small enough V will be a neighborhood of x_0 as well. □

We like to point out that Maxwell's system (1.83) is overdetermined. One can check that the first and the third equation imply

$$4\pi\partial_t \varrho - div\, \mathbf{j} = 0 .$$

Of course, this is only true when ε is time-independent. The question is whether uniqueness of continuation still holds when we do not include the third equation. This is an important question for certain applicatio ns [11].

We can give a satisfactory answer only for $\mathbf{H}$ in the case of $\sigma = 0$. This means we consider the reduced Maxwell system

$$\begin{aligned} \varepsilon\partial_t \mathbf{E} - curl\, \mathbf{H} &= \mathbf{j} \\ \mu\partial_t \mathbf{H} + curl\, \mathbf{E} &= 0 \\ div(\mu\mathbf{H}) &= 0 . \end{aligned} \tag{1.91}$$

Again, we consider the homogeneous system. Because of the first equation we can deduce a simpler version of (1.85)

$$\mu\varepsilon\partial_t^2 \mathbf{H} - \ Delta\mathbf{H} - \nabla\varepsilon \times \partial_t \mathbf{E} - \nabla\left(\frac{\nabla\mu}{\mu}\mathbf{H}\right) = 0 .$$

and use the first equation in (1.91) to eliminate $\partial_t \mathbf{E}$. Hence, we have

$$\mu\varepsilon\partial_t^2 \mathbf{H} - \Delta\mathbf{H} - \nabla\varepsilon \times \frac{curl\, \mathbf{H}}{\varepsilon} - \nabla\left(\frac{\nabla\mu}{\mu}\mathbf{H}\right) = 0 .$$

This is now a system of three equations for $\mathbf{H}$ only and the principal part of each equation is again the wave operator (1.86). U sing the same argument as in the proof of theorem 49 we obtain uniqueness of continuation for $\mathbf{H} \in \mathcal{D}'((0,T), H^1(\Omega'))$.

Then we can apply elliptic regularity theory and show that the uniqueness of continuation property holds for $\mathbf{H} \in \mathcal{D}'((0,T), L^2(\Omega'))$ as well.

1.6.2 The equations of linear elasticity

The second system which we consider is the system of linear elasticity in the isotropic case. An earlier result here is due to Isakov [6].

Let $\rho(x')$ denote the density function and $\lambda(x')$ and $\mu(x')$ the Lam e parameters. These three functions are time independent and positive. Let $\mathbf{u} =$

(u_1, u_2, u_3) denote the elastic displacement vector and **F** is the external force. The equations of linear elasticity are

$$\rho \partial_t^2 \mathbf{u} - \mu(\Delta \mathbf{u} + \nabla div\, \mathbf{u}) - \nabla(\lambda\, div\, \mathbf{u}) - \sum_{j=1}^{3} \nabla\mu(\nabla u_j + \partial_j u)\mathbf{e}_j = \; bfF\,. \quad (1.92)$$

Again, the isotropic case is treated in chapter 11 of [7].

The homogeneous system can be written as

$$\partial_t^2 \mathbf{u} = \frac{\mu}{\rho}\Delta \mathbf{u} + \frac{\mu+\lambda}{\rho}\nabla\, div\, \mathbf{u} + \frac{\nabla\lambda}{\rho} div\, \mathbf{u} + \sum_{j=1}^{3} \frac{\nabla\mu}{\rho}(\nabla u_j + \partial_j u)\mathbf{e}_j = 0\,. \quad (1.93)$$

This system consisting of three equations is coupled even in the highest order terms. We will change that by introducing two more function $v = div\, u$ and $\mathbf{w} = curl\, u$. For these two functions we will find differential equations simply by applying the divergence and the rotation to the whole system. A similar approach was suggested by G. Nakamura.

That leads to the following system of seven equations

$$\begin{aligned}
\partial_t^2 \mathbf{u} &= \frac{\mu}{\rho}\Delta \mathbf{u} + \frac{\mu+\lambda}{\rho}\nabla v + \frac{\nabla\lambda}{\rho} v + \sum_{j=1}^{3} \nabla\mu(\nabla u_j + \partial_j u)\mathbf{e}_j = 0 \\
\partial_t^2 v &= \frac{2\mu+\lambda}{\rho}\Delta v + \left(\frac{2\nabla\mu + \nabla\lambda}{\rho} + \nabla\frac{2\mu+\lambda}{\rho}\right)\nabla v \\
&\quad - \left(\frac{\nabla\mu}{\rho} + \nabla\frac{\mu}{\rho}\right)\cdot curl\, \mathbf{w} + div\frac{\nabla\lambda}{\rho} v + \sum_{j=1}^{3} \nabla\partial_j\mu(\nabla u_j + \partial_j u) \\
\partial_t^2 \mathbf{w} &= \frac{\mu}{\rho}\Delta \mathbf{w} + \left(\nabla\frac{2\mu+\lambda}{rho} + \frac{\nabla\lambda}{\rho}\right)\times\nabla v + \left(\frac{\nabla\mu}{\rho}\cdot\nabla\right)\mathbf{w} - \nabla\frac{\mu}{\rho}\times curl\, \mathbf{w} \\
&\quad + curl\frac{\nabla\lambda}{\rho} div\, \mathbf{u} + l(\nabla \mathbf{u})
\end{aligned}$$

Here $l(\mathbf{u})$ denotes a term which depends only on the matrix of first derivatives of $\mathbf{u}$ and the first and second derivatives of μ and ρ with its first derivatives.

The advantage of this representation is that the principal part

of this system is decoupled and consist of wave operators. The principal part of the first three and last three equations is

$$\Box_{\mu\rho} = \partial_t^2 - \frac{\mu}{\rho}\Delta \quad (1.94)$$

and the principal part of the fourth equation is

$$\Box_{\mu\rho\lambda} = \partial_t^2 - \frac{2\mu+\lambda}{\rho}\Delta\,. \quad (1.95)$$

In complete analogy with theorem 49 we obtain the following result.

Theorem 51 *Let $\Omega = (0,T) \times \Omega'$ and let S be an oriented C^2-surface in Ω that is non-characteristic at x_0 with respect to the hyperbolic operators (1.94) and (1.95). Furthermore, assume $\lambda, \mu \in C^2(\bar{\Omega}')$ and $\rho \in C^1(\bar{\Omega}')$. Then the uniqueness of continuation property holds with respect to the elasticity system across S at x_0 for $\mathbf{u} \in \mathcal{D}'((0,T), H^1(\Omega'))$.*

We like to emphasize that we ne ed to start the proof with higher regularity $\mathbf{u} \in \mathcal{D}'((0,T), H^2(\Omega'))$ since we need to work with *div*$\mathbf{u}$ and *curl*$\mathbf{u}$ as well. After that the regularity can be lowered using the same technique as in corollary 50.

References

[1] J. B. Conway *Functions of one complex variable* Springer-Verlag, Berlin 1973

[2] D. Gilbarg, N. Trudinger *Elliptic partial differential equations of second order* Springer-Verlag, Berlin 1983

[3] L. Hörmander *Linear partial differential operators* Springer-Verlag, Berlin 1966

[4] L. Hörmander *The analysis of linear partial differential operators I-IV* Springer-Verlag, Berlin 1983

[5] L. Hörmander *A uniqueness of continuation theorem for second order hyperbolic differential equations* Comm. part. diff. equations **17** pp. 699-714 (1992)

[6] V. Isakov *A non-hyperbolic Cauchy problem for $\Box_b\Box_c$ and its applications to elasticity theory* Comm. pure and applied math. **39** pp. 747-767 (1986)

[7] R. Leis *Initial boundary value problems in mathematical physics* Wiley, New York 1986

[8] M. Renardy, R. Rogers *An introduction to partial differential equations* Springer-Verlag, Berlin 1993

[9] L. Robbiano *Théorème d'unicité adapté au contrôle des solutions des problèmes hyperboliques* Comm. part. diff. equations **16** pp. 789-900 (1991)

[10] D. Tataru *Unique continuation for solutions to PDE's; between Hörmander's and Holmgren's theorem* Comm. part. diff. equations **20** pp. 855-884 (1995)

[11] M. Yamamoto *On an inverse problem of determining source terms in Maxwell's equations with a single measurement* preprint, 1997

DETERMINATION OF A DISTRIBUTED INHOMOGENEITY IN A TWO-LAYERED WAVEGUIDE FROM SCATTERED SOUND

Robert P. Gilbert *

Department of Mathematical Sciences
University of Delaware
Newark, DE 19716

Christopher Mawata † and Yongzhi Xu ‡

Department of Mathematics
University of tennessee at Chattanooga
Chattanooga, TN 37403

*This author's research was supported in part by NSF Grant BES-9402539.
†This author's research was supported in part by grants from CECA of University of Tennessee at Chattanooga
‡This author's research was supported in part by NSF Grant BES-9402539 and by grants from CECA of University of Tennessee at Chattanooga

R.P. Gilbert et al.(eds.), Direct and Inverse Problems of Mathematical Physics, 107–124.

Abstract: This paper considers the determination of a distributed inhomogeneity in a two-layered waveguide from scattered sound. Assuming that we know the acoustic properties of the waveguide, we determine the unknown inhomogeneity by sending in incident waves from point sourses in given locations, and detecting the total waves along a line. In this paper we consider the case that wavenumber k is small. In this case we obtain the representation, uniqueness, and existence of the direct scattering problem, and the uniqueness of inverse scattering problem. Numerical examples are also presented.

1.1 INTRODUCTION

The inverse problem of the localization and identification of a *passive object* in a body of water has important practical applications (identification of seamounts, mineral deposits, submarines, submerged wreckages and navigational obstacles, etc.), and has received fairly considerable attention. Only recently have attempts been made to deviate from the usual (generally non-automatized) signature–recognition strategy to that of the achievement of full-fledged (automatized) parameter-estimation (the parameters being those associated with the location and shape of the object, assuming the sources and underwater environment to be known.

This paper considers the determination of a distributed inhomogeneity in a two-layered waveguide from scattered sound. Assuming that we know the acoustic properties of the waveguide, we determine the unknown inhomogeneity by sending in incident waves from point sourses in given locations, and detecting the total waves along a line. In this paper we consider the case that wavenumber k is small. In this case we obtain the representation, uniqueness, and existence of the direct scattering problem, and the uniqueness of inverse scattering problem. Numerical examples are also presented.

1.2 ACOUSTIC MODEL IN A TWO-LAYERED WAVEGUIDE AND THE GREEN'S FUNCTION

Consider a 2-D acoustic model in a two-layered waveguide:

$$M_1 = \{(x, z) \mid 0 < z < d, -\infty < x < \infty\},$$

$$M_2 = \{(x, z) \mid d < z < h, -\infty < x < \infty\},$$

$$\Gamma = \{(x, z) \mid z = d, -\infty < x < \infty\},$$

$$R_2^h = \{(x, z) \mid 0 < z < h, -\infty < x < \infty\} = R \times (0, h).$$

Here d and h are constants, and $h > d > 0$. The acoustic field

$$u(\xi, z) = \begin{cases} u_1(\xi, z), & (\xi, z) \in M_1 \\ u_2(\xi, z), & (\xi, z) \in M_2 \end{cases}$$

satisfies

$$\Delta u_1 + k_1^2 u_1 = -\delta(x - x_s)\delta(z - z_s) \text{ in } \mathrm{M}_1 \tag{1.1}$$

$$\Delta u_2 + k_2^2 u_2 = 0 \text{ in } \mathrm{M}_2 \tag{1.2}$$

$$\rho_1 u_1 = \rho_2 u_2 \text{ on } \Gamma \tag{1.3}$$

$$\frac{\partial u_1}{\partial \nu} = \frac{\partial u_2}{\partial \nu} \text{ on } \Gamma \tag{1.4}$$

$$u_1(x, 0) = 0 \tag{1.5}$$

$$\frac{\partial u_2}{\partial z}(x, h) = 0 \tag{1.6}$$

where k_1, k_2 are known wavenumbers in M_1, M_2 respectively, ρ_1, ρ_2 are known densities in M_1, M_2 respectively, u_1, u_2 are the total acoustic fields in M_1, M_2, respectively, and the acoustic source at $(x_s, z_s) \in M_1$.

Following we construct the outgoing Green's function, i.e., the solution of the equations (2.1)-(2.6) takes the form

$$u(x, z) = \sum_{n=1}^{\infty} u_n(z) e^{ik\xi_n |x|}, \ \textit{for large } |x|, \tag{1.7}$$

where $u_n(x)$ and ξ_n will be specified later. We call the above condition the outgoing radiation condition.

We will denote the outgoing Green's function by $G(x, z, x_s, z_s)$. Note that $G(x, z, x_s; z_s) = G(|x - x_s|, z, z_s)$, which can be represented by (setting $x_s = 0$ for simplicity)

$$G(x, z; 0, z_s) = \frac{1}{2\pi} \int_{-\infty}^{\infty} \widehat{G}(\xi, z; 0, z_s) e^{i|x|\xi} d\xi, \tag{1.8}$$

where

$$\widehat{G}(\xi, z; 0, z_s) = \begin{cases} \widehat{G}_1(\xi, z; 0, z_s), & 0 \leq z < d \\ \widehat{G}_2(\xi, z; 0, z_s), & d < z < h \end{cases} \tag{1.9}$$

and $\widehat{G}_1$, $\widehat{G}_2$ satisfy

$$\frac{\partial^2}{\partial z^2} \widehat{G}_1 + \tau_1^2 \widehat{G}_1 = -\delta(z - z_s), \ 0 < z < d, \tag{1.10}$$

$$\frac{\partial^2}{\partial z^2}\widehat{G}_2 + \tau_2^2\widehat{G}_2 = 0,\ \ d < z < h, \tag{1.11}$$

$$\widehat{G}_1\left(\xi, 0\right) = 0, \tag{1.12}$$

$$\rho_1\widehat{G}_1\left(\xi, d\right) = \rho_2\widehat{G}_2\left(\xi, d\right), \tag{1.13}$$

$$\frac{\partial \widehat{G}_1}{\partial z}\left(\xi, d\right) = \frac{\partial \widehat{G}_2}{\partial z}\left(\xi, d\right), \tag{1.14}$$

$$\frac{\partial \widehat{G}_2}{\partial z}\left(\xi, h\right) = 0. \tag{1.15}$$

Here we use the notation $\tau_1 = \sqrt{k_1^2 - \xi^2}$, and $\tau_2 = \sqrt{k_2^2 - \xi^2}$.

We construct $\widehat{G}\left(\xi, z; 0, z_s\right)$ in the following way: first we construct two linearly independent solutions of the homogeneous equations corresponding to equations (2.10), (2.11) that both satisfy (2.13) and (2.14), and that one solution satisfies (2.12), and another satisfies (2.15), respectively. We shall denote these solutions by $\phi_1\left(\xi, z\right)$ and $\phi_2\left(\xi, z\right)$, respectively. Then

$$\widehat{G}\left(\xi, z; 0, z_s\right) = -\frac{\phi_1\left(\xi, z_<\right)\phi_2\left(\xi, z_>\right)}{W\left(\phi_1, \phi_2\right)\left(\xi, z_s\right)},$$

where

$$z_< = \begin{cases} z, & \text{if } 0 < z < z_s \\ z_s, & \text{if } z_s < z < d \end{cases},$$

$$z_> = \begin{cases} z_s, & \text{if } 0 < z < z_s \\ z, & \text{if } z_s < z < d \end{cases},$$

and

$$W\left(\phi_1, \phi_2\right) = \det\begin{vmatrix} \phi_1' & \phi_2' \\ \phi_1 & \phi_2 \end{vmatrix}.$$

$\phi_1\left(\xi, z\right)$ and $\phi_2\left(\xi, z\right)$ can be obtained explicitly:

$$\phi_1\left(\xi, z\right) = \begin{cases} \sin\left(\tau_1 z\right), & 0 \le z < d \\ A_1\cos\left(\tau_2 z\right) + B_1\sin\left(\tau_2 z\right), & d < z < h \end{cases} \tag{1.16}$$

where

$$\begin{pmatrix} A_1 \\ B_1 \end{pmatrix} = \begin{pmatrix} \cos\left(\tau_2 d\right) & -\sin\left(\tau_2 d\right) \\ \sin\left(\tau_2 d\right) & \cos\left(\tau_2 d\right) \end{pmatrix}\begin{pmatrix} \frac{\rho_1}{\rho_2}\sin\left(\tau_1 d\right) \\ \frac{\tau_1}{\tau_2}\cos\left(\tau_1 d\right) \end{pmatrix},$$

$$= \begin{pmatrix} \frac{\rho_1}{\rho_2}\cos\left(\tau_2 d\right)\sin\left(\tau_1 d\right) - \frac{\tau_1}{\tau_2}\cos\left(\tau_1 d\right)\sin\left(\tau_2 d\right) \\ \frac{\rho_1}{\rho_2}\sin\left(\tau_2 d\right)\sin\left(\tau_1 d\right) + \frac{\tau_1}{\tau_2}\cos\left(\tau_1 d\right)\cos\left(\tau_2 d\right) \end{pmatrix}.$$

$$\phi_2(\xi,z) = \begin{cases} A_2 \cos(\tau_1 z) + B_2 \sin(\tau_1 z), & 0 \le z < d \\ \cos(\tau_2 (h - z)), & d < z < h \end{cases} \tag{1.17}$$

where

$$\begin{pmatrix} A_2 \\ B_2 \end{pmatrix} = \begin{pmatrix} \frac{\rho_2}{\rho_1} \cos(\tau_1 d) \cos(\tau_2 (h-d)) - \frac{\tau_2}{\tau_1} \sin(\tau_1 d) \sin(\tau_2 (h-d)) \\ \frac{\rho_2}{\rho_1} \sin(\tau_2 d) \cos(\tau_2 (h-d)) + \frac{\tau_2}{\tau_1} \cos(\tau_1 d) \sin(\tau_2 (h-d)) \end{pmatrix}$$

$W(\phi_1, \phi_2)$ does not depend on z_s. We shall therefore denote it by $W(\phi_1, \phi_2)(\xi)$.

$$\begin{aligned} W(\phi_1, \phi_2)(\xi) &= \phi_1'(\xi, 0)\, \phi_2(\xi, 0) - \phi_1(\xi, 0)\, \phi_2'(\xi, 0) \\ &= \phi_1'(\xi, 0)\, \phi_2(\xi, 0) \\ &= \tau_1 \cos(\tau_1 0)\left(A_2 \cos(\tau_1 0) + B_2 \sin(\tau_1 0)\right) = \tau_1 A_2 \\ &= \tau_1 \left[\frac{\rho_2}{\rho_1} \cos(\tau_1 d) \cos(\tau_2 (h-d)) - \frac{\tau_2}{\tau_1} \sin(\tau_1 d) \sin(\tau_2 (h-d))\right]. \end{aligned}$$

The Green's function has representation

$$G(|x|, z, z_s) = -\frac{1}{2\pi} \int_{-\infty}^{\infty} \frac{\phi_1(\xi, z_<)\, \phi_2(\xi, z_>)}{W(\phi_1, \phi_2)} e^{i|x|\xi}\, d\xi. \tag{1.18}$$

The integral (2.18) is considered as a contour integral which has poles at the zeros of $W(\phi_1, \phi_2)(\xi)$. We denote the zeros by ξ_n. At the zeros, $\phi_1(\xi_n, z)$ and $\phi_2(\xi_n, z)$ are linearly dependent functions, i.e. there exists a constant c_n such that

$$\phi_2(\xi_n, z) = c_n \phi_1(\xi_n, z).$$

By the residue theorem,

$$\begin{aligned} G(|x|, z, z_s) &= -\frac{1}{2\pi} \int_C \frac{\phi_1(\xi, z_<)\, \phi_2(\xi, z_>)}{W(\phi_1, \phi_2)} e^{i|x|\xi}\, d\xi \\ &= -\frac{1}{2\pi} \cdot 2\pi i \sum_{n=1}^{\infty} \frac{\phi_1(\xi_n, z_<)\, \phi_2(\xi_n, z_>)}{\frac{\partial}{\partial \xi} W(\phi_1, \phi_2)|_{\xi=\xi_n}} e^{i|x|\xi_n}, \\ &= -i \sum_{n=1}^{\infty} \frac{\phi_1(\xi_n, z_<)\, c_n \phi_1(\xi_n, z_>)}{\frac{\partial}{\partial \xi} W(\phi_1, \phi_2)|_{\xi=\xi_n}} e^{i|x|\xi_n}, \\ &= -i \sum_{n=1}^{\infty} \frac{\Phi_n(z_<)\, \Phi_n(z_>)}{W_n(\phi_1, \phi_2)} e^{i|x|\xi_n}, \end{aligned} \tag{1.19}$$

where

$$\Phi_n(z) = \frac{\phi_1(\xi_n, z)}{||\phi_1(\xi_n, \cdot)||},$$

$$\|\phi_1(\xi_n,\cdot)\|^2 = (\phi_1(\xi_n,\cdot),\phi_1(\xi_n,\cdot)) := \int_0^h \phi_1^2(\xi_n,z)dz,$$

and

$$W_n(\phi_1,\phi_2) = \frac{1}{c_n\|\phi_1(\xi_n,\cdot)\|^2}\frac{\partial}{\partial\xi} W(\phi_1,\phi_2)|_{\xi=\xi_n}.$$

Figure 1(a) shows the propagating wave in a two-layered ocean from a point acoustic sourse located at (50, 30). For comparison Figure 1(2) shows the propagating wave in a homogeneous ocean from a point acoustic sourse located at (50, 30). We have used the following parameters.

$c_1 = 1500,\ c_2 = 900,\ \rho_1 = 1000,\ \rho_2 = 1000,\ f = 50,\ h = 100,\ d0 = 80,\ c2 = 1800.$

1.3 DIRECT SCATTERING PROBLEM

Now we consider the case that there exists a given inhomogeneity in the layer M_1. We assume that the inhomogeneity is contained in a bounded domain Ω with C^2 boundary having outward normal vector. The propagating solution

$$u(x,z) = \begin{cases} u_1(x,z), & \text{if (x,z)} \in \mathrm{M}_1 \setminus \overline{\Omega} \\ u_2(x,z), & \text{if (x,z)} \in \mathrm{M}_2 \\ u_3(x,z), & \text{if (x,z)} \in \Omega \end{cases}, \tag{1.1}$$

satisfying

$$\Delta u_1 + k_1^2 u_1 = -\delta(x-x_s)\delta(z-z_s) \text{ in } \mathrm{M}_1 \setminus \overline{\Omega} \tag{1.2}$$

$$\Delta u_3 + k_3^2(x,z)u_3 = 0 \text{ in } \Omega \tag{1.3}$$

$$\Delta u_2 + k_2^2 u_2 = 0 \text{ in } \mathrm{M}_2 \tag{1.4}$$

$$\rho_1 u_1 = \rho_2 u_2 \text{ on } \Gamma \tag{1.5}$$

$$\frac{\partial u_1}{\partial \nu} = \frac{\partial u_2}{\partial \nu} \text{ on } \Gamma \tag{1.6}$$

$$u_1(x,0) = 0 \tag{1.7}$$

$$\frac{\partial u_2}{\partial z}(x,h) = 0 \tag{1.8}$$

$$\rho_1 u_1 = \rho_3 u_3 \text{ on } \partial\Omega \tag{1.9}$$

$$\frac{\partial u_1}{\partial \nu} = \frac{\partial u_3}{\partial \nu} \text{ on } \partial\Omega \tag{1.10}$$

and u satisfies the outgoing radiation condition (2.7). In this paper we assume the $k_3 \in C(\Omega)$.

Figure 1.1 Acoustic wave from a point sourse

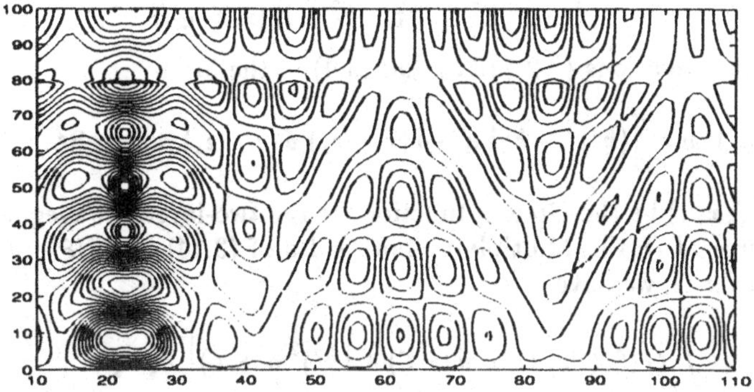

(a) In a two-layered ocean

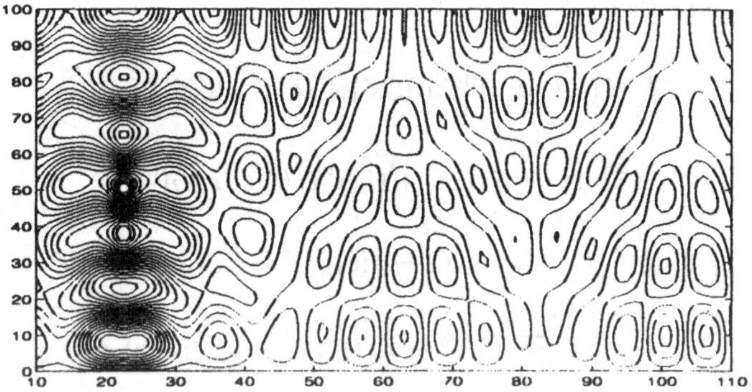

(b) In a homogeneous ocean

Denote

$$k^{o}(x,z)=\begin{cases} k_1, & \text{if } (x,z)\in M_1 \\ k_2, & \text{if } (x,z)\in M_2 \end{cases}, \tag{1.11}$$

$$k(x,z)=\begin{cases} k_1, & \text{if } (x,z)\in M_1\setminus\Omega \\ k_2, & \text{if } (x,z)\in M_2 \\ k_3(x,z), & \text{if } (x,z)\in\Omega \end{cases}. \tag{1.12}$$

Therefore,

$$\widehat{k}^2 := (k^0)^2 - k^2 = \begin{cases} 0, & \text{if } (x,z) \notin \Omega \\ k_1^2 - k_3^2, & \text{if } (x,z) \in \Omega \end{cases} . \tag{1.13}$$

We rewrite the equations (3.2), (3.3) and (3.4) in the form

$$\Delta u + (k^o)^2 u = \left[(k^o)^2 - k^2\right] u - \delta(x - x_s)\,\delta(z - z_s) \text{ a.e. in } \mathrm{R}_h^2. \tag{1.14}$$

Let $G(\xi,\zeta;x,z)$ be the Green's function for the two-layered waveguide with acoustic source at (ξ,ζ). Multiplying both sides of (3.14) by $G(\xi,\zeta;x,z)$ and integrating over Ω, we have

$$\int_\Omega G(\xi,\zeta;x,z)\left[\Delta u(\xi,\zeta) + (k^o)^2 u(\xi,\zeta)\right] d\xi\, d\zeta$$

$$= \int_\Omega G(\xi,\zeta;x,z)\left[\widehat{k}^2 u(\xi,\zeta) - \delta(\xi - x_s)\,\delta(\zeta - z_s)\right] d\xi\, d\zeta. \tag{1.15}$$

If $(x,z) \in R_h^2 \setminus \overline{\Omega}$, then

$$\int_\Omega G(\xi,\zeta;x,z)\left[\Delta u(\xi,\zeta) + (k^o)^2 u(\xi,\zeta)\right]$$

$$-u(\xi,\zeta)\left[\Delta G(\xi,\zeta;x,z) + (k^o)^2 G(\xi,\zeta;x,z)\right] d\xi\, d\zeta$$

$$= \int_\Omega G(\xi,\zeta;x,z)\,\Delta u(\xi,\zeta) - u(\xi,\zeta)\,\Delta G(\xi,\zeta;x,z)\, d\xi\, d\zeta$$

$$= \int_{\partial\Omega} G(\xi,\zeta;x,z)\frac{\partial u_-}{\partial n}(\xi,\zeta) - u_-(\xi,\zeta)\frac{\partial G}{\partial n}(\xi,\zeta;x,z)\, ds \tag{1.16}$$

where $\frac{\partial u_-}{\partial n}(\xi,\zeta)$ and $u_-(\xi,\zeta)$ are the limits of $\frac{\partial u}{\partial n}(\xi,\zeta)$ and $u(\xi,\zeta)$ as (ξ,ζ) approaches $\partial\Omega$ from the interior of Ω. Similarly we use $\frac{\partial u_+}{\partial n}(\xi,\zeta)$ and $u_+(\xi,\zeta)$ to denote the limits of $\frac{\partial u}{\partial n}(\xi,\zeta)$ and $u(\xi,\zeta)$ as (ξ,ζ) approaches $\partial\Omega$ from the exterior of Ω. For $(x,z) \notin \partial\Omega$, $G(\xi,\zeta;x,z)$ and $\frac{\partial G}{\partial n}(\xi,\zeta;x,z)$ are continuous across $\partial\Omega$. Since u satisfies $\Delta u + (k^o)^2 u = 0$ for $(x,z) \in R_h^2 \setminus \overline{\Omega}$, we have by Green's formula

$$u(x,z) = \int_{\partial\Omega}\left[u_+(\xi,\zeta)\frac{\partial G}{\partial n}(\xi,\zeta;x,z) - \frac{\partial u_+}{\partial n}(\xi,\zeta)\,G(\xi,\zeta;x,z)\right] d\xi\, d\zeta \tag{1.17}$$

From (3.15), (3.16), (3.17) and by (3.9), (3.10), we obtain the representation

$$u(x,z) = \int_{\partial\Omega}\left[u_+\frac{\partial G}{\partial n} - \frac{\partial u_+}{\partial n}G\right] ds$$

$$= \int_{\partial\Omega} \left[u_- \frac{\partial G}{\partial n} - \frac{\partial u_-}{\partial n} G \right] ds + \int_{\partial\Omega} (u_+ - u_-) \frac{\partial G}{\partial n} ds$$

$$= \int_{\Omega} (u\Delta G - G\Delta u)\, d\xi\, d\zeta + \int_{\partial\Omega} (u_+ - u_-) \frac{\partial G}{\partial n} ds$$

$$= \int_{\Omega} \left(u \left[\Delta G + (k^o)^2 G \right] - G \left[\Delta u + (k^o)^2 u \right] \right) d\xi\, d\zeta + \int_{\partial\Omega} (u_+ - u_-) \frac{\partial G}{\partial n} ds$$

$$= - \int_{\Omega} G \left[\Delta u + (k^o)^2 u \right] d\xi\, d\zeta + \int_{\partial\Omega} (u_+ - u_-) \frac{\partial G}{\partial n} ds$$

$$= - \int_{\Omega} G(\xi, \zeta; x, z) \widehat{k}^2 u(\xi, \zeta)\, d\xi\, d\zeta + G(x_s, z_s; x, z) + \int_{\partial\Omega} (u_+ - u_-) \frac{\partial G}{\partial n} ds,$$

$$(x, z) \in R_h^2 \setminus \overline{\Omega}. \tag{1.18}$$

If $(x, z) \in \Omega$, then, using the notation

$$D_\epsilon = \left\{ (\xi, \zeta) \mid (\xi - x)^2 - (\zeta - z)^2 \leq \epsilon \right\},$$

and ∂D_ϵ^- for the boundary of D_ϵ with a clockwise orientation, we have

$$\int_{\Omega \setminus D_\epsilon} G(\xi, \zeta; x, z) \left[\Delta u(\xi, \zeta) + (k^o)^2 u(\xi, \zeta) \right]$$

$$-u(\xi, \zeta) \left[\Delta G(\xi, \zeta; x, z) + (k^o)^2 G(\xi, \zeta; x, z) \right] d\xi\, d\zeta$$

$$= \int_{\partial\Omega} G(\xi, \zeta; x, z) \frac{\partial u_-}{\partial n}(\xi, \zeta) - u_-(\xi, \zeta) \frac{\partial G}{\partial n}(\xi, \zeta; x, z)\, ds$$

$$+ \int_{\partial D_\epsilon^-} G(\xi, \zeta; x, z) \frac{\partial u}{\partial n}(\xi, \zeta) - u(\xi, \zeta) \frac{\partial G}{\partial n}(\xi, \zeta; x, z)\, ds. \tag{1.19}$$

Since both $u(\xi, \zeta)$ and $G(\xi, \zeta; x, z)$ satisfy $\Delta u + (k^o)^2 u = 0$ on $R_h^2 \setminus \overline{\Omega}$, when $(x, z) \in \Omega$, we know

$$\int_{\partial\Omega} G(\xi, \zeta; x, z) \frac{\partial u_+}{\partial n}(\xi, \zeta) - u_+(\xi, \zeta) \frac{\partial G}{\partial n}(\xi, \zeta; x, z)\, ds = 0. \tag{1.20}$$

Moreover, writing $r = \sqrt{(\xi - x)^2 + (\zeta - z)^2}$, $G(\xi, \zeta; x, z)$ has a singular approximation

$$G(\xi, \zeta; x, z) = -\frac{1}{2\pi} \log r + O(1), \text{for r} \to 0. \tag{1.21}$$

Hence

$$\lim_{\epsilon\to 0}\int_{\partial D_\epsilon^-} G(\xi,\zeta;x,z)\frac{\partial u}{\partial n}(\xi,\zeta)-u(\xi,\zeta)\frac{\partial G}{\partial n}(\xi,\zeta;x,z)\,ds$$

$$=\lim_{\epsilon\to 0}-\int_0^{2\pi} u(x+\epsilon\cos\theta,z+\epsilon\sin\theta)\left[\frac{1}{2\pi\epsilon}+O(1)\right]\epsilon\,d\theta$$

$$=-u(x,z) \tag{1.22}$$

Now from (3.15), (3.19), (3.20),(3.22) and by (3.9), (3.10), letting $\epsilon\to 0$, we obtain

$$-u(x,z)+\int_{\partial\Omega}\left[G\frac{\partial u_-}{\partial n}-u_-\frac{\partial G}{\partial n}\right]ds$$

$$=-u(x,z)+\int_{\partial\Omega}\left[G\frac{\partial u_+}{\partial n}-u_+\frac{\partial G}{\partial n}\right]ds+\int_{\partial\Omega}(u_+-u_-)\frac{\partial G}{\partial n}ds$$

$$=-u(x,z)+\int_{\partial\Omega}(u_+-u_-)\frac{\partial G}{\partial n}ds$$

$$=\lim_{\epsilon\to 0}\int_{\Omega\setminus D_\epsilon} G\left[\Delta u+(k^o)^2u\right]-u\left[\Delta G+(k^o)^2G\right]d\xi\,d\zeta$$

$$=\int_{\Omega\setminus D_\epsilon} G\left[\Delta u+(k^o)^2u\right]d\xi\,d\zeta$$

$$=\int_\Omega G(\xi,\zeta;x,z)\widehat{k}^2u(\xi,\zeta)\,d\xi\,d\zeta-G(x_s,z_s;x,z).$$

Hence

$$u(x,z)=-\int_\Omega G(\xi,\zeta;x,z)\widehat{k}^2u(\xi,\zeta)\,d\xi\,d\zeta$$

$$+G(x_sz_s;x,z)+\int_{\partial\Omega}(u_+-u_-)\frac{\partial G}{\partial n}ds,\ (x,z)\in\Omega. \tag{1.23}$$

Combining (3.18) and (3.22), we get the representation of the wave field in R_h^2.

Now we deduce the integral equations that determine $u(x,z)$. Let

$$\phi(\xi,\zeta)=u_+(\xi,\zeta)-u_-(\xi,\zeta). \tag{1.24}$$

Equation (3.22) becomes

$$u(x,z)+\int_\Omega G(\xi,\zeta;x,z)\widehat{k}^2u(\xi,\zeta)\,d\xi\,d\zeta-\int_{\partial\Omega}\phi(\xi,\zeta)\frac{\partial G}{\partial n}(\xi,\zeta;x,z)\,ds$$

$$=G(x_s,z_s;x,z),\ (x,z)\in\Omega. \tag{1.25}$$

Using (3.9) and from (3.18), (3.22), we have for $(x,z) \in \partial\Omega$

$$0 = \rho_1 u_+ (x,z) - \rho_3 u_- (x,z)$$

$$= \rho_1 \left[\frac{1}{2}\phi(x,z) + \int_{\partial\Omega} \phi(\xi,\zeta) \frac{\partial G}{\partial n}(\xi,\zeta;x,z)\, ds\right]$$

$$-\rho_3 \left[-\frac{1}{2}\phi(x,z) + \int_{\partial\Omega} \phi(\xi,\zeta) \frac{\partial G}{\partial n}(\xi,\zeta;x,z)\, ds\right]$$

$$+(\rho_1 - \rho_3)\left[G(x_s,z_s;x,z) - \int_{\Omega} G(\xi,\zeta;x,z)\widehat{k}^2 u(\xi,\zeta)\, d\xi\, d\zeta\right],$$

so

$$\phi(x,z) + \frac{2(\rho_1-\rho_3)}{\rho_1+\rho_3}\int_{\partial\Omega} \phi(\xi,\zeta)\frac{\partial G}{\partial n}(\xi,\zeta;x,z)\, ds$$

$$-\frac{2(\rho_1-\rho_3)}{\rho_1+\rho_3}\int_{\Omega} G(\xi,\zeta;x,z)\widehat{k}^2 u(\xi,\zeta)\, d\xi\, d\zeta = -\frac{2(\rho_1-\rho_3)}{\rho_1+\rho_3} \mathrm{G}(\mathrm{x_s},\mathrm{z_s};\mathrm{x},\mathrm{z}),$$

$$(x,z) \in \partial\Omega. \tag{1.26}$$

Here we have used the facts that

$$G(x_s,z_s;x,z) - \int_{\Omega} G(\xi,\zeta;x,z)\widehat{k}^2 u(\xi,\zeta)\, d\xi\, d\zeta$$

is continuous across $\partial\Omega$, and

$$v(x,y) = \int_{\partial\Omega} \phi(\xi,\zeta)\frac{\partial G}{\partial n}(\xi,\zeta;x,z)\, ds$$

satisfies jump conditions

$$v^+(x,z) = -\frac{1}{2}\phi(x,z) + \int_{\partial\Omega} \phi(\xi,\zeta)\frac{\partial G}{\partial n}(\xi,\zeta;x,z)\, d\xi\, d\zeta,$$

$$v^-(x,z) = -\frac{1}{2}\phi(x,z) + \int_{\partial\Omega} \phi(\xi,\zeta)\frac{\partial G}{\partial n}(\xi,\zeta;x,z)\, d\xi\, d\zeta,$$

$$v^+(x,z) - v^-(x,z) = \phi(x,z),\ (x,z) \in \partial\Omega.$$

From the above analysis, and similar to the proof of Theorem 3.1 of [], we have the following theorem.

Theorem 3.1 If (u,ϕ) satisfies the direct scattering problem (3.1)-(3.9), then (u,ϕ) satisfies the integral equaitons (3.25) and (3.26).

Conversely, if $(u,\phi) \in C(\Omega) \times C(\partial\Omega)$ is a solution of the integral equaitons (3.25) and (3.26), then (u,ϕ) is a solution of the direct scattering problem. □

Theorem 3.2 If $k_0 := max\{\widehat{k}\}$ and $|\rho_1 - \rho_3|$ are small enough, then the system of integral equations (3.25) and (3.26) has a unique solution.

Proof: Denote (3.25) in the form

$$u + \mathbf{T}u = f, \tag{1.27}$$

where

$$\mathbf{T}u(x,z) = \int_\Omega G(\xi,\zeta;x,z)\widehat{k}^2 u(\xi,\zeta)\, d\xi\, d\zeta$$

and

$$f(x,z) = \int_{\partial\Omega} \phi(\xi,\zeta)\frac{\partial G}{\partial n}(\xi,\zeta;x,z)\, ds + G(x_s,z_s;x,z).$$

If k_0 is small enough, then the operator $\mathbf{I}+\mathbf{T}$ has bounded inverse operator $(\mathbf{I}+\mathbf{T})^{-1}$, where $\mathbf{I}$ denotes the identical operator in $L^2(\Omega)$. Substituting $u = (\mathbf{I}+\mathbf{T})^{-1}f$ into (3.26), we obtain

$$\phi(x,z) + \frac{2(\rho_1-\rho_3)}{\rho_1+\rho_3}\mathbf{S}\phi(x,z) - \frac{2(\rho_1-\rho_3)}{\rho_1+\rho_3}\mathbf{T}\circ(\mathbf{I}+\mathbf{T})^{-1}f(x,z)$$

$$= -\frac{2(\rho_1-\rho_3)}{\rho_1+\rho_3}G(x_s,z_s;x,z), \tag{1.28}$$

where

$$\mathbf{S}\phi(x,z) = \int_{\partial\Omega} \phi(\xi,\zeta)\frac{\partial G}{\partial n}(\xi,\zeta;x,z)\, ds.$$

The operators $\mathbf{S}$ and $\mathbf{T}\circ(\mathbf{I}+\mathbf{T})^{-1}$ are bounded in $L^2(\partial\Omega)$ and $L^2(\Omega)$, respectively. So do their composites. If $|\rho_1 - \rho_3|$ is small enough, (3.28) has a unique solution $\phi \in L^2(\partial\Omega)$. Substituting the ϕ into (3.27), we determine uniquely u.
□

For given Ω, $\widehat{k}^2 = (k^o)^2 - k^2$, ρ_1, ρ_2, ρ_3 and source location $(x_s,\ z_s)$, we can determine $u(x,z)$ on Ω and $\phi = u^+ - u^-$ on $\partial\Omega$ by integral equations (3.25) and (3.26). Hence we can compute the wave field by (3.18) and (3.22).

In case that $\rho_1 = \rho_3$ (3.24) becomes

$$\phi(x,z) = 0,\ (x,z) \in \partial\Omega. \tag{1.29}$$

The system of integral equations reduces to a single integral equation

$$u(x,z) + \int_\Omega G(\xi,\zeta;x,z)\widehat{k}^2 u(\xi,\zeta)\, d\xi\, d\zeta = G(x_s,z_s;x,z). \tag{1.30}$$

Following we develop an iterative algorithm for integral equation (3.30). Under the assumptions the wave field can be determined in R_h^2 if we solve

$u(x,z) \equiv u(x,z,x_s,z_s)$ such that

$$u(x,z) = G(x_s,z_s;x,z) - \int_\Omega G(\xi,\zeta;x,z)\widehat{\mathrm{k}}^2 \mathrm{u}(\xi,\zeta;\mathrm{x_s},\mathrm{z_s})\,d\xi\,d\zeta \qquad (1.31)$$

We use the following algorithm: Let

$$u(\xi,\zeta;x_s,z_s) = G(\xi,\zeta;x,z),\ (\xi,\zeta) \in \Omega \qquad (1.32)$$

and for $n = 1,2,3,\cdots$, and $(x,z) \in \Omega$, let

$$u_{n+1}(x,z) = G(x_s,z_s;x,z) - \int_\Omega G(\xi,\zeta;x,z)\widehat{\mathrm{k}}^2 \mathrm{u_n}(\xi,\zeta;\mathrm{x_s},\mathrm{z_s})\,d\xi\,d\zeta \qquad (1.33)$$

This algorithm converges if $\int_\Omega \widehat{k}^2 d\xi\, d\zeta := \widehat{k}^2 |\Omega|$ is small. The integral operator

$$T_\Omega u_n(x,z;x_s,z_s) = \int_\Omega G(\xi,\zeta;x,z)\widehat{\mathrm{k}}^2 \mathrm{u_n}(\xi,\zeta;\mathrm{x_s},\mathrm{z_s})\,d\xi\,d\zeta,\ (\mathrm{x},\mathrm{z}) \in \Omega \qquad (1.34)$$

is a singular integral. The integral kernel

$$G(\xi,\zeta;x,z) = -\frac{1}{2\pi}\log\sqrt{(x-\xi)^2+(z-\zeta)^2} + O(1) \text{ as } (\mathrm{x},\mathrm{z}) \to (\xi,\zeta) \qquad (1.35)$$

We split $T_\Omega u_n(x,z;x_s,z_s)$ into two integrals

$$T_\Omega u_n(x,z;x_s,z_s) = \int_{D_\epsilon} G(\xi,\zeta;x,z)\widehat{\mathrm{k}}^2 \mathrm{u_n}(\xi,\zeta;\mathrm{x_s},\mathrm{z_s})\,d\xi\,d\zeta \qquad (1.36)$$

$$+\int_{\Omega\setminus D_\epsilon} G(\xi,\zeta;x,z)\widehat{\mathrm{k}}^2 \mathrm{u_n}(\xi,\zeta;\mathrm{x_s},\mathrm{z_s})\,d\xi\,d\zeta \qquad (1.37)$$

where

$$\mathrm{D}_\epsilon = \left\{(\xi,\zeta) \mid (\mathrm{x}-\xi)^2 + (\mathrm{z}-\zeta)^2 \le \epsilon^2\right\}. \qquad (1.38)$$

For small $\epsilon > 0$,

$$\int_{D_\epsilon} G(\xi,\zeta;x,z)\widehat{\mathrm{k}}^2 \mathrm{u_n}(\xi,\zeta;\mathrm{x_s},\mathrm{z_s})\,d\xi\,d\zeta$$

$$= -\widehat{k}^2 \int_{D_\epsilon} \frac{\log(\rho)}{2\pi} u_n(x,z;x_s,z_s)\,\rho d\rho\, d\theta + \mathrm{O}(\epsilon^2)$$

$$-\widehat{k}^2 u_n(x,z;x_s,z_s)\left[-\frac{\rho^2}{2}\log(\rho)\Big|_{0+}^{\epsilon} + \int_0^\epsilon \rho\, d\rho\right] + O(\epsilon^2) \qquad (1.39)$$

$$-\frac{\widehat{k}^2}{2}\epsilon^2 \log(\epsilon)\ u_n(x,z;x_s,z_s) + O(\epsilon^2). \qquad (1.40)$$

The implementation of the above algorithm has been carried out on a PC using Matlab.

1.4 INVERSE SCATTERING PROBLEM: A NUMERICAL EXPERIMENT

The inverse problem considered is as follows: let Γ_d be a subset of $\Gamma_1 := \{(x, z_1) | -\infty < x < \infty,\ z_1 = constant\}$, and Γ_s be a subset of $\Gamma_2 := \{(x, z_2) | -\infty < x < \infty,\ z_2 = constant\}$. Given $u(x, z; x_s, z_s)$ for $(x, z; x_s, z_s) \in \Gamma_d \times \Gamma_s$, determine the inhomogeneity $k_3(x, z)$.

Here we assume that Γ_1 and Γ_2 are strictly above the inhomogeneity Ω; i.e., $\max_{(x,z)\in\Omega} \{z\} < \min\{z_1, z_2\}$..

In the three dimensional waveguide case, we can prove that the inverse scattering problem for given data on two planes has a unique solution. (See Appendix A). However, the uniqueness is still an open problem in the two dimensional case, even if we take $\Gamma_d = \Gamma_1$ and $\Gamma_s = \Gamma_2$. Therefore, we consider this section as a numerical experiment for the inverse problem.

In the following we reformulate the inverse problem as an overdetermined linear system, and use a nonlinear optimization scheme to solve the regularized nonlinear least squares problem. We consider only the case that $\rho_1 = \rho_3$ in this paper. Moreover, we assume that $\widehat{k}^2 \in C(M_1)$.

From (3.18) we can represent the acoustic field detected on Γ_d with sourses on Γ_s as

$$u(x, z; x_s, z_s) = -\mathbf{F}(\widehat{k}^2 u)(x, z; x_s, z_s) + G(x_s, z_s; x, z),\ (x, z) \in \Gamma_d,$$
$$(x_s, z_s) \in \Gamma_s. \tag{1.1}$$

where

$$\mathbf{F}(\widehat{k}^2 u)(x, z; x_s, z_s) := \int_\Omega G(\xi, \zeta; x, z)\, \widehat{k}^2 u(\xi, \zeta; x_s, z_s)\, d\xi\, d\zeta.$$

$u(\xi, \zeta; x_s, z_s)$, $(\xi, \zeta) \in \Omega$ satiisfies (3.27) with $\phi = 0$; i.e.,

$$u + \mathbf{T}u = G, \tag{1.2}$$

where

$$\mathbf{T}(\widehat{k}^2 u)(x, z) = \int_\Omega G(\xi, \zeta; x, z)\, \widehat{k}^2 u(\xi, \zeta)\, d\xi\, d\zeta$$

and

$$G = G(x_s, z_s; x, z).$$

For given measured data $u^* = u^*(x, z; x_s, z_s)$, $(x, z; x_s, z_s) \in \Gamma_d \times \Gamma_s$, we formulate the determination of inhomogeneity as the minimization problem: find $\widehat{k}^2 = k_3^2(x, z)$, $(x, z) \in \Omega$ that minimize $\mathcal{J}_\epsilon(\widehat{k}^2)$ for some suitable chosen ϵ, where

$$\mathcal{J}_\epsilon(\widehat{k}^2) := \|u^* - \mathbf{F}(\widehat{k}^2(\mathbf{I} + \mathbf{T})^{-1}G) - G\|^2_{L^2(\Gamma_d)} + \epsilon \|\nabla \widehat{k}^2\|^2_{L^2(\Omega)}. \tag{1.3}$$

Since $max|\{\widehat{k^2}\}|$ is small, we have approximation

$$(\mathbf{I}+\mathbf{T})^{-1} \sim I - \mathbf{T} + \mathbf{T}^2 - \mathbf{T}^3,$$

which allows us to solve the inverse problem by minimizing

$$\mathcal{J}_\epsilon(\widehat{k^2}) := \|u^* - \mathbf{F}(\widehat{k^2}(I - \mathbf{T} + \mathbf{T}^2 - \mathbf{T}^3)G) - G\|^2_{L^2(\Gamma_d)} + \epsilon\|\nabla \widehat{k^2}\|^2_{L^2(\Omega)}. \quad (1.4)$$

The difference between the minimization of (2.32) and the minimization of (2.33) is that no integral equation needs to be solved in (2.33).

Following are some numerical examples of the inverse probelm.

In the examples we assume that

$$c_1 = 1500,\ \rho_1 = 1000,\ c_2 = 1800,\ \rho_2 = 1800,\ f = 75,\ h = 100,\ d_0 = 80,$$

$$(x_s, z_s) = (20, 30),\ k_2 = \frac{2\pi f}{c_1} = 0.314,\ k_2 = \frac{2\pi f}{c_2} = 0.262,$$

The distributed inhomogeneity is contained in a rectangle $\{(x,z)|50 < x < 70,\ 50 < z < 60\}$. The measured data are from $\Gamma_d = \{(10 + 0.5m, 30)|m = 0, 1, 2, ..., 200\}$. The input data for the inverse problem are obtained by solving the direct problem (2.27) accurately. (We use iteration (2.29) until $|u_{n+1} - u_n| < 10^{-12}$.) The minimization problem is solved by the nonlinear leastsquares algorithm from the Matlab Optimization Toolbox.

In Figure 1.2, the graphs in the right colomn are the originals, while those in the left are the reconstructions. The graphs in row 2 through row 4 are the z-sections for $z = 50, 52.5, 55$, respectively.

Figure 1.2 Identification of unknown objects

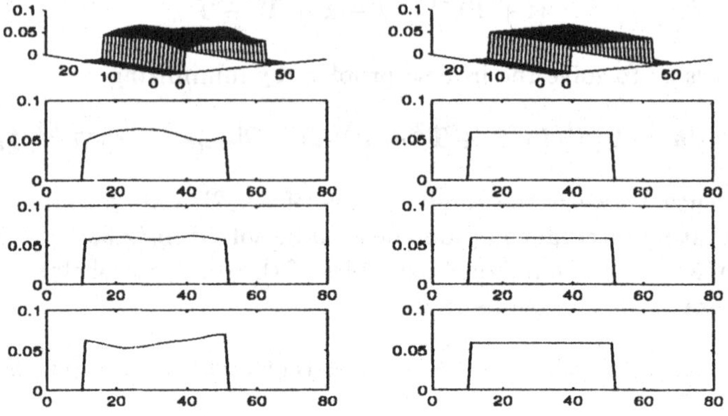

(a) Example 1

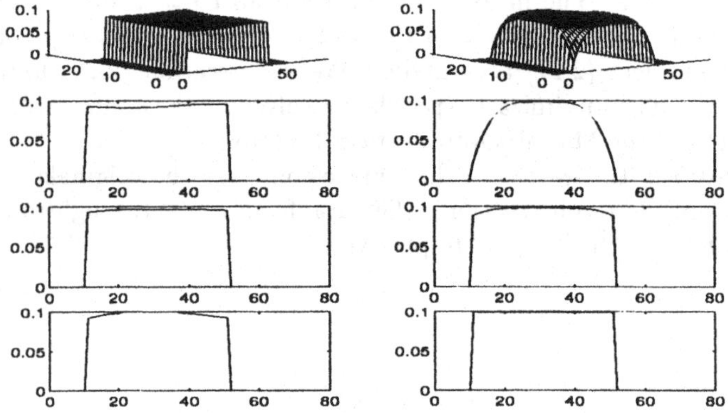

(b) Example 2

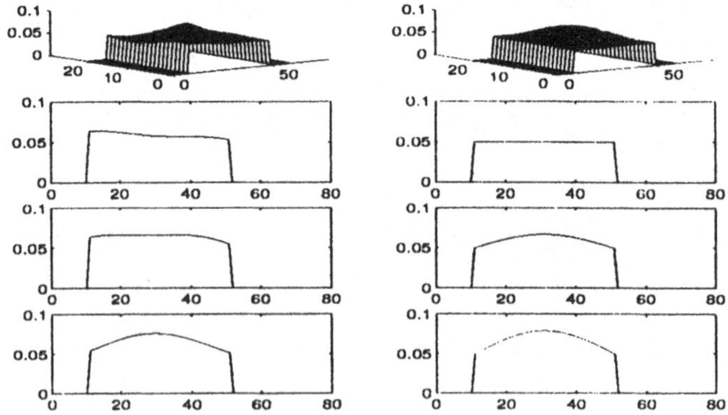

(c) Example 3

References

[1] Ahluwalia,D. and Keller, J.: *Exact and asymptotic representations of the sound field in a stratified ocean* in *Wave Propagation and Underwater Acoustics*, Lecture Notes in Physics 70, Springer, Berlin, 1977.

[2] T. S. Angell, R. E. Kleinman, C. Rozier and D. Lesselier, *Uniqueness and complete families for an acoustic waveguide problem*, submitted to Wave Motion (1996).

[3] P. Carrion and G. Boehm, *Tomographic imaging of opaque and low-contrast objects in range-independent waveguides*, J. Acoust. Soc. Am., **91**, 1440-1446, (1992).

[4] Colton, David and Kress, Rainer: *Inverse Acoustic and Electromagnetic Scattering Theory*, Springer-Verlag, (1993).

[5] Gilbert, R.P. and Xu, Y.: *Generalized Herglotz functions and inverse scattering problems in finite depth oceans* in *Inverse Problems*, SIAM, (1992).

[6] Gilbert, R.P., Xu, Y.: *An inverse problem for harmonic acoustics in stratified oceans*, J. Math. Anal. Appl. 17(1) (1993), 121-137.

[7] R. P. Gilbert and Y. Xu, *The seamount problem*, in SIAM special issue on the occasion of Prof. I. Stakgold's 70th birthday, Nonlinear Problems in Applied Mathematics, T. Angell et al. Eds., SIAM, Philadelphia, pp. 140-149, (1996).

[8] Gilbert, R.P. and Zhongyan Lin: *On the conditions for uniqueness and existence of the solution to an acoustic inverse problem: I Theory*, J. computational Acoustics, Vol.1, No.2(1993) 229-247

[9] Gilbert, R.P. and Zhongyan Lin: *An acoustic inverse problem: Numerical experiments*, J. Computational Acoustics,(to appear 1995).

[10] F. B. Jensen, W. A. Kuperman, M. B. Porter and H. Schmidt, COMPUTATIONAL OCEAN ACOUSTICS, AIP, New York, (1994).

[11] D. Lesselier and B. Duchene, *Wavefield inversion of objects in stratified environments. From backpropagation schemes to full solutions*, in REVIEW OF RADIO SCIENCE 1993–1995, R. Stone, ed., Oxford U. Press, Oxford, pp. 235–268, 1996.

[12] N. C. Makris, F. Ingenito and W. A. Kuperman, *Detection of a submerged object insonified by surface noise in an ocean waveguide*, J. Acoust. Soc. Am., **96**, 1703-1724, (1994).

[13] C. Rozier, D. Lesselier and T. Angell, *Optimal shape reconstruction of a perfect target in shallow water*, in PROC. 3RD EUROP. CONF. UNDERWATER ACOUSTICS, Heraklion (1996).

[14] D. Rozier, D. Lesselier, T. Angell and R. Kleinman, *Reconstruction of an impenetrable obstacle immersed in a shallow water acoustic waveguide*, in PROC.CONF. PROBLEMES INVERSES PROPAGATION ET DIFFRACTION D'ONDES, Aix-les-Bains (1996).

[15] C. Rozier, D. Lesselier, T. Angell and R. Kleinman, *Shape retrieval of an obstacle immersed in shallow water from single-frequency farfields using a complete family method*, submitted to Inverse Probs. (1996).

[16] Xu, Y.: *Direct and Inverse Scattering in Shallow Oceans*, Ph.D. Thesis, University of Delaware, 1990.

[17] Y. Xu and Y. Yan, *Boundary integral equation method for source localization with a continuous wave sonar*, J. Acoust. Soc. Am. **92**, 995-1002, (1992).

[18] Y. Xu, T. C. Poling and T. Brundage, *Direct and inverse scattering of harmonic acoustic waves in an inhomogeneous shallow ocean*, in Computational Acoustics, D. Lee et al. Eds., IMACS 91, vol 2., North-Holland, Amsterdam, pp. 21-43, (1993).

DIFFERENTIABILITY WITH RESPECT TO PARAMETERS OF INTEGRATED SEMIGROUPS

Min He

Department of Mathematics and Computer Science
Kent State University-Trumbull Campus
Warren, OH 44483

Abstract: This paper is concerned with a family of non-densely defined operators $A(\varepsilon)$ and studies differentiability with respect to a multi-parameter ε of integrated semigroup $S(t,\varepsilon)$ generated by $A(\varepsilon)$. The obtained results are conveniently employed for some hyperbolic and parabolic partial differential equation.

1.1 INTRODUCTION

The operators of abstract Cauchy problems often generate integrated semigroups when the Cauchy problems are formulated from hyperbolic or parabolic initial and boundary value problems which are posed on spaces of continuous functions. When considering solutions of the abstract Cauchy problems, one of the natural properties is differentiability with respect to parameters of integrated semigroups. So far as we know, there is one result [1] on differentiability with respect to parameters of integrated semigroups. This result is applied to the case when the domains of generators of integrated semigroups are independent of parameters.

R.P. Gilbert et al.(eds.), Direct and Inverse Problems of Mathematical Physics, 125–135.

Consider the hyperbolic initial and boundary value problem:

$$\begin{aligned} u_{tt} &= u_{xx}, \\ u(x,0) &= u_0(x), \quad u_t(x,0) = u_1(x) \quad \text{for} \quad x \in [0,1], \\ \alpha u_t(0,t) &- u_x(0,t) = 0 \\ u_t(1,t) &+ \beta u_x(1,t) = 0, \alpha, \ \beta \geq 0. \end{aligned} \tag{1.1}$$

The reformulated abstract Cauchy problem obtained by taking $v = u_t$, $w = u_x$, and $z = (v, w)^t$ and is given by:

$$\frac{dz(t)}{dt} = A(\varepsilon)z(t) \tag{1.2}$$

$$\text{on} \quad X = \prod_{i=1}^{2}(C[0,1], \|\cdot\|_\infty)$$

$$z(0) = z_0$$

where

$$\begin{aligned} A(\varepsilon) &= \begin{pmatrix} 0 & \partial x \\ \partial x & 0 \end{pmatrix}, \quad \varepsilon = (\alpha, \beta) \in R_+^2, \\ D(A(\varepsilon)) &= \left\{ \begin{pmatrix} v \\ w \end{pmatrix} \in C^1[0,1] \times C^1[0,1] \middle| \begin{array}{l} \alpha v(0) = w(0), \\ v(1) = -\beta w(1), \end{array} \quad \alpha, \beta \geq 0 \right\} \end{aligned}$$

Note that the domain of $A(\varepsilon)$ is dependent on ε. This example motivates our study on differentiability with respect to parameters of integrated semigroups concerning the case when the domains of generators are dependent on parameters.

After providing some needed background material in Section 2, we proceed to prove the main result, Theorem 3.4 in Section 3. This result gives convenient conditions for determining differentiability with respect to parameters of integrated semigroups. In the last section, we shall illustrate the obtained results in application to Equation (1.1).

1.2 PRELIMINARIES

We begin with giving some basic terminology, definitions, and results that will be needed in the sequel.

Definition 1.2.1 (2) $\{S(t)\}_{t\geq 0} \subset \mathcal{B}(X)$ *is called an integrated semigroup iff*

(a) $S(0) = 0$,

(b) $S(t)S(\tau)x = \int_0^t (S(\tau + r) - S(r))x \, dr$, *for every* $x \in X$,

(c) *For any* $x \in X$, $S(t)x : [0,\infty)$ *is continuous.*

Definition 1.2.2 (2) *An integrated semigroup* S *is called non-degenerate if* $S(t)x = 0$ *for all* $t \geq 0$ *implies that* $x = 0$.

Definition 1.2.3 (2) *The generator* A *of a non-degenerate integrated semigroup* S *is defined by: Let* $x, y \in X$. *Then* $x \in D(A)$ *and* $Ax = y$ *iff*

$$S(\cdot)x \in C^1([0,\infty), X) \text{ and } S'(t)x - x = S(t)y, \quad t \geq 0.$$

Definition 1.2.4 (2) *An integrated semigroup* S *is said to be of type* (M, ω) *where* $M \geq 1$, *iff for* $t, \tau \geq 0$,

$$\|S(t+r) - S(t)\| \leq M \int_t^{t+r} e^{\omega s}\, ds.$$

Definition 1.2.5 *Let* X, Y *be Banach spaces and* Ω *be an open set in* X. *A mapping* $f : \Omega \to Y$ *is said to be (Frechét) differentiable at a point* x *in* Ω *if there exists* $Df \in \mathcal{B}(X, Y)$ *(the space of bounded linear operators taking* $X \to Y$*) such that for every* $h \in X$ *with* $x + h \in \Omega$,

$$\|f(x+h) - f(x) - Df(x)h\|_Y = o(\|h\|_X, x),$$

where $o(\|h\|_X, x)$ *satisfies* $o(\|h\|_X, x)/\|h\|_X \to 0$ *as* $\|h\|_X \to 0$. *The linear operator* $Df(x)$ *is called the derivative of* f *at* x *and* $Df(x)h$ *is the differential of* f *at* x.

In the sequel we use "A is a Hille-Yosida operator" to mean that there exist $M \geq 1$ and $\omega \in R$ such that $\lambda > \omega$ implies $\lambda \in \rho(A)$ (the resolvent set of A) and

$$\|(\lambda I - A)^{-n}\| \leq \frac{M}{(\lambda - \omega)^n} \qquad \text{for } \lambda > \omega \text{ and } n = 1, 2, \ldots.$$

Note that the boundedness of the resolvent $(\lambda I - A)^{-1}$ implies that A is closed.

Theorem 1.2.6 (3) *The following two statements are equivalent:*
(a) A *is the generator of a non-degenerate semigroup* S *of type* (M, ω).
(b) A *is a Hille-Yosida operator.*

Theorem 1.2.7 (2) *Define* $X_0 = \overline{D(A)}$. *Define the part* A_0 *of* A *as*

$$A_0 = A \text{ on } D(A_0) = \{u \in D(A) : Au \in X_0\}.$$

Assume that $(\lambda I - A)^{-1} \in \mathcal{B}(X)$ *for all* $\lambda > 0$ *large and that*

$$\limsup_{\lambda \to \infty} \lambda \|(\lambda I - A)^{-1}\| < +\infty.$$

Then $D(A_0)$ is dense in X_0, and if A_0 generates a C_0-semigroup on X_0, then A generates a non-degenerate integrated semigroup of type (M,ω) on X.

Theorem 1.2.8 (2) *If A is a Hille-Yosida operator, then the part A_0 of A in X_0 generates a C_0-semigroup on X_0 satisfying*

$$\|T(t)\| \le Me^{\omega t}.$$

Lemma 1.2.9 *Let $\{Z(\varepsilon)\}_{\varepsilon\in P} \subset \mathcal{B}(X,Y)$, and let C be a compact subset of X. Assume that*
(2.1) *$Z(\varepsilon)x$ is continuous in ε for each $x \in C$, i.e.*

$$\|Z(\varepsilon+h)x - Z(\varepsilon)x\| = o(1) \qquad as\ |h| \to 0.$$

(2.2) *$\{Z(\varepsilon)\}_{\varepsilon\in P}$ is uniformly bounded, i.e. there is an $H > 0$ such that*

$$\|Z(\varepsilon)\| \le H \qquad for\ all \varepsilon \in P.$$

Then $Z(\varepsilon)$ restricted to C is uniformly continuous with respect to ε , i.e.

$$\lim_{|h|\to 0} \sup_{x\in C} \|Z(\varepsilon+h)x - Z(\varepsilon)x\| = o(1). \tag{2.3}$$

Proof 1.2.10 *The proof is standard, and is omitted here.*

1.3 INTEGRATED SEMIGROUPS DEPENDENT ON PARAMETERS

In this section, we study the operators $A(\varepsilon)$ with the domains dependent on parameter ε and present the main results on differentiability with respect to parameter ε of integrated semigroup $S(t,\varepsilon)$ generated by $A(\varepsilon)$. We start with a result concerning the case when the domains of $A(\varepsilon)$ are independent of ε.

Let X be a Banach space with norm $\|\cdot\|$. Let P be an open subset of a finite-dimensional normed linear space $mathcalP$ with norm $|\cdot|$. For each $\varepsilon \in P$, let $S(t,\varepsilon)$ be the integrated semigroup generated by the operator $A(\varepsilon)$.

Theorem 1.3.1 (1) *Assume that*
(3.1) *$D(A_0(\varepsilon)) = D_0$ for all $\varepsilon \in P$.*
(3.2) *There are constants $M \ge 1$ and $\omega \in R$ such that*

$$\|(\lambda I - A(\varepsilon))^{-n}\| \le \frac{M}{(\lambda-\omega)^n} \quad for\ \lambda > \omega,\ n \in N,\ and\ all\ \varepsilon \in P.$$

(3.3) *For each $x \in D_0$, $A_0(\varepsilon)x$ is continuously (Frechét) differentiable with respect to ε on P.*
Then for each $x \in X_0 = \overline{D_0}$, the integrated semigroup $S(t,\varepsilon)x$ is continuously (Frechét differentiable with respect to ε uniformly on $[0,t_0]$ for any $t_0 > 0$.

Now consider the case of domains of $A(\varepsilon)$ dependent on the parameter ε.

ASSUMPTION Q.

Let $\varepsilon_0 \in P$ be given. Then for each $\varepsilon \in P$ there exists bounded operators $Q_1(\varepsilon), Q_2(\varepsilon) : X \to X$ with bounded inverses $Q_1^{-1}(\varepsilon)$ and $Q_2^{-1}(\varepsilon)$, such that $A(\varepsilon) = Q_1(\varepsilon)A(\varepsilon_0)Q_2(\varepsilon)$.

Note that if $A(\varepsilon_1) = Q_1(\varepsilon_1)A(\varepsilon_0)Q_2(\varepsilon_1)$, then

$$\begin{aligned} A(\varepsilon) &= Q_1(\varepsilon)A(\varepsilon_0)Q_2(\varepsilon) \\ &= Q_1(\varepsilon)Q_1^{-1}(\varepsilon_1)Q_1(\varepsilon_1)A(\varepsilon_0)Q_2(\varepsilon_1)Q_2^{-1}(\varepsilon_1)Q_2(\varepsilon) \\ &= \tilde{Q}_1(\varepsilon)A(\varepsilon_1)\tilde{Q}_2(\varepsilon). \end{aligned}$$

Thus, having such a relationship for some ε_0 implies a similar relationship at any other $\varepsilon_1 \in P$. Without loss of generality then, we may just consider the differentiability of integrated semigroups $S(t,\varepsilon)$ at $\varepsilon = \varepsilon_0 \in P$.

Let $\tilde{A}(\varepsilon) = Q_2(\varepsilon)A(\varepsilon)Q_2^{-1}(\varepsilon)$ and $\tilde{X}_0(\varepsilon_0) = \overline{D(\tilde{A}_0(\varepsilon_0))}$.

Lemma 1.3.2 *Assume that Assumption Q and (3.2) are satisfied and suppose that*

(3.4) $Q_i(\varepsilon)x (i = 1, 2)$ *and* $Q_2^{-1}(\varepsilon)x$ *are continuously (Frechét) differentiable with respect to* ε *for* $x \in X$.

Then for each $\varepsilon \in P$, $\tilde{A}(\varepsilon)$ *generates a integrated semigroup* $\tilde{S}(t,\varepsilon)$. *Furthermore, for each* $x \in \tilde{X}_0(\varepsilon_0)$, $\tilde{S}(t,\varepsilon)x$ *is continuously (Frechét differentiable with respect to* ε *uniformly on* $[0, t_0]$ *for any* $t_0 > 0$.

Proof 1.3.3 *First note that*

$$\tilde{A}(\varepsilon) = Q_2(\varepsilon)Q_1(\varepsilon)A(\varepsilon_0). \tag{3.4}$$

It is clear that (3.4) implies $\tilde{A}_0(\varepsilon) = Q_2(\varepsilon)Q_1(\varepsilon)A_0(\varepsilon_0)$. *Hence* $D(\tilde{A}_0(\varepsilon)) = D(A_0(\varepsilon))$ *which indicates that* $\tilde{A}_0(\varepsilon)$ *satisfies Hypothesis (3.1).*

Secondly, we see that (3.4) implies that there exists $\delta > 0$ *so that* $Q_2(\varepsilon)$ *and* $Q_2^{-1}(\varepsilon)$ *are uniformly bounded on* $B_\delta(\varepsilon_0) = \{\varepsilon \mid |\varepsilon - \varepsilon_0| \le \delta\}$. *Since*

$$(\lambda I - \tilde{A}(\varepsilon))^n = Q_2(\varepsilon)(\lambda I - A(\varepsilon))^n Q_2^{-1}(\varepsilon),$$

we have

$$(\lambda I - \tilde{A}(\varepsilon))^{-n} = Q_2(\varepsilon)(\lambda I - A(\varepsilon))^{-n} Q_2^{-1}(\varepsilon).$$

It follows from (3.2) that $\tilde{A}(\varepsilon)$ *satisfies (3.2). From Theorem 2.7 and Theorem 2.8, we have that for each* $\varepsilon \in P$, $\tilde{A}(\varepsilon)$ *generates an integrated semigroup* $\tilde{S}(t,\varepsilon)$. *Thirdly,* $\tilde{A}_0(\varepsilon)$ *satisfies (3.3) because* $Q_1(\varepsilon)$ *and* $Q_2(\varepsilon)$ *are continuously (Frechét differentiable) with respect to* ε. *Now the desired result directly follows from Theorem 3.1.*

Q.E.D.

Lemma 1.3.4 *Assume that Assumption Q and (3.2) are satisfied and suppose that*

(3.6) $Q_i(\varepsilon)x (i = 1, 2)$ *and* $Q_2^{-1}(\varepsilon)x$ *are continuous in* ε *for* $x \in X$.

Then

$$S(t,\varepsilon) = Q_2^{-1}(\varepsilon)\tilde{S}(t,\varepsilon)Q_2(\varepsilon).$$

Proof 1.3.5 *First note that for each $\varepsilon \in P$,*

$$\begin{aligned}(\lambda I - A_0(\varepsilon))^{-1} &= (\lambda I - A(\varepsilon))^{-1} \text{ on } X_0(\varepsilon) = \overline{D(A(\varepsilon))} \\ (\lambda I - \tilde{A}_0(\varepsilon))^{-1} &= (\lambda I - \tilde{A}(\varepsilon))^{-1} \text{ on } \tilde{X}_0(\varepsilon) = \overline{D(\tilde{A}(\varepsilon))}.\end{aligned}$$

Also for $x \in D(\tilde{A}(\varepsilon)) \subseteq \tilde{X}_0(\varepsilon), Q_2^{-1}(\varepsilon)x \in D(A(\varepsilon)) \subseteq X_0(\varepsilon)$.
Hence, for each $x \in D(\tilde{A}_0(\varepsilon)) \subseteq \tilde{X}_0(\varepsilon)$,

$$\begin{aligned}(\lambda I - \tilde{A}_0(\varepsilon))^{-1} &= (\lambda I - A(\varepsilon))^{-1} \text{ on } X_0(\varepsilon) = \overline{D(A(\varepsilon))} \\ (\lambda I - \tilde{A}_0(\varepsilon))^{-1} &= (\lambda I - \tilde{A}(\varepsilon))^{-1} \text{ on } \tilde{X}_0(\varepsilon) = \overline{D(\tilde{A}(\varepsilon))}. \qquad (3.5)\end{aligned}$$

Since $\tilde{X}_0(\varepsilon) = \overline{D(\tilde{A}_0(\varepsilon))}$ and all operators in (3.5) are bounded, we have (3.5) holds for every $x \in \tilde{X}_0(\varepsilon)$. Hence

$$\tilde{A}_0(\varepsilon) = Q_2(\varepsilon)A_0(\varepsilon)Q_2^{-1}(\varepsilon) \text{ on } D(\tilde{A}_0(\varepsilon)).$$

From the proof of Lemma 3.2, we see that $\tilde{A}_0(\varepsilon)$ generates a C_0-semigroup $\tilde{T}(t,\varepsilon)$. It is clear that $T(t,\varepsilon) = Q_2^{-1}(\varepsilon)\tilde{T}(t,\varepsilon)Q_2(\varepsilon)$.

Note that the relation between a C_0-semigroup $T(t,\varepsilon)$ and the related integrated semigroup (t,ε) is as follows: for $\lambda > \omega$ and $x \in X$,

$$S(t,\varepsilon)x = \lambda S_0(t,\varepsilon)(\lambda I - A(\varepsilon))^{-1}x - T(t,\varepsilon)(\lambda I - A(\varepsilon))^{-1}x + (\lambda I - A(\varepsilon))^{-1}x$$

where

$$S_0(t,\varepsilon) = \int_0^t T(s,\varepsilon)\,ds.$$

Using (3.8), one can easily show that $S(t,\varepsilon) = Q_2^{-1}(\varepsilon)\tilde{S}(t,\varepsilon)Q_2(\varepsilon)$. Q.E.D.

Theorem 1.3.6 *Assume that Assumption Q, (3.2), and (3.4) are satisfied, then for each $x \in X_0(\varepsilon_0) = \overline{D(A(\varepsilon_0))}$, the integrated semigroup $S(t,\varepsilon)x$ is continuously (Frechét differentiable) at $\varepsilon = \varepsilon_0$ for $\varepsilon_0 \in P$. In particular,*

$$\begin{aligned}[D_\varepsilon S(t,\varepsilon)x\Big|_{\varepsilon=\varepsilon_0}] &= [D_\varepsilon Q_2^{-1}(\varepsilon)\tilde{S}(t,\varepsilon_0)x\Big|_{\varepsilon=\varepsilon_0}] + Q_2^{-1}(\varepsilon_0)[D_\varepsilon\tilde{S}(t,\varepsilon)x\Big|_{\varepsilon=\varepsilon_0}] \\ &+ Q_2^{-1}(\varepsilon_0)\tilde{S}(t,\varepsilon_0)[D_\varepsilon Q_2(\varepsilon)x\Big|_{\varepsilon=\varepsilon_0}]\end{aligned}$$

Proof 1.3.7 *Let*

$$\begin{aligned}I_1 &= \frac{1}{|h|}\|[(Q_2^{-1}(\varepsilon_0+h)\tilde{S}(t,\varepsilon_0+h) - Q_2^{-1}(\varepsilon_0))\tilde{S}(t,\varepsilon_0)][D_\varepsilon Q_2(\varepsilon)x\Big|_{\varepsilon=\varepsilon_0}]h\| \\ I_2 &= \frac{1}{|h|}\|[Q_2^{-1}(\varepsilon_0+h) - Q_2^{-1}(\varepsilon_0)][D_\varepsilon\tilde{S}(t,\varepsilon)x\Big|_{\varepsilon=\varepsilon_0}]h\|.\end{aligned}$$

Claim. $I_1 \to 0$ *and* $I_2 \to 0$ *as* $|h| \to 0$.
In fact, It is clear that $Q_2^{-1}(\varepsilon)\tilde{S}(t,\varepsilon)$ *is continuous in* ε. *let* $C = \{[D_\varepsilon Q_2(\varepsilon)x]p \mid p \in P, |p| = 1\}$. *Since* $\{p \in P \,|\, |p| = 1\}$ *is bounded and closed, and* $\mathcal{P}$ *is a finite-dimensional normed linear space, it is compact. Also,* $D_\varepsilon Q_2(\varepsilon)x \in \mathcal{B}(\mathcal{P}, X)$ *implies that* C *is compact. Now applying Lemma 2.9, we obtain that*

$$I_1 \leq \| (Q_2^{-1}(\varepsilon_0+h)\tilde{S}(t,\varepsilon_0+h) - Q_2^{-1}(\varepsilon_0))\tilde{S}(t,\varepsilon_0)] \| \left\| [D_\varepsilon Q_2(\varepsilon)x]\frac{h}{|h|} \right\|.$$

Hence $I_1 \to 0$ *as* $|h| \to 0$. *Using the similar discussion, we can show that* $I_2 \to 0$ *as* $|h| \to 0$.

¿From Lemma 3.3, we see that $A_0(\varepsilon_0) = Q_2^{-1}(\varepsilon_0)\tilde{A}_0(\varepsilon_0)Q_2(\varepsilon_0)$. *Note that* $Q_2(\varepsilon_0) = I$. *Hence* $A_0(\varepsilon_0) = Q_2^{-1}(\varepsilon_0)\tilde{A}_0(\varepsilon_0)$ *implies* $D(A_0(\varepsilon_0)) = D(\tilde{A}_0(\varepsilon_0))$, *and thus* $X_0(\varepsilon_0) = \tilde{X}_0(\varepsilon_0)$. *By Lemma 3.3, we also have*

$$S(t,\varepsilon) = Q_2^{-1}(\varepsilon)\tilde{S}(t,\varepsilon)Q_2(\varepsilon).$$

For $x \in X_0(\varepsilon_0)$ *and* $h \in \mathcal{P}$ *with* $\varepsilon_0 + h \in B_\delta(\varepsilon_0)$,

$$\begin{aligned}
& \frac{1}{|h|}\|S(t,\varepsilon_0+h)x - S(t,\varepsilon_0)x - [D_\varepsilon S(t,\varepsilon)x\Big|_{\varepsilon=\varepsilon_0}]h\| \\
= \; & \frac{1}{|h|}\|Q_2^{-1}(\varepsilon_0+h)\tilde{S}(t,\varepsilon_0+h)Q_2(\varepsilon_0+h)x - Q_2^{-1}(\varepsilon_0)\tilde{S}(t,\varepsilon_0)Q_2(\varepsilon_0)x \\
- \; & \{[D_\varepsilon Q_2^{-1}(\varepsilon)\tilde{S}(t,\varepsilon_0)x\Big|_{\varepsilon=\varepsilon_0}] + Q_2^{-1}(\varepsilon_0)[D_\varepsilon \tilde{S}(t,\varepsilon)x\Big|_{\varepsilon=\varepsilon_0}] \\
+ \; & Q_2^{-1}(\varepsilon_0)\tilde{S}(t,\varepsilon_0)[D_\varepsilon Q_2(\varepsilon)x\Big|_{\varepsilon=\varepsilon_0}]\}h\| \\
\leq \; & \frac{1}{|h|}\|Q_2^{-1}(\varepsilon_0+h)\tilde{S}(t,\varepsilon_0+h)\|\|Q_2(\varepsilon_0+h)x - Q_2(\varepsilon_0)x - [D_\varepsilon Q_2(\varepsilon)x\Big|_{\varepsilon=\varepsilon_0}]h\| \\
+ \; & \frac{1}{|h|}\|Q_2^{-1}(\varepsilon_0+h)\|\|\tilde{S}(t,\varepsilon_0+h)x - \tilde{S}(t,\varepsilon_0)x - [D_\varepsilon \tilde{S}(t,\varepsilon)x\Big|_{\varepsilon=\varepsilon_0}]h\| \\
+ \; & \frac{1}{|h|}\|[Q_2^{-1}(\varepsilon_0+h) - Q_2^{-1}(\varepsilon_0)]\tilde{S}(t,\varepsilon_0)x - [D_\varepsilon Q_2^{-1}(\varepsilon)\tilde{S}(t,\varepsilon_0)x\Big|_{\varepsilon=\varepsilon_0}]h\| \\
+ \; & \frac{1}{|h|}\|[Q_2^{-1}(\varepsilon_0+h)\tilde{S}(t,\varepsilon_0+h) - Q_2^{-1}(\varepsilon_0))\tilde{S}(t,\varepsilon_0)][D_\varepsilon Q_2(\varepsilon)x\Big|_{\varepsilon=\varepsilon_0}]h\| \\
+ \; & \frac{1}{|h|}\|[Q_2^{-1}(\varepsilon_0+h) - Q_2^{-1}(\varepsilon_0)][D_\varepsilon \tilde{S}(t,\varepsilon)x\Big|_{\varepsilon=\varepsilon_0}]h\|.
\end{aligned}$$

It is easy to show that $\|Q_2^{-1}(\varepsilon_0+h)\tilde{T}(t,\varepsilon_0+h)\|$, $\|Q_2^{-1}(\varepsilon_0+h)\|$ *are uniformly bounded. Hence, the desired result follows from Lemma 3.2, the Claim, and (3.4).* Q.E.D.

1.4 APPLICATION TO A WAVE EQUATION

Consider the wave equation with damping boundary conditions

$$\begin{aligned} u_{tt} &= u_{xx}, \\ u(x,0) &= u_0(x), \quad u_t(x,0) = u_1(x) \qquad \text{for } x \in [0,1], \\ \alpha u_t(0,t) &- u_x(0,t) = 0, \\ u_t(1,t) &+ \beta u_x(1,t) = 0, \qquad \alpha, \beta \geq 0. \end{aligned} \tag{4.6}$$

As we discussed in Section 1, Equation (4.7) can be formulated as an abstract Cauchy problem (1.2).

First we see that, for each ε,

$$\overline{D(A(\varepsilon))} = X_0(\varepsilon) = \left\{ \begin{pmatrix} v \\ w \end{pmatrix} \in X = \prod_{i=1}^{2} C[0,1] \middle| \begin{matrix} \alpha v(0) = w(0), \\ v(1) = -\beta w(1), \end{matrix} \quad \alpha, \beta \geq 0 \right\}.$$

Thus, the operator $A(\varepsilon)$ is not densely defined. Also the domain of $A(\varepsilon)$ is dependent on ε, and so is $X_0(\varepsilon) = \overline{D(A(\varepsilon))}$.

Next we show that all hypotheses of Theorem 3.4 are satisfied.

Making the change of variables

$$\bar{z} = \begin{pmatrix} \bar{v} \\ \bar{w} \end{pmatrix} = U \begin{pmatrix} v \\ w \end{pmatrix} \tag{4.7}$$

where

$$U = \begin{pmatrix} \frac{1}{2} & \frac{1}{2} \\ \frac{1}{2} & -\frac{1}{2} \end{pmatrix},$$

the transformed equation is

$$\frac{d\bar{z}(t)}{dt} = A(\varepsilon)\bar{z}(t) \tag{4.8}$$

$$\text{on} \quad X = \prod_{i=1}^{2} (C[0,1], \|\cdot\|_\infty)$$

$$\bar{x}(0) = \bar{z}_0$$

where

$$\begin{aligned} A_1(\varepsilon) &\equiv UA(\varepsilon)U^{-1} = \begin{pmatrix} \partial x & 0 \\ 0 & -\partial x \end{pmatrix}, \\ D(A_1(\varepsilon)) &= \left\{ \begin{pmatrix} v \\ w \end{pmatrix} \in C^1[0,1] \times C^1[0,1] \middle| \begin{matrix} (1-\alpha)v(0) = (1+\alpha)w(0), \\ (1+\beta)v(1) = -(1-\beta)w(1), \end{matrix} \right\}. \end{aligned}$$

Defining

$$\bar{\alpha} = \frac{1-\alpha}{1+\alpha}, \qquad \bar{\beta} = \frac{1-\beta}{1+\beta},$$

$A_1(\varepsilon)$ has the domain

$$D(A_1(\varepsilon)) = \left\{ \begin{pmatrix} v \\ w \end{pmatrix} \in C^1[0,1] \times C^1[0,1] \left| \begin{array}{l} \bar{\alpha}v(0) = w(0), \\ v(1) = -\bar{\beta}w(1), \end{array} \right. \quad \bar{\alpha}, \bar{\beta} \in (-1,1] \right\}.$$

Since $A_1(\varepsilon) = U(A(\varepsilon)U^{-1}$, we have

$$(\lambda I - A(\varepsilon))^{-1} = U^{-1}(\lambda I - A_1(\varepsilon))^{-1}U.$$

Because U, U^{-1} are bounded and independent of ε, it suffices to show that the operator $A_1(\varepsilon)$ satisfies all hypotheses of Theorem 3.4

For $\varepsilon = (\bar{\alpha}, \bar{\beta})$, with $\bar{\alpha}, \bar{\beta} \in (-1, 1]$

$$A_1(\varepsilon) = \frac{1}{1 + \bar{\alpha}\bar{\beta}} Q(\varepsilon) A_1(0) Q(\varepsilon)$$

where

$$Q(\varepsilon) = \begin{pmatrix} 1 & \bar{\beta} \\ -\bar{\alpha} & 1 \end{pmatrix}.$$

It is clear that Assumption Q and (3.4) are satisfied for any $\varepsilon_0 = (\bar{\alpha}_0, \bar{\beta}_0)$ in $(-1, 1]$.

Now the remaining thing is to show that $A_1(\varepsilon)$ satisfies (3.2).

For any $(f, g)^t \in X$, consider

$$(\lambda I - A_1(\varepsilon)) \begin{pmatrix} v \\ w \end{pmatrix} = \begin{pmatrix} f \\ g \end{pmatrix}, \quad \text{i.e.} \quad \begin{array}{l} \lambda v - v' = f \\ \lambda w + w' = g \end{array} \tag{4.9}$$

with the boundary conditions $\bar{\alpha}v(0) = w(0), v(1) = -\bar{\beta}w(1)$.

We want to estimate $(\lambda I - A_1(\varepsilon))^{-1}(f, g)^t$ for a given $(f, g)^t \in X$ and $\lambda > 0$.

Let

$$\begin{pmatrix} v \\ w \end{pmatrix} = (\lambda I - A_1(\varepsilon))^{-1} \begin{pmatrix} f \\ g \end{pmatrix}.$$

<u>Case 1</u>. $\left\| \begin{pmatrix} v \\ w \end{pmatrix} \right\| = \|v\|_\infty$.

If $\|v\|_\infty = |v(x_0)|$ where $x_0 \in (0, 1)$, then $v'(x_0) = 0$ and $\lambda v(x_0) = f(x_0)$. Thus,

$$\|v\|_\infty \le \frac{1}{\lambda}|f(x_0)| \le \frac{1}{\lambda}\|f\|_\infty \le \frac{1}{\lambda}\left\| \begin{pmatrix} f \\ g \end{pmatrix} \right\|$$

implies

$$\left\| \begin{pmatrix} v \\ w \end{pmatrix} \right\| = \|v\|_\infty \le \frac{1}{\lambda}\left\| \begin{pmatrix} f \\ g \end{pmatrix} \right\|.$$

If $\|v\|_\infty = |v(0)|$, we may assume that $v(0) > 0$. Then $v'(0) \le 0$ and $0 < \lambda v(0) \le f(0)$. Thus,

$$\left\| \begin{pmatrix} v \\ w \end{pmatrix} \right\| = \|v\|_\infty = v(0) \le \frac{1}{\lambda} f(0) \le \frac{1}{\lambda}\left\| \begin{pmatrix} f \\ g \end{pmatrix} \right\|.$$

If $\|v\|_\infty = |v(1)|$, then $v(1) = -\bar{\beta}w(1)$ implies $|v(1)| \leq |\bar{\beta}||w(1)| \leq |w(1)|$ (since $|\bar{\beta}| \leq 1$). Also

$$\|w\|_\infty \leq \left\| \begin{pmatrix} v \\ w \end{pmatrix} \right\| = \|v\|_\infty = |v(1)| \leq |w(1)| \leq \|w\|_\infty$$

implies

$$\left\| \begin{pmatrix} v \\ w \end{pmatrix} \right\| = \|w\|_\infty = |w(1)|.$$

We may assume that $w(1) > 0$. Then $w'(1) \geq 0$, and $0 < \lambda w(1) \leq g(1)$. Thus,

$$\left\| \begin{pmatrix} v \\ w \end{pmatrix} \right\|_\infty = \|w\|_\infty = |w(1)| \leq \frac{1}{\lambda} g(1) \leq \frac{1}{\lambda} \left\| \begin{pmatrix} f \\ g \end{pmatrix} \right\|.$$

<u>Case 2</u>. $\left\| \begin{pmatrix} v \\ w \end{pmatrix} \right\| = \|w\|_\infty$.

For the case of $\|w\|_\infty = |w(x_0)|$ where $x_0 \in (0,1)$ and the case of $\|w\|_\infty = |w(1)|$, the argument is similar to that in Case 1.

For the case of $\|w\|_\infty = |w(0)|$, $w(0) = \bar{\alpha}v(0)$ implies $|w(0)| \leq |\bar{\alpha}||v(0)| \leq |v(0)|$ (since $|\bar{\alpha}| \leq 1$). Also

$$\|v\|_\infty \leq \left\| \begin{pmatrix} v \\ w \end{pmatrix} \right\| = \|w\|_\infty = |w(0)| \leq |v(0)| \leq \|v\|_\infty$$

implies

$$\left\| \begin{pmatrix} v \\ w \end{pmatrix} \right\| = \|v\|_\infty = |v(0)|.$$

Using the similar argument in Case 1, we have

$$\left\| \begin{pmatrix} v \\ w \end{pmatrix} \right\| = \|v\|_\infty = |v(0)| \leq \frac{1}{\lambda} f(0) \leq \frac{1}{\lambda} \left\| \begin{pmatrix} f \\ g \end{pmatrix} \right\|.$$

In summary, we have

$$\left\| (\lambda I - A_1(\varepsilon))^{-1} \begin{pmatrix} f \\ g \end{pmatrix} \right\| \leq \frac{1}{\lambda} \left\| \begin{pmatrix} f \\ g \end{pmatrix} \right\|.$$

Thus, the operator $A_1(\varepsilon)$ satisfies the "uniform Hille-Yosida" condition (3.2).

Now all hypotheses of Theorem 3.4 are satisfied, and we have strong differentiability of the integrated semigroup $S(t, \varepsilon)$ which is generated by $A(\varepsilon)$.

Remarks: The theorems in [2] and [4] show that the classical or integral solution of (1.2) can be expressed in terms of the integrated semigroup $S(t, \varepsilon)$ generated by $A(\varepsilon)$. Hence, it can be shown that the existence of the integrated semigroup $S(t, \varepsilon)$ and the strong differentiability with respect to parameter ε of $S(t, \varepsilon)$ imply the existence of an unique classical or integral solution of (1.2) and the continuity in parameter ε of such a solution. Same results are obviously true for Equation (1.1).

References

[1] Grimmer, R. and M. He. (1997). *Differentiability with Respect to Parameters of Semigroups*, Semigroup Forum.

[2] Thieme, H.R. (1990). *Integrated Semigroups and Integrated Solutions to Abstract Cauchy Problems*, J. Math. Anal. Appl., Vol. 152, (pages 416–447).

[3] Arendt, W. (1987). *Vector-valued Laplace Transforms and Cauchy Problems*, Isreal J. Math., Vol. 54, (pages 327–352).

[4] Da Prato, G. and E. Sinestrari. (1987). *Differential Operators with Non-dense Domains*, Ann. Scuola Norm. Sup. Pisa Cl. Sci., Vol. 14, (pages 285–344).

BOUNDEDNESS OF PSEUDO-DIFFERENTIAL OPERATORS ON HÖRMANDER SPACES

G.M. Iancu*

Department of Mathematics and Statistics
York University
4700 Keele Street, North York, Ontario
M3J 1P3 CANADA

ysma9303@@mathstat.yorku.ca

Abstract: We prove the boundedness of a class of pseudo-differential operators on the Hörmander spaces $\dot{H}^{0,p}, 1 < p < \infty$. Examples are given to show that the classical pseudo-differential operators fail to be bounded on these spaces. Fourier integral operators and their global mapping properties are studied in this setting. Applications to the regularity of solutions of semilinear pseudo-differential equations on $\mathbb{R}^n$ are presented.

1.1 INTRODUCTION

We define for any function φ in the Schwartz space $\mathcal{S}$ the Fourier transform $\widehat{\varphi}$ by

$$\widehat{\varphi}(\xi) = (2\pi)^{-\frac{n}{2}} \int_{\mathbb{R}^n} e^{-ix\cdot\xi}\varphi(x)dx, \quad \xi \in \mathbb{R}^n.$$

The Fourier transform, first defined on the Schwartz space $\mathcal{S}$, can be extended to the space $\mathcal{S}'$ of tempered distributions. To wit, the Fourier transform $\widehat{u}$ of a tempered distribution u on $\mathbb{R}^n$ is defined by

$$\widehat{u}(\varphi) = u(\widehat{\varphi}), \quad \varphi \in \mathbb{R}^n.$$

*This is an expanded version of a lecture given at the ISAAC97 Conference, University of Delaware, June 3–7, 1997.

R.P. Gilbert et al.(eds.), Direct and Inverse Problems of Mathematical Physics, 137–147.

For $s \in \mathbb{R}$ and $1 \leq p < \infty$, we define the spaces $\dot{H}^{s,p}$ by

$$\dot{H}^{s,p} = \{u \in \mathcal{S}' : \langle\cdot\rangle^s \hat{u} \in L^p\},$$

where $\langle\cdot\rangle$ is the function on $\mathbb{R}^n$ given by

$$\langle\xi\rangle = (1 + |\xi|^2)^{\frac{1}{2}}, \quad \xi \in \mathbb{R}^n.$$

It is easy to see that $\dot{H}^{s,p}$ is a Banach space with the norm $\|\cdot\|_{s,p}$ given by

$$\|u\|_{s,p} = \|\langle\cdot\rangle^s \hat{u}\|_p, \quad u \in \dot{H}^{s,p},$$

where $\|\cdot\|_p$ is the usual norm in $L^p(\mathbb{R}^n)$. This space coincides with the space $B_{p,k}$ introduced by Hörmander in Section 10.1 of the book [5] if the function k is taken to be the function $\langle\cdot\rangle^s$. It has been proved by Pi and Wong in the paper [7] that, under appropriate conditions, the space $\dot{H}^{s,p}$ is a Banach algebra, generalizig a classical theorem of Schauder. More precisely, Theorem 1.1 in the above-mentioned paper states that

$$u, v \in \dot{H}^{s,p}, 1 \leq p < \infty, s > \frac{n}{p'}, \text{ where } \frac{1}{p} + \frac{1}{p'} = 1 \Rightarrow uv \in \dot{H}^{s,p}$$

and there exists a positive constant C, depending on s, p and n only, such that

$$\|uv\|_{s,p} \leq C\|u\|_{s,p}\|v\|_{s,p}, \quad u, v \in \dot{H}^{s,p}.$$

We define S^m, $m \in \mathbb{R}$, to be the set of all functions $\sigma(x,\xi)$ in $C^\infty(\mathbb{R}^n \times \mathbb{R}^n)$ such that for any two multi-indices α and β, there exists a positive constant $C_{\alpha,\beta}$, depending on α and β only, such that

$$\left|\left(D_x^\alpha D_\xi^\beta \sigma\right)(x,\xi)\right| \leq C_{\alpha,\beta}(1 + |\xi|)^{m-|\beta|}, \quad x, \xi \in \mathbb{R}^n,$$

We call any function σ in $\bigcup_{m\in\mathbb{R}} S^m$ a classical symbol.

Let σ be a continuous function defined on $\mathbb{R}^n \times \mathbb{R}^n$. Then the pseudo-differential operator T_σ on the Schwartz space $\mathcal{S}$ associated to σ is defined by

$$(T_\sigma\varphi)(x) = (2\pi)^{-\frac{n}{2}} \int_{\mathbb{R}^n} e^{ix\cdot\xi}\sigma(x,\xi)\hat{\varphi}(\xi)d\xi, \quad \varphi \in \mathcal{S}.$$

The aim of this paper is to answer the question of whether the pseudo-differential operator with classical symbols in S^0 is bounded on the space $\dot{H}^{0,p}$, $1 \leq p < \infty$, and in case the answer is negative, to find another class of symbols for which the associated pseudo-differential operator is bounded. It has been shown in the paper [7] by Pi and Wong that for $1 \leq p < \infty$, $s \in \mathbb{R}$, the pseudo-differential operator with "constant coefficient " symbols in S^m, $m \in \mathbb{R}$, is bounded from $\dot{H}^{s,p}$ into $\dot{H}^{s-m,p}$. Less recent results can be found in the papers [11] and [12] by Zaidman, where the pseudo-differential operators with "variable coefficient" symbols are very restrictive, and the $\dot{H}^{0,p}$ boundedness, $1 \leq p < \infty$, is relying havily on the Young's inequality and properties of the

Fourier transform of $L^1(\mathbb{R}^n)$ functions. The answer is that for general classical symbols in S^0, the associated pseudo-differential operator fails to be bounded on $\dot{H}^{0,p}$, $1 < p < \infty$, $p \neq 2$. The proof will be provided in Section 2.

For now, our attention will be restricted to the following class of symbols:

Let $\dot{S}_q^m$, $m, q \in \mathbb{R}$, denote the set of $C^\infty(\mathbb{R}^n \times \mathbb{R}^n)$ functions with the property that for any two multi-indices α and β, there exists a positive constant $C_{\alpha,\beta}$, depending on α and β only, such that

$$\left|\left(D_x^\alpha D_\xi^\beta \sigma\right)(x,\xi)\right| \leq C_{\alpha,\beta}(1+|x|)^{q-|\alpha|}(1+|\xi|)^m, \quad x,\xi \in \mathbb{R}^n.$$

We denote for simplicity the symbol class $\dot{S}_0^m$ by $\dot{S}^m$, $m \in \mathbb{R}$.

The main result in this paper is the following theorem.

Theorem 1.1.1 *Let $\sigma \in \dot{S}^m$. Then the pseudo-differential operator T_σ : $\dot{H}^{s,p} \to \dot{H}^{s-m,p}$ is a bounded linear operator for $1 < p < \infty$ and $s \in \mathbb{R}$.*

We study the basic pseudo-differential theory for the newly introduced class of operators in Section 3. The proof of Theorem 1.1.1 is given in Section 4. We then show that Fourier integral operators are best studied in this new setting. Applications to the regularity of solutions of semilinear pseudo-differential equations on $\mathbb{R}^n$ are presented in Section 5.

The impetus for the study of pseudo-differential operators on Hörmander spaces comes from the desire of developing new results involving pseudo-differential theory on spaces with useful properties in applications. It should be acknowledged the importance and ramifications of Schauder's theorem to nonlinear wave equations, as presented in the book [3] by Beals.

1.2 MOTIVATION

The decaying condition in the x variable introduced for the special class of operators $\dot{S}_q^m$, is necessary, as the following example shows:

Assume that $\sigma \in S^0$ is a function of x, only. Then, by the Fourier inversion formula,

$$(T_\sigma\varphi)(x) = \sigma(x)\varphi(x), \quad \varphi \in \mathcal{S},$$

for all $x \in \mathbb{R}^n$. Thus, by taking the Fourier transform,

$$\widehat{T_\sigma\varphi}(\xi) = \widehat{\sigma\varphi}(\xi) = \left(T_{\widetilde{\sigma}}\widehat{\varphi}\right)(\xi), \quad \varphi \in \mathcal{S},$$

for all $\xi \in \mathbb{R}^n$, where $\widetilde{\sigma}(x) = \sigma(-x)$, $x \in \mathbb{R}^n$. It is well-known that only the boundedness of the symbol σ and all its derivatives, does not guarantee the L^p , $1 < p < \infty$, $p \neq 2$, boundedness of the pseudo-differential operator $T_{\widehat{\sigma}}$. Moreover, by following the same path of thought, we can prove that the partial-differential operator associated to symbols of the form

$$\sigma(x,\xi) = \sum_{|\alpha| \leq m} a_\alpha(x)\xi^\alpha, \quad x,\xi \in \mathbb{R}^n,$$

with $a_\alpha \in S^0$, fail to be bounded from $\dot{H}^{s,p}$ into $\dot{H}^{s-m,p}$, $s \in \mathbb{R}$ and $1 < p < \infty$, $p \neq 2$. The reader should consult the paper [1] by Beals for further reference.

1.3 PSEUDO-DIFFERENTIAL OPERATOR CALCULUS

In this section we will present the most important theorems and results involving pseudo-differential operators associated to symbols in $\dot{S}^m_q$. Proofs are not provided, what we are interested in, is the precise statement of these results. We begin with one of the main notions, that is, asymptotic expansion of a symbol.

Definition 1.3.1 *Let $\sigma \in \dot{S}^m_q$. Suppose we can find $\sigma_j \in \dot{S}^m_{q_j}$, where $q = q_0 > q_1 > \cdots > q_j \to -\infty$, $j \to \infty$, such that*

$$\sigma - \sum_{j=0}^{N-1} \sigma_j \in \dot{S}^m_{q_N}$$

for every positive integer N. Then we call $\sum_{j=0}^{\infty} \sigma_j$ an asymptotic expansion of σ and write

$$\sigma \sim \sum_{j=0}^{\infty} \sigma_j.$$

An important result in this connection is the following theorem. The proof can be modelled by using the pattern of any such theorem in any introductory book in pseudo-differential theory.

Theorem 1.3.2 *Let $q_0 > q_1 > \cdots > q_j \to -\infty$ as $j \to \infty$. Suppose $\sigma_j \in \dot{S}^m_{q_j}$. Then there exists a symbol $\sigma \in \dot{S}^m_{q_0}$ such that $\sigma \sim \sum_{j=0}^{\infty} \sigma_j$. Moreover, if τ is another symbol with the same asymptotic expansion, then $\sigma - \tau \in \bigcap_{q \in \mathbb{R}} \dot{S}^m_q$.*

From now on we will be concerned only with symbols in $\bigcup_{m \in \mathbb{R}} \dot{S}^m$. The next two results though standard, need to be mentioned .

Lemma 1.3.3 *Let σ and τ be two symbols in $\dot{S}^m$ such that $T_\sigma = T_\tau$. Then $\sigma = \tau$.*

Lemma 1.3.4 *Let σ be a symbol in $\dot{S}^m$. Then the pseudo-differential operator T_σ maps the Schwartz space into itself.*

Next, we need to know the composition rules for the newly introduced class of pseudo-differential operators. The theorem that is showing that the product of two pseudo-differential operators is again a pseudo-differential operator, with the bonus of giving an asymptotic expansion for the symbol of the product, is given next.

Theorem 1.3.5 *Let $\sigma \in \dot{S}^{m_1}$ and $\tau \in \dot{S}^{m_2}$. Then the product $T_\sigma T_\tau$ is again a pseudo-differential operator T_λ, where $\lambda \in \dot{S}^{m_1+m_2}$ and has the following asymptotic expansion:*

$$\lambda \sim \sum_{\mu} \frac{(-i)^{|\mu|}}{\mu!} (\partial_\xi^\mu \sigma)(\partial_x^\mu \tau). \qquad (3.1)$$

Here (3.1) *means that*

$$\lambda - \sum_{|\mu|<N} \frac{(-i)^{|\mu|}}{\mu!}(\partial_\xi^\mu \sigma)(\partial_x^\mu \tau) \in \dot{S}_{-N}^{m_1+m_2}$$

for every positive integer N.

We end this section with the notion of formal adjoint. The following theorem is showing that the formal adjoint of a pseudo-differential operator in this new class exists, and it is a pseudo-differential operator in the same class. Moreover, we obtain a useful asymptotic expansion for its symbol.

Theorem 1.3.6 *Let* $\sigma \in \dot{S}^m$. *Then the formal adjoint* T_σ^* *of the pseudo-differential operator* T_σ *is again a pseudo-differential operator* T_{σ^*}, *where* $\sigma^* \in \dot{S}^m$ *and has the following asymptotic expansion*

$$\sigma^* \sim \sum_\mu \frac{(-i)^{|\mu|}}{\mu!}(\partial_\xi^\mu \partial_x^\mu \overline{\sigma}). \tag{3.2}$$

Here (3.2) *means that*

$$\sigma^* - \sum_{|\mu|<N} \frac{(-i)^{|\mu|}}{\mu!}(\partial_\xi^\mu \partial_x^\mu \overline{\sigma}) \in \dot{S}_{-N}^m$$

for every positive integer N.

Remark 1.3.7 *The formal adjoint allows us to extend the pseudo-differential operator* T_σ, *defined initially on the Schwartz space* $\mathcal{S}$, *to a linear mapping defined on the space* $\mathcal{S}'$ *of tempered distributions. To wit, take any* $u \in \mathcal{S}'$ *and define* $T_\sigma u$ *by*

$$(T_\sigma u)(\varphi) = u\left(\overline{T_\sigma^* \overline{\varphi}}\right), \quad \varphi \in \mathcal{S}.$$

1.4 THE PROOF OF THEOREM 1.1.1

The proof of Theoren 1.1.1 is indirect, and for this we introduce further notations.

Let σ be a symbol in $\dot{S}^m$. Then we associate to σ the pseudo-differential operator $\widetilde{T}_\sigma$ on the Schwartz space $\mathcal{S}$, defined by

$$\left(\widetilde{T}_\sigma \varphi\right)(\xi) = (2\pi)^{-\frac{n}{2}} \int_{\mathbb{R}^n} e^{ix\cdot\xi} \sigma(x,\xi)\widehat{\varphi}(x)dx, \quad \xi \in \mathbb{R}^n.$$

To prove Theorem 1.1.1 we first show the following:

Lemma 1.4.1 *Let* $\sigma \in \dot{S}^0$. *Then* $\widehat{T_\sigma \varphi} = \widetilde{T}_{\widetilde{\overline{\sigma}}}^* \widehat{\varphi}$, *for all* $\varphi \in \mathcal{S}$, *where* $\widetilde{\sigma}(x,\xi) = \sigma(-x,\xi)$, $x, \xi \in \mathbb{R}^n$, *and by* $\widetilde{T}_\sigma^*$ *we understand the formal adjoint of the pseudo-differential operator* $\widetilde{T}_\sigma$.

Proof. For any pair of functions φ and ψ in $\mathcal{S}$, we define (φ, ψ) by

$$(\varphi, \psi) = \int_{\mathbb{R}^n} \varphi(x)\overline{\psi(x)}dx.$$

Now, by the Fourier inversion formula, the adjoint formula and by Fubini's theorem, we compute

$$\begin{aligned}
\left(\widehat{T_\sigma\varphi}, \psi\right) &= \int_{\mathbb{R}^n} (T_\sigma\varphi)(x)\, \overline{\widehat{\psi}}(x)dx \\
&= \int_{\mathbb{R}^n} \left\{(2\pi)^{-\frac{n}{2}} \int_{\mathbb{R}^n} e^{ix\cdot\xi}\sigma(x,\xi)\widehat{\varphi}(\xi)d\xi\right\} \overline{\widehat{\psi}}(x)dx \\
&= \int_{\mathbb{R}^n} \left\{(2\pi)^{-\frac{n}{2}} \int_{\mathbb{R}^n} e^{ix\cdot\xi}\sigma(x,\xi)\overline{\widehat{\psi}}(x)dx\right\} \widehat{\varphi}(\xi)d\xi \\
&= \int_{\mathbb{R}^n} \left\{(2\pi)^{-\frac{n}{2}} \overline{\int_{\mathbb{R}^n} e^{ix\cdot\xi}\, \overline{\widetilde{\sigma}}(x,\xi)\widehat{\psi}(x)dx}\right\} \widehat{\varphi}(\xi)d\xi \\
&= \left(\widehat{\varphi}, \widetilde{T}_{\overline{\widetilde{\sigma}}}\psi\right) = \left(\widetilde{T}^*_{\overline{\widetilde{\sigma}}}\widehat{\varphi}, \psi\right), \quad \varphi, \psi \in \mathcal{S}. \qquad (4.1)
\end{aligned}$$

Since (4.1) is true for all ψ in $\mathcal{S}$, the conclusion of the lemma follows. □

Remark 1.4.2 *It is now obvious, based on the proof of Lemma 1.4.1, that for* $\sigma \in \dot{S}^0$, *we have* $\widehat{T_\sigma\varphi} = \widetilde{T}_\tau\varphi$, *for all* $\varphi \in \mathcal{S}$, *where* τ *is a symbol in* $\dot{S}^0$ *and has the following asymptotic expansion*

$$\tau \sim \sum_\mu \frac{(-i)^{|\mu|}}{\mu!} \left(\partial_x^\mu \partial_\xi^\mu \widetilde{\sigma}\right), \qquad (4.2)$$

Here (4.2) *means that*

$$\tau - \sum_{|\mu|<N} \frac{(-i)^{|\mu|}}{\mu!} \left(\partial_x^\mu \partial_\xi^\mu \widetilde{\sigma}\right)$$

is a symbol in $\dot{S}^0_{-N}$ *for every positive integer* N.

We can now give the proof of Theorem 1.1.1.

Proof of Theorem 1.1.1. We already saw in Lemma 1.4.1 that for $\sigma \in \dot{S}^0$, $\widehat{T_\sigma\varphi} = \widetilde{T}_\tau\widehat{\varphi}$, for $\varphi \in \mathcal{S}$, where $\tau \in \dot{S}^0$. Thus, we cam employ the L^p, $1 < p < \infty$, boundedness theorem for pseudo-differential operators with classical symbols to get a positive constant C depending on n and p only such that

$$||\widehat{T_\sigma\varphi}||_p = ||\widetilde{T}_\tau\widehat{\varphi}||_p \leq C\, ||\widehat{\varphi}||_p, \quad \varphi \in \mathcal{S},$$

which proves the $\dot{H}^{0,p}$ boundedness of the pseudo-differential operator T_σ by a standard density argument.

Next, we are going to show that if $\sigma \in \dot{S}^m$, then T_σ is a bounded linear operator from $\dot{H}^{m,p}$ into $\dot{H}^{0,p}$. For this, we cosider the bounded linear operators

$$\begin{aligned} J_{-s} &: \dot{H}^{s,p} \to \dot{H}^{0,p}, \\ T_\sigma J_m &: \dot{H}^{0,p} \to \dot{H}^{0,p}, \\ J_{s-m} &: \dot{H}^{0,p} \to \dot{H}^{s-m,p}, \end{aligned}$$

where we denote by J_t, $t \in \mathbb{R}$ the pseudo-differential operator with classical symbol $\lambda(\xi) = \langle\xi\rangle^{-t}$, $\xi \in \mathbb{R}^n$. The first and third operators are obviously bounded, while the second is bounded since its symbol is equal to $\tau(x,\xi) = \sigma(x,\xi)\langle\xi\rangle^{-m}$ and it belongs to $\dot{S}^0$. To see this, we compute for any two multi-indices α and β,

$$\begin{aligned} \left|\left(\partial_x^\alpha \partial_\xi^\beta \tau\right)(x,\xi)\right| &= \left|\sum_{\delta \le \beta} \binom{\beta}{\delta} \partial_\xi^\delta \{\langle\xi\rangle^{-m}\} \left(\partial_\xi^{\beta-\delta}\partial_x^\alpha \sigma\right)(x,\xi)\right| \\ &\le \sum_{\delta \le \beta} \binom{\beta}{\delta} C_\delta (1+|\xi|)^{-m-|\delta|} C_{\alpha,\beta,\delta}(1+|x|)^{-|\alpha|}(1+|\xi|)^m \\ &= \sum_{\delta \le \beta} \binom{\beta}{\delta} C_{\alpha,\beta,\delta}(1+|\xi|)^{-|\delta|}(1+|x|)^{-|\alpha|} \\ &\le C_{\alpha,\beta}(1+|x|)^{-|\alpha|}, \quad x,\xi \in \mathbb{R}^n. \end{aligned}$$

Thus the product $J_{s-m}T_\sigma J_{m-s}$ is a bounded linear operator from $\dot{H}^{s,p}$ into $\dot{H}^{s-m,p}$. Since J_{m-s} and J_{s-m} are isometric and onto, it follows immediately that for $1 < p < \infty$ and $s \in \mathbb{R}$, $T_\sigma : \dot{H}^{m,p} \to \dot{H}^{0,p}$ is a bounded linear operator.

For the conclusion, we note that in view of Theorem 1.3.5, the product $J_{m-s}T_\sigma$ is a pseudo-differential operator with symbol in $\dot{S}^s$. Hence, by the previous calculations, we can find a positive constant C depending on n and p, only, such that

$$||T_\sigma u||_{s-m,p} = ||J_{m-s}T_\sigma u||_{0,p} \le C||u||_{s,p}, \quad u \in \dot{H}^{s,p},$$

and the proof of the theorem is over. □

Remark 1.4.3 *Lemma 1.4.1 provides us with the necessary insight into generalizing the symbols in* $\dot{S}^m$. *We define* $\dot{S}^m_{1,\delta}$, $0 \le \delta < 1$, *to be the set of all* $C^\infty(\mathbb{R}^n \times \mathbb{R}^n)$ *functions such that for any two multi-indices* α *and* β, *there exists a positive constant* $C_{\alpha,\beta}$ *such that*

$$\left|\left(D_x^\alpha D_\xi^\beta \sigma\right)\right| \le C_{\alpha,\beta}(1+|x|)^{-|\alpha|+\delta|\beta|}(1+|\xi|)^m, \quad x,\xi \in \mathbb{R}^n. \tag{4.3}$$

Then, the conclusion of Theorem 1.1.1 is valid for the symbols in $\dot{S}^m_{1,\delta}$. *Further generalizations of* (4.3) *which guarantees the validity of Theorem 1.1.1 should be mentioned. First we can consider symbols with singularities at* $x = 0$ *and secondly* (4.3) *is required only for some derivatives, e.g.,* $|\alpha|$, $|\beta| \le N$, *where* N *depends only on the dimension of the underlying space* $\mathbb{R}^n$.

The Fourier integral operators that are natural in our setting are given by

$$\left(T_{(\sigma,\phi)}\varphi\right)(x) = \int_{\mathbb{R}^n} e^{ix\cdot\xi+i\phi(x,\xi)}\sigma(x,\xi)\widehat{\varphi}(\xi)d\xi, \ \ \varphi \in \mathcal{S}, \tag{4.4}$$

where ϕ is the real-valued phase function. In other words, Fourier integral operators can be viewed as pseudo-differential operators with the symbol $e^{i\phi}\sigma$. Various choices of symbols σ and various restrictions for the real phase function ϕ have lead mathematicians to studying the mapping properties of (4.4) mostly in L^p and Besov spaces. We refer to Chapter 8 Section 6.2 in the book [10] by Treves and Section 2.5.2 in the book [6] by Hörmander for the L^2 setting, to the article [2] by Beals for the L^p setting, $1 < p < \infty$, and to the articles [8, 9] by Qing-jiu for Besov spaces.

We are going to show that the theory developed so far guarantees good global mapping properties of operators $T_{(\sigma,\phi)}$ under minimal restrictions on σ and ϕ. In order to fix ideas and motivate our next, more general, considerations, let σ be the "variable coefficient" polynomial symbol $\sigma(x,\xi) = \sum_{|\alpha|\le m} a_\alpha(x)\xi^\alpha$, where a_α have the property that for any multi-index β, there exists a positive constant $C_{\alpha,\beta}$ such that

$$\left|\left(D^\beta a_\alpha\right)(x)\right| \le C_{\alpha,\beta}(1+|x|)^{-|\beta|}, \ \ x \in \mathbb{R}^n.$$

Obviously, σ can be viewed as both a classical symbol in S^m and an element of $\dot{S}^m$.

Let ϕ be a real-valued, independent of x, classical symbol in S^1. Then, after an easy computation we conclude that $e^{i\phi}\sigma$ is a symbol in $\dot{S}^m$, and so, by Theorem 1.1.1, the Fourier integral operator $T_{(\sigma,\phi)}$ is bounded from $\dot{H}^{s,p}$ into $\dot{H}^{s-m,p}$, for $1 < p < \infty$ and $s \in \mathbb{R}$.

All considerations made above lead us to define S_m as the set of all $C^\infty(\mathbb{R}^n\times\mathbb{R}^n)$ functions with the property that for all multi-indices α and β there exists a positive constant $C_{\alpha,\beta}$ such that

$$\left|\left(D_x^\alpha D_\xi^\beta \sigma\right)(x,\xi)\right| \le C_{\alpha,\beta}(1+|x|)^{-|\alpha|}(1+|\xi|)^{m-|\beta|}, \ \ x,\xi \in \mathbb{R}^n.$$

The following theorem is a restatement, in a particular case, of Theorem 1.1.1.

Theorem 1.4.4 *Let* σ *be a symbol in* S_m *and* ϕ *a real-valued, independent of* x, *symbol in* S^1. *Then the Fourier integral operator* $T_{(\sigma,\phi)}$ *is bounded from* $\dot{H}^{s,p}$ *into* $\dot{H}^{s-m,p}$ *for* $1 < p < \infty$ *and* $s \in \mathbb{R}$.

Remark 1.4.5 *Results similar to Theorem 1.4.4 can be obtained for* $\phi \in S_1$ *symbol in the both variables* x *and* ξ, *but, in this case,* σ *has to belong to* S_{-N+m}, *where* N *is as in Remark 1.4.3.*

1.5 REGULARITY OF SOLUTIONS OF SEMILINEAR PSEUDO-DIFFERENTIAL EQUATIONS

In the theory of regularity of solutions of partial differential equations, an important rôle is played by a special class of pseudo-differential operators called elliptic operators. They are important in applications because they have approximate inverses that are also pseudo-differential operators. We have to make all these concepts precise for our class of operators.

A symbol $\sigma \in \dot{S}^m$ is said to be elliptic if there exist positive constants C and R such that

$$|\sigma(x,\xi)| \geq C(1+|\xi|)^m, \quad |x|+|\xi| \geq R.$$

Of course, a pseudo-differential operator T_σ is said to be elliptic if its symbol is elliptic. The importance of ellipticity is revealed by the following theorem.

Theorem 1.5.1 *Let σ be an elliptic symbol in $\dot{S}^m$. Then there exists a symbol τ in $\dot{S}^{-m}$ such that*

$$T_\tau T_\sigma = I + S$$

and

$$T_\sigma T_\tau = I + \widetilde{S},$$

where S and $\widetilde{S}$ are pseudo-differential operators with symbols in $\bigcap_{k\in\mathbb{R}} \dot{S}^0_k$, and I is the identity operator.

The proof of Theorem 1.5.1 can be obtained, with some modifications due to the particularities of our class of symbols, by using the methods in, e.g., Section 1, Chapter 2 in the book [4] by Cordes.

More useful to us are elliptic operators with symbols in S_m. A minor modification of the proof of Theorem 1.5.1 leads to the following result.

Theorem 1.5.2 *Let σ be an elliptic symbol in S_m. Then there exists a symbol τ in S_{-m} such that*

$$T_\tau T_\sigma = I + S$$

and

$$T_\sigma T_\tau = I + \widetilde{S},$$

where S and $\widetilde{S}$ are pseudo-differential operators with symbols in $\bigcap_{k\in\mathbb{R}} \dot{S}^k$, and I is the identity operator.

The basic result on the regularity of solutions of linear pseudo-differential equations on $\mathbb{R}^n$ is provided by the following lemma.

Lemma 1.5.3 *Let $\sigma \in S_m$, $m > 0$, be an elliptic symbol, and $f \in \dot{H}^{s,p}$. Then any solution u in $\bigcup_{t\in\mathbb{R}} \dot{H}^{t,p}$ of the equation $T_\sigma u = f$ is in $\dot{H}^{s+m,p}$.*

Proof. Since σ is elliptic, it follows from Theorem 1.5.2 that there exists $\tau \in S_{-m}$ and $\kappa \in \bigcap_{k\in\mathbb{R}} \dot{S}^k$ such that

$$T_\tau T_\sigma = I + T_\kappa,$$

where I is the identity operator. Hence

$$u = T_\tau f - T_\kappa u.$$

By Theorem 1.1.1, $T_\tau f \in \dot{H}^{s+m,p}$. Since $\kappa \in \bigcap_{k\in\mathbb{R}} \dot{S}^k$, it follows from Theorem 1.1.1 again that $T_\kappa u \in \dot{H}^{s+m,p}$ as well. □

We consider next the semilinear pseudo-differential equation $T_\sigma u = P(u)$ on $\mathbb{R}^n$, where $\sigma \in S_m$, $m > 0$, is an elliptic symbol, and $P(u)$ is a polynomial in u. We end this section with the following regularity theorem. It should be acknowledged that this theorem is a generalization of Theorem 4.4 in the paper [7] by Pi and Wong.

Theorem 1.5.4 *Let $u \in \dot{H}^{s,p}$ be a solution of $T_\sigma u = P(u)$, where $1 < p < \infty$, $s > \frac{n}{p'}$ and $\frac{1}{p} + \frac{1}{p'} = 1$. Then $u \in C^\infty(\mathbb{R}^n)$ after modification of the function u on a set of measure zero.*

Proof. Since $u \in \dot{H}^{s,p}$, by the Banach algebra property of Hörmander spaces, $P(u) \in \dot{H}^{s,p}$. Hence by Lemma 1.5.3, $u \in \dot{H}^{s+m,p}$. By the Banach algebra property of the Hörmander spaces applied again, $u \in \dot{H}^{s+m,p}$. If we apply the argument repeatedly, we conclude that $u \in \bigcap_{t\geq s} \dot{H}^{t,p}$, and the proof is complete. □

References

[1] Beals, M. (1979). *L^p and Hölder estimates for pseudodifferential operators: Necessary conditions*, Harmonic analysis in Euclidean spaces, Proc. Symp. Pure Math., Vol. 35(II), Providence, Amer. Math. Soc., (pages 153–157).

[2] Beals, M. (1982). *L^p boundedness of Fourier integral operators*, Mem. Amer. Math. Soc., Vol. 38(264).

[3] Beals, M. (1989). *Propagation and Interaction of Singularities in Nonlinear Hiperbolic Problems*, Birkhauser.

[4] Cordes, H. O. (1995). *The Technique of Pseudodifferential Operators*, Cambridge University Press.

[5] Hörmander, L. (1983). *The Analysis of Linear Partial Differential Operators* II, Springer-Verlag.

[6] L. Hörmander, L. (1983/83). *The Analysis of Linear Partial Differential Operators* IV, Springer-Verlag.

[7] Pi, L. and M.W. Wong. (1992). *On a generalization of Schauder's theorem*, in *Proceedings of Iinternational Conference on Nonlinear Partial Differential Equations, Zhejang University, R.P. China, June 1992, (eds. G. Dong and F. Lin)*, International Academic Press, (1993), (pages 222–232).

[8] Qing-jiu, Q. (1985). *The Besov space boundedness for certain Fourier integral operators*, Acta Math. Scientia, Vol. 5, (pages 167–174).

[9] Q. Qing-jiu, Q. (1986). *On L^p-estimates for certain Fourier integral operators*, Scientia Sinica, Ser. *A*, Vol. 29, (pages 350–362).

[10] Treves, F. (1980). *Introduction to Pseudodifferential and Fourier Integral Operators* II, Plenum Press.

[11] Zaidman, S. (1970). *Certaines classes d' operateurs pseudo-differentiels*, J. Math. Anal. Appl., Vol. 30, (pages 522–563).

[12] Zaidman, S. (1972). *Pseudo-differential operators*, Ann. Mat. Pura. Appl., Vol. 92, (pages 345–399).

COEFFICIENT IDENTIFICATION IN ELLIPTIC DIFFERENTIAL EQUATIONS

Ian Knowles *

Department of Mathematics
University of Alabama at Birmingham
Birmingham, AL 35294

Abstract: An outline is given for new variational approach to the problem of computing the (possibly discontinuous) coefficient functions p, q, and f in elliptic equations of the form $-\nabla \cdot (p(x)\nabla u) + \lambda q(x)u = f, \quad x \in \Omega \subset \mathbb{R}^n$, from a knowledge of the solutions u.

1.1 INTRODUCTION

Consider the differential equation

$$Lv = -\nabla \cdot (p(x)\nabla v) + \lambda q(x)v = f(x), \quad x \in \Omega, \tag{1.1}$$

where Ω is an open bounded set in $\mathbb{R}^n$,

$$f \in \mathcal{L}^2(\Omega), \quad q \in \mathcal{L}^\infty(\Omega), \quad \text{and} \quad p \in \mathcal{L}^\infty(\Omega), \tag{1.2}$$

*Supported in part by US National Science Foundation grant DMS-9505047. This is an expanded version of a lecture given at the ISAAC97 Conference, University of Delaware, June 3–7, 1997

R.P. Gilbert et al.(eds.), Direct and Inverse Problems of Mathematical Physics, 149–160.

are real and p satisfies

$$p(x) \geq \nu > 0, \quad x \in \Omega. \tag{1.3}$$

In addition, we assume that the real constant λ, and q, are chosen so that the homogeneous Dirichlet operator $L = L_{p,q}$ (i.e. L acting on $W_0^{1,2}(\Omega)$) satisfies

$$L \text{ is a positive operator in } \mathcal{L}^2(\Omega); \tag{1.4}$$

for later use we note that as $q \in \mathcal{L}^\infty(\Omega)$, (1.4) is true for all $|\lambda|$ small enough. It is known [6, Chapter 8] that the generalized Dirichlet problems associated with (1.1) are uniquely solvable, and that the solutions v lie in the Sobolev space $W^{1,2}(\Omega)$. Our main concern here is the corresponding collection of inverse problems: given v (for one or more values of λ), find one or more of the coefficient functions p, q, and f.

These inverse problems arise in connection with groundwater flow (and oil reservoir simulation); in such cases the flow in the porous medium is governed by the diffusion equation:

$$\nabla \cdot [p(x)\nabla w(x,t)] = S(x)\frac{\partial w}{\partial t} - R(x,t), \tag{1.5}$$

in which w represents the piezometric head, p the hydraulic conductivity (or sometimes, for a two dimensional aquifer, the transmissivity), R the recharge, and S the storativity of the aquifer (see, for example, [2, p. 214]). It has long been recognized among hydro-geologists [1, Chapter 8] that the inability to obtain reliable values for the coefficients in (1.5) is a serious impediment to the confident use of such models. Methods that have been employed range from educated guesswork (referred to as "trial and error calibration" in the hydrology literature – the method preferred by most practitioners at this time [1, p. 226]) to various attempts at "automatic calibration" (see [3, 4, 5, 10, 11, 12] as part of the extensive literature on these inverse problems).

In the steady-state case we have that $\frac{\partial w}{\partial t} = 0$, and $R = R(x)$; if R is presumed known, the inverse problem reduces to finding p from a knowledge of w. In [7, 9] a new approach to this reduced problem was given in terms of finding unique minima for certain convex functionals. This approach was shown to be effective in the presence of mild discontinuities in the coefficients, an important practical consideration in view of the fact that fractures in the porous media are commonly encountered. We present here an extension of these ideas to cover the general equation (1.1). The basic value of such an extension can be seen by looking at (1.5) in the case $R = R(x)$. In this situation, (1.5) can be transformed to (1.1) by (for example) applying a Laplace transform to the time variable. The raw data needed consists of head measurements taken over both space and time, as well as hydraulic conductivity (or transmissivity) values on the spatial boundary; such data is in general readily available, and reasonably accurate. In principle, one can then use an appropriate functional (of the type discussed below) to recover p, S, and R. We note in passing that while there are other methods to obtain p (mainly from steady-state data on the heads), it has been observed ([1, p. 152 and p. 197]) that there are essentially no universally applicable methods for estimating R and S and most practitioners use quite rough estimates of these parameters. This leads to instabilities in the model, especially when transient simulations are involved.

The functionals used in [7, 9] may be generalized as follows. Let a solution u of (1.1) be given for which P, Q, and F are the coefficients corresponding to p, q, and f, respectively, that we seek to compute. For functions p, q, f satisfying (1.2, 1.3, 1.4), let $v = u_{p,q,f}$ denote the solution of the boundary value problem determined by (1.1) and

$$v|_{\partial\Omega} = u|_{\partial\Omega}. \tag{1.6}$$

Thus $u = u_{P,Q,F}$. Define

$$\mathcal{D}_G = \{(p,q,f) : p, q, f \text{ satisfy } (1.2), (1.3), (1.4) \text{ and } p|_\Gamma = P|_\Gamma\}$$

where Γ is a hypersurface in Ω transversal to ∇u. It is convenient to take Γ to be the boundary of the bounded region Ω, and we henceforth assume this to be so. For (p,q,f) in D_G define

$$G(p,q,f) = \int_\Omega p(x)(|\nabla u|^2 - |\nabla u_{p,q,f}|^2) + \lambda q(x)(u^2 - u^2_{p,q,f}) - 2f(x)(u - u_{p,q,f})\, dx \tag{1.7}$$

The functional studied in [8, 9] corresponds to $p = 1$ and $f = 0$, while that in [7] corresponds to $q = f = 0$.

1.2 PROPERTIES OF THE FUNCTIONAL G

Some of the properties of G are summarized in the following theorem:

Theorem 1

(a) For any $c = (p,q,f)$ *in* $\mathcal{D}_G$,

$$G(c) = \int_\Omega p(x)|\nabla(u - u_c)|^2 + \lambda q(x)(u - u_c)^2\, dx = (L_{p,q}(u - u_c), u - u_c). \tag{1.8}$$

(b) $G(c) \geq 0$ *for all* $c = (p,q,f)$ *in* $\mathcal{D}_G$, *and* $G(c) = 0$ *if and only if* $u = u_c$.

(c) For $c_1 = (p_1, q_1, f_1)$ and $c_2 = (p_2, q_2, f_2)$ in $\mathcal{D}_G$ we have

$$\begin{aligned} G(c_1) - G(c_2) &= \int_\Omega (p_1 - p_2)\left(|\nabla u|^2 - \nabla u_{c_1} \cdot \nabla u_{c_2}\right) + \\ &\qquad + \lambda (q_1 - q_2)(u^2 - u_{c_1} u_{c_2}) - \\ &\qquad\qquad - 2(f_1 - f_2)(u - \frac{u_{c_1} + u_{c_2}}{2}). \end{aligned} \tag{1.9}$$

(d) The first Gâteaux differential for G is given by

$$\begin{aligned} G'(p, q, f)[h_1, h_2, h_3] &= \int_\Omega \left(|\nabla u|^2 - |\nabla u_c|^2\right) h_1 + \\ &\qquad + \lambda (u^2 - u_c^2) h_2 - \\ &\qquad\qquad - 2(u - u_c) h_3, \end{aligned} \tag{1.10}$$

for $h_1, h_2 \in \mathcal{L}^\infty(\Omega)$ with $h_1|_{\partial\Omega} = 0$, and $h_3 \in \mathcal{L}^2(\Omega)$, and $G'(p, q, f) = 0$ if and only if $u = u_c$.

(e) The second Gâteaux differential of G is given by

$$G''(p, q, f)[h, k] = 2\left(L_{p,q}^{-1}(e(h)), e(k)\right), \tag{1.11}$$

where $h = (h_1, h_2, h_3)$, $k = (k_1, k_2, k_3)$, and the functions $h_1, h_2, k_1, k_2 \in \mathcal{L}^\infty(\Omega)$, with $h_1|_{\partial\Omega} = k_1|_{\partial\Omega} = 0$, $h_3, k_3 \in \mathcal{L}^2(\Omega)$,

$$e(h) = -\nabla \cdot (h_1 \nabla u_{p,q,f}) + \lambda h_2 u_{p,q,f} - h_3,$$

and $(\cdot, \cdot)$ denotes the usual inner product in $\mathcal{L}^2(\Omega)$.

Proof. If v is a solution of the generalized Dirichlet problem (1.1,1.6), by the standard theory (see for example [6, Chapter 8]) we have

$$(Lv, \phi) = \int_\Omega p(x) \nabla v \cdot \nabla \phi + \lambda q(x) v \phi \, dx, \tag{1.12}$$

and hence, by (1.1),

$$\int_\Omega \phi \nabla \cdot (p(x) \nabla v) \, dx = -\int_\Omega p(x) \nabla v \cdot \nabla \phi \, dx, \tag{1.13}$$

for any function $v \in W^{1,2}(\Omega)$ and any $\phi \in W_0^{1,2}(\Omega)$. The latter formula is essentially Green's formula for this situation ("integration by parts") and will be much used in the rest of this proof.

In the sequel, it will be convenient to set $c = (p, q, f)$. Observe that

$$\begin{aligned}
G(c) &= \int_\Omega p|\nabla(u-u_c)|^2 + 2p\nabla u_c \cdot \nabla(u-u_c) + \lambda q(u^2 - u_c^2) - 2f(u-u_c) \\
&= \int_\Omega p|\nabla(u-u_c)|^2 - 2(u-u_c)\nabla\cdot(p\nabla u_c) + \lambda q(u^2-u_c^2) - 2f(u-u_c) \\
&\qquad \text{from (1.13), using } \phi = u - u_c \in W_0^{1,2}(\Omega), \\
&= \int_\Omega p|\nabla(u-u_c)|^2 - 2(u-u_c)(\lambda q u_c - f) + \lambda q(u^2-u_c^2) - 2f(u-u_c),
\end{aligned}$$

from (1.1), and the first result in (1.8) follows after some rearrangement; the remaining part of (1.8) follows by a further integration by parts, and the results in (b) follow from this and (1.4).

For $c_1 = (p_1, q_1, f_1)$ and $c_2 = (p_2, q_2, f_2)$ in $\mathcal{D}_G$ we have from (1.7) that

$$\begin{aligned}
&G(c_1) - G(c_2) \\
&= \int_\Omega p_1(|\nabla u|^2 - |\nabla u_{c_1}|^2) - p_2(|\nabla u|^2 - |\nabla u_{c_2}|^2) + \lambda q_1(u^2 - u_{c_1}^2) - \\
&\qquad -\lambda q_2(u^2 - u_{c_2}^2) - 2f_1(u-u_{c_1}) + 2f_2(u-u_{c_2}) \\
&= \int_\Omega p_1\nabla(u_{c_2}+u_{c_1})\cdot\nabla(u_{c_2}-u_{c_1}) + \lambda q_1(u_{c_2}^2 - u_{c_1}^2) + \\
&\qquad +(p_1-p_2)(|\nabla u|^2 - |\nabla u_{c_2}|^2) + \lambda(q_1-q_2)(u^2-u_{c_2}^2) - \\
&\qquad -2f_1(u-u_{c_1}) + 2f_2(u-u_{c_2}) \\
&= \int_\Omega (u_{c_1}-u_{c_2})(\nabla\cdot(p_1\nabla(u_{c_1}) + \nabla\cdot(p_1\nabla u_{c_2})) + \lambda q_1(u_{c_2}^2 - u_{c_1}^2) + \\
&\qquad +(p_1-p_2)(|\nabla u|^2 - |\nabla u_{c_2}|^2) + \lambda(q_1-q_2)(u^2-u_{c_2}^2) - \\
&\qquad -2f_1(u-u_{c_1}) + 2f_2(u-u_{c_2}) \\
&= \int_\Omega (u_{c_1}-u_{c_2})\{\nabla\cdot p_1\nabla u_{c_1} + \nabla\cdot p_2\nabla u_{c_2} + \nabla\cdot((p_1-p_2)\nabla u_{c_2})\} + \\
&\qquad +(p_1-p_2)(|\nabla u|^2 - |\nabla u_{c_2}|^2) + \lambda q_1(u_{c_2}^2 - u_{c_1}^2) \\
&\qquad +\lambda(q_1-q_2)(u^2-u_{c_2}^2) - 2f_1(u-u_{c_1}) + 2f_2(u-u_{c_2}) \\
&= \int_\Omega (u_{c_1}-u_{c_2})(\lambda q_1 u_{c_1} + \lambda q_2 u_{c_2} - f_1 - f_2) + \\
&\qquad -(p_1-p_2)\nabla u_{c_2}\cdot\nabla(u_{c_1}-u_{c_2}) + (p_1-p_2)(|\nabla u|^2 - |\nabla u_{c_2}|^2) + \\
&\qquad +\lambda q_1(u_{c_2}^2 - u_{c_1}^2) + \lambda(q_1-q_2)(u^2-u_{c_2}^2) - \\
&\qquad -2f_1(u-u_{c_1}) + 2f_2(u-u_{c_2}),
\end{aligned}$$

by (1.1) and an integration by parts, using $u_{c_1} - u_{c_2} \in W_0^{1,2}(\Omega)$. Part (c) now follows after some rearrangement.

In order to prove (d) and (e) we need two ancillary results. First we note that, for c and h as above (and fixed),

$$\lim_{\epsilon \to 0} u_{c+\epsilon h} = u_c \tag{1.14}$$

in $W^{1,2}(\Omega)$. To see this, we subtract the equations

$$-\nabla \cdot (p \nabla u_c) + \lambda q u_c = f \tag{1.15}$$

$$-\nabla \cdot ((p + \epsilon h_1) \nabla u_{c+\epsilon h}) + \lambda (q + \epsilon h_2) u_{c+\epsilon h} = f + \epsilon h_3 \tag{1.16}$$

to obtain

$$L_{p,q}(u_{c+\epsilon h} - u_c) = \epsilon (\nabla \cdot (h_1 \nabla u_{c+\epsilon h}) - \lambda h_2 u_{c+\epsilon h} + h_3) \tag{1.17}$$

If this equation is multiplied on both sides by $u_{c+\epsilon h} - u_c$ and integrated over Ω we arrive (after the usual integration by parts) at

$$\begin{aligned}
&\int_\Omega p|\nabla(u_{c+\epsilon h} - u_c)|^2 + \lambda q (u_{c+\epsilon h} - u_c)^2 \\
&= \epsilon \int_\Omega h_1 \nabla u_{c+\epsilon h} \cdot \nabla (u_{c+\epsilon h} - u_c) - \lambda h_2 u_{c+\epsilon h}(u_{c+\epsilon h} - u_c) + h_3 (u_{c+\epsilon h} - u_c) \\
&= \epsilon \int_\Omega h_1 |\nabla (u_{c+\epsilon h} - u_c)|^2 + h_1 \nabla u_c \cdot \nabla (u_{c+\epsilon h} - u_c) - \\
&\qquad - \lambda h_2 (u_c (u_{c+\epsilon h} - u_c) + (u_{c+\epsilon h} - u_c)^2) + h_3 (u_{c+\epsilon h} - u_c) \\
&\leq \epsilon \int_\Omega |h_1||\nabla (u_{c+\epsilon h} - u_c)|^2 + |h_1/2|(|\nabla u_c|^2 + |\nabla (u_{c+\epsilon h} - u_c)|^2) + \\
&\qquad + |\lambda||h_2/2|(u_c^2 + (u_{c+\epsilon h} - u_c)^2) + |\lambda||h_2|(u_{c+\epsilon h} - u_c)^2 + \\
&\qquad\qquad + (1/2)(|h_3|^2 + (u_{c+\epsilon h} - u_c)^2),
\end{aligned}$$

after repeated use of the inequality $ab \leq (a^2 + b^2)/2$.

Now, for $|\lambda|$ small enough (this is where the need for assumptions (1.3,1.4) becomes apparent), the term on the left of the above inequality is bounded below by a constant multiple of $||u_{c+\epsilon h} - u_c||_{W^{1,2}(\Omega)}$. For ϵ small enough all the terms in $u_{c+\epsilon h} - u_c$ on the right side of the inequality can be moved to the left, and the resulting left side then can be bounded below by a (smaller) constant multiple of $||u_{c+\epsilon h} - u_c||_{W^{1,2}(\Omega)}$. As the remaining terms on the right are $O(\epsilon)$, the result (1.14) now follows.

We also need to know that for any function $\eta \in \mathcal{L}^\infty(\Omega)$

$$||\nabla \cdot (\eta \nabla u_{c+\epsilon h})||_{W^{1,2}(\Omega)} \leq K \tag{1.18}$$

where the constant K does not depend on ϵ. To see this, note that the functional F defined on $W_0^{1,2}(\Omega)$ by $F(\phi) = \int_\Omega \eta \nabla u_{c+\epsilon h} \cdot \nabla \phi$ satisfies

$$|F(\phi)| \leq K||\phi||_{W^{1,2}(\Omega)}, \tag{1.19}$$

where it follows from (1.14) that the constant does not depend on ϵ. Consequently, $F \in (W_0^{1,2}(\Omega))^*$. If we use the Riesz representation theorem to identify $(W_0^{1,2}(\Omega))^*$ with $W_0^{1,2}(\Omega)$, F is identified with a unique element of $W_0^{1,2}(\Omega)$ which we may take to be $\nabla \cdot (\eta \nabla u_{c+\epsilon h})$, and $||F|| = ||\nabla \cdot (\eta \nabla u_{c+\epsilon h})||_{W^{1,2}(\Omega)}$; the estimate (1.18) then follows from (1.19).

Now, from (1.7) and some algebra,

$$\begin{aligned}
&(G(c+\epsilon) - G(c))/\epsilon \\
= &\int_\Omega h_1(|\nabla u|^2 - |\nabla u_{c+\epsilon h}|^2) + \lambda h_2(u^2 - u_{c+\epsilon h}^2) - 2h_3(u - u_{c+\epsilon h}) \\
&+\epsilon^{-1} \int_\Omega p(|\nabla u_c|^2 - |\nabla u_{c+\epsilon h}|^2) + \\
&\qquad +\lambda q(u_c^2 - u_{c+\epsilon h}^2) - 2f(u_c - u_{c+\epsilon h})
\end{aligned} \tag{1.20}$$

By (1.14) it is sufficient to show that the second integral expression above tends to zero as $\epsilon \to 0$. But,

$$\begin{aligned}
&\epsilon^{-1} \int_\Omega p(|\nabla u_c|^2 - |\nabla u_{c+\epsilon h}|^2) \\
= \ &\epsilon^{-1} \int_\Omega p\nabla(u_c - u_{c+\epsilon h}) \cdot \nabla(u_c + u_{c+\epsilon h}) \\
= \ &\epsilon^{-1} \int_\Omega (u_{c+\epsilon h} - u_c)\nabla \cdot (p\nabla(u_c + u_{c+\epsilon h})), \\
&\text{after an integration by parts,} \\
= \ &\epsilon^{-1} \int_\Omega (u_{c+\epsilon h} - u_c)\{(\lambda q u_c - f) + \\
&\qquad +\lambda(q + \epsilon h_2)u_{c+\epsilon h} - (f + \epsilon h_3) - \epsilon \nabla \cdot (h_1 \nabla u_{c+\epsilon h})\}, \\
&\text{using (1.15) and (1.16),} \\
= \ &\int_\Omega (u_{c+\epsilon h} - u_c)\{-\nabla \cdot (h_1 \nabla u_{c+\epsilon h}) + \lambda h_2 u_{c+\epsilon h} - h_3\} + \\
&+\epsilon^{-1} \int_\Omega \lambda q(u_{c+\epsilon h}^2 - u_c^2) - 2f(u_{c+\epsilon h} - u_c)
\end{aligned}$$

Consequently, the second integral expression in (1.20) equals

$$\int_\Omega (u_{c+\epsilon h} - u_c)\{-\nabla\cdot(h_1\nabla u_{c+\epsilon h}) + \lambda h_2 u_{c+\epsilon h} - h_3\} \qquad (1.21)$$

and this tends to zero as $\epsilon \to 0$ by (1.14) and (1.18); (d) is thus established.

Finally, the second Gâteaux differential is given by

$$G''(c)[h,k] = \lim_{\epsilon\to 0} \frac{G'(c+\epsilon h)[k] - G'(c)[k]}{\epsilon}$$

From (1.9) and some algebra

$$\begin{aligned}
&(G'(c+\epsilon h)[k] - G'(c)[k])/\epsilon \\
= \;& \epsilon^{-1}\int_\Omega (|\nabla u_c|^2 - |\nabla u_{c+\epsilon h}|^2)k_1 + \lambda(u_c^2 - u_{c+\epsilon h}^2)k_2 - 2(u_c - u_{c+\epsilon h})k_3 \\
= \;& \epsilon^{-1}\int_\Omega k_1\nabla(u_c - u_{c+\epsilon h})\cdot\nabla(u_c + u_{c+\epsilon h}) + \\
& \qquad\qquad + \lambda(u_c^2 - u_{c+\epsilon h}^2)k_2 - 2(u_c - u_{c+\epsilon h})k_3 \\
= \;& \epsilon^{-1}\int_\Omega (u_{c+\epsilon h} - u_c)\{\nabla\cdot(k_1\nabla(u_c + u_{c+\epsilon h})) - \lambda(u_c + u_{c+\epsilon h})k_2 + 2k_3\}, \\
& \text{after an integration by parts,} \\
= \;& \int_\Omega L_{p,q}^{-1}(-\nabla\cdot(h_1\nabla u_{c+\epsilon h}) + \lambda h_2 u_{c+\epsilon h} - h_3)\{-\nabla\cdot(k_1\nabla(u_c + u_{c+\epsilon h})) + \\
& \qquad\qquad + \lambda(u_c + u_{c+\epsilon h})k_2 - 2k_3\}, \\
& \text{by (1.17),} \\
= \;& 2\int_\Omega L_{p,q}^{-1}(-\nabla\cdot(h_1\nabla u_c) + \lambda h_2 u_c - h_3)\{-\nabla\cdot(k_1\nabla u_c) + \lambda u_c k_2 - k_3\} + \\
& + \int_\Omega L_{p,q}^{-1}(-\nabla\cdot(h_1\nabla(u_{c+\epsilon h} - u_c)) + \lambda h_2(u_{c+\epsilon h} - u_c)) \times \\
& \qquad\qquad \times\{-\nabla\cdot(k_1\nabla(u_c + u_{c+\epsilon h})) + \lambda(u_c + u_{c+\epsilon h})k_2 - 2k_3\} + \\
& + \int_\Omega L_{p,q}^{-1}(-\nabla\cdot(h_1\nabla u_c) + \lambda h_2 u_c - h_3)\times \\
& \qquad\qquad \times\{-\nabla\cdot(k_1\nabla(u_{c+\epsilon h} - \nabla u_c)) + \lambda(u_{c+\epsilon h} - u_c)k_2\} \qquad (1.22)
\end{aligned}$$

It remains to show that the second and third integrals in (1.22) tend to zero as $\epsilon \to 0$. As the operator $L_{p,q}^{-1}$ is self-adjoint, if we set

$$w_\epsilon = -\nabla\cdot(k_1\nabla(u_c + u_{c+\epsilon h})) + \lambda(u_c + u_{c+\epsilon h})k_2 - 2k_3$$

the second integral may be rewritten as

$$\int_\Omega (-\nabla \cdot (h_1 \nabla (u_{c+\epsilon h} - u_c)) + \lambda h_2 (u_{c+\epsilon h} - u_c)) L_{p,q}^{-1} w_\epsilon$$

$$= \int_\Omega h_1 \nabla (u_{c+\epsilon h} - u_c) \cdot \nabla (L_{p,q}^{-1} w_\epsilon) + \lambda h_2 (u_{c+\epsilon h} - u_c)) L_{p,q}^{-1} w_\epsilon$$

Now, from (1.18), w_ϵ is uniformly bounded in ϵ in $\mathcal{L}^2(\Omega)$, and as $L_{p,q}^{-1}$ may be extended uniquely as a bounded linear operator from $\mathcal{L}^2(\Omega)$ to $W^{1,2}(\Omega)$, $L_{p,q}^{-1} w_\epsilon$ is bounded independently of ϵ in $W^{1,2}(\Omega)$. From the boundedness of ∇ on $W^{1,2}(\Omega)$ to $\mathcal{L}^2(\Omega) \times \mathcal{L}^2(\Omega)$ it follows that $|\nabla (L_{p,q}^{-1} w_\epsilon)|$ is bounded independently of ϵ in $\mathcal{L}^2(\Omega)$. From (1.14) it now follows that the second integral in (1.22) tends to zero with ϵ. Finally, note that $L_{p,q}^{-1}(-\nabla \cdot (h_1 \nabla u_c) + \lambda h_2 u_c - h_3)$ lies in $W^{1,2}(\Omega)$; that the third integral vanishes as $\epsilon \to 0$ follows via (1.14) after an integration by parts. This completes the proof of the theorem □

Some comments are in order. First, the differentials listed in (d) and (e) are actually Fréchet differentials, but we omit the proofs. From (1.4) and (1.11) the functional G is convex, but not necessarily strictly convex (see below). The functional introduced in [9] corresponds to the case $p = 1, f = 0$, and $h_1 = h_3 = k_1 = k_3 = 0$, while the functional in [7] corresponds to the case $q = f = 0$ and $h_2 = h_3 = k_2 = k_3 = 0$. As can be seen from the theorem, most of the properties that were important in the special cases are also present in the general result. In particular, part (c) generalizes [8, eq. (2.17)].

One notable exception concerns property (b). As in the earlier cases, the non-negativity of G is a direct consequence of the Dirichlet principle for the elliptic equation. But the uniqueness connection between the minimum of G and the coefficients p, q, f is the subject of on-going work. It will be shown elsewhere that if the coefficient f is presumed to be known, and if one is given solutions $u_1 = u_{P,Q,\lambda_1}$ and $u_2 = u_{P,Q,\lambda_2}$, where $\lambda_1 \neq \lambda_2$, then under certain conditions the functional $H(p,q) = G(p,q,\lambda_1) + G(p,q,\lambda_2)$ satisfies $H(p,q) = 0$ if and only if $p = P$ and $q = Q$; furthermore, H is strictly convex in that for any p, q the second differential H'' satisfies $H''(p,q)[h_1, h_2] = 0$ if and only $h_1 = h_2 = 0$. It is conjectured that in the general case if one had three solutions of (1.1), corresponding to different values of λ, then one could form a functional H containing three terms, and recover, uniquely, by minimization of this H, all three coefficients p, q, f.

1.3 SOME APPLICATIONS

While the recovery of multiple coefficients has not as yet been tested computationally, the recovery of single coefficients has been extensively investigated

Figure 1.1 Computed examples; $q = f = 0$

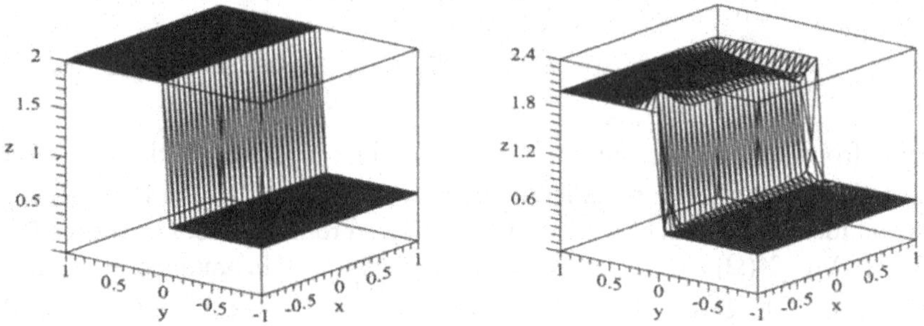

(a) $z = P_1(x, y)$, 31 × 31 grid

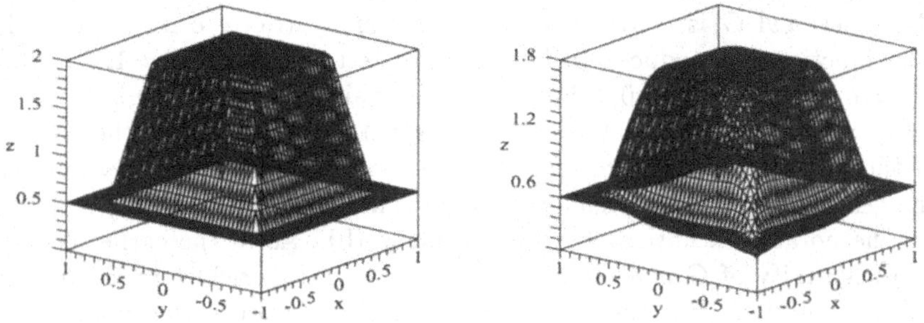

(b) $z = P_2(x, y)$, 49 × 49 grid

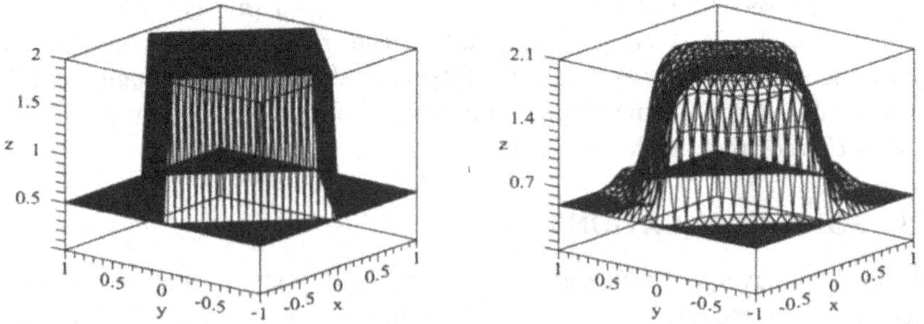

(c) $z = P_3(x, y)$, 31 × 31 grid

in [7, 9]. The method is remarkably effective, especially so in the case of coefficients (including principal coefficients) with mild discontinuities. Some results additional to those shown in [7] are given in Figure 1.1, with various choices of the function p, listed below:

$$P_1(x,y) = \begin{cases} 2, & \text{if } y > 0, \\ 0.5, & \text{otherwise;} \end{cases}$$

$$P_2(x,y) = \begin{cases} -6y+5, & \text{if } .5 \le y \le .75 \text{ and } |x| \le y, \\ 6x+5, & \text{if } -.75 \le x \le -.5 \text{ and } |y| \le -x, \\ 6y+5, & \text{if } -.75 \le y \le -.5 \text{ and } |x| \le -y, \\ -6x+5, & \text{if } .5 \le x \le .75 \text{ and } |y| \le x, \\ 2, & \text{if } |y| < .5 \text{ and } |x| < .5, \\ .5, & \text{otherwise;} \end{cases}$$

$$P_3(x,y) = \begin{cases} 2, & \text{if } |y| < x + .95 \text{ and } |x-y| < .95 \\ 0.5, & \text{otherwise.} \end{cases}$$

In each case, the correct p is on the left, and the p computed with a preconditioned steepest descent algorithm (see [7] for details) using the functional G (with $q = f = 0$) appears on the right. The function P_1 was recovered after about 100 steepest descent iterations, while P_2 and P_3 were recovered after 65 and 40 iterations, respectively.

References

[1] Mary P. Anderson and William W. Woessner. *Applied Groundwater Modeling.* Academic Press, New York, 1992.

[2] J. Bear. *Dynamics of Fluids in Porous Media.* American Elsevier, New York, 1972.

[3] J. Carrera. State of the art of the inverse problem applied to the flow and solute equations. In E. Custodio, editor, *Groundwater Flow and Quality Modeling,* pages 549–583. D. Reidel Publ. Co., 1988.

[4] J. Carrera and S.P. Neumann. Adjoint state finite element estimation of aquifer parameters under steady-state and transient conditions. In *Proceedings of the 5th International Conference on Finite Elements in Water Resources.* Springer-Verlag, 1984.

[5] R.L. Cooley and R.L. Naff. Regression modeling of groundwater flow. In *Techniques of Water-Resources Investigations*, number 03-B4. USGS, 1990.

[6] David Gilbarg and Neil S. Trudinger. *Elliptic Partial Differential Equations of Second Order.* Springer-Verlag, New York, 1977.

[7] Ian Knowles. Parameter identification for elliptic problems with discontinuous principal coefficients. *preprint*, 1997.

[8] Ian Knowles and Robert Wallace. A variational method for numerical differentiation,. *Numerische Mathematik*, 70:91–110, 1995.

[9] Ian Knowles and Robert Wallace. A variational solution of the aquifer transmissivity problem. *Inverse Problems*, 12:953–963, 1996.

[10] W. Menke. *Geophysical Data Analysis: Discrete Inverse Theory*. Academic Press, New York, 1989.

[11] A. Peck, S.M. Gorelick, G. De Marsily, S. Foster, and V. Kovalevsky. *Consequences of Spatial Variability in Aquifer Properties and Data Limitations for Groundwater Modeling Practice*. Number 175. International Association of Hydrologists, 1988.

[12] William W-G. Yeh. Review of parameter identification procedures in groundwater hydrology: The inverse problem. *Water Resources Research*, 22(2):95–108, 1986.

QUASI-EXPONENTIAL SOLUTIONS FOR SOME PDE WITH COEFFICIENTS OF LIMITED REGULARITY.

Alexander Panchenko

University of Delaware
Department of Mathematical Sciences
Newark, DE 19716

Abstract: Let $\Omega \subset \mathbf{R}^3$ be a bounded Lipshitz domain, and let $\zeta \in \mathbf{C}^3$, $\zeta \cdot \zeta = 0$. In Ω, consider an elliptic equation

$$\operatorname{div}(a\nabla u) + b \cdot \nabla u + cu = 0$$

with $a \in C^1(\bar{\Omega})$, $b \in L^\infty(\Omega)^3$,$c \in L^\infty(\Omega)$. Assume also that a is real valued and has a positive lower bound. We prove that for $|\zeta|$ sufficiently large, this equation has special quasi-exponential solutions of the form

$$u = e^{-\frac{1}{2}i\zeta \cdot x}(1 + w(x, \zeta))$$

depending on parameter ζ and such that $\| w \|_{L^2(\Omega)} = O(|\zeta|^{-\alpha})$, for any $\alpha \in (0, 1)$.

1.1 INTRODUCTION.

Since the appearance of the papers [SU], [N1] and [HN], construction of the quasi-exponential solutions (QES) has been extensively used in uniqueness proofs for inverse problems (see [I1] and references therein.) The origins of the idea can be traced back to Calderón [C] who observed that the function

R.P. Gilbert et al.(eds.), Direct and Inverse Problems of Mathematical Physics, 161–184.

$\exp(\zeta \cdot x)$ is harmonic in x provided $\zeta \in \mathbf{C}^n$ satisfies $\zeta \cdot \zeta = 0$. Taking products of the form $u_1 u_2 = \exp((\zeta_1 + \zeta_2) \cdot x)$ in a bounded domain Ω one obtains a family of exponentials which is dense in $L^2(\Omega)$.

Sylvester and Uhlmann in [SU] proposed to look for solutions of the Schrödinger equation:

$$\Delta u + qu = 0. \tag{1.1}$$

with potential $q \in L^\infty(\Omega)$ in the form of perturbed harmonis functions, or QES

$$u(x, \zeta) = e^{-\frac{1}{2} i\zeta \cdot x}(1 + w(x, \zeta))$$

where the influence of q is described by a function w that tends to zero in the appropriate norm as $|\zeta| \to \infty$. Substitution of u into (1.1) yields the equation

$$\Delta w - i\zeta \cdot \nabla w + qw = -q. \tag{1.2}$$

We denote the operator

$$\Delta - i\zeta \cdot \nabla$$

by P. The reciprocal $1/(-\xi^2 + \zeta \cdot \xi)$ of its symbol $P(\xi)$ is locally integrable in $\mathbf{R}^n, n \geq 2$ and hence defines a tempered distribution. Taking the inverse Fourier transform of $1/P$ one obtains a fundamental solution E. Then w can be found from the integral equation

$$w + Eqw = -Eq.$$

Denote by L^2_δ the space of functions on $\mathbf{R}^n$ square integrable with weight $(1 + |x|^2)^\delta$ with the norm

$$\| f \|_{L^2_\delta} = \left(\int_{\mathbf{R}^n} (1 + |x|^2)^\delta |f(x)|^2 dx\right)^{1/2}.$$

Let $\delta \in (-1, 0)$ and $|\zeta| > \beta > 0$. The estimate

$$\| Ef \|_{L^2_\delta} \leq \frac{C(\beta, \delta)}{|\zeta|} \| f \|_{L^2_{\delta+1}} \tag{1.3}$$

obtained in [SU] makes the operator Eq a contraction in L^2_δ for large $|\zeta|$. By the Banach contraction principle, the solution exists. The estimate above also implies existence of $\beta > 0$ such that

$$\| w \|_{L^2_\delta} \leq \frac{c}{|\zeta|} \tag{1.4}$$

for all ζ satisfying $|\zeta| \geq \beta$. Moreover, for dimensions three and greater, it is possible to choose ζ_1 and ζ_2 both depending on a large parameter $\gamma = O(|\zeta|)$ such that the limiting equality

$$\lim_{\gamma \to \infty} \int f(x) u(x, \zeta_1) u(x, \zeta_2) dx = 0 \tag{1.5}$$

implies $f = 0$ a.e.

The rate of decay from the estimate (1.3), corresponding estimate for w (1.4), and the density property (1.5) are of quite general nature, as was shown by Isakov [I]. He proved existence of QES for equations of the form:

$$Qu + cu = 0,$$

where Q is an arbitrary constant coefficient differential operator and $c \in L^\infty(\Omega)$. As the next step, one could try to generalize the method to the operators containing first order perturbations of a given constant coefficient operator. It turns out that the direct extension is not possible. To explain the difficulty, consider the equation

$$\Delta u + b \cdot \nabla u = 0$$

with $b \in C^\infty(\Omega)$. Substituting $u = e^{-\frac{1}{2}i\zeta\cdot x}(1+w)$ one obtains and equation for w:

$$Pw + b \cdot \nabla u - \frac{1}{2}i\zeta \cdot bw = \frac{1}{2}i\zeta \cdot b.$$

Taking $w = Ev$ transforms this into the integral equation:

$$v + b \cdot \nabla Ev - \frac{1}{2}i\zeta \cdot bEv = \frac{1}{2}i\zeta \cdot b. \tag{1.6}$$

Now the estimate (1.3) does not imply that $-1/2i\zeta \cdot bE$ is a contraction for large $|\zeta$. Moreover, the norm of ∇E is only bounded by a constant (see, for instance, [S]) which means that $b \cdot \nabla E$ is not a contraction either. The solution of (1.6) can be shown to exist, but the crucial estimate (1.4) will not hold in general. Hence, a different method is needed.

When the coefficients of the operator are smooth, one could replace the fixed point theorem by the pseudodifferential calculus of parameter-dependent symbols developed by Shubin ([Sh], ch.II, section 9). Using this approach, Nacamura and Uhlmann [NU] proved existence of QES for equations containing smooth first order perturbations of the Laplacian. In the case of less regular coefficients available results include the one by Sun [S] who constructed slightly modified QES for the Schrödinger equation with magnetic potential

$$\Delta u + b \cdot \nabla u + cu = 0$$

under the assumptions that $b \in W^{2,\infty)}(\Omega)$, $c \in L^\infty(\Omega)$, and $\|$ curl$b \|_{L^\infty(\Omega)}$ is small. Recently, Tolmasky [To] obtained a general result in this direction. Combining the method of Nacamura and Uhlmann with the symbol smoothing technique (see [T], ch.1, section 1.3), he proved existence of QES for equations with nonsmooth first order perturbations of the Laplacian. By the proposition 1.3.B from [T], the minimal regularity required for symbol smoothing is C^s, $s > 0$. This makes extension of QES-methods to the case of discontinuous coefficients an open problem.

In this work we consider the second-order linear differential equation

$$\text{div }(a\nabla u) + b \cdot \nabla u + cu = 0 \tag{1.7}$$

in a bounded Lipshitz domain Ω under assumptions: $a \in C^1(\bar{\Omega})$, $b \in L^\infty(\Omega)^3$, $c \in L^\infty(\Omega)$. We also assume that a is real valued and has a positive lower bound. The dimension of the underlying space is taken to be three. This was done mainly for notational simplicity, and generalization of the method to $n > 3$ is straightforward. The case $n = 2$ is exceptional and should be treated differently.

Substitution of $u = e^{-\frac{1}{2}i\zeta\cdot x}(1 + w)$ and division by a yields an equation for w:

$$Pw + \frac{1}{a}[(\nabla a + b)\cdot \nabla w - \frac{1}{2}i\zeta \cdot (\nabla a + b)w + cw] = \frac{1}{2}i\zeta \cdot \frac{\nabla a + b}{a} + \frac{c}{a}. \tag{1.8}$$

To prove existence of w satisfying an appropriate decay estimate, we use fixed point iteration combined with what is known in numerical analysis as preconditioning. Loosely speaking, to improve convergence of an iterative scheme one needs a good initial guess. We propose to look for w of the form $w = Nv$ where N is a special parametrix of the operator P. Recall that by definition N satisfies $PN = \delta - PF$ where δ denotes the delta distribution and F is a smoothing operator. Substituting $w = Nv$ into (1.8) we obtain

$$v - PFv + QNv = f, \tag{1.9}$$

where Q denotes the first order perturbation of P in the left hand side of (1.8), and f stands for the right hand side of (1.8). For a good choice of N, the operator $A = PF - QN$ will be a contraction on $L^2(\Omega)$ for sufficiently large $|\zeta|$. Then v could be written as the sum of the convergent series

$$v = \sum_{j=0}^{\infty} A^j f,$$

so that the solution w of (1.8) would have the form

$$w = N\sum_{j=0}^{\infty} A^j f.$$

Notice also that the decay of the norm of w is controlled by the decay of the norm of N.

To carry out this program, we need a parametrix which would be "more contracting" for large $|\zeta|$ than the fundamental solution E. A possible way to achieve this is to cut off the kernel of E and make the size of a support of a cut-off function shrink as $|\zeta| \to \infty$. Let us fix a cut-off function ϕ identically equal to one in a fixed neighborhood X_0 of zero. Combining techniques from [H], th 10.3.7, [I] and [SU], we can obtain that for large enough $|\zeta|$

$$\| (\phi E) * f \|_{L^2(\mathbf{R}^3)} \leq \frac{C}{|\zeta|} \| f \|_{L^2(\mathbf{R}^3)} . \tag{1.10}$$

Suppressing the dependence on $\zeta/|\zeta|$ from now on, we will write $E(x,\gamma)$ instead of $E(x,\zeta)$, where $\gamma = O(|\zeta|)$ is a generic large parameter. Next, we observe that E has a specific invariance with respect to dilations:

$$E(\gamma^{-\alpha}x, \gamma) = \gamma^{\alpha}E(x, \tilde{\gamma}), \tag{1.11}$$

where $\tilde{\gamma}$ denotes $\gamma^{1-\alpha}$, and α is any real positive number. Now fix $\alpha \in (0,1)$ and consider the operator with the kernel $\phi_{\tilde{\gamma}}E(x,\gamma)$ where $\phi_{\tilde{\gamma}}(x) = \phi(\gamma^{\alpha}x)$. We show that there is a real $\beta > 1$ such that for all $\gamma \geq \beta$ the following estimate holds:

$$\| (\phi_{\tilde{\gamma}}E) * f \|_{L^2(\mathbf{R}^3)} \leq C\gamma^{-1-\alpha} \| f \|_{L^2(\mathbf{R}^3)}, \tag{1.12}$$

where C depends only on ϕ, α, β. This implies that the operator $Q(\phi_{\tilde{\gamma}}E)$ will be a contraction for large enough γ.

The most difficult problem is to show that the norm of the corresponding operator $PF = P((1-\phi_{\tilde{\gamma}})E(x,\gamma))$ tends to zero as $\gamma \to \infty$. For an arbitrary ϕ, the kernel of PF is given by

$$-P(\phi_{\tilde{\gamma}}E) = -EP\phi_{\tilde{\gamma}} - 2\nabla\phi_{\tilde{\gamma}} \cdot \nabla E - iE\zeta \cdot \nabla\phi_{\tilde{\gamma}}.$$

Since derivatives of $\phi_{\tilde{\gamma}}$ blow up as γ grows, decay of the norm of $P(\phi_{\tilde{\gamma}}E)$ is far from obvious. Using the estimates for E and the estimates

$$|\partial_j^k \phi_{\tilde{\gamma}}(x)| \leq C_{kj}\gamma^{k\alpha}$$

one can show boundedness of $P(\phi_{\tilde{\gamma}}E)$ in $L^2(\mathbf{R}^3)$. But it seems to be quite difficult to show that for unspecified real valued cut-off function ϕ the norm of this operator tends to zero as γ grows.

We bypass this problem by constructing a special parameter-dependent complex valued cut-off function $\psi(x,\gamma)$. This function is identically equal to one in X_0 and oscillates rapidly on a set where it is nonconstant. More precisely, in section 2 we prove that there exists a cut-off function ψ with the gradient given by

$$\nabla\psi(x,\gamma) = e^{i\tilde{\gamma}S(x)}(h_0(x) + \frac{1}{\tilde{\gamma}}h_1(x)),$$

where S is a smooth real valued phase function satisfying a number of condition listed in section 2 below, and h_0 and h_1 are smooth compactly supported outside of X_0 vector fields. Using the function ψ we define a special parametrix of $P = \Delta - i\zeta \cdot \nabla$:

Definition 1.1.1 *The near field parametrix N of the operator P is defined as the tempered distribution with the Schwartz kernel*

$$N(x,\gamma) = \psi(\gamma^{\alpha}x, \gamma)E(x,\gamma),$$

The terminology seems to be natural. If E is the wave field generated by the point source at the origin, then N represents the field in the zone close to the source, outside of which one can use far-field asymptotics. The smoothing operator $E - N$ is denoted by F. Due to the special choice of the cut -off function, the kernel of PF contains a rapidly oscillating factor $e^{i\tilde{\gamma}S(x)}$. The nature of critical points of S determines dependence of the norm of PF on γ.

In section 4, th.4 we show that for a proper choice of S,

$$\| PF \| \leq C\tilde{\gamma}^{-1/6} \tag{1.13}$$

where C is independent of γ. The estimate shows that the rate of decay of $N = \psi(\gamma^\alpha x, \gamma)E(x, \gamma)$ and PF are closely connected. To make the norm of the N decay faster, one would like to take α as large as possible. But for $\alpha = 1$ the estimate above implies just boundedness (recall that $\tilde{\gamma} = \gamma^{1-\alpha}$) which makes this value of α unacceptable.

The main result of the paper is the following

Theorem 1 *Let $\Omega \subset \mathbf{R}^3$ be a bounded Lipshitz domain.— Suppose that $a \in C^1(\bar{\Omega})$, $b \in L^\infty(\Omega)^3$, $c \in L^\infty(\Omega)$, a is real valued and has a positive lower bound. Then there is $\beta > 1$ such that a quasi-exponential solution $u(x, \zeta)$ of (1.7) exists for all ζ satisfying $\zeta \cdot \zeta = 0$, $|\zeta| \geq \beta$. The function w has the explicit representation:*

$$w(x, \zeta) = N \sum_{j=0}^{\infty} A^j f(x, \zeta),$$

where N is the near field parametrix,

$$A = PF - QN,$$

$$Qw = \frac{1}{a}[(\nabla a + b) \cdot \nabla w - \frac{1}{2} i\zeta \cdot (\nabla a + b)w + cw] \tag{1.14}$$

$$f(x, \zeta) = \frac{1}{2} i\zeta \cdot \frac{\nabla a + b}{a} + \frac{c}{a}. \tag{1.15}$$

Moreover, the estimates

$$\| w \|_{L^2(\Omega)} \leq C_1 |\zeta|^{-1-\alpha} \| f \|_{L^2(\Omega)}$$

and

$$\| \nabla w \|_{L^2(\Omega)^3} \leq C_2 |\zeta|^{-\alpha} \| f \|_{L^2(\Omega)}$$

hold for any $\alpha \in (0, 1)$ with C_1, C_2 independent of $|\zeta|$.

The following should be noted here.

1. Since the right hand side f has an order $O(\gamma)$, the actual rate of decay of the perturbation w is $O(\gamma^{-\alpha})$. The parameter α can be chosen to be arbitrary close to one. This means that the best decay of the norm of w is almost as good as $O(\gamma^{-1})$ obtained in [SU] and [I] for equations with zero order perturbations.
2. The theorem can be easily generalized to the case when Q is not a linear differential operator. The essential requirement for $Q : L^2(\Omega) \to L^2(\Omega)$ is the following: there is a constant c independent of γ such that

$$\| e^{\frac{1}{2}i\zeta \cdot x} Q(e^{-\frac{1}{2}i\zeta \cdot x} \bullet) \| \leq c\gamma^k$$

with $k < 2$.
The paper is organized as follows. In section 2 we construct the cut-off function ψ. In section 3 we prove a number of estimates for convolution operators with kernels given by the fundamental solution E times various cut-off functions. The main result of this section is the theorem 3 where we prove the estimates

$$\| N * f \|_2 \leq C_1 \gamma^{-1-\alpha} \| f \|_2$$

and

$$\| \nabla N * f \|_2 \leq C_2 \gamma^{-\alpha} \| f \|_2$$

Then, in section 4, theorem 4, the estimate (1.13) for the norm of PF is obtained. Finally, in section 5, the main theorem is proved. The proof of the main theorem follows quickly from the estimates above and the Banach contraction principle. Appendix 1 contains some results on asymptotics and oscillatory integrals. In appendix 2 we give a proof of a technical result about the fundamental solution E needed in section 3, and also obtain an explicit asymptotics of E with respect to the large parameter $\gamma|x|$. This theorem together with the theorem 2 from section 2 allows one to prescribe the nature of critical points of phase functions, which in turn leads to parameter dependent estimates from the theorem 4 and proposition 3.5.

1.2 CONSTRUCTION OF THE PARAMETRIX.

Let X_0 be a fixed neighborhood of the origin. Fix a compact X containing X_0 and such that $\text{dist}(\partial X, X_0) > 0$. The aim of this section is to construct a special (complex-valued) function $\psi(x,\gamma)$ subject to the following conditions:

1) $\psi \in C_0^\infty(\mathbf{R}^3)$, $X_0 \subset \text{supp}\psi \subset X$;

2) $\text{supp}\psi$ is independent of γ;

3) for all $x \in X_0$, $\psi(x,\gamma) = 1$.

Let $T_{\partial_j \psi} : L^2(\mathbf{R}^3) \to L^2(\mathbf{R}^3)$ be the convolution operators associated with the first partial derivatives of ψ. The purpose of the construction is to make the norm of $T_{\partial_j \psi}$ decreasing as $\gamma \to \infty$. Typically, such decay is a consequence of rapid oscillations of the kernel $\partial_j \psi$ (see, for instance, [S], ch 8,9). Motivated by this, we look for a function ψ with gradient of the form

$$\nabla \psi(x) = c(\gamma) e^{i\tilde{\gamma} S(x)} (h_0(x) + \frac{1}{\tilde{\gamma}} h_1(x)), \tag{2.16}$$

where $h_0, h_1 : \mathbf{R}^3 \to \mathbf{R}^3$ are smooth functions supported on a compact specified by conditions 1)-3).

Theorem 2 *There exists a function $\psi(x,\gamma)$ with $\nabla\psi$ defined by (2.16) satisfying conditions 1)-3). Moreover, S can be chosen to satisfy:*

i) $\det \partial^2_{jk} S \neq 0$ *in* X;

Proof. Denote the right-hand side of (2.16) by u. The necessary and sufficient condition for existence of a scalar function ψ with $\nabla\psi = u$ is

$$\text{curl} u = 0.$$

Taking *curl* of u one obtains:

$$\text{curl} u = e^{i\tilde{\gamma}S}[i\tilde{\gamma}A_S h_0 + \tilde{\gamma}^0(\text{curl} h_0 + iA_S h_1) + \tilde{\gamma}^{-1}\text{curl} h_1].$$

Here A_S is the matrix of the form

$$\begin{pmatrix} 0 & \partial_3 S & -\partial_2 S \\ -\partial_3 S & 0 & \partial_1 S \\ \partial_2 S & -\partial_1 S & 0 \end{pmatrix}$$

depending on the phase function S. We try to choose S, h_0, and h_1 such that

$$A_S h_0 = 0, \tag{2.17}$$

$$\text{curl} h_0 + iA_S h_1 = 0,$$

and

$$\text{curl} h_1 = 0.$$

One can easily check that $\det A_S = 0$ identically for any choice of S, so the kernel of A_S is nontrivial. First, we choose a smooth real valued function S satisfying i). This condition is crucial for the decay of the norm of the operator PF. Its role will be explained in detail below in the proof of the theorem 4. Let x be a point in $X \backslash X_0$ such that $\nabla S(x) \neq 0$. Then there exists a partial derivative $\partial_j S$ not equal to zero at x. Direct computation shows that $h_0(x)$ must be of the form

$$h_0(x) = \frac{f(x)}{\partial_j S(x)} \nabla S(x),$$

where $f(x)$ is an arbitrary C_0^∞- scalar function supported in $X \backslash X_0$.

Next, we look for h_1 satisfying the second equation of (2.17). Since

$$\text{curl} h_0 = \text{curl}(\frac{f}{\partial_j S}\nabla S) = A_S \nabla(\frac{f}{\partial_j S}),$$

we have

$$h_1 = i\nabla(\frac{f}{\partial_j S}).$$

For this choice of h_1, the third equation of (2.17) is automatically satisfied. When $\nabla\psi$ is constructed, we define the function itself. Since $\nabla S \neq 0$, there is a partial derivative of ψ not identically equal to zero. Suppose, it is $\partial_1\psi$. Integrating it we obtain:

$$\psi(x) = c(\gamma) \int^{x_1} e^{i\tilde{\gamma}S(y_1,x')} h_0^{(1)}(y_1, x') + \frac{1}{\tilde{\gamma}} h_1^{(1)}(y_1, x') dy_1,$$

where the superscript (1) denotes the first component of a corresponding vector field. By the choice of h_0 and h_1, the function ψ must be identically equal to a constant k in X_0. The value of this constant (possibly depending on γ,) equals, for instance,

$$k(\gamma) = \int^{x_1=0} e^{i\tilde{\gamma}S(y_1,0)} h_0^{(1)}(y_1,0) + \frac{1}{\tilde{\gamma}} h_1^{(1)}(y_1,0) dy_1.$$

If we normalize ψ by taking

$$c(\gamma) = \frac{1}{k(\gamma)},$$

then for all $x \in X_0$ we will have

$$\psi(x,\gamma) = 1,$$

Of course, $k(\gamma)$ should not be equal to zero. Moreover, for construction to be useful, we need to make decay of $k(\gamma)$ as slow as possible, so $c(\gamma)$ would be only slightly increasing as γ grows. This can be achieved by imposing the following additional condition on S: $S(x_1, 0, 0)$ has one degenerate critical point x_0 of index 2 on the interval $\mathrm{supp} h_0 \cap (-\infty, 0]$. Under this condition we have

$$\partial_1 S(x_0) = \partial_1^2 S(x_0) = 0, \tag{2.18}$$

$$\partial_1^3 S(x_0) \neq 0.$$

To estimate $k(\gamma)$, we use the following proposition ([F], ch.3, th. 1.5):

Proposition 1.2.1 *Let $I = [x_0, x_0 + \delta] \subset \mathbf{R}^1$ be a finite interval, $\delta > 0$. Suppose that the functions $f(x), S(x) \in C^\infty(I)$, $d^k f/dx^k(x_0 + \delta) = 0$ for $k = 0, 1, 2....$ Let x_0 be the only critical point of S on I, $m \geq 2$ integer, and*

$$\frac{d^k S}{dx^k}(x_0) = 0, \quad 1 \leq k \leq m-1,$$

$$\frac{d^m S}{dx^m}(x_0) \neq 0.$$

Then, as $\lambda \to \infty$,

$$\int_{x_0}^{x_0+\delta} e^{i\lambda S(x)} f(x) dx \sim \lambda^{-1/m} e^{i\lambda S(x_0)} \sum_{k=0}^{\infty} a_k \lambda^{-k/m}, \tag{2.19}$$

where coefficients a_k can be computed by the formula:

$$a_k = \frac{1}{mk!} \Gamma\left(\frac{k+1}{m}\right) \exp\left(\frac{i\pi \, sgn S^{(m)}(x_0)(k+1)}{2m}\right) \times$$

$$\left(\frac{d}{dx}\right)^k [f(x)(-sgn S^{(m)}(x_0))(S(x) - S(x_0)))^{-\frac{k+1}{m}} (x - x_0)^{k+1}]_{|x=x_0}.$$

The proposition immediately implies that if $\tilde{\beta} > 1$ is fixed and $\tilde{\gamma} \geq \tilde{\beta}$, then

$$|c(\gamma)| \leq K\tilde{\gamma}^{1/3},$$

where K depends only on S_0, h_0 and β. Also, the proposition shows that $k(\gamma) \neq 0$ provided $h_0^{(1)}(x_0) \neq 0$ which can be achieved by the choice of the function h_0.

The proof is complete.

1.3 ESTIMATES FOR N.

We start with the

Proposition 1.3.1 *For any real $\alpha > 0$ the following identities hold:*

$$E(\gamma^{-\alpha}x, \gamma) = \gamma^{\alpha} E(x, \gamma^{1-\alpha}), \tag{3.20}$$

$$\nabla E(\gamma^{-\alpha}x, \gamma) = \gamma^{2\alpha} \nabla E(x, \gamma^{1-\alpha}). \tag{3.21}$$

Proof. Change variables in the oscillatory integrals

$$E(x, \gamma) = (2\pi)^{-3} \int e^{ix\cdot\xi} \frac{d\xi}{-\xi^2 + \zeta \cdot \xi}$$

and

$$\nabla E(x, \gamma) = (2\pi)^{-3} \int e^{ix\cdot\xi} \frac{i\xi d\xi}{-\xi^2 + \zeta \cdot \xi},$$

respectively.

The proof is complete.

Consider a function $\tilde{P}(\xi$ defined by

$$\tilde{P}(\xi)^2 = \sum_{|k|\leq 2} |\partial^k P(\xi)|^2.$$

Let $B_{\infty,\tilde{P}}$ be a Banach space of tempered disributions f such that

$$\| f \|_{\infty,\tilde{P}} = \text{ess sup}(\tilde{P}|\hat{f}|) < \infty$$

For theory of this and similar spaces, see [H] ch.10. First, we state the following technical result proven in the appendix 2.

Proposition 1.3.2 *The distribution $\phi E \in B_{\infty,\tilde{P}}$. If $\gamma \geq \beta > 0$, there exist a constant C such that*

$$\| \phi E \|_{\infty,\tilde{P}} \leq C.$$

Let $\phi(x)$ be a fixed cut off function. First, we prove the estimate (1.12) from the introduction and the corresponding estimate for the gradient. These estimates are similar to the estimates obtained in [I]. Cutting off the kernel was implicitly present in [I] via the method used in [H] to prove th. 10.3.7.

Proposition 1.3.3 *Let $\gamma \geq \beta > 0$. The estimates*

$$\| (\phi E) * f \|_2 \leq C_1 \gamma^{-1} \| f \|_2,$$

$$\| (\phi \partial_j E) * f \|_2 \leq C_2 \| f \|_2$$

hold with C_1, C_2 depending only on ϕ and β.

Proof. First, we estimate $|\mathcal{F}(\phi E)(\xi)|$. By the previous proposition we have

$$|\mathcal{F}(\phi E)| = \frac{1}{\tilde{P}}(\tilde{P}|\mathcal{F}(\phi E)|) \leq C(\phi, \beta)\frac{1}{\inf \tilde{P}}.$$

Since $\inf \tilde{P} \geq \gamma$, the first estimate is proved. To prove the second estimate, we observe that the following inequality holds:

$$\frac{|\xi|}{\tilde{P}} \leq c(\beta)$$

The proof is complete.

Remark. The inclusion from the proposition 3.2 does not hold for E even though it holds for the regular fundamental silution of P consrtucted by Hormander and used by Isakov to obtain his parameter-dependent estimates. Thus, presence of a cut-off function in the kernel is crucial for the validity of the propositions 3.2 and 3.3.

Next, we fix $\alpha \in (0, 1)$ and let $\phi_{\tilde{\gamma}}$ denote the scaled function $\phi(\gamma^\alpha x)$. Consider the convolution operators with the kernels $\phi_{\tilde{\gamma}} E$ and $\phi_{\tilde{\gamma}} \partial_j E$. The following proposition shows that shrinking the cut-off function one obtains better decay with respect to parameter.

Proposition 1.3.4 *Let $\gamma \geq \beta > 0$. The estimates*

$$\| (\phi_{\tilde{\gamma}} E) * f \|_2 \leq C_1 \gamma^{-1-\alpha} \| f \|_2,$$

$$\| (\phi_{\tilde{\gamma}} \partial_j E) * f \|_2 \leq C_2 \gamma^{-\alpha} \| f \|_2$$

hold for any $\alpha \in (0, 1)$. The constants C_1, C_2 depend only on $\psi(x)$ and β.

Proof. Scaling and using proposition 3.1 we obtain

$$(\phi_{\tilde{\gamma}} E) * f(\gamma^{-\alpha} y) = \gamma^{-2\alpha} \int \psi(x) E(\gamma^{-\alpha} x, \gamma^{1-\alpha}) f(\gamma^{-\alpha}(y - x)) dx.$$

Denote the left hand side by $(\phi_{\tilde{\gamma}}E)f_\gamma$ and $f(\gamma^{-\alpha}x)$ by f_γ. The cut-off function $\psi(y)$ is parameter independent, and by the proposition 3.3

$$\| (\phi_{\tilde{\gamma}}E) * f_\gamma \|_2 \leq C\gamma^{-2\alpha}\gamma^{-1+\alpha} \| f_\gamma \|_2 .$$

Scaling L^2-norms of $(\phi_{\tilde{\gamma}}E)f_\gamma$ and f_γ, one obtains the estimate claimed. The estimate for gradient follows from (3.21) in the same manner.

The proof is complete.

Consider now the operator $N_{\tilde{\gamma}} = \psi(y,\gamma)E_{\tilde{\gamma}}$, where $E_{\tilde{\gamma}} = E(y,\tilde{\gamma})$ and ψ is the parameter-dependent cut-off function constructed in section 2 (theorem 2). The following proposition shows that the estimates from the proposition 3.3 remain true in this case.

Proposition 1.3.5

$$\| N_{\tilde{\gamma}}f \|_2 \leq C_1\tilde{\gamma}^{-1} \| f \|_2, \tag{3.22}$$

$$\| \nabla N_{\tilde{\gamma}}f \|_2 \leq C_2 \| f \|_2 \tag{3.23}$$

hold with C_1, C_2 independent of γ.

Proof. Using partition of unity, we write

$$N_{\tilde{\gamma}} = T_{\tilde{\gamma}} + \rho E_{\tilde{\gamma}},$$

where ρ is a cut-off function supported in X_0, identically equal to one near zero and independent of γ. By the proposition 3.3,

$$\| \rho E_{\tilde{\gamma}}f \|_2 \leq c\tilde{\gamma}^{-1} \| f \|_2, \tag{3.24}$$

with c in dependent of γ. The proposition will be proved if we show that the norm of $T_{\tilde{\gamma}}$ decays faster than $\tilde{\gamma}^{-1}$. Since

$$\| T_{\tilde{\gamma}}f \|_2 = \| \hat{T}_{\tilde{\gamma}}\hat{f} \|_2 \leq \sup |\hat{T}_{\tilde{\gamma}}| \; \| f \|_2,$$

it is sufficient to estimate supremum of $|\hat{T}_{\tilde{\gamma}}|$. By the choice of ρ we have $\text{dist}(0, \text{supp}T_{\tilde{\gamma}}) > d > 0$ for some fixed d. Hence, the fundamental solution can be well approximated by its asymptotics from the theorem 5 from appendix 2. Moreover, since multiplication of the kernel by $e^{\frac{1}{2}i\tilde{\gamma}y_2}$ is equivalent to the shift of its Fourier transform which preserves the supremum, we can disregard the corresponding factor in the expression for the asymptotics for E. Hence, it is sufficient to estimate $I(\xi) = \hat{T}_{\tilde{\gamma}}(\xi - \frac{1}{2}\tilde{\gamma}e_2$, where $e_2 = (0,1,0)$. By the theorem 5 from appendix 2,up to a constant factor,

$$I(\xi) = c(\gamma)\int e^{-iy\cdot\xi}F(y)\int^{y_1} e^{i\tilde{\gamma}S(x_1,y')}h_0^{(1)}(x_1,y')dx_1dy + \frac{c(\gamma)}{\tilde{\gamma}}R(\xi,\tilde{\gamma})$$

where

$$F(y) = \int_{S^1} e^{\frac{1}{2}iy\cdot\theta}\tilde{E}(y,\theta)ds(\theta),$$

where

$$\tilde{E}(y,\theta) = (1-\rho)\chi(y,\theta)(y_1 + i\theta\cdot y')^{-1}.$$

The cut-off function χ is defined in appendix 2. The remainder R has the same form as the main term with $\tilde{E}$ and $h_0^{(1)}$ replaced by functions $r_1(y,\gamma)$ and $h_1(y)$. These functions can be explicitly calculated and estimated by the theorem 5 and theorem 2. The integral with respect to y is taken over the support of $h_0^{(1)}$. Interchanging the order of integration one obtains:

$$I(\xi) = c(\gamma)\int_{S^1}\int e^{-iy'\cdot\xi'}e^{\frac{1}{2}i\tilde{\gamma}y'\cdot\theta}$$

$$\int e^{i\tilde{\gamma}S(x_1,y')}h_0^{(1)}(x_1,y')\int_{x_1} e^{-iy_1\xi_1}\tilde{E}(y)dy_1 ds(\theta)dx_1 dy' + \frac{c(\gamma)}{\tilde{\gamma}}R(\xi,\tilde{\gamma}).$$

Denote by $\hat{F}(x_1,y',\theta,\xi_1)$ the function

$$\int_{x_1} e^{-iy_1\xi_1}\tilde{E}(y)dy_1.$$

Then, replacing x_1 by y_1 we have:

$$I(\xi) = c(\gamma)\int_{S^1}\int e^{-iy'\cdot\xi'}e^{i\tilde{\gamma}(S(y)-\frac{1}{2}y'\cdot\theta)}h_0^{(1)}(y)\hat{F}(y,\theta,\xi_1)dyds(\theta) + \frac{c(\gamma)}{\tilde{\gamma}}R(\xi,\tilde{\gamma}).$$

Without changing supremum, we can replace ξ' by $\tilde{\gamma}\xi'$. Then

$$\sup|I(\xi)| \le |c(\gamma)\int_{S^1}\int e^{i\tilde{\gamma}(S_0(y)+\frac{1}{2}y'\cdot\theta-y'\cdot\xi')}\hat{F}(y,\theta,\xi_1)dyds(\theta)| + |\frac{c(\gamma)}{\tilde{\gamma}}R(\xi,\tilde{\gamma})|.$$

To estimate the main term, we apply the stationary phase method to the integral with respect to y. The integral is taken over the support of $h^{(1)}$. Hence, when $-1/2\theta+\xi'$ does not belong to the set $\{\nabla S(y)|y \in \mathrm{supp} h_0^{(1)}\}$, the phase function $\Phi(y,\theta,\xi)$ does not have critical points. In this case the integral decays faster than any power of $\tilde{\gamma}$ by the asymptotic localization lemma. Hence, for large $\tilde{\gamma}$ it is sufficient to estimate the integral assuming that

$$-\frac{1}{2}\theta + \xi' \in K = \{\nabla S(y)|y \in \mathrm{supp} h_0^{(1)}\}.$$

The nondegeneracy condition $\det\partial^2_{jk}S_0 \ne 0$ implies that for any vector from the set K, Φ has a unique nondegenerate critical point $y_0(\theta,\xi)$. By the proposition...from the appendix, the integral with respect to y decays as $O(\tilde{\gamma}^{-3/2})$. More precisely, we obtain the estimate:

$$\sup|I(\xi)| \le c(\gamma)\sup_{1/2\theta-\xi'\in K}|c\tilde{\gamma}^{-3/2}\int_{S^1} e^{i\tilde{\gamma}\Phi(y_0,\theta,\xi)}h^{(1)}(y_0)\hat{F}(y_0,\theta,\xi_1)ds(\theta)|+$$

$$|c(\gamma)\tilde{\gamma}^{-1}R(\xi,\tilde{\gamma})|.$$

Estimating the last integral one obtains:

$$\sup |I(\xi)| \leq Ac(\gamma)\tilde{\gamma}^{-3/2} \sup |h_0^{(1)}| \sup |\hat{F}(y_0, \theta, \xi_1| + c(\gamma)\tilde{\gamma}^{-1}|R|.$$

where A is independent of γ. Since the remainder R has the same structure as the main term, repeating the process, one obtains:

$$|R(\xi_1, \tilde{\gamma}\xi'| \leq c\tilde{\gamma}^{-3/2} \sup |h_1(y_0)| \sup |\int_{y_1} r_1(x_1, y', \tilde{\gamma})e^{-ix_1\xi_1} dx_1 \,_{|y=y_0}|.$$

The function r_1 is the remainder of an asymptotic expansion. By the theorem 5 it is bounded from above by a constant. Moreover, by the theorem 2, $c(\gamma) \leq c\tilde{\gamma}^{1/3}$. Then it follows that

$$\sup |\hat{T}_{\tilde{\gamma}}(\xi)| \leq c\tilde{\gamma}^{-7/6}.$$

The proof is complete.

Theorem 3 *Let $\psi(y, \gamma)$ be the cut off function constructed in the theorem 2 and let $N : L^2(\mathbf{R}^3) \to L^2(\mathbf{R}^3)$ be the operator with the kernel $\psi(\gamma^\alpha y, \gamma)E(y, \gamma)$. Then the estimates*

$$\| Nf \|_2 \leq C_1\gamma^{-1-\alpha} \| f \|_2$$

and

$$\| (\partial_j N)f \|_2 \leq C_2\gamma^{-\alpha} \| f \|_2$$

hold with C_1, C_2 independent of γ.

Proof. Scaling and using proposition 3.1 we obtain:

$$Nf(\gamma^{-\alpha}y) = \gamma^{-2\alpha} \int \psi(x, \gamma)E(\gamma^{-\alpha}x, \tilde{\gamma})f(\gamma^{-\alpha}(y - x))dx.$$

Hence, the estimate for Nf will follow if we prove that $\| N_{\tilde{\gamma}} \| \leq c\tilde{\gamma}^{-1}$, where $N_{\tilde{\gamma}} : L^2(\mathbf{R}^3) \to L^2(\mathbf{R}^3)$ is the convolution operator with the kernel $\psi(y, \gamma)E(y, \tilde{\gamma})$. The last estimate holds by the proposition 3.5.

The estimate for $(\partial_j N)f$ follows from the second estimates in propositions 3.3-3.5.

The proof is complete.

1.4 ESTIMATE FOR PF.

Since

$$PF = P[(1 - \psi(\gamma^\alpha x, \gamma)E] = -EP\psi(\gamma^\alpha x, \gamma) - 2\nabla\psi(\gamma^\alpha x, \gamma) \cdot \nabla E,$$

application of lemma... yields the following scaling property:

$$PF(x, \gamma) = -\gamma^{3\alpha}[E(\tilde{x}, \tilde{\gamma})P_{\tilde{\gamma}}\psi(\tilde{x}, \tilde{\gamma}) - 2\nabla\psi \cdot \nabla E(\tilde{x}, \tilde{\gamma})], \qquad (4.25)$$

where $\tilde{x} = \gamma^{-\alpha}, \tilde{\gamma} = \gamma^{1-\alpha}$, differentiations are with respect to $\tilde{x}$, and

$$P_{\tilde{\gamma}} = \Delta - i\gamma^{-\alpha}\zeta \cdot \nabla.$$

Theorem 4 *Let The estimate*

$$\| PFf \|_2 \leq C\tilde{\gamma}^{-1/6} \| f \|_2, \tag{4.26}$$

holds with C independent of γ.

Proof. Scaling as in the proof of the proposition 3.1 shows that it is sufficient to prove the estimate for the operator $T_{\tilde{\gamma}}$ with the kernel

$$K_{\tilde{\gamma}} = -P\psi(y,\gamma)E(y,\tilde{\gamma}) + 2\nabla\psi(y,\gamma) \cdot \nabla E(y,\tilde{\gamma}).$$

All cut-off functions in this expression are supported away from zero. Hence, E and ∇E can be replaced by its asymptotic expansions from the theorem 5 from appendix 2. The remainder estimate implies that it is sufficient to consider only the main term of the corresponding asymptotic expansion. Substituting expressions for the derivatives of ψ one can represent T as a finite sum of operators T_j with kernels K_j defined by:

$$K_j(y,\gamma) = c_j c(\gamma)\tilde{\gamma}e^{i\tilde{\gamma}S(y)}e^{\frac{1}{2}i\tilde{\gamma}y_2}\int_{S^1} e^{\frac{1}{2}i\tilde{\gamma}y'\cdot\theta}a_j(y,\theta)ds(\theta),$$

where c_j are constants, a_j are smooth cut-off functions supported in $X\backslash X_0$ for all $\theta \in S^1$. Hence, if we prove the estimate (4.26) for the operator T_j : $L^2(\mathbf{R}^3) \to L^2(\mathbf{R}^3)$ with the kernel K_j, the theorem will follow.

The rest of the proof is a modification of the proof of the proposition 5.3 from appendix 1 given in [S], ch.9, proposition 1.1. Consider the composition operator $T_j^*T_j$ of T_j and its dual. If we prove that the norm of this operator is bounded by $c\tilde{\gamma}^{-1/3}$, the required bound for T_j will follow. Up to the factor $|c(\gamma)|^2\tilde{\gamma}^2$, the kernel $A(x,z)$ of $T_j^*T_j$ is given by the integral:(from now on we can drop the subscript):

$$A(x,z) = e^{\frac{1}{2}i\tilde{\gamma}(z_2-x_2)}$$

$$\int e^{i\tilde{\gamma}(S(y-z)-S(y-z))}\int e^{\frac{1}{2}i\tilde{\gamma}(y-x)\cdot\theta_1-(y-z)\cdot\theta_2}a_1(y-x)\bar{a}_2(y-z)ds_1ds_2dy,$$

where subscripts of a and ds denote dependence on θ_1 and θ_2 respectively. Interchange the order of integration so the inner integral is taken with respect to y, denote this integral by $I_{\tilde{\gamma}}(\theta_1,\theta_2)$ and consider it separately. Assume for the moment that $|x-z| \geq t > 0$, where t is a fixed positive number to be specified later and estimate the absolute value of

$$I_{\tilde{\gamma}}(\theta_1,\theta_2) = \int e^{i\tilde{\gamma}[(S(y-x)-S(y-z)+\frac{1}{2}y\cdot k}a(y-x,\theta_1)\bar{a}(y-z,\theta_2)dy,$$

where k denotes the vector $\frac{1}{2}(\theta_1-\theta_2)$. Denote the phase function in brackets by $\Phi(x,y,z,\theta_1-\theta_2$, and let $M(y,x)$ be a matrix of the second partial derivatives of S evaluated at (y,x). We notice that

$$\nabla_y\Phi = M(y,x)(x-z) + k + O(|x-z|^2).$$

By the theorem 2, M is invertible. We multiply $\nabla\Phi$ by the vector $b = M^{-1}(\frac{x-z+M^{-1}k}{|x-z+M^{-1}k|})$. Then

$$b \cdot (M(y,x)(x-z)+k) = |x-z+M^{-1}k|.$$

It follows that

$$|b \cdot \nabla_y \Phi| \geq c|x-z+M^{-1}k|,$$

provided the support of a is sufficiently small. This requirement can always be satisfied if we use an appropriate partition of unity. Let D denote the differential operator

$$D = b \cdot \nabla_y.$$

Since

$$\frac{1}{i\tilde{\gamma}|x-z+M^{-1}k|} D(e^{i\tilde{\gamma}\Phi}) = e^{i\tilde{\gamma}\Phi},$$

we can integrate by parts N times. Then we have:

$$I_{\tilde{\gamma}}(\theta_1,\theta_2) = \int \frac{1}{i\tilde{\gamma}|x-z+M^{-1}k|^N} e^{i\tilde{\gamma}\Phi} A_N(y,\theta_1,\theta_2),$$

where A_N is a smooth function with the same y-support as a depending on $D^t(M^{-1})$ and $D^t a$. Next, consider

$$|x-z+M^{-1}k|^2 \geq |x-z|(|x-z|+2M^{-1}(\frac{x-z}{|x-z|})\cdot k).$$

This inequality shows that for large enough t (the assumed lower bound for $|x-z|$) depending on the norm of M^{-1}, there will be a constant $B > 0$ such that

$$|x-z+M^{-1}k|^2 \geq B|x-z|^2.$$

This implies that there is a constant C' such that

$$|A(x,z)| \leq C'|c(\gamma)|^2\tilde{\gamma}^2(\tilde{\gamma}|x-z|)^{-N}$$

when $|x-z| \geq t$. Since $\frac{1}{|c(\gamma)|^2\tilde{\gamma}^2}|A(x,z)|$ is bounded independent of $\tilde{\gamma}$ for small $|x-z|$, one obtains the following estimate:

$$|A(x,z)| \leq C|c(\gamma)|^2\tilde{\gamma}^2(1+\tilde{\gamma}|x-z|)^{-N}.$$

Therefore,

$$\int |A(x,z)|dx \leq C|c(\gamma)|^2\tilde{\gamma}^2\tilde{\gamma}^{-3}\int (1+|x-z|)^N dx,$$

with analogous bound for the integral with respect to z. By the Shur's lemma, the norm of $T_j^*T_j$ is bounded by

$$C_1|c(\gamma)|^2\tilde{\gamma}^{-1}$$

which, together with the estimate for $c(\gamma)$ from the theorem 2 implies

$$\| T_j \| \leq C\tilde{\gamma}^{-1/6}.$$

The proof is complete.

1.5 PROOF OF THE MAIN THEOREM.

Proof. Substitute u of the form

$$u(x,\zeta) = e^{-\frac{1}{2}ix\cdot\zeta}(1 + w(x,\zeta))$$

into (1.7). Since a is continuous and positive, division by a is possible. Dividing one obtains the equation for w:

$$Pw + Qw + f,$$

with Q and F defined by (1.14) and (1.15) respectively. Taking $w = Nv$ and using $N = E - F$ we obtain an equation for v:

$$v = Av + f$$

with $A = -PF + QN$, a bounded linear operator on $L^2(\Omega)$. The operator T on $L^2(\Omega)$ defined by $Tu = Au + f$ is a contraction for sufficiently large $|\zeta|$. To show this, we observe that for large enough $|\zeta|$, PF is a contraction by the theorem 4, and by the theorem 3, the norm of QN is bounded by $c\gamma^{-\alpha}$ where c depends on the constant from the theorem 3 and respective norms of the coefficients a, b, c. Therefore, for any fixed $\alpha \in (0,1)$ we have the following estimate:

$$\| A \| \leq c\gamma^{-t},$$

where $t = \min(\frac{1}{6}(1-\alpha), \alpha)$. Thus, if $\beta > 1$ is sufficiently large, A will be a contraction for all ζ satisfying $|\zeta| \geq \beta$. By the Banach fixed point theorem, there exists a unique solution v that can be written as the sum of the convergent Neumann series for A. Moreover we have the estimates

$$\| Nv \|_{L^2(\Omega)} \leq c\gamma^{-1-\alpha} \frac{\| f \|_2}{1 - \| A \|}$$

and

$$\| \nabla Nv \|_{L^2(\Omega)^3} \leq c\gamma^{-\alpha} \frac{\| f \|_2}{1 - \| A \|}$$

which imply the estimates for w and its gradient.

The proof is complete.

Appendix 1.

In this appendix we state the basic facts concerning the stationary phase method (see, for instance, [H], section 7.7.). The following two proposition appear as theorems 7.7.1 and 7.7.5, respectively.

Proposition 1.5.1 *Let $K \subset bfR^n$ be a compact set, X and open neighborhood of K and j, k non-negative integers. If $u \in C_0^k(K)$, $f \in C^{k+1}(X)$ and $Imf \geq 0$ in X, then*

$$\omega^{j+k} \left| \int u(x)(\mathrm{Im} f(x))^j e^{i\omega f(x)} dx \right| \leq$$

$$C \sum_{|\alpha \leq k|} \sup |D^\alpha u|(|f'|^2 + \operatorname{Im} f)^{|\alpha|/2-k}$$

$\omega > 0$. *Here C is bounded when f stays in a bounded set in $C^{k+1}(X)$. When f is real valued, the estimate above reduces to*

$$\omega^k | \int u(x) e^{i\omega f(x)} dx | \leq C \sum_{|\alpha \leq k|} \sup |D^\alpha u| |f'|^{|\alpha|-2k},$$

$\omega > 0$.

Proposition 1.5.2 *Let $K \subset bfR^n$ be a compact set, X and open neighborhood of K and k positive integer. If $u \in C_0^{2k}(K)$, $f \in C^{3k+1}(X)$ and $Imf(x_0) = 0$,$f'(x_0) = 0$, $\det f''(x_0) \neq 0$ in $K \backslash x_0$ then*

$$| \int u(x) e^{i\omega f(x)} dx - e^{i\omega f(x_0)} (\det \omega f''(x_0/2\pi i))^{\mp 2} \sum_{j<k} \omega^{-j} L_j u| \leq$$

$$C\omega^{-k} \sum_{|\alpha| \leq 2k} \sup |D^\alpha u|,$$

$\omega > 0$. *Here C is bounded when f stays in a bounded set in $C^{3k+1}(X)$ and $|x - x_0|/|f'(x)|$ has a uniform bound. With*

$$g_{x_0}(x) = f(x) - f(x_0) - (f''(x_0)(x - x_0), x - x_0)/2$$

which vanishes of third order at x_0 we have

$$L_j u = \sum_{\nu - \mu = j} \sum_{2\nu \geq 3\mu} i^{-j} 2^{-\nu} (f''(x_0)^{-1} D, D)^\nu (g_{x_0}^\mu u)(x_0)/\mu!\nu!.$$

This is a differential operator og order $2j$ acting on u at x_0. The coefficients are rational functions of degree $-j$ in $F''(x_0)$, ..., $f^{(2j+2)}(x_0)$ with denominator $(\det f'')^{3j}$. In every term the total number of derivatives of u and of f'' is at most $2j$.

Appendix 2.

In this appendix we list some results about the fundamental solution E. First, we prove the proposition 3.2 from section 3, and then we prove the theorem which gives the explicit asymptotic expansion for E with respect to the large parameter $|x|\gamma$. Essentially, the asymptotics can be written as the inverse Fourier transform of an explicitly given distribution (which depends also on x) supported on a set of zeros of symbol $P(\xi)$. Asymptotics of this type were obtained by Vainberg [V] when a set of zeros of symbol has codimension one and consists of components with nonzero curvature. In the present case the set of zeros of $P(\xi)$ has codimension two which requires somewhat different techniques.

Proof of the proposition 3.3. We need to estimate the absolute value of

$$\tilde{P}(\xi)\int \phi(\xi-\xi')\frac{1}{P(\xi')}d\xi'$$

¿From the Taylor theorem it follows that:

$$\tilde{P}(\xi) \le (1+c|\xi-\xi'|^2)\tilde{P}(\xi')$$

with c independent of γ. Hence,

$$\sup \tilde{P}(\xi)|\hat{\phi} * \hat{E}|(\xi) \le \sup(|\hat{\phi}(\xi)|(1+c|\xi|^2)^m)\int \frac{\tilde{P}(\xi')}{|P(\xi')|}(1+c|\xi'-\xi|^2)^{-m}d\xi'$$

with some positive integer $m > 3/2$. To estimate the last integral we utilize the techniques introduced in [SU]. Let $N_\epsilon(S)$ be an ϵ-tubular neighborhood of S and let $\{U_j\}_{j=2}^3$ be a covering of S defined by:

$$U_2 = S \cap \{\xi|(\xi_2 - 1/\sqrt{2}\gamma)^2 > \frac{\gamma^2}{12}\},$$

$$U_3 = S \cap \{\xi|\xi_3^2 > \frac{\gamma^2}{12}\}.$$

We define a covering $\{V_j\}$ of $\mathbf{R}^3$ by

$$V_1 = \mathbf{R}^3 \backslash N_{\epsilon/2}(S),$$

$$V_j = N_\epsilon(U_j), j = 2,3$$

with $\frac{\beta}{4} < \epsilon < \frac{\gamma}{2\sqrt{2}}$. Consider first the integral

$$\int_{V_1} (1+c|\xi'-\xi|^2)^{-m}\frac{\tilde{P}(\xi')}{|P(\xi')|}d\xi'.$$

On V_1, $|P(\xi)|$ is bounded from below by $\frac{\beta\gamma}{4\sqrt{2}}$ (see proof of the proposition 3.6 in [SU]). Moreover we have $|\xi| \ge \epsilon/2 \ge \beta/2$ on V_1. Estimating the numerator from above and denominator from below we obtain

$$\frac{\tilde{P}(\xi')}{|P(\xi')|} = (1+\frac{4(\xi'^2 - \xi'\cdot k) + 2\gamma^2 + 12}{|P(\xi'|^2})^{1/2} \le$$

$$(1+\frac{A'\xi'^2\gamma + B'\gamma^2 + C}{\gamma^2})^{1/2} \le (1 + A\xi'^2 + B)^{1/2}$$

with A, B depending only on β. Hence,

$$\sup \int_{V_1} (1+c|\xi'-\xi|^2)^{-m}\frac{\tilde{P}(\xi')}{|P(\xi')|}d\xi' \le C_1,$$

where C_1 depends only on m and β.

Consider the integrals over V_j, $j = 2, 3$. Since $\xi^2 - \xi \cdot k = -\mathrm{Re}P(\xi)$ And V_j lies inside the ball $B(3\gamma)$, there exists a constant C'_j such that $\tilde{P}^2 - |P|^2$ is bounded from above by $A\gamma^2 + B$ with A, B depending only on β. Using the diffeomorphism φ defined in the proof of the proposition 3.6 in [SU], one obtains:

$$\int_{V_j} (1 + c|\xi - \xi'|^2)^{-m} \frac{\tilde{P}(\xi')}{|P(\xi')|} d\xi' \leq$$

$$\sup \int (1 + \frac{A\gamma^2 + B\gamma + C}{\gamma^2(\eta_1^2 + \eta_2^2)})^{1/2} (1 + c|\varphi^{-1}(\eta') - \varphi^{-1}(\eta)|^2)^{-m} |(\varphi^{-1})'(\eta)| d\eta.$$

Since $|(\varphi^{-1})'|$ and $|\varphi'|$ are uniformly bounded on $\mathbf{R}^3$ by a constant b independent of γ,

$$|\varphi^{-1}(\eta') - \varphi^{-1}(\eta)| \geq \frac{1}{b}|\eta' - \eta|,$$

and

$$\int_{V_j} \frac{\tilde{P}(\xi')}{|P(\xi')|} (1 + c|\xi - \xi\prime^2)^{-m} \leq$$

$$\sup_{\eta'} b \int (1 + \frac{A_j}{\eta_1^2 + \eta_2^2})^{1/2} (1 + \frac{c}{b}|\eta - \eta'|^2)^{-m} d\eta \leq C_j$$

with C_j depending only on β and m.

The proof is complete.

Let

$$S = \{\xi | -\xi^2 + \gamma\xi_2 + i\gamma\xi_1 = 0\},$$

and $\theta = \frac{\xi'}{|\xi'|}$. To formulate the next theorem we need to define the following smooth cut-off functions:

- $h(y)$ is identically zero when $|y_1| < 1$ or $|y'| \geq 2$, and identically equal to one when $|y_1| \geq 2$ or $|y'| < 1$;
- $\chi(\xi, y)$ is supported in a tubular neighborhood of the set S for all y, equal to one for ξ such that $\frac{y'}{|y'|} \cdot \theta \geq 1/2$, where and equal to zero for ξ such that $\frac{y'}{|y'|} \cdot \theta \leq 1/4$;
- $\chi_1(\xi)$ is supported in a tubular neighborhood of S and identically equal to one on ξ-support of ξ.

Theorem 5 *Let M be a positive integer. Then the function*

$$F_M(y) = \frac{\gamma}{4\pi^2} e^{\frac{1}{2}i\gamma y_2} \int_S e^{i\gamma y \cdot \xi} \sum_{j=0}^{M} A_j (\chi + h(1 - \chi)) d\xi, \tag{5.27}$$

with

$$A_j = [-\gamma(y_1 + i\theta \cdot y']^{-j-1} (\theta \cdot \nabla_{\xi'} + i\partial_{\xi_1})^j,$$

satisfies

$$E(y) - F_M(y) = O((\gamma|y|)^{-M-1})$$

in the closure of $\Omega_1 \backslash B(1)$.

The proof of the theorem also provides an explicit expression for the remainder of the asymptotic expansion.

Corollary 1.5.3 *The remainder of the expansion (3.20) has the form:*

$$E(y) - F_M(y) = R_1(y) + R_2(y) + R_3(y), \tag{5.28}$$

where

$$R_1(y) = \frac{1}{8\pi^3}\gamma \int e^{i\gamma\xi\cdot y} \frac{(1-\chi_1)(\gamma\xi)}{-\xi^2 + \xi_2 + i\xi_1} d\xi,$$

and the integral is understood as an oscillatory integral,

$$R_2(y) = \frac{1}{8\pi^3}\gamma e^{\frac{1}{2}i\gamma y_2} \int e^{i\gamma(\xi_1 y_1 + r y\prime \cdot \theta)} \frac{r(1-h)\chi_1(1-\chi)(\gamma\xi)}{-\xi^2 + \frac{1}{4} + i\xi_1} dr d\xi_1 d\theta,$$

$$R_3(y) = \frac{1}{8\pi^3}\gamma e^{\frac{1}{2}i\gamma y_2} \int e^{i\gamma(\xi_1 y_1 + r y\prime \cdot \theta)} \frac{1}{r + i\xi_1 - \frac{1}{2}} \frac{D^{j+1}}{(\gamma(y_1 + i y\prime \cdot \theta))^{j+1}} F dr d\xi_1 d\theta$$

where

$$F = \frac{r\chi_1(\gamma\xi)(\chi + h(1-\chi))(\gamma\xi, y)}{r + i\xi_1 + \frac{1}{2}}.$$

and

$$D = -\theta \cdot \nabla_{\xi\prime} - i\partial_{\xi_1}.$$

Proof. In $\mathbf{R}^3 \backslash B(1)$, the fundamental solution E is defined by an oscillatory integral

$$E(y) = (2\pi)^{-3} \int e^{i\eta\cdot y} \frac{1}{-\eta^2 + \gamma\eta_2 + i\gamma\eta_1} d\eta$$

in aligned with ζ coordinates. Introduce a new variable $\xi = \gamma^{-1}\eta$. Then

$$E(y) = (2\pi)^{-3}\gamma \int e^{i\gamma\xi\cdot y} \frac{1}{-\xi^2 + \xi_2 + i\xi_1} d\xi.$$

Let χ_1 be cut-off function supported in the neighborhood of the set S (see above for more precise definition). We have:

$$E(y) = (2\pi)^{-3}\gamma \int e^{i\gamma\xi\cdot y} \frac{\chi_1(\xi)}{-\xi^2 + \xi_2 + i\xi_1} d\xi + R_1(y), \tag{5.29}$$

where R_1 is defined in Corollary 1 in section 3. Observing that an oscillatory integral for E restricted to the exterior of $B(1)$ can be replaced by a convolution with the Fourier transform of a smooth compactly supported function f , and making use of the theorem 7.7.1 in [H], we see that $R_1(y)$ is asymptotically small.

Shifting variables in the first term in (5.29), one obtains:

$$E(y) = L + O((\gamma|y|)^{-\infty}),$$

where

$$L = (2\pi)^{-3}\gamma e^{\frac{1}{2}i\gamma y_2} \int e^{i\gamma\xi\cdot y} \frac{\chi_1(\xi + \frac{1}{2}\gamma e_2)}{-\xi^2 + \frac{1}{4} + i\xi_1} d\xi.$$

Since

$$-\xi^2 + \frac{1}{4} + i\xi_1 = (\frac{1}{2} + i\xi_1 - |\xi'|)(\frac{1}{2} + i\xi_1 + |\xi'|),$$

we have:

$$L = (2\pi)^{-3}\gamma e^{\frac{1}{2}i\gamma y_2} \int e^{i\gamma\xi\cdot y} \frac{F(\xi)}{\frac{1}{2} + i\xi_1 - |\xi'|} d\xi,$$

where

$$F(\xi) = \frac{\chi_1(\xi + \frac{1}{2}\gamma e_2)}{\frac{1}{2} + i\xi_1 + |\xi'|}.$$

Introduce polar coordinates $\xi' = r\theta$, $\theta \in S^2$. Then

$$L = -(2\pi)^{-3}\gamma e^{\frac{1}{2}i\gamma y_2} \int e^{i\gamma(-\xi_1 y_1 + r\theta\cdot y'} \frac{rF(r\theta, -\xi_1)}{-\frac{1}{2} + i\xi_1 - r} d\xi_1 dr d\theta.$$

Using functions χ, h defined in section 3, we split M into three parts:

$$L = L(\chi, 1) + L(1 - \chi, h) + R_2(y),$$

where $L(a, b)$ denotes L with the product ab inserted under the integral. Again, the theorem 7.7.1 from [H] implies that $R_2(y)$ is asymptotically small. Asymptotic expansions for the first two terms are obtained via multiple integration by parts. We start with the identity:

$$\frac{2}{\gamma(y_1 + iy' \cdot \theta)} \bar{\partial} e^{i\gamma(-\xi_1 y_1 + r\theta\cdot y')} = e^{i\gamma(-\xi_1 y_1 + r\theta\cdot y')}, \tag{5.30}$$

where $\bar{\partial} = \frac{1}{2}(\partial_r + i\partial_{\xi_1})$. Moreover, since $-\frac{1}{\pi z}$ is a fundamental solution of $\bar{\partial}$, we have

$$f(x_1, x_2) = -\frac{1}{\pi} \int \frac{\bar{\partial} f(r, \xi_1)}{z - x_1 - ix_2} d\xi_1 dr \tag{5.31}$$

where $z = r + i\xi_1$ and $f \in C_0^1(\mathbf{R}^2)$. Combining (5.30), (5.31), and integrating M times by parts one obtains:

$$L(\chi, 1) + L(1 - \chi, h) = \frac{1}{4\pi^2} e^{\frac{1}{2}i\gamma y_2} \int_S e^{i\gamma\xi\cdot y} \sum_{j=0}^{L} A_j(\chi + h(1 - \chi)) d\xi + R_3(y),$$

where A_j are defined in the formulation of the theorem, and the expression for R_3 is given in the corollary.

The proof is complete.

References

[BU] Brown R.M. and Uhlmann G. *Uniqueness in the Inverse Conductivity Problem for Nonsmooth Conductivities in Two Dimensions.* To appear in Comm. in PDE.

[C] Calderón A.P. *On an Inverse Boundary Value Problem.* Seminar on Numerical Analysis and Its Applications to Continuum Physics, Rio de Janeiro,
1980, 65-73.

[ER] Eskin G. and Ralston J. *Inverse Scattering Problem for the Scrödinger equation with Magnetic Potential at a Fixed Energy.* Comm.Math.Phys., 1995, (173), 173-199.

[F] Fedoryuk M.V. *Asymptotics: Integrals and Series.*, Nauka, Moscow, 1987. (Russian).

[HeN] Henkin G.M. and Novikov R.G. *The ∂-equation in the Multidimensional Inverse Scattering Problem.* Russian Math. Serveys, 1987,**42**, 101-180.

[H] Hörmander L. *Analysis of Linear Partial Differential Operators.*, v.I,II. Springer-Verlag.

[I] Isakov V.*Completeness of Products of Solutions and Some Inverse Problems for PDE.* J. Diff. Equations, 1991,**92**, 305-317.

[I1] Isakov V. *Inverse Problems for PDE.* Springer-Verlag, 1997.

[N] Nachman A. *Reconstruction from Boundary Measurements.* Ann. Math., 1988, **128**, 531-577.

[NU] Nakamura G. and Uhlmann G. *Global Uniqueness for an Inverse Boundary Value Problem Arising in Elasticity.* Invent. Math., 1994,**118**, 457-474.

[S] Stein E.M. *Harmonic Analysis: Real-variable Methods, Orthogonality, and Oscillatory Integrals.* Prinnceton University Press, 1993.

[Sh] Shubin M.A. *Pseudodifferential Operators and Spectral Theory.* Springer-Verlag, 1987.

[SU] Sylvester J. and Uhlmann G. *Global Uniqueness Theorem for an Inverse Boundary Value Problem.* Ann. Math., 1987,**125**, 153-169.

[Su] Sun Z. *An Inverse Boundary Value Problem for Schrodinger Operator with Vector Potentials.* Trans. AMS, 1993, **338**, 2, 953-969.

[T] Taylor M.E.*Pseudeodifferential Operators and Nonlinear PDE.* Birkhauser, Boston, 1991.

[To] Tolmasky C.F. *Exponentially Growing Solutions for Non-smooth First Order Perturbations of the Laplacian.* To appear in SIAM J. Math. Anal.

[V] Vaiberg B.R. *Asymptotic Methods in Equations of Mathematical Physics.* Gordon and Breach, 1989.

ANALYTICALLY SMOOTHING EFFECT FOR SCHRÖDINGER TYPE EQUATIONS WITH VARIABLE COEFFICIENTS

Kunihiko Kajitani and Seiichiro Wakabayashi

Institute of Mathematics
University of Tsukuba
305 Tsukuba Ibaraki Japan

1.1 INTRODUCTION

We shall investigate analytically smoothing effects of the solutions to the Cauchy problem for Schrödinger type equations. We shall prove that if the initial data decay exponentially then the solutions become analytic with respect to the space variables. Let $T > 0$. We consider the following Cauchy problem,

$$(1)\quad \frac{\partial}{\partial t}u(t,x) - ia(x,D)u(t,x) - b(t,x,D)u(t,x) = f(t,x), t \in [-T,T], x \in R^n,$$

$$(2)\qquad u(0,x) = u_0(x), x \in R^n,$$

where

$$(3)\qquad a(x,D)u = \sum_{j,k=1}^{n} D_j(a_{jk}(x)D_k u),$$

R.P. Gilbert et al.(eds.), Direct and Inverse Problems of Mathematical Physics, 185–219.

$$b(t,x,D)u = \sum_{j=1}^{n} b_j(t,x)D_j u + b_0(t,x)u, \tag{4}$$

and $D_j = -i\frac{\partial}{\partial x_j}$. We assume that the coefficients $a_{jk}(z)$ and $b_j(t,z)$ are bounded and holomorhpic in a complex domain $\Gamma_{\tau_0,\varepsilon_0} = \{z \in C^n; |Imz| \leq \tau_0|Rez| + \varepsilon_0\}$ for some $\tau_0 > 0$ and $\varepsilon_0 > 0$ and continuous in $t \in [-T,T]$, and that the principal part $a(x,\xi) = \sum a_{ij}(x)\xi_i\xi_j$ is real valued and elliptic for $x, \xi \in R^n$, that is, there is $c > 0$ such that

$$|a(x,\xi)| \geq c|\xi|^2, \tag{5}$$

for $x, \xi \in R^n$. Moreover we assume the coefficients of the first order term $b_j(t,x)(j = 1,...,n)$ satisfy,

$$\lim_{|x|\to\infty} Reb_j(t,x) = 0, \text{uniformly in } t \in [-T,T]. \tag{6}$$

Furthermore we suppose that there is a function $\theta(z,\zeta)$ holomorphic in a complex domain $\Gamma_{\tau_\theta,\varepsilon_\theta} \times \Gamma_{\tau'_\theta,\varepsilon'_\theta}(\tau_\theta, \varepsilon_\theta, \tau'_\theta, \varepsilon'_\theta > 0)$ and $c > 0$ such that

$$H_a\theta(x,\xi) \geq c|\xi|, \tag{7}$$

for $x, \xi \in R^n$ and $< z >^{-1} \theta(z,\zeta)$ is bounded in $\Gamma_{\tau_\theta,\varepsilon_\theta} \times \Gamma_{\tau'_\theta,\varepsilon'_\theta}$, where denote by H_a the Hamilton vector field of a and $< z >= (1 + z_1^2 + \cdots + z_n^2)^{\frac{1}{2}}$. For example we can choose $\theta = \sum z_j\zeta_j < \zeta >^{-1}$ when a is Laplacian in R^n. See Doi [3] for a criterion to existence of the function θ which satisfies the condition (7). For $\rho \geq 0$ let define a exponential operator $e^{\rho<D>}$ as follows,

$$e^{\rho<D>}u(x) = \int_{R^n} e^{ix\xi+\rho<\xi>}\hat{u}(\xi)\bar{d}\xi$$

where $\hat{u}(\xi)$ stands for a Fourier transform of u and $\bar{d}\xi = (2\pi)^{-n}d\xi$ and for $\varepsilon \in R$ denote $\phi_\varepsilon = x\xi - i\varepsilon\vartheta(x,\xi)$ and we define

$$I_{\phi_\varepsilon}(x,D)u(x) = \int_{R^n} e^{i\phi_\varepsilon(x,\xi)}\hat{u}(\xi)d\xi.$$

Then our main theorem follows.

Theorem. *Assume (5)-(7) are valid and there are $\rho_0 > 0$ and $\delta_0 > 0$ such that $I_{\phi_{\pm\delta_0}}u_0 \in L^2(R^n)$ and $I_{\phi_{\pm\delta_0}}e^{\rho_0<D>}f(t,x) \in C([-T,T];L^2(R^n))$. Then there exists a solution of (1)-(2) satisfying that there are $C > 0, \rho > 0$ and $\delta > 0$ such that*

$$|\partial_x^\alpha u(t,x)| \leq C(\rho|t|)^{-|\alpha|}|\alpha|!e^{\delta<x>}, \tag{8}$$

for $(t,x) \in [-T,T]\backslash 0 \times R^n, \alpha \in N^n$.

Remark. *(i) If $e^{\delta_1<x>}u_0(\delta_1 > 0)$ is in $L^2(R^n)$, then we have $\delta_0 > 0$ such that $I_{\phi_{\pm\delta_0}}u_0$ belongs to $L^2(R^n)$ (See Lemma 2.7 below). (ii) In the case of $a = -\Delta$, A. Jensen in [6] and Hayashi,Nakamitsu & Tsutsumi in [5] showed that if $< x >^k u_0(x) \in L^2(R^n)$, the solution u of (1)-(2) belongs to H^k_{loc} for $t \neq 0$, Hayashi & Saitoh in [4] proved that if $e^{\delta<x>^2}u_0$ $(\delta > 0)$ is in $L^2(R^n)$, the solution u is analytic in x for $t \neq 0$ and De Bouard, Hayashi & Kato in [1], Kato & Taniguti in [10] show that if u_0 satisfies $\|(x\nabla)^j u_0\| \leq C^{j+1}j!^2$ for j=0,1.2..., then the solution is analytic in x for $t \neq 0$. Theorem 1 is proved by Kajitani in [8], when $a(x,\xi) = |\xi|^2$ and $b_j(t,x) = 0$.*

The idea in the proof of the above Theorem is based on the methods introduced in [8] in which is investigated the well posedness in Gevrey classes of the Cauchy problem for Schrödinger type equations and in [12] in which is studied the analytically microlocal properties of hyperfunctions by use of peudodifferential oprators of infinite order. Roughly speaking we transform an unknown function u to w by pseudodifferential operators of infinite order such as $w = I_{\phi_\epsilon}(x,D)E(t)u$ which satisfies a Schrödinger type equation to be well posed in L^2, where ϵ, ρ small parameters and $E(t) = e^{t\rho p(x,D)}$ $(p(x,\xi) = (1+a(x,\xi))^{1/2})$ is the fundamental solution of the equation $E'(t) = \rho p(x,D)E(t), E(0) = I$. The L^2- well posed Cauchy problem for Schrödinger type equations with variable coefficients are studied in [7] and [3]. In §1 we introduce Sobolev spaces with exponential weights in which act pseudodifferential operators of infinite order. In §2 and §3 we shall study pseudodifferental operators of infinite order and Fourier integral operators, especially in §3 we construct the fundamental solution $E(t)$ and the inverse operator of $I_{\phi_\epsilon}(x,D)$ as a pseudodifferental operator of infinite order respectively. See [14] for the microlocal properties of pseudodifferentia! ! ! l operators of infinite order and Fourier integral operators. In §4 we consider L^2- well posed Cauchy problem for Schrödinger type equation. Finally in §5 we shall give the proof of our main Theorem.

1.2 WEIGHTED SOBOLEV SPACES

We introduce some Sobolev spaces with weights. Let ρ, δ be real numbers. Define

$$\hat{H}_\delta = \{u \in L^2_{loc}(R^n); e^{\delta<x>}u(x) \in L^2(R^n)\}.$$

For $\rho \geq 0$ let define

$$H_\rho = \{u \in L^2(R^n); Fu(\xi) \in \hat{H}_\rho(R^n_\xi)\},$$

where Fu stands for the Fourier transform of u. For $\rho < 0$ we define H_ρ as the dual space of $H_{-\rho}$. Then the Fourier transform F becomes bijective from H_ρ to $\hat{H}_\rho$. We define the operator $e^{\rho<D>}$ mapping continuously from H_{ρ_1} to $H_{\rho_1-\rho}$ as follows;

$$e^{\rho<D>}u(x) = F^{-1}(e^{\rho<\xi>}Fu(\xi))(x),$$

for $u \in H_{\rho_1}$ and $e^{\delta<x>}$ maps continuously from $\hat{H}_{\delta_1}$ to $\hat{H}_{\delta_1-\delta}$. We define for $\delta \geq 0$ and $\rho \in R$

$$H_{\rho,\delta} = \{u \in H_\rho; e^{\rho<D>}u \in \hat{H}_\delta\}. \tag{1.1}$$

For $\delta < 0$ we define $H_{\rho,\delta}$ as the dual space of $H_{-\rho,-\delta}$. We note that $H_{\rho,0} = H_\rho, H_{0,\delta} = \hat{H}_\delta$ and $H_{0,0} = L^2(R^n)$. Furthermore we define for $\rho \geq 0$ and $\delta \in R$

$$\tilde{H}_{\rho,\delta} = \{u \in \hat{H}_\delta; e^{\delta<x>}u \in H_\rho\} \tag{1.2}$$

and for $\rho < 0$ define $\tilde{H}_{\rho,\delta}$ as the dual spase of $\tilde{H}_{-\rho,-\delta}$. Denote by H' the dual space of a topological space H. Then $H'_{\rho,\delta} = H_{-\rho,-\delta}$ and $\tilde{H}'_{\rho,\delta} = \tilde{H}_{-\rho,-\delta}$ hold for any ρ and $\delta \in R$. We shall prove $H_{\rho,\delta} = \tilde{H}_{\rho,\delta}$ later on (see Proposition 2.11).

The exponential map $e^{\rho<D>}$ (resp. $e^{\delta<x>}$) operates continuously ¿from $\hat{H}_\delta$ into $H_{-\rho,\delta} = H'_{\rho,-\delta}$ (resp. from H_ρ into $\tilde{H}_{\rho,-\delta} = \tilde{H}'_{-\rho,\delta}$). In fact, if $\delta \geq 0$ then $\hat{H}_\delta \subset L^2$ and consequently $\hat{H}_\delta$ is contained in the definition domain of the operator $e^{\rho<D>}$. Hence we have $e^{\rho<D>}u \in H_{-\rho,\delta}$ for $u \in \hat{H}_\delta$. If $\delta < 0$, we can define $e^{\rho<D>}$ as a map from $\hat{H}_\delta$ into $H_{-\rho,\delta}$. In fact, for $u \in \hat{H}_\delta$ and $\varphi \in H'_{-\rho,\delta} = H_{\rho,-\delta}$,

$$< e^{\rho<D>}u, \varphi >=< u, e^{\rho<D>}\varphi >$$

because $-\delta > 0$ implies $e^{\rho<D>}\varphi \in \hat{H}_{-\delta}$. In similar way, $e^{\delta<x>}$ can be extended as a map from H_ρ into $\tilde{H}_{\rho,-\delta}$. Thus we get

$$e^{-\rho<D>}\hat{H}_\delta \subset H_{\rho,\delta}, \tag{1.3}$$

$$e^{-\delta<x>}H_\rho \subset \tilde{H}_{\rho,\delta}. \tag{1.4}$$

Lemma 1.1. *Let $\rho, \delta \in R$. Then*

(i) $$H_{\rho,\delta} = e^{-\rho<D>}e^{-\delta<x>}L^2 = e^{-\rho<D>}\hat{H}_\delta.$$

(ii) $$\tilde{H}_{\rho,\delta} = e^{-\delta<x>}e^{-\rho<D>}L^2 = e^{-\delta<x>}H_\rho.$$

Proof. When $\delta \geq 0$, (i) is trivial. Let $\delta < 0$. It suffices to show $H_{\rho,\delta} \subset e^{-\rho<D>}\hat{H}_\delta$. Let $T \in H_{\rho,\delta} = H'_{-\rho,-\delta}$. Then it follows from Riesz Theorem we have $u \in H_{-\rho,-\delta}$ such that for any $\phi \in H_{-\rho,-\delta}$

$$(1.5) \qquad < T, \bar{\phi} >= (u, \phi)_{H_{-\rho,-\delta}} = (e^{-\delta<x>}e^{-\rho<D>}u, e^{-\delta<x>}e^{-\rho<D>}\phi)_{L^2},$$

where $\bar{\phi}$ stands for the complex conjugate of ϕ. Since $-\delta > 0$, we have $w \in \hat{H}_{-\delta}$ such that $u = e^{\rho<D>}w$. Hence we get by (1.5)

$$< T, \bar{\phi} >= (e^{-\delta<x>}w, e^{-\delta<x>}e^{-\rho<D>}\phi)_{L^2} = (e^{-2\delta<x>}w, e^{-\rho<D>}\phi)_{L^2} =$$
$$< e^{-\rho<D>}e^{-2\delta<x>}w, \bar{\phi} >,$$

for any $\phi \in H_{-\rho,-\delta}$. Therefore $T = e^{-\rho<D>}e^{-2\delta<x>}w \in e^{-\rho<D>}\hat{H}_\delta$. We can prove (ii) in a similar way. Q.E.D.

Lemma 1.2 *Let* $1 > \rho > 0, \delta \in R$ *and* $u \in \tilde{H}_{\rho,\delta}$. *Then*

$$(1.6) \qquad |D_x^\alpha u(x)| \le C_n(1-\epsilon)^{-n/2}\|u\|_{\tilde{H}_{\rho,\delta}}(\epsilon\rho)^{-|\alpha|}|\alpha|!e^{\delta<x>}$$

for $x \in R^n, \alpha \in N^n$ *and* $0 < \epsilon < 1$.

Proof. Put $v(x) = e^{\rho<D>}e^{\delta<x>}u$ and $w = e^{-\rho<D>}v(x)$. Then v belongs to $L^2(R^n)$ and hence $w \in H_\rho$ satisfying for any $\epsilon \in (0,1)$

$$(1.7) \qquad |D_x^\alpha w(x)| \le C(1-\epsilon)^{-n/2}\|w\|_{H_\rho}(\epsilon\rho)^{-|\alpha|}|\alpha|!,$$

which implies (1.6) because of $u(x) = e^{\delta<x>}w(x)$. Q.E.D.

1.3 PSEUDODIFFERENTIAL OPERATORS

In this section we shall treat pseudodifferential operators with symbol of infinite order.

Definition 2.1. *We say* $a(x,\zeta) \in A_{0,0}$, *if* $a(x,\zeta) \in C^\infty(R^{2n})$ *and there are* $C_a > 0, \varepsilon_a > 0$ *such that*

$$(2.1) \qquad |a_{(\beta)}^{(\alpha)}(x,\zeta)| \le C_a\varepsilon_a^{-|\alpha+\beta|}|\alpha+\beta|!,$$

for $x, \zeta \in R^n, \alpha, \beta \in N^n$. *Let* $\rho, \delta \in R$. *We say* $a(x,\zeta) \in A_{\rho,\delta}$ *if there is* $\tilde{a}(x,\zeta) \in A_{0,0}$ *suth that*

$$(2.2) \qquad a(x,\zeta) = e^{\rho<\zeta>+\delta<x>}\tilde{a}(x,\zeta),$$

where we denote $< x >= (1 + x_1^2 + ... + x_n^2)^{\frac{1}{2}}$.

We define the product of two symbols $a_i \in A_{\rho_i,\delta_i}, i = 1,2$ as follows;

$$(2.3) \qquad (a_1 \circ a_2)(x,\zeta) = os - \int\int_{R^{2n}} e^{-iy\eta}a_1(x,\zeta+\eta)a_2(x+y,\zeta)dy\bar{d}\eta,$$

$$= \lim_{\epsilon\to 0} \int\int_{R^{2n}} e^{-iy\eta-\epsilon(|y|^2+|\eta|^2)} a_1(x,\zeta+\eta)a_2(x+y,\zeta)dyd\bar{\eta},$$

where $d\bar{\eta} = (2\pi)^{-n}d\eta$. Then we can show the proposition below.

Proposition 2.3. *(i) Let $a_i \in A_{\rho_i,\delta_i}, i = 1,2$. Then there is $\epsilon_0 > 0$ such that if $|\rho_1|, |\delta_2| \le \epsilon_0$, the product $a_1 \circ a_2$ belongs to $A_{\rho_1+\rho_2,\delta_1+\delta_2}$.*

(ii) Let $a_i \in A_{\rho_i,\delta_i}, i = 1,2,3$. Then if $|\rho_i|(i = 1,2), |\delta_i|(i = 2,3) \le \epsilon_0/2$, we have $(a_1 \circ a_2) \circ a_3 = a_1 \circ (a_2 \circ a_3)$.

Proof. (i) Put $\tilde{a}_i = e^{-\rho_i<\zeta>-\delta_i<x>}a_i(x,\zeta)$ and $\tilde{a} = e^{-(\rho_1+\rho_2)<\zeta>-(\delta_1+\delta_2)<x>}a_1\circ a_2$. Then (2.3) implies

$$(2.4)\quad \tilde{a}(x,\zeta) = \lim_{\epsilon\to+0} \int\int_{R^{2n}} e^{-iy\eta-\epsilon(|y|^2+|\eta|^2)} e^{\rho_1(<\zeta+\eta>-<\zeta>)+\delta_2(<x+y>-<x>)}$$

$$\times\tilde{a}_1(x,\zeta+\eta)\tilde{a}_2(x+y,\zeta)dyd\bar{\eta} = \lim_{\epsilon\to+0} \tilde{a}_\epsilon(x,\zeta).$$

We shall show that $a_\epsilon(x,\zeta)$ converges in $A_{0,0}$ tending $\epsilon \to +0$. Noting

$$<x+y> - <x> = \omega(x,y)y,$$

where

$$(2.5)\qquad \omega(x,y) = (2x+y)(<x+y> + <y>)^{-1},$$

we can rewrite from (2.4)

(2.6)

$$a_\epsilon(x,\zeta) = \int\int_{R^{2n}} e^{-iy\eta-\epsilon(|y|^2+|\eta|^2)} e^{\rho_1\omega(\zeta,\eta)\eta+\delta_2\omega(x,y)y}\tilde{a}_1(x,\zeta+\eta)\tilde{a}_2(x+y,\zeta)dyd\bar{\eta}$$

$$= \int\int_{R^{2n}} e^{-i(y+i\rho_1\omega(\zeta,\eta))(\eta+i\delta_2\omega(x,y))-i\rho_1\delta_2\omega(x,y)\omega(\zeta,\eta)-\epsilon(|y|^2+|\eta|^2)}\tilde{a}_1(x,\zeta+\eta)\tilde{a}_2(x+y,\zeta)dyd\bar{\eta}$$

$$= \int\int_C e^{-iuv}b_\epsilon(x,\zeta;u,v)\frac{\partial(\Phi,\Psi)}{\partial(u,v)}dud\bar{v},$$

where

(2.7)

$$b_\epsilon(x,\zeta;u,v) = e^{-\epsilon(|y|^2+|\eta|^2)-i\rho_1\delta_2\omega(x,y)\omega(\zeta,\eta)}\tilde{a}_1(x,\zeta+\eta)\tilde{a}_2(x+y,\zeta) = e^{-\epsilon(|y|^2+|\eta|^2)}b(x,\zeta;u,v)$$

and $(y,\eta) = (\Phi,\Psi)(x,\zeta;u,v)$ is a solution of the following equation

$$(2.8)\qquad y + i\rho_1\omega(\zeta,\eta) = u,$$

$$\eta + i\delta_2\omega(x,y) = v,$$

and

$$(2.9) \qquad C = C(x,\zeta) = \{(y + i\rho_1\omega(\zeta,\eta), \eta + i\delta_2\omega(x,y)); (y,\eta) \in R^{2n}\},$$

and $\frac{\partial(\Phi,\Psi)}{\partial(y,\eta)}$ is the Jacobian. Then by the implicit function theorem we can solve the equation (2.8) and there is $\varepsilon_0 > 0$ such that for $|\rho_1 + \delta_2| \le \varepsilon_0$ and $x, \zeta \in R^n$ the solution $(\Phi,\Psi)(x,\zeta;u,v)$ is holomorphic in $(u,v) \in K(x,\zeta) = \{(y + i\rho_1 t\omega(\zeta,\eta), \eta + i\delta_2 t\omega(x,y)); (y,\eta) \in R^{2n}, 0 \le t \le 1\}$ and the Jacobian $\frac{\partial(\Phi,\Psi)}{\partial(y,\eta)}$ is bounded in $K(x,\zeta)$. Therefore $b_\epsilon(z,\zeta;u,v)$ and $b(z,\zeta;u,v)$ are bounded and holomorphic in $(u,v) \in K(x,\zeta)$. Moreover, since $b_\epsilon(z,\zeta;u,v)$ is integrable in $(u,v) \in K(x,\zeta)$ if $\epsilon > 0$, applying Stokes formula to (2.6) and integrating by part, we get

$$\tilde{a}_\epsilon(x,\zeta) = \int\int_{R^{2n}} e^{-iy\eta} b_\epsilon(x,\zeta;y,\eta)\frac{\partial(\Phi,\Psi)}{\partial(y,\eta)} dy\bar{d}\eta$$

$$= \int\int_{R^{2n}} e^{-iy\eta}(1-\Delta_\eta)^\ell(1+|y|^2)^{-\ell}(1-\Delta_y)^\ell\{(1+|\eta|^2)^{-\ell} b_\epsilon(x,\zeta;y,\eta)\frac{\partial(\Phi,\Psi)}{\partial(y,\eta)}\}dy\bar{d}\eta.$$

Hence choosing $\ell > n/2$, we have

$$(2.10) \qquad \tilde{a}(x,\zeta) = \lim_{\epsilon\to+o} \tilde{a}_\epsilon(x,\zeta)$$

$$= \int\int_{R^{2n}} e^{-iy\eta}(1-\Delta_\eta)^\ell(1+|y|^2)^{-\ell}(1-\Delta_y)^\ell\{(1+|\eta|^2)^{-\ell} b(x,\zeta;y,\eta)\frac{\partial(\Phi,\Psi)}{\partial(y,\eta)}\}dy\bar{d}\eta.$$

Therefore we can see easily that $\tilde{a}(x,\zeta)$ belongs to $A_{0,0}$. (ii) follows from Fubini's theorem. Q.E.D.

Now we want to define a pseudo differential operator $a(x,D)$ for a symbol $a(x,\xi) \in A_{\rho,\delta}$, which operates from $H_{\rho',\delta'}$ to $H_{\rho'-\rho,\delta'-\delta}$. When ρ and δ are non positive, since $A_{\rho,\delta}$ is contained in the usual symbol class $S^0_{0,0}$(denote by $S^m_{\rho,\delta}$ the Hörmander's class), we can define

$$(2.11) \qquad a(x,D)u(x) = \int e^{ix\xi} a(x,\xi)\hat{u}(\xi)\bar{d}\xi,$$

for $u \in L^2(R^n)$ and for $a \in A_{\rho,\delta}$. Moreover for $a_i \in A_{\rho_i,\delta_i}, i = 1,2$ (ρ_i and δ_i non positive) the symbol $\sigma(a_1(x,D)a_2(x,D))(x,\xi)$ of the product of $a_1(x,D)$ and $a_2(x,D)$ can be written as follows,

$$(2.12) \qquad \sigma(a_1(x,D)a_2(x,D))(x,\xi) = (a_1 \circ a_2)(x,\xi)$$

and we have

$$a_1(x,D)(a_2(x,D)u)(x) = (a_1 \circ a_2)(x,D)u(x) \tag{2.13}$$

for $u \in L^2(R^n)$, where $a_1 \circ a_2$ is defined by (2.3). Next we shall show that (2.12) and (2.13) are valid for any ρ_i, δ_i. To do so, we need some preparations. Let $a \in A_{\rho,\delta}$ and $u \in H_\rho$. Then we can define $a(x,D)u(x)$ which belongs to $\hat{H}_\delta$. In fact, from the definition of $A_{\rho,\delta}$ we have $\tilde{a}(z,\zeta) \in A_{0,0}$ satisfying (2.2). Since $e^{\rho<\xi>}\hat{u}(\xi) \in L^2$ and $\tilde{a} \in S^0_{0,0}$, we can define

$$e^{-\delta<x>}a(x,D)u(x) = \int e^{ix\xi}\tilde{a}(x,\xi)e^{\rho<\xi>}\hat{u}(\xi)\bar{d}\xi, \tag{2.14}$$

which is in L^2, that is, $a(x,D)u \in \hat{H}_\delta$. For $\epsilon > 0$ we denote $\chi_\epsilon(x) = e^{-\epsilon<x>^2}$ and $\chi_\epsilon(D) = e^{-\epsilon<D>^2}$.

Lemma 2.4. *(i) Let $a \in A_{\rho,\delta}(\rho,\delta \in R), u \in L^2$ and $\epsilon_0 > 0$ chosen in Proposition 2.3. Then for any $\epsilon > 0$*

$$a(x,D)(\chi_\epsilon(D)\chi_\epsilon(x)u)(x) = (a(x,\xi)\chi_\epsilon(\xi)) \circ \chi_\epsilon(x))(x,D)u(x) \tag{2.15}$$

Proof. *(i) When $\epsilon > 0$, $\chi_\epsilon(D)\chi_\epsilon(x)u(x) \in H_{\rho'}$ for any $\rho' \in R$. Therefore we can use (2.14) and consequently we get*

(2.19)

$$a(x,D)(\chi_\epsilon(D)\chi_\epsilon(x)u(x)) = \int e^{ix\xi}a(x,\xi)\chi_\epsilon(\xi)\{\int \hat{\chi}_\epsilon(\xi-\eta)\hat{u}(\eta)\bar{d}\eta\}\bar{d}\xi =$$

$$\int e^{ix\eta}a_\epsilon(x,\eta)\hat{u}(\eta)\bar{d}\eta,$$

where $$a_\epsilon(x,\eta) = \int e^{ix(\xi-\eta)}a(x,\xi)\chi_\epsilon(\xi)\hat{\chi}_\epsilon(\xi-\eta)\bar{d}\xi,$$

and $\hat{\chi}_\epsilon$ is a Fourier transform of χ_ϵ . By Fubini's theorem and the definition of product of symbols we obtain

$$a_\epsilon(x,\eta) = \int e^{ix(\xi-\eta)}a(x,\xi)\chi_\epsilon(\xi)\{\int e^{-iy(\xi-\eta)}\chi_\epsilon(y)dy\}\bar{d}\xi$$

$$= \int\int e^{-iy\xi}a(x,\xi+\eta)\chi_\epsilon(\xi+\eta)\chi_\epsilon(x+y)dy\bar{d}\xi = (a\chi_\epsilon(\eta)) \circ \chi_\epsilon(x),$$

which implies (2.15) together with (2.19). Since $a\chi_\epsilon(\xi) \in A_{\rho',\delta}$ for any ρ' and $\chi_\epsilon(x) \in A_{0,\delta'}$ for any δ', it follows from Proposition 2.3 that $a\chi_\epsilon(\xi) \circ \chi_\epsilon(x)$ belongs to $A_{\rho+\rho',\delta+\delta'}$ if $|\rho'| \le \varepsilon_0, |\delta'| \le \varepsilon_0$. Hence we get (2.16) if we take $\rho' = -\epsilon_0$ and $\delta' = -\epsilon_0$.

(ii) Put $w(x) = e^{-\delta<x>-\epsilon<x>^2}e^{-\epsilon<D>^2}u(x)$. We shall prove there is $\epsilon_1 > 0$ independent of $\epsilon > 0$ such that $w \in H_{\rho'}$ for $|\rho'| \le \epsilon_1$. In fact, the Fourier image of w is give by

$$\hat{w}(\xi) = \int \hat{\chi}_{\epsilon,\delta}(\xi - \eta)e^{-\epsilon<\eta>^2}\hat{u}(\eta)\bar{d}\eta,$$

where

$$\hat{\chi}_{\epsilon,\delta}(\xi) = \int e^{ix\xi}e^{-\delta<x>-\epsilon<x>^2}dx.$$

Noting that the Cauchy integral formula shows that there are $C_0 > 0$ and $\rho_0 > 0$ independent of ϵ such that

$$|D_x^\alpha(e^{-\delta<x>-\epsilon<x>^2})| \leq C_0\rho_0^{-|\alpha|}|\alpha|!e^{2|\delta|<x>-\epsilon/2<x>^2}, x \in R^n,$$

we can see that there are $C_{\epsilon,\delta} > 0$ and $\epsilon_1 > 0$ (independent of ϵ) such that

$$|\hat{\chi}_{\epsilon,\delta}(\xi)| \leq C_{\epsilon,\delta}e^{-\epsilon_1<\xi>},$$

for $\xi \in R^n$. Hence we can express for $\rho < \epsilon_1$,

$$e^{-\rho<D>}w(x) = \int e^{ix\xi-\rho<\xi>}\hat{w}(\xi)\bar{d}\xi$$

$$= \int e^{ix\xi-\rho<\xi>}\{\int \hat{\chi}_{\epsilon,\delta}(\xi-\eta)e^{-\epsilon<\eta>^2}\hat{u}(\eta)\bar{d}\eta\}\bar{d}\xi = \int e^{ix\eta}a_\epsilon(x,\eta)\hat{u}(\eta)\bar{d}\eta,$$

where we put

$$a_\epsilon(x,\eta) = \int e^{ix(\xi-\eta)-\rho<\xi>}\hat{\chi}_{\epsilon,\delta}(\xi-\eta)e^{-\epsilon<\eta>^2}\bar{d}\xi$$

$$= \int e^{ix(\xi-\eta)-\rho<\xi>-\epsilon<\eta>^2}\{\int e^{-iy(\xi-\eta)-\epsilon<y>^2-\delta<y>}dy\}\bar{d}\xi$$

$$= \int e^{ix\xi-\rho<\xi+\eta>-\epsilon<\eta>^2}\{\int e^{-iy\xi-\epsilon<y>^2-\delta<y>}dy\}\bar{d}\xi$$

$$= \lim_{\vartheta\to+0}\int e^{-iy\xi-\vartheta(|y|^2+|\xi|^2)-\rho<\xi+\eta>}\chi_\epsilon(x+y)e^{-\delta<x+y>}\chi_\epsilon(\eta)dy\bar{d}\xi$$

$$= e^{-\rho<\xi>}\circ(\chi_\epsilon(x)e^{-\delta<x>}\chi_\epsilon(\xi))(x,\eta).$$

Therefore we get (2.17). On the other hand since $e^{-\rho<\zeta>}$ belongs to $A_{-\rho,0}$ and $e^{-\epsilon<x>^2-\delta<x>-\epsilon<\zeta>^2}$ is in $A_{\rho',\delta'}$ for any $\rho',\delta' \in R$, it follows from Proposition 2.3 that we get $e^{-\rho<\xi>}\circ(\chi_\epsilon(x)e^{-\delta<x>}\chi_\epsilon(\xi))(z,\zeta)$ belongs to $A_{-\rho+\rho',-\delta+\delta'}$ for $|\rho|,|\delta'| \leq \epsilon_0$. Hence we get (2.18) if we choose $\rho' = -\epsilon_0, \delta' = -\epsilon_0$. Q.E.D.

Remark. *We note that ϵ_1 depends on the holomorphic domain of $< x >$. If we replace the definition of $< x >$ by $< x >_h = (h^2 + |x|^2)^{1/2}(h > 0)$, then ϵ_1 can be taken arbitrarily large by choosing $h > 0$ suitably.*

Lemma 2.5. *Let $u \in H_{\rho,\delta}$ and $|\rho|,|\delta| \leq \epsilon_0/2$ (ϵ_0 is given in Proposition 2.3). Then for any $\epsilon > 0$ there is $u_\epsilon \in H_{\epsilon_0/2,\epsilon_0/2}$ such that*

(2.20) $$\|u-u_\epsilon\|_{H_{\rho,\delta}}<\epsilon.$$

Proof. *Put* $w=e^{\delta<x>}e^{\rho<D>}u$, *which is in* L^2. *Then it follows from (ii) of Lemma 2.4 that for* $\mu>0$ $u_\mu=e^{-\rho<D>}e^{-\delta<x>}e^{-\mu<x>^2}e^{-\mu<D>^2}w=:a_\mu(x,D)w$ *belongs to* $H_{\epsilon_0/2,\epsilon_0/2}$. *In fact we have* $u_\mu\in L^2$ *because* $a_\mu(x,\xi)=e^{-\rho<\xi>}\circ(e^{-\delta<x>-\mu<x>^2-\mu<\xi>^2})\in A_{0,0}\subset S^0_{0,0}$ *for* $|\rho|\le\epsilon_0,\rho<\varepsilon_1$ *from (ii) of Lemma 2.4 and* $w\in L^2$. *Hence we have* $e^{\epsilon<x>/2}e^{\epsilon_0<D>/2}u_\mu=e^{\epsilon_0/2<x>}e^{\epsilon_0/2<D>}a_\mu w$ *in* $H_{-\epsilon_0/2,-\epsilon_0/2}$. *On the other hand it follows from (ii) of Proposition 2.3 and (i) of Lemma 2.4 that we have the symbol of* $e^{\epsilon_0/2<x>}e^{\epsilon_0/2<D>}a_\mu=(e^{\epsilon_0<x>/2}e^{\epsilon_0<\xi>/2-\rho<\xi>})\circ(e^{-\delta<x>-\mu<\xi>^2})\circ e^{-\mu<x>^2}\in A_{-\epsilon_0/2-\rho,-\epsilon_0/2-\delta}\subset S^0_{0,0}$ *for ! ! !* $|\rho|,|\delta|\le\epsilon_0/2$ *and consequently* $e^{\epsilon_0/2<x>}e^{\epsilon_0/2<D>}a_\mu w$ *belongs to* L^2 *because of* $w\in L^2$. *Hence* $u_\mu\in H_{\epsilon_0/2,\epsilon_0/2}$ *and satisfies*

$$\|u-u_\mu\|_{H_{\rho,\delta}}=\|e^{-\mu<x>^2}e^{-\mu<D>^2}w-w\|_{L^2}<\epsilon,$$

if we choose $\mu>0$ *suitably. Q.E.D.*

Lemma 2.6. *Let* $a\in A_{\rho,\delta}, 0<\epsilon_0',\tilde\epsilon_0\le\epsilon_0$ *(*ϵ_0 *is given in Proposition 2.3) and* $u\in H_{\epsilon_0',\tilde\epsilon_0}$. *Then there is* $\epsilon_2>0$ *independent of* a,ρ *and* δ *such that* $a(x,D)u(x)$ *belongs to* $H_{\epsilon_0'-\rho,\tilde\epsilon_0-\delta}$ *if* $0<\epsilon_0'-\rho\le min\{\epsilon_0,\epsilon_2\epsilon_a\}$ *and* $0<\tilde\epsilon_0-\delta\le\epsilon_0$, *where* ϵ_a *is given in Definition 2.1.*

Proof. *Put* $w=e^{\tilde\epsilon_0<x>}e^{\epsilon_0'<D>}u\in L^2$, *that is,* $u=e^{-\epsilon_0'<D>}e^{-\tilde\epsilon_0<x>}w$. *Then since* $e^{\epsilon_0'<D>}u\in\hat H_{\tilde\epsilon_0}\subset L^2$, *we have for* $\rho\le\epsilon_0'$

$$a(x,D)u(x)=\int e^{ix\xi}a(x,\xi)\hat u(\xi)\bar d\xi,$$

which is contained in $\hat H_{-\delta}$. *On the other hand since*

$$\hat u(\xi)=e^{-\epsilon_0'<\xi>}\int e^{-iy\xi-\tilde\epsilon_0<y>}w(y)dy=e^{-\epsilon_0'<\xi>}\int\int e^{-iy\xi-\tilde\epsilon_0<y>}e^{iy\eta}\hat w(\eta)dy\bar d\eta,$$

we have

$$a(x,D)u(x)=\int e^{ix\eta}\tilde a(x,\eta)\hat w(\eta)\bar d\eta, \tag{2.21}$$

where

(2.22)
$$\tilde a(x,\eta)=\int\int e^{-iy\xi}a(x,\xi+\eta)e^{-\epsilon_0'<\xi+\eta>-\tilde\epsilon_0<x+y>}dy\bar d\xi=\{(ae^{-\epsilon_0'<\xi>})\circ e^{-\tilde\epsilon_0<x>}\}(x,\xi).$$

Therefore it follows from Proposition 2.3

$$(2.23) \qquad \tilde{a}(x,\xi) \in A_{\rho-\epsilon_0',\delta-\tilde{\epsilon}_0},$$

if $|\rho-\epsilon_0'| \le \epsilon_0$ *and* $\tilde{\epsilon} \le \epsilon_0$. *Moreover if* $\rho-\epsilon_0' < 0$ *and* $\delta-\tilde{\epsilon}_0 < 0$, *Cauchy integral formula yields*

$$|D_x^\alpha \tilde{a}(x,\xi)| \le C_1^{|\alpha|+1}\epsilon_a^{-|\alpha|}|\alpha|! e^{(\rho-\epsilon_0')<\xi>+(\delta-\tilde{\epsilon}_0)<x>)/2},$$

for $x,\xi \in R^n$ *and consequently there is* $\epsilon_2>0$ *and* $C>0$, *where* ϵ_2 *depends only on* n, *such that* $\hat{\tilde{a}}(\eta,\xi)$ *the Fourier image of* $\tilde{a}(x,\xi)$ *with respect to* x *satisfies*

$$(2.24) \qquad |\hat{\tilde{a}}(\eta,\xi)| \le Ce^{-\epsilon_2\epsilon_a<\eta>-(\epsilon_0'-\rho)<\xi>/2} <\eta>^{-(n+1)},$$

for $\xi,\eta \in R^n$ *and consequently since* $e^{\rho'<\xi>}\hat{au}(\xi)$ *belongs to* $L^1(R^n_\xi)$ *when* $|\rho'| \le min\{\epsilon_2\epsilon_{a_2},\epsilon_0'-\rho\}$,*we get from (2.21) and Fubini's theorem*

$$(2.25) \qquad e^{\rho'<D>}(a(x,D)u)(x) = \int e^{ix\xi+\rho'<\xi>}\widehat{a(x,D)u}(\xi)\bar{d}\xi$$

$$= \int e^{ix\xi+\rho'<\xi>}\left(\int\int e^{-iy\xi+iy\eta}\tilde{a}(y,\eta)\hat{w}(\eta)\bar{d}\eta\right)dy\bar{d}\xi$$

$$= \int\int e^{ix\xi+\rho'<\xi>}\hat{\tilde{a}}(\xi-\eta,\eta)\hat{w}(\eta)\bar{d}\xi\bar{d}\eta = \int e^{ix\eta}a_1(x,\eta)\hat{w}(\eta)\bar{d}\eta,$$

where

$$(2.26) \quad a_1(x,\eta) = \int e^{ix(\xi-\eta)+\rho'<\xi>}\hat{\tilde{a}}(\xi-\eta,\eta)\bar{d}\xi = \int e^{ix\xi+\rho'<\xi+\eta>}\hat{\tilde{a}}(\xi,\eta)\bar{d}\xi$$

$$= \lim_{\epsilon\to+0}\int e^{ix\xi+\rho'<\xi+\eta>}\hat{\tilde{a}}(\xi,\eta)e^{-\epsilon<\xi>^2}\bar{d}\xi = \lim_{\epsilon\to+0}\int e^{ix\xi+\rho'<\xi+\eta>}\int e^{-iy\xi-\epsilon<\xi>^2}\tilde{a}(y,\eta)dy\bar{d}\xi$$

$$= os - \int\int e^{-iy\eta+\rho'<\xi+\eta>}\tilde{a}(x+y,\eta)dy\bar{d}\xi = (e^{\rho'<\xi>}\circ\tilde{a})(x,\eta),$$

which belongs to $A_{\rho-\epsilon_0'-\rho',\delta-\tilde{\epsilon}_0}$ *for* $|\rho'| \le \epsilon_0, |\tilde{\epsilon}_0-\delta| \le \epsilon_0$ *from (2.23) and from Proposition 2.3. Thus we obtain* $a_1(x,\eta) = (e^{\rho'<\xi>}\circ\tilde{a})(x,\eta)$ *and moreover if* $\epsilon_0'-\rho \le \epsilon_2\epsilon_a$ *we can choose* $\rho' = \epsilon_0'-\rho$ *and we get from (2.25)*

$$e^{-(\delta-\tilde{\epsilon}_0)<x>}(e^{-(\rho-\epsilon_0')<D>}a(x,D)w)(x) = e^{-(\delta-\tilde{\epsilon}_0)<x>}a_1(x,D)w(x)$$

which is in L^2, *because of* $e^{-(\delta-\tilde{\epsilon}_0)<x>}a_1(x,\xi) \in S^0_{0,0}$ *and* $w \in L^2$. *Therefore* $a(x,D)u(x)$ *belongs to* $H_{\epsilon_0'-\rho,\tilde{\epsilon}_0-\delta}$ *if* $0 < \epsilon_0'-\rho \le min\{\epsilon_0,\epsilon_2\epsilon_a\}$ *and* $0 < \tilde{\epsilon}_0-\delta \le \epsilon_0$. *Q.E.D.*

Remark. *We may choose* $\epsilon_0 = \epsilon_2$ *without loss of generality.*

Lemma 2.7. *Let $a_i \in A_{\rho_i,\delta_i} (i = 1,2)$ and $u \in H_{\epsilon_0',\tilde{\epsilon}_0}(\epsilon_0', \tilde{\epsilon}_0 > 0)$. Then if $|\rho_1| \le \epsilon_0, |\delta_2| \le \epsilon_0, 0 < \epsilon_0' - \rho_2 \le \epsilon_0 min\{1, \epsilon_{a_2}\}, 0 < \tilde{\epsilon}_0 - \delta_2 \le \epsilon_0, 0 < \epsilon_0' - \rho_2 - \rho_1 \le \epsilon_0 min\{1, \epsilon_{a_1}\}$ and $0 < \tilde{\epsilon}_0 - \delta_2 - \delta_1 \le \epsilon_0$ are valid (ϵ_0 is given in Proposition 2.3), we have*

$$a_1(x,D)(a_2(x,D)u)(x) = (a_1 \circ a_2)(x,D)u(x), \tag{2.27}$$

which is in $H_{\epsilon_0'-\rho_1-\rho_2,\tilde{\epsilon}_0-\delta_1-\delta_2}$.

Proof. *Lemma 2.6 implies that $a_2 u \in H_{\epsilon_0'-\rho_2,\tilde{\epsilon}_0-\delta_2}$ if $0 < \epsilon_0' - \rho_2 \le \epsilon_0 min\{1, \epsilon_{a_2}\}, 0 < \tilde{\epsilon}_0 - \delta_2 \le \epsilon_0$ (here we take $\epsilon_2 = \epsilon_0$) and $a_1(x,D)(a_2(x,D)u) \in H_{\epsilon_0'-\rho_2-\rho_1,\tilde{\epsilon}_0-\delta_2-\delta_1}$ if $0 < \epsilon_0' - \rho_2 - \rho_1 \le \epsilon_0 min(1.\epsilon_{a_1}), 0 < \tilde{\epsilon}_0 - \delta_2 - \delta_1 \le \epsilon_0$. We must show (2.27) holds. Since (2.24) with $a = a_2$ holds, $a_1(x,\xi)\hat{a_2 u}(\xi)$ belongs to $L^1(R^n)$. Hence we can write*

$$a_1(x,D)(a_2(x,D)u)(x) = \int e^{ix\xi} a_1(x,\xi)\widehat{a_2 u}(\xi) \bar{d}\xi = \int e^{ix\xi} a_1(x,\xi)$$

$$(\int e^{-iy\xi} a_2(y,D)u)(y)dy)\bar{d}\xi = \int\int e^{ix\xi-iy\xi} a_1(x,\xi)(\int e^{-iy\eta} a_2(y,\eta)\hat{u}(\eta)\bar{d}\eta)dy\bar{d}\xi$$

$$= \lim_{\epsilon\to+0} \int\int\int e^{ix\xi-iy\xi+iy\eta-\epsilon<x-y>^2-\epsilon<\xi>^2} a_1(x,\xi)a_2(y,\eta)\hat{u}(\eta)dy\bar{d}\xi\bar{d}\eta$$

$$= \int e^{ix\eta}\{\lim_{\epsilon\to+0}\int\int e^{-iy\xi-\epsilon<x-y>^2-\epsilon<\xi-\eta>^2} a_1(x,\xi+\eta)a_2(x+y,\eta)dy\bar{d}\xi\}\hat{u}(\eta)\bar{d}\eta$$

$$= (a_1 \circ a_2)(x,D)u(x),$$

because of $a_1 \circ a_2 \in A_{\rho_1+\rho_2,\delta_1+\delta_2}$ if $|\rho_1| \le \epsilon_0, |\delta_2| \le \epsilon_0$ from Proposition 2.3. Q.E.D.

Let $a \in A_{\rho,\delta}(|\rho|, |\delta| \le \epsilon_0/4), u \in H_{\epsilon_0/2,\epsilon_0/2}$ and $|\rho_1|, |\delta_1| < \epsilon_0/4$. Put $w = e^{\delta_1<x>}e^{\rho_1<D>}u$, which is in $H_{\epsilon_0/2-\rho_1,\epsilon_0/2-\delta_1}$. Since we can write $u = e^{-\rho_1<D>}(e^{-\delta_1<x>}w)$, we get by use of Lemma 2.7 with $\epsilon_0' = \epsilon_0/2 - \rho_1, \tilde{\epsilon}_0 = \epsilon_0/2 - \delta_1, a_1 = a(x,\xi)e^{-\rho_1<\xi>}$ and $a_2 = e^{-\delta_1<x>}, \epsilon_{a_2} = 1$.

$$a(x,D)u(x) = a(x,D)(e^{-\rho_1<D>}(e^{-\delta_1<x>}w)$$
$$= ((a(x,\xi)e^{-\rho_1<\xi>})o e^{-\delta_1<x>})(x,D)w(x),$$

Noting that $a_1(x,\xi) := (e^{(\delta_1-\delta)<x>}e^{(\rho_1-\rho)<\xi>}) \circ (a(x,\xi)e^{-\rho_1<\xi>}) \circ e^{-\delta_1<x>} \in A_{0,0}$, we obtain

$$||au||_{H_{\rho_1-\rho,\delta_1-\delta}} = ||a_1(x,D)w||_{L^2} \le C||w||_{L^2} = C||u||_{H_{\rho_1,\delta_1}} \tag{2.28}$$

for any $u \in H_{\epsilon_0/2,\epsilon_0/2}$. Since $H_{\epsilon_0/2,\epsilon_0/2}$ is dense in H_{ρ_1,δ_1} from Lemma 2.5, we get the following theorem.

Theorem 2.8 *Let $a \in A_{\rho,\delta}(|\rho|,|\delta| \le \epsilon_0/4), |\rho_1|,|\delta_1| < \epsilon_0/4$, where ϵ_0 are given in Proposition 2.3. Then $a(x,D)$ maps from H_{ρ_1,δ_1} to $H_{\rho_1-\rho,\delta_1-\delta}$ and satisfies the following inequality*

$$(2.29) \qquad \|au\|_{H_{\rho_1-\rho,\delta_1-\delta}} \le C\|u\|_{H_{\rho_1,\delta_1}}$$

for any $u \in H_{\rho_1,\delta_1}$.

For $a \in A_{\rho,\delta}$, we difine

$$(2.30) \qquad a^t(x,\xi) = os - \int\int e^{-iy\eta} a(x+y,-\xi-\eta)dy\bar{d}\eta,$$

and $a^*(x,\xi) = a^t(\bar{x},-\xi)$. Then we can prove the following lemma, by the same way as that of the proof (i) of Proposition 2.3.

Lemma 2.9. *Let $a \in A_{\rho,\delta}$ and $|\rho|,|\delta| \le \epsilon_0$. Then $a^t(x,\xi)$ defined in (2.29) belongs to $A_{\rho,\delta}$. Moreover it holds*

$$(2.31), \qquad (a^t(x,D)u,\varphi)_{L^2} = (u, a(\bar{x},D)\varphi)_{L^2},$$
$$(a^*(x,D)u,\varphi)_{L^2} = (u, a(x,D)\varphi)_{L^2},$$

for any $u,\varphi \in H_{\epsilon_0}$.

The relation (2.31) and the inequality (2.29) yield

$$|(a^t u,\varphi)| \le \|u\|_{H_{\rho-\rho_1,\delta-\delta_1}} \|\bar{a}\varphi\|_{H_{\rho_1-\rho,\delta_1-\delta}} \le C\|u\|_{H_{\rho-\rho_1,\delta-\delta_1}} \|\varphi\|_{H_{\rho_1,\delta_1}},$$

if $|\rho|,|\delta| \le \epsilon_0/4$ and $|\rho_1|,|\delta_1| < \epsilon_0/4$. Therefore taking account that $H_{\epsilon_0/2,\epsilon_0/2}$ is dense in H_{ρ_1,δ_1}, we get from (2.31)

$$(2.32) \qquad \|a^t u\|_{H_{-\rho_1,-\delta_1}} \le C\|u\|_{H_{\rho-\rho_1,\delta-\delta_1}},$$

for any $u \in H_{\rho_1,\delta_1}$. Thus we get the following proposition.

Propostion 2.10. *Let $a \in A_{\rho,\delta}$ and $|\rho|,|\delta| \le \epsilon_0/4$ and $|\rho_1|,|\delta_1| < \epsilon_0/4$. Then the pseudodifferential operators $a^t(x,D)$ and $a^*(x,D)$ satisfy (2.32).*

Noting that $(e^{\delta<x>}e^{\rho<D>})^t = e^{\rho<D>}e^{\delta<x>}$, we have for $u \in H_{\rho,\delta}$

$$e^{\rho<D>}e^{\delta<x>}u(x) = (e^{\delta<x>}e^{\rho<D>})^t(e^{-\rho<D>}e^{-\delta<x>}e^{\delta<x>}e^{\rho<D>}u)(x)$$
$$= (e^{\delta<x>}e^{\rho<D>})^t \circ (e^{-\delta<x>}e^{-\rho<D>})^t e^{\delta<x>}e^{\rho<D>}u(x).$$

Moreover we can see from Proposition 2.3 and Lemma 2.9 that $(e^{\delta<x>}e^{\rho<\xi>})^t \circ (e^{-\delta<x>}e^{-\rho<\xi>})^t$ *is in* $A_{0,0}$. *Hence we obtain the fact below.*

Proposition 2.11. *Let* $|\rho|, |\delta| \leq \epsilon_0/4$. *Then* u *belongs to* $H_{\rho,\delta}$ *if and only if* $u \in \tilde{H}_{\rho,\delta}$.

The following result on the multiple symbols of pseudodifferential operators is a special case of Lemma 2.2 of Chapter 7 in Kumanogo's book [12].

Lemma 2.12. *Let* $r_j(x,\zeta) \in A_{0,0} (j = 1, 2, ..., v)$ *and put*

$$q_v(x, D) = r_1(x, D)r_2(x, D) \cdots r_v(x, D).$$

Then the symbol $q_v(x,\zeta)$ *belongs to* $A_{0,0}$ *and satisfies*

$$|q_{v(\beta)}^{(\alpha)}(x,\zeta)| \leq C^v \prod_{j=1}^{v} C_{r_j} \bar{\varepsilon}_v^{-|\alpha+\beta|}|\alpha+\beta|!, \tag{2.33}$$

for $(x,\zeta) \in R^{2n}, \alpha, \beta \in N^n$, *where* C *is independent of* v *and* $\bar{\varepsilon}_v = min\{\varepsilon_{r_j}/4\}$.

Proof. *It follows from the product formula (2.3) that we have*

$$q_v(x,\zeta) = os - \int\int_{R^{2n(v-1)}} e^{-iy^{(v-1)}\eta^{(v-1)}} r_1(x,\zeta+\eta^1)r_2(x+\bar{y}^1,\zeta+\eta^2)$$

$$\cdots r_{v-1}(x+\bar{y}^{v-2},\zeta+\eta^{v-1})r_v(x+\bar{y}^{v-1},\zeta)dy^{(v-1)}\bar{d}\eta^{(v-1)},$$

where $y^{(v-1)} = (y^1, y^2, ...y^{v-1}) \in R^{n(v-1)}, y^j \in R^n$, *and* $y^{(v-1)}\eta^{(v-1)} = y^1\eta^1 + \cdots y^{v-1}\eta^{v-1}$ *and* $\bar{y}^j = y^1 + \cdots + y^j$. *Since* $e^{-iy^j\eta^j} =< y^j >^{-2}< D_{\eta^j} >^2 e^{-iy^j\eta^j}$ *holds, we have integrating by part,*

$$q_v(x,\zeta) = os-\int\int e^{-iy^{(v-1)}\eta^{(v-1)}} \prod_{j=1}^{v-1} < y^j >^{-l_0}< D_{\eta^j} >^{l_0} r_1(x,\zeta+\eta^1)r_2(z+\bar{y}^1,\zeta+\eta^2)$$

$$\times \cdots r_{v-1}(z+\bar{y}^{v-2},\zeta+\eta^{v-1})r_v(x+\bar{y}^{v-1},\zeta)dy^{(v-1)}\bar{d}\eta^{(v-1)}$$

Changing the variables y *to* w *as* $w^j = \bar{y}^j (j = 1, \cdots, v-1)$ *and noting* $y^j = w^j - w^{j-1}$ *and*

$$y^{(v-1)}\eta^{(v-1)} = w^1(\eta^1 - \eta^2) + \cdots + w^{v-2}(\eta^{v-2} - \eta^{v-1}) + w^{v-1}\eta^{v-1},$$

we get using integration by part

$$q_v(x,\zeta) = \int\int e^{-i(w^1(\eta^1-\eta^2)+\cdots+w^{v-2}(\eta^{v-2}-\eta^{v-1})+w^{v-1}\eta^{v-1})}$$

$$\times J(w^{(v-1)}, \eta^{(v-1)}; x, \zeta) dw^{(v-1)} \bar{d}\eta^{(v-1)}, \tag{2.34}$$

where

$$J(w^{(v-1)}, \eta^{v-1}); x, \zeta) = \prod_{j=1}^{v-1} < \eta^j - \eta^{j+1} >^{-\ell_0} < D_{w^j} >^{\ell_0}$$

$$\times \prod_{j=1}^{v-1} < w^j - w^{j-1} >^{-\ell_0} < D_{\eta^j} >^{\ell_0} r_1(x, \zeta + \eta^1) r_2(x + w^1, \zeta + \eta^2)$$

$$\cdots r_{v-1}(x + w^{v-2}, \zeta + \eta^{v-1}) r_v(x + w^{v-1}, \zeta)(w^0 = \eta^v = 0)$$

Since r_j are in $A_{0,0}$, we have from (2.1)

$$|r^{(\alpha)}_{j(\beta)}(x, \zeta)| \le C_{r_j} \varepsilon_{r_j}^{-|\alpha+\beta|} |\alpha + \beta|!,$$

for $(x, \zeta) \in R^{2n}, \alpha, \beta \in N^n$. Therefore we obtain

$$|J(w^{(v-1)}, \eta^{(v-1)}; x, \zeta)| \le C^v \prod_{j=1}^{v-1} sup_{R^{2n}, |\alpha+\beta| \le 4\ell_0} |r^{(\alpha)}_{j(\beta)}(\cdot, \cdot)|$$

$$\times \prod_{j=1}^{v-1} < w^j - w^{j-1} >^{-\ell_0} < \eta^j - \eta^{j+1} >^{-\ell_0}, (w^0 = \eta^v = 0).$$

Integrating this we obtain (2.33) with $\alpha = \beta = 0$. For $(\alpha, \beta) \ne 0$, we have

$$q^{(\alpha)}_{v(\beta)}(x, \zeta) = \sum \binom{\alpha}{\alpha^1} \binom{\alpha - \bar{\alpha}^1}{\alpha^2} \cdots \binom{\alpha - \bar{\alpha}^{v-1}}{\alpha^v} \binom{\beta}{\beta^1} \binom{\beta - \bar{\beta}^1}{\beta^2} \cdots \binom{\beta - \bar{\beta}^{v-1}}{\beta^v}$$

$$\times \int \int e^{-i(w^1(\eta^1 - \eta^2) + \cdots + w^{v-2}(\eta^{v-2} - \eta^{v-1}) + w^{v-1}\eta^{v-1})} J^{\alpha^{(v)}}_{\beta^{(v)}}(w^{(v-1)}, \eta^{(v-1)}; x, \zeta) dw^{(v-1)} \bar{d}\eta^{(v-1)},$$

where $\bar{\alpha}^j = \alpha^1 + \cdots + \alpha^j, \alpha^{(v)} = (\alpha^1, \cdots, \alpha^v)$, the summation ranges over $\alpha^j \le \alpha - \bar{\alpha}^{j-1}, \beta^j \le \beta - \bar{\beta}^{j-1}$, and

$$J^{\alpha^{(v)}}_{\beta^{(v)}}(w^{(v-1)}, \eta^{v-1}); x, \zeta) = \prod_{j=1}^{v-1} < \eta^j - \eta^{j+1} >^{-\ell_0} < D_{w^j} >^{\ell_0}$$

$$\times \prod_{j=1}^{v-1} < w^j - w^{j-1} >^{-\ell_0} < D_{\eta^j} >^{\ell_0} r^{(\alpha^1)}_{1(\beta^1)}(x, \zeta + \eta^1) r^{(\alpha^2)}_{2(\beta^2)}(x + w^1, \zeta + \eta^2)$$

$$\cdots r^{(\alpha^{v-1)})}_{v-1(\beta^{v-1})}(x + w^{v-2}, \zeta + \eta^{v-1}) r^{(\alpha - \bar{\alpha}^{v-1})}_{v(\beta - \bar{\beta}^{v-1})}(x + w^{v-1}, \zeta)(w^0 = \eta^v = 0)$$

Since we can estimate

$$|J^{\alpha^{(v)}}_{\beta^{(v)}}(w^{(v-1)}, \eta^{v-1}); x, \zeta)| \le C^v \prod_{j=1}^{v-1} sup_{|\delta+\gamma| \le 4\ell_0} |r^{(\alpha^j + \gamma)}_{j(\delta + \beta^j)}(\cdot, \cdot)|$$

$$\times \prod_{j=1}^{v-1} < \omega^j - \omega^{j-1} >^{-2\ell_0} < \eta^j - \eta^{j-1} >^{-\ell_0} sup_{|\delta+\gamma|\le 4\ell_0} |r^{(\alpha-\bar{\alpha}^{v-1}+\gamma)}_{j(\beta-\bar{\beta}^{v-1}+\delta)}(\cdot,\cdot)|$$

$$\le C^v [\prod_{j=1}^{v-1} C_{r_j} \varepsilon_{r_j}^{-|\alpha^j+\beta^j|-4\ell_0}](|\alpha^j+\beta^j|+4\ell_0)!(|\alpha-\bar{\alpha}^{v-1}+\beta-\bar{\beta}^{v-1}|+4\ell_0)!$$

$$\times \prod_{j=1}^{v-1} < \omega^j - \omega^{j-1} >^{-\ell_0} < \eta^j - \eta^{j-1} >^{-\ell_0}$$

$$\le (2^{4(\ell_0+2}(4\ell_0)!C)^v (2\bar{\varepsilon}_v^{-1})^{|\alpha+\beta|} [\prod_{j=1}^{v-1} C_{r_j} |\alpha^j+\beta^j|!] |\alpha-\bar{\alpha}^{v-1}+\beta-\bar{\beta}^{v-1}|!$$

$$\times \prod_{j=1}^{v-1} < \omega^j - \omega^{j-1} >^{-\ell_0} < \eta^j - \eta^{j-1} >^{-\ell_0},$$

we obtain (2.33) integrating this and using the inequality below,

$$\sum \binom{\alpha}{\alpha^1} \cdots \binom{\alpha-\bar{\alpha}^{v-1}}{\alpha^v} \binom{\beta}{\beta^1} \cdots \binom{\beta-\beta^{v-1}}{\beta^v} \prod_{j=1}^{v-1} |\alpha^j+\beta^j|!|\alpha-\bar{\alpha}^{v-1}|!|\beta-\bar{\beta}^{v-1}|!$$

$$\le 2^{|\alpha+\beta|+2v} |\alpha+\beta|!.$$

Q.E.D.

We can prove easily the following lemma as a corollary of Lemma 2.12, by using the Neumann series method.

Lemma 2.13. *Let $r(x,\xi)$ be in $A_{0,0}$. If $C_r > 0$ is sufficiently small, then there is the inverse $(I + r(x,D))^{-1}$ which is a pseudodifferential operator with its symbol contained in $A_{0,0}$.*

Finally we mention the following lemma which will be used in the next section.

Lemma 2.14.(Boutet de Monvel and Krée [2]) *Assume that $p_k(x,\xi)$ satisfies*

$$|p^{(\alpha)}_{k(\beta)}(x,\xi)| \le C_0 r_0^{-|\alpha+\beta|-k} (|\alpha+\beta|+k)! < x >^{-|\beta|} < \xi >^{m-|\alpha|} (< x >< \xi >)^{\ell-k},$$

for $x,\xi \in R^n, \alpha,\beta \in N^n, k = 0,1,2,\cdots$. Then there exist $p(x,\xi), C > 0$ and $\varepsilon_1 > 0$ such that

$$|\partial_\xi^\alpha D^\beta (p(x,\xi) - \sum_{k=0}^{N-1} p_k(x,\xi))| \le C_1 \varepsilon_1^{-|\alpha+\beta|-N} (|\alpha+\beta|+N)! < x >^{m-|\beta|-N} < \xi >^{\ell-|\alpha|-N},$$

for $x, \xi \in R^n, \alpha, \beta \in N^n, N = 0, 1, 2, \cdots$.

Refer Matsumoto's note [13] for the detail of proof of this Lemma.

1.4 FOURIER INTEGRAL OPERATORS

We introduce some symbol classes. For a Riemann metric $g = <x>^{-2} dx^2 + <\xi>^{-2} d\xi^2$ *and a weight* $m(x,\xi)$ *we denote by* $S(m,g)$ *the set of functions* $a(x,\xi) \in C^\infty(R^{2n})$ *satisfying*

$$(3.1) \qquad |a^{(\alpha)}_{(\beta)}(x,\xi)| \leq C_{\alpha\beta} m(x,\xi) <x>^{-|\beta|}<\xi>^{-|\alpha|},$$

for $x, \xi \in R^n, \alpha, \beta \in N^n$ *and denote by* $AS(m,g)$ *the set of functions* $a(x,\xi) \in S(m,g)$ *satisfying*

$$(3.2), \qquad |a^{(\alpha)}_{(\beta)}(x,\xi)| \leq C_a \varepsilon_a^{-|\alpha+\beta|} m(x,\xi)|\alpha+\beta|! <x>^{-|\beta|}<\xi>^{-|\alpha|},$$

for $x, \xi \in R^n, \alpha, \beta \in N^n$, *where* $C_a > 0, \varepsilon_a > 0$ *are independent of* x, ξ, α, β.

Proposition 3.1 *Let* $a_i \in AS(<x>^{m_i}<\xi>^{\ell_i}, g), i = 1, 2$. *Then* $a_1 \circ a_2$ *belongs to* $S(<x>^{m_1+m_2}<\xi>^{\ell_1+\ell_2}, g)$ *and moreover we can decompose*

$$(3.3) \qquad a_1 \circ a_2(x,\xi) = p(x,\xi) + r(x,\xi),$$

where $p(x,\xi) \in AS(<x>^{m_1+m_2}<\xi>^{\ell_1+\ell_2}, g)$ *satisfies that there are* $C > 0$ *and* $\varepsilon_0 >$ *such that*

(3.4)

$$p(x,\xi) - \sum_{|\gamma|<N} \gamma!^{-1} a_1^{(\gamma)}(x,\xi) a_{2(\gamma)}(x,\xi)) \in AS(C^{1+N} N! <x>^{m_1+m_2-N}<\xi>^{\ell_1+\ell_2-N}, g),$$

for any non negative integer N, *and* $r(x,\xi)$ *belongs to* $A_{-\varepsilon_0,-\varepsilon_0}$.

Proof. *Put for* $j = 0, 1, \cdots$,

$$p_j(x,\xi) = \sum_{|\gamma|=j} \gamma!^{-1} a_1^{(\gamma)}(x,\xi) a_{2(\gamma)}(x,\xi).$$

Then it follows from Lemma 2.14 that there is $p(x,\xi)$ *which satisfies (3.4). We prove that there is* $\varepsilon_0 > 0$ *such that* $r(x,\xi) = (a_1 \circ a_2)(x,\xi) - p(x,\xi)$ *is in* $A_{-\varepsilon_0,-\varepsilon_0}$. *Let* $|x| \leq |\xi|, N$ *a positive integer and* $\alpha, \beta \in N^n$. *From (2.3) we have*

$$a_1 \circ a_2(x,\xi) = os - \int\int e^{-iy\eta} a_1(x, \xi+\eta) a_2(x+y, \xi) dy d\bar{\eta}$$

$$= os - \int\int e^{-iy\eta} a_1(x,\xi+\eta) a_2(x+y,\xi) \chi_{N+|\alpha+\beta|}(|\eta|/<\xi>) dy d\bar{\eta}$$

$$+ os - \int\int e^{-iy\eta} a_1(x,\xi+\eta) a_2(x+y,\xi)(1 - \chi_{N+|\alpha+\beta|}(|\eta|/<\xi>)) dy d\bar{\eta}$$

$$\equiv b_1(x,\xi)+b_2(x,\xi),$$

where $\chi_N(t)=1$, *if* $|t|\le 1/4, \chi_N(t)=0$, *if* $|t|\ge 1/2$ *and*

$$|D_t^{k+j}\chi_N(t)|\le C_jC^{1+N}k!,$$

if $k\le N$ *and* $j=0,1.\cdots$. *Then we have by Taylor expansion,*

$$b_1(x,\xi)=\sum_{|\gamma|<N}\gamma!^{-1}a_1^{(\gamma)}(x,\xi)a_{2(\gamma)}(x,\xi)+r_N(x,\xi),$$

where

$$r_N(x,\xi)=os-\sum_{|\gamma|=N}N\gamma!^{-1}\int\int\int_0^1(1-t)^{N-1}e^{-iy\eta}\partial_\eta^\gamma\{a_1(x,\xi+\eta)a_{2(\gamma)}(x+ty,\xi)$$

$$\times\chi_{N+|\alpha+\beta|}(|\eta|/<\xi>)\}dy\bar{d}\eta dt.$$

Integrating by part we get

$$r_{N(\beta)}^{(\alpha)}(x,\xi)=\sum_{|\gamma|=N}N\gamma!^{-1}\sum_{\alpha'}\sum_{\beta'}\binom{\alpha}{\alpha'}\binom{\beta}{\beta'}\int\int\int_0^1(1-t)^{N-1}e^{-iy\eta}<\eta>^{-\ell_0}<D_y>^{\ell_0}$$

$$\times<y>^{-\ell_0}<D_\eta>^{\ell_0}\partial_\eta^\gamma[a_{1(\beta')}^{(\alpha')}(x,\xi+\eta)\partial_\xi^{\alpha-\alpha'}\{a_{2(\gamma+\beta-\beta')}(x+ty,\xi)\chi_{N+|\alpha+\beta|}(|\eta|/<\xi>)\}]dy\bar{d}\eta dt$$

$$=\sum_{|\gamma|=N}N\gamma!^{-1}\sum_{\alpha'}\sum_{\beta'}\binom{\alpha}{\alpha'}\binom{\beta}{\beta'}\int\int\int_0^1(1-t)^{N-1}F_{\gamma,\alpha,\beta,\alpha',\beta'}(x,\xi;y,\eta)dy\bar{d}\eta dt,$$

where we put

$$F_{\gamma,\alpha,\beta,\alpha',\beta'}(x,\xi;y,\eta)=<\eta>^{-\ell_0}<D_y>^{\ell_0}<y>^{-\ell_0}<D_\eta>^{\ell_0}$$

$$\times\partial_\eta^\gamma[a_{1(\beta')}^{(\alpha')}(x,\xi+\eta)\partial_\xi^{\alpha-\alpha'}\{a_{2(\gamma+\beta-\beta')}(x+ty,\xi)\chi_{N+|\alpha+\beta|}(|\eta|/<\xi>)\}]$$

Then since $|\eta|\le<\xi>/2$, *on* $supp\chi_{N+|\alpha+\beta|}$, *we can estimate*

$$|F_{\gamma,\alpha,\beta,\alpha',\beta'}(x,\xi;\eta)|\le C_{a_1}C_{a_2}C_0r_0^{-2N-|\alpha+\beta|}(N+|\alpha'+\beta'|)!(N+|\alpha-\alpha'+\beta-\beta'|)!$$

$$\times<x>^{m_1+m_2}<\xi>^{\ell_1+\ell_2-N}<\eta>^{-\ell_0},$$

where $\ell_0=max\{[(n+|m_2|)/2+1,[(n+|\ell_1|)/2+1]\}$ *and* $r_0>0, C_0>0$ *depend only on* n,ε_{a_1}. *Therefore integrating this term we can estimate*

$$(3.5)\quad |r_{N(\beta)}^{(\alpha)}|\le C_{a_1}C_{a_2}C_1r_1^{-|\alpha+\beta|-N}N!|\alpha+\beta|!<x>^{m_1+m_2}<\xi>^{\ell_1+\ell_2-N}.$$

On the other hand we have

$$b_{2(\beta)}^{(\alpha)}(x,\xi)=\sum_{\alpha'}\sum_{\beta'}\binom{\alpha}{\alpha'}\binom{\beta}{\beta'}\int\int e^{-iy\eta}<\eta>^{-\ell_0}<D_y>^{\ell_0}<y>^{-\ell_0}<D_\eta>^{\ell_0}$$

$$\times a^{(\alpha')}_{1(\beta')}(x,\xi+\eta)\partial_\xi^{\alpha-\alpha'}\{a_{2(\beta-\beta')}(x+y,\xi)(1-\chi_{N+|\alpha+\beta|}(|\eta|/<\xi>))\}dy\bar{d}\eta.$$

We put

$$F_{\alpha,\beta,\alpha',\beta'}(x,\xi;y,\eta)=<\eta>^{-\ell_0}<D_y>^{\ell_0}<y>^{-\ell_0}<D_\eta>^{\ell_0}$$

$$\times a^{(\alpha')}_{1(\beta')}(x,\xi+\eta)\partial_\xi^{\alpha-\alpha'}\{a^{(\alpha-\alpha')}_{2(\gamma+\beta-\beta')}(x+ty,\xi)(1-\chi_{N+|\alpha+\beta|}(|\eta|/<\xi>))\},$$

$$F_{\alpha,\beta,\alpha',\beta'}(x,\xi;\eta)=\int e^{-iy\eta}F_{\alpha,\beta,\alpha',\beta'}(x,\xi;y,\eta)dy.$$

Then we have

$$|D_y^\lambda F_{\alpha,\beta,\alpha',\beta'}(x,\xi;y,\eta)|\le C_{a_1}C_{a_2}C_2r_2^{-|\alpha+\beta+\lambda|}<x>^{m_1+m_2}<\xi>^{\ell_1+\ell_2}$$

$$\times<y>^{-\ell_0+|m_2|}<\eta>^{-\ell_0+|\ell_1|}(|\alpha'+\beta'|)!(|\alpha-\alpha'+\beta-\beta'+\lambda|)!,$$

for any $\lambda\in N^n$. Therefore there is $\varepsilon_0>0$ such that we obtain

$$|F_{\alpha,\beta,\alpha',\beta'}(x,\xi;\eta)|\le C_{a_1}C_{a_2}C_3r_3^{-|\alpha+\beta|}<x>^{m_1+m_2}<\xi>^{\ell_1+\ell_2}$$

$$\times<y>^{-\ell_0+|m_2|}<\eta>^{-\ell_0}(|\alpha'+\beta'|)!(|\alpha-\alpha'+\beta-\beta')!e^{-\varepsilon_0(<\xi>+<x>)},$$

because of $|\eta|\ge<\xi>/2$ and $|x|\le|\xi|$ and consequently we get
(3.6)

$$|b^{(\alpha)}_{2(\beta)}(x,\xi)|\le C_{a_1}C_{a_2}C_4r_4^{-|\alpha+\beta|}<x>^{m_1+m_2}<\xi>^{\ell_1+\ell_2}|\alpha+\beta|)!e^{-\varepsilon_0(<\xi>+<x>)}.$$

Thus we get

$$|r^{(\alpha)}_{(\beta)}(x,\xi)|=|D_x^\beta\partial_\xi^\alpha(a_1\circ a_2(x,\xi)-p(x,\xi))|\le|D_x^\beta\partial_\xi^\alpha(r_N(x,\xi)-b_2(x,\xi))|$$

$$\le C_{a_1}C_{a_2}C_1r_1^{-|\alpha+\beta|-N}N!|\alpha+\beta|!<x>^{m_1+m_2}<\xi>^{\ell_1+\ell_2-N}$$

$$+C_{a_1}C_{a_2}C_4r_4^{-|\alpha+\beta|}<x>^{m_1+m_2}<\xi>^{\ell_1+\ell_2}|\alpha+\beta|)!e^{-\varepsilon_0(<\xi>+<x>)},$$

for any positive integer N. Taking the minimum with respect to N, we can see that $r(x,\xi)\in A_{-\varepsilon_0,-\varepsilon_0}$ for some $\varepsilon_0>0$, if $|x|\le|\xi|$. We can prove this similarly in the case of $|x|\ge|\xi|$. Q.E.D.

By use of Proposition 3.1, Lemma 2.13 and Lemma 2.14 we can prove the following lemma.

Lemma 3.2. *Let $j(x,\xi)\in AS(\varepsilon_1,g)$. Then if $\varepsilon_1>0$ is small enough, there are $k_1(x,\xi)\in AS(\varepsilon_1<x>^{-1}<\xi>^{-1},g),\varepsilon_0>0$ independent of ε_1 and $r_\infty(x,\xi)\in A_{-\varepsilon_0,-\varepsilon_0}$ such that $(I+j(x,D))^{-1}=k(x,D)+k_1(x,D)+r_\infty(x,D)$, where $k(x,\xi)=(1+j(x,\xi))^{-1}$.*

For $\vartheta\in AS(\rho_\vartheta<\xi>+\delta_\vartheta<x>,g)(\rho_\vartheta,\delta_\vartheta\ge0)$, we denote

$$\phi(x,\xi) = x\xi - i\vartheta(x,\xi).$$

For $a \in A_{0,0}$ we define a Fourier integral operator with a phase function $\phi(x,\xi)$ as follows,

$$a_\phi(x,D)u(x) = \int_{R^n} e^{i\phi(x,\xi)} a(x,\xi)\hat{u}(\xi)\bar{d}\xi, \tag{3.7}$$

for $u \in H_{\epsilon_0,\epsilon_0}$. Putting $p(x,\xi) = a(x,\xi)e^{\vartheta(x,\xi)}$, we can see $p(x,\xi) \in A_{\rho_\vartheta,\delta_\vartheta}$. Therefore we can regard $a_\phi(x,D)$ as a pseudo differential operator with its symbol $p = ae^\vartheta$ defined in §2 and consequently it follows from Theorem 2.8 that $a_\phi(x,D)$ acts continuously from $H_{\rho,\delta}$ to $H_{\rho-\rho_\vartheta,\delta-\delta_\vartheta}$. However in order to construct the inverse operator of $p(x,D)$ it is better to regard $p(x,D)$ as a Fourier integral operator. In paticular for $a = 1$ we denote

$$I_\phi(x,D)u(x) = \int e^{i\phi(x,\xi)}\hat{u}(\xi)\bar{d}\xi, \tag{3.8}$$

$$I_\phi^R(x,D)v(x) = \int e^{ix\xi}\bar{d}\xi \int e^{\phi(y,\xi)}v(y)dy. \tag{3.9}$$

Lemma 3.3. *Let $a(x,\xi) \in AS(<x>^m<\xi>^\ell, g)$ and $\vartheta \in AS(\rho_\vartheta <\xi> + \delta_\vartheta <x>, g)(\rho_\vartheta, \delta_\vartheta \geq 0)$. Put $\phi = x\xi - i\vartheta(x,\xi)$ and $\tilde{a}(x,D) = a_\phi(x,D)I_{-\phi}^R(x,D)$. If ρ_ϑ and δ_ϑ are sufficiently small, then $\tilde{a}(x,\xi)$ belongs to $S(<x>^m<\xi>^\ell, g)$ and moreover satisfies*

$$\tilde{a}(x,\xi) = \tilde{p}(x,\xi) + r(x,\xi), \tag{3.10}$$

for $x,\xi \in R^n$, and

$$\tilde{p}(x,\xi) - \sum_{|\gamma|<N} \gamma!^{-1} D_y^\gamma \partial_\eta^\gamma \{a(x,\Phi(x,y,\eta))J(x,y,\eta)\}_{y=0,\eta=\xi}$$

$$\in AS(C^{1+N}N! <x>^{m-N}<\xi>^{\ell-N}, g) \tag{3.11}$$

for any N, where $\Phi(x,y,\xi)$ is a solution of the following equation,

$$\Phi(x,y,\xi) - i\tilde{\nabla}_x\vartheta(x,y,\Phi(x,\xi)) = \xi, \tag{3.12}$$

$$\tilde{\nabla}_x\vartheta(x,y,\xi) = \int_0^1 \nabla\vartheta(x+ty,\xi)dt, \tag{3.13}$$

$J(x,y,\xi)=\frac{D\Phi(x,y,\xi)}{D\xi}$ *is the Jacobian of* $\Phi, r(x,\xi)\in A_{-\varepsilon_0,-\varepsilon_0}$, *and* $C>0, \varepsilon_0>0$ *are independent of* $\rho_\vartheta, \delta_\vartheta$.

Proof. *It follows from (3.7) and (3.9) that we have*

(3.14)
$$\begin{aligned}\tilde a(x,\xi) &= os-\int\int e^{-ix\xi+iy\xi+i(\phi(x,\eta)-\phi(y,\eta))}a(x,\eta)dyd\bar\eta\\ &= os-\int\int e^{-iy(\xi-\eta+i\tilde\nabla_x\vartheta(x,y,\eta))}a(x,\eta)dyd\bar\eta\\ &= os-\int\int_{R^n\times\gamma} e^{-iy(\xi-\omega)}a(x,\Phi(x,y,\omega))J(x,y,\omega)dyd\bar\omega,\end{aligned}$$

where $\gamma=\{\eta-i\tilde\nabla_x\vartheta(x,y,\eta);\eta\in R^n\}$. *Since* $\nabla_x\vartheta(x,\xi)$ *is in* $AS(\delta_\vartheta+\rho_\vartheta<\xi><x>^{-1},g)$ *and so it holds as follows,*

(3.15)
$$|\partial_\eta^\alpha D_x^\beta D_y^\lambda\tilde\nabla_x\vartheta(x,y,\eta)|\le(\rho_\vartheta+\delta_\vartheta)r_\vartheta^{-|\alpha+\beta+\lambda|-1}<\eta>^{1-|\alpha|}|\alpha+\beta+\lambda|!\int_0^1<x+ty>^{-|\beta+\lambda|}dt,$$

the implicit function theorem assures the following estimates

(3.16)
$$|\partial_\omega^\alpha D_x^\beta D_y^\lambda(\Phi(x,y,\omega)-\omega)|\le C_\Phi\varepsilon_\Phi^{-|\alpha+\beta+\lambda|}|\alpha+\beta+\lambda|!<\omega>^{1-|\alpha|}\int_0^1<x+ty>^{-\beta+\lambda|}dt,$$

(3.17)
$$|\partial_\omega^\alpha D_x^\beta D_y^\lambda(J(x,y,\omega)-1)|\le C_J\varepsilon_J^{-|\alpha+\beta+\lambda|}|\alpha+\beta+\lambda|!<\omega>^{-|\alpha|}\int_0^1<x+ty>^{-\beta+\lambda|}dt,$$

for $x,y,\omega\in R^n, \alpha,\beta,\lambda\in N^n$, *if* $\rho_\vartheta+\delta_\vartheta$ *is small eneugh. By use of Stokes formula we can see from (3.14)*

$$\tilde a(x,\xi)==os-\int\int_{R^{2n}} e^{-iy(\xi-\eta)}a(x,\Phi(x,y,\eta))J(x,y,\eta)dyd\bar\eta.$$

Put for $k=0,1,\cdots,$

$$\tilde p_k(x,\xi)=\sum_{|\gamma|=k}\gamma!^{-1}D_y^\gamma\partial_\eta^\gamma\{a(x,\Phi(x,y,\eta))J(x,y,\eta)\}_{y=0,\eta=\xi}.$$

Then by virtue of (3.16) and (3.17) we can see that $\tilde p_k(x,\xi)$ *satisfies the condition of Lemma 2.14. Hence there is* $\tilde p(x,\xi)$ *which satisfies (3.11). We can prove by a quite simillar way as the proof of Proposition 3.1 that there is a* $\varepsilon_0>0$ *independent of* $\rho_\vartheta,\delta_\vartheta$ *such that* $\tilde a-\tilde p\in A_{-\varepsilon_0,-\varepsilon_0}$, *taking account of (3.16) and (3.17). Q.E.D.*

Lemma 3.4. *Let* $a(x,\xi)$ *and* ϑ *be satisfied with the same condition as one of Lemma 3,3. For* $\phi=x\xi-i\vartheta(x,\xi)$ *put* $a'(x,\xi)=I_{-\phi}^R(x,D)a_\phi(x,D)$. *Then if* ρ_ϑ *and* δ_ϑ *are sufficiently small,* $a'(x,\xi)$ *belongs to* $S(<x>^m<\xi>^\ell,g)$ *and moreover satisfies*

$$(3.18) \qquad a'(x,\xi) = p'(x,\xi) + r'(x,\xi),$$

$$(3.19) \qquad p'(x,\xi) - \sum_{|\gamma|<N} \gamma^{-1} D_y^\gamma \partial_\eta^\gamma \{a(\Phi'(y,\xi,\eta),\xi) J'(y,\xi,\eta)\}_{y=x,\eta=0}$$

$$\in AS(C^{1+N} N! < x >^{m-N} < \xi >^{\ell-N}, g),$$

for any non negative integer N, where $\Phi'(y,\xi,\eta)$ is a solution of the equation

$$(3.20) \qquad \Phi'(y,\xi,\eta) - i\tilde{\nabla}_\xi \vartheta(\Phi'(y,\xi,\eta),\xi,\eta) = y,$$

$$(3.21) \qquad \tilde{\nabla}_\xi \vartheta(y,\xi,\eta) = \int_0^1 \nabla_\xi \vartheta(y,\xi + t\eta) dt,$$

and $J'(y,\xi,\eta) = \frac{D\Phi'(y,\xi,\eta)}{Dy}$, and $r'(x,\xi) \in A_{-\varepsilon_0,-\varepsilon_0}$ ($\varepsilon_0 > 0$ is independent of $\rho_\vartheta, \delta_\vartheta$.

Proof. *Since it holds*

$$a'(x,\xi) = os - \int\int e^{-ix(\xi-\eta)+i(\phi(y,\xi)-\phi(y,\eta))} a(y,\xi) dy \bar{d}\eta,$$

$$= os - \int\int e^{-i(x-y)(\xi-\eta)+(\vartheta(y,\xi)-\vartheta(y,\eta)} a(y,\xi) dy \bar{d}\eta,$$

$$= os - \int\int e^{-i(x-y+i\tilde{\nabla}_\xi \vartheta(y,\xi,\eta))(\eta} a(y,\xi) dy \bar{d}\eta,$$

$$= os - \int\int e^{-i(x-y)\eta} a(\Phi'(y,\xi,\eta),\xi) J(y,\xi,\eta) dy \bar{d}\eta,$$

we can show $a' - p' \in A_{-\varepsilon_0,-\varepsilon_0}$ by the same way as the proof of Proposition 3.1 and Lemma 3.4. Q.E.D.

Lemma 3.5. *Let $\vartheta(x,\xi) \in AS(\rho_\vartheta < \xi > +\delta_\vartheta < x >, g)$. If ρ_ϑ and δ_ϑ are sufficently small, there is the inverse of $I_\phi(x,D)$, which maps continuously from H_{ρ_1,δ_1} to $H_{\rho_1-\rho_\vartheta,\delta_1-\delta_\vartheta}$ for $|\rho_1|, |\delta_1|$ small enough and satisfies*

$$(3.22) \qquad I_\phi(x,D)^{-1} = I^R_{-\phi}(x,D)(I + j(x,D))^{-1} = (I + j'(x,D))^{-1} I^R_{-\phi}(x,D)$$

$$= I^R_{-\phi}(x,D)(k(x,D) + k_1(x,D) + r(x,D)) = (k'(x,\xi) + k'_1(x,D) + r'(x,D))) I^R_{-\phi}(x,D),$$

where $j(x,\xi) = J(x,0,\xi) - 1 + r_1(x,\xi)$, $j'(x,\xi) = J'(x,\xi,0) - 1 + r_2(x,\xi)$, $k(x,\xi) = J(x,0,\xi)^{-1}$, $k'(x,\xi) = J'(x,\xi,0)^{-1}$ and $k_1, k'_1 \in AS(< x >^{-1} < \xi >^{-1}, g)$ and $r, r' \in A_{-\varepsilon_0,-\varepsilon_0}$.

Proof. *It follows from (3.10) and (3.18) with $a = 1$ that $I_\phi(z,D) I^R_{-\phi} = J(x,x,D) + r(x,D)$ and $I^R_{-\phi} I_\phi(x,D) = J'(x,D,0) + r'(x,D)$. Hence we obtain* *the first two equality in (3.22) from Lemma 2.13, if we choose $\rho_\vartheta, \delta_\vartheta$ sufficiently small. Moreover we get the last two equlity of (3.22) from Lemaa 3.2. Q.E.D.*

Lemma 3.6. *Let $a(x,\xi)$ and ϑ be satisfied with the same condition as one of Lemma 3,3. Let $\phi = x\xi - i\vartheta$. Then we have*

$$\sigma(I_\phi(x,D)a(x,D))(x,\xi) = I_\phi \circ a(x,\xi) = e^{\vartheta(x,\xi)}(q(x,\xi) + r(x,\xi)), \tag{3.23}$$

$$\sigma(a(x,D)I_\phi(x,D)(x,\xi) = a \circ I_\phi(x,\xi) = e^{\vartheta(x,\xi)}(q'(x,\xi) + r'(x,\xi)), \tag{3.24}$$

where r, r' is in $A_{-\varepsilon_0,-\varepsilon_0}$, if $\rho_\vartheta, \delta_\vartheta$ is sufficiently small, and q, q' satisfies

$$q(x,\xi) - \sum_{|\gamma|<N} \gamma!^{-1} D_y^\delta \partial_\eta^\gamma \{a(x+y-i\tilde\nabla_\xi \vartheta(x,\xi,\eta),\xi)\}_{y=\eta=0} \tag{3.25}$$
$$\in AS(C^{1+N}N! <x>^{m-N}<\xi>^{\ell-N}, g),$$

$$q'(x,\xi) - \sum_{|\gamma|<N} \gamma^{-1} D_y^\gamma \partial_\eta^\gamma \{a(x,\xi+\eta-i\tilde\nabla_x \vartheta(x,y,\xi))\}_{y=\eta=0} \tag{3.26}$$
$$\in AS(C^{1+N}N! <x>^{m-N}<\xi>^{\ell-N}, g),$$

for any positive integer N, and $C>0$ and $\varepsilon_0>0$ are independent of $\rho_\vartheta, \delta_\vartheta$.
Proof. *We have*

$$b(x,\xi) \equiv e^{-\vartheta(x,\xi)} I_\varphi \circ a(x,\xi) = os - \int\int e^{-iy\eta+\vartheta(x,\xi+\eta)-\vartheta(x,\xi)} a(x+y,\xi) dy\bar d\eta$$
$$= os - \int\int e^{-i(y+i\tilde\nabla_\xi\vartheta(x,\xi,\eta))\eta} a(x+y,\xi) dy\bar d\eta$$
$$= os - \int\int e^{-iz\eta} a(x+z-i\tilde\nabla_\xi\vartheta(x,\xi,\eta)),\xi) dz\bar d\eta.$$

Since $a(x+z-i\tilde\nabla_\xi\vartheta(x,\xi,\eta)),\xi)$ is holomorphic in $z \in \Gamma = \{z \in C^n; |Imz - Re\tilde\nabla_\xi\vartheta(x,\xi,\eta)| \le \tau|Rez + x + Im\tilde\nabla_\xi\vartheta(x,\xi,\eta)| + \varepsilon\}$ for some $\tau>0, \varepsilon>0$, for a positive integer $N, \alpha, \beta \in N^n$ we get by Stokes formula,

$$b(x,\xi) := os - \int\int_C e^{-iz\eta} a(x+z-i\tilde\nabla_\xi\vartheta(x,\xi,\eta)),\xi) dz\bar d\eta$$

$$= os - \int\int_C e^{-iz\eta} a(x+z-i\tilde\nabla_\xi\vartheta(x,\xi,\eta)),\xi)\chi_{N+|\alpha+\beta|}(|z|/<x>) dz\bar d\eta$$

$$+ os - \int\int_C e^{-iz\eta} a(x+z-i\tilde\nabla_\xi\vartheta(x,\xi,\eta)),\xi)(1-\chi_{N+|\alpha+\beta|}(|z|/<x>)) dz\bar d\eta$$

$$=: b_1(x,\xi) + b_2(x,\xi),$$

where $C = \{z \in \Gamma; Imz = 0, |Re\tilde\nabla_\xi\vartheta(x,\xi,\eta)| \le \tau|Rez + x + Im\tilde\nabla_\xi\vartheta(x,\xi,\eta)| + \varepsilon\} \cap \{z \in \partial\Gamma'; |Re\tilde\nabla_\xi\vartheta(x,\xi,\eta)| > \tau|Rez + x + Im\tilde\nabla_\xi\vartheta(x,\xi,\eta)| + \varepsilon = |Imz -$

$Re\tilde{\nabla}_\xi\vartheta(x,\xi,\eta)|\}$ and $\partial\Gamma'$ is a connected component of $\partial\Gamma$ which intersects R^n. Put for $k = 0, 1, \cdots$,

$$q_k(x,\xi) = \sum_{|\gamma|=k} \gamma!^{-1} D_y^\gamma \partial_\eta^\gamma \{a(x + y - i\tilde{\nabla}_\xi\vartheta(x,\xi,\eta),\xi)\}_{y=\eta=0}.$$

Then it follows from Lemma 2.14 that there is $q(x,\xi)$ satisfying (3.25). Moreover we have using again Stokes formula,

$$os-\int\int_C e^{-iz\eta} \sum_{|\gamma|<N} D_z^\gamma\{a(x+z-i\tilde{\nabla}_\xi\vartheta(x,\xi,\eta)),\xi)\chi_{N+|\alpha+\beta|}(|z|/<x>)\}_{z=0}(iz)^\gamma\gamma!^{-1}dz d\bar{\eta}$$

$$= \sum_{k=0}^{N-1} q_k(x,\xi).$$

Therefore we get

$$r(x,\xi) = b(x,\xi) - q(x,\xi) = b_1(x,\xi) + b_2(x,\xi) - q(x,\xi)$$

$$= os-\int\int_C e^{-iz\eta}N \sum_{|\gamma|=N} \gamma!^{-1}D_z^\gamma\partial_\eta^\gamma\{a(x+z-i\tilde{\nabla}_\xi\vartheta(x,\xi,\eta)),\xi)(1-\chi_{N+|\alpha+\beta|}(|z|/<x>))\}dz d\bar{\eta}$$

$$+\sum_{k=0}^{N-1} q_k(x,\xi) - q(x,\xi) + b_2(x,\xi)$$

$$\equiv r_1(x,\xi) + r_2(x,\xi) + b_2(x,\xi).$$

(3.25) yields that r_2 belongs to $AS(C^{1+N}N! < x >^{m-N}< \xi >^{\ell-N}, g)$. We can prove by the same way as one of Lemma 3.1 that r_1 and b_2 satisfy the simillar estimate to (3.5) and to (3.6) respectively. Since N is arbitrary, there is $\varepsilon_0 > 0$ such that $r \in A_{-\varepsilon_0,-\varepsilon_0}$. Q.E.D.

Summing up Lemma 3.2-Lemma 3.6, we obtain the following theorem.

Theorem 3.7. *Let $a \in AS(< x >^m< \xi >^\ell, g), \vartheta \in AS(\rho_\vartheta < \xi > +\delta_\vartheta < x >, g)$ and $\phi = x\xi - i\vartheta(x,\xi)$. Then if $\rho_\vartheta, \delta_\vartheta$ are sufficiently small, $\tilde{a}(x,D) = I_\phi(x,D)a(x,D)I_\phi^{-1}$ and $\tilde{a}'(x,D) = I_\phi(x,D)^{-1}a(x,D)I_\phi(x,D)$ are pseudodifferential operators of which symbols are given by*

$$\tilde{a}(x,\xi) = p(x,\xi) + r(x,\xi), \tag{3.27}$$

$$a'(x,\xi) = p'(x,\xi) + r'(x,\xi), \tag{3.28}$$

where

$$p(x,\xi) - a(x - i\nabla_\xi\vartheta(x,\Phi), \xi + i\nabla_x\vartheta(x,\Phi)) \in AS(< x >^{m-1}< \xi >^{\ell-1}, g), \tag{3.29}$$

(3.30)
$$\tilde{p}'(x,\xi) - a(x + i\nabla_\xi \vartheta(\Phi',\xi), \xi - i\nabla_x \vartheta(\Phi',\xi)) \in AS(<x>^{m-1}<\xi>^{\ell-1}, g),$$

where $\Phi = \Phi(x,0,\xi)$ and $\Phi' = \Phi'(x,\xi,0)$ are given by (3.12) and (3.20) respectively and r, r' belong to $A_{-\varepsilon_0,-\varepsilon_0}$ for an $\varepsilon_0 > 0$ independent of $\rho_\vartheta, \delta_\vartheta$.

Let $p_0(x,\xi)$ be satisfied with

(3.31)
$$|p_{0(\beta)}^{(\alpha)}(x,\xi)| \le C_{p_0}\varepsilon_{p_0}^{-|\alpha+\beta|}(<x> + <\xi>) <x>^{-|\beta|}<\xi>^{-|\alpha|}\ |\alpha+\beta|!,$$

for $x, \xi \in R^n, \alpha, \beta \in N^n$. For $T > 0$ and $\rho \in R$ we consider the following Cauchy problem,

$$\frac{d}{dt}E(t) = \rho p_0(x,D)E(t), t \in (-T,T), E(0) = I. \tag{3.32}$$

Then we can construct the fundamental solution $E(t)$ as a pseudo-differential operator following the method of Kumanogo in [12].

Proposition 3.8. *Let $T > 0$. Then there are $\rho_0 > 0$ and $\epsilon_0 > 0$ such that there exists the fundamental solution $E(t)$ of (3.32) which is given by,*

$$E(t) = I_{\phi(t)}(x,D)(e(t,x,D) + r(t,x,D)), \tag{3.33}$$

where $r(t,x,\xi)$ belongs to $A_{-\epsilon_0,-\epsilon_0}$ uniformly in $t \in [-T,T]$, $|\rho| \le \rho_0$ and $e(t,x,\xi)$ satisfies that there are $C_e >, \varepsilon_e > 0$ such that

$$|e_{(\beta)}^{(\alpha)}(t,x,\xi)| \le C_e\varepsilon_e^{-|\alpha+\beta|}e^{C_e|\rho t|} <x>^{-|\beta|}<\xi>^{-|\alpha|}\ |\alpha+\beta|!, \tag{3.34}$$

for $x, \xi \in R^n, \alpha, \beta \in N^n$ and $|t| \le T$, where $\phi(t,x,\xi) = x\xi - i\vartheta(t,x,\xi)$ and ϑ is holomorphic in $(z,\zeta) \in \Gamma^2_{\tau_\vartheta,\varepsilon_\vartheta}$ uniformly in $t \in [-T,T]$ for some $\tau_\vartheta > 0, \varepsilon_\vartheta > 0$ and satisfies the following equation

$$\partial_t \vartheta(t,z,\zeta) = \rho p_0(z, \zeta - i\nabla_z \vartheta(t,z,\zeta)), \vartheta(0,z,\zeta) = 0, \tag{3.35}$$

for $z, \zeta \in \Gamma_{\tau_\vartheta,\varepsilon_\vartheta}, |\rho| \le \rho_0$ and $|t| \le T$.

Proof. *Since p_0 satisfies (3.31), there are $\tau' > 0, \varepsilon' > 0$ such that p_0 has a holomorphic extension $p_0(z,\zeta)$ satisfying*
(3.31)'
$$|p_{0(\beta)}^{(\alpha)}(z,\zeta)| \le C_p\varepsilon_p^{-|\alpha+\beta|}(<Rez> + <Re\zeta>) <Rez>^{-|\beta|}<Re\zeta>^{-|\alpha|}\ |\alpha+\beta|!$$

for $z, \zeta \in \Gamma_{2\tau',2\varepsilon'} = \{\zeta \in C^n; |Im\zeta| \le 2\tau|Re\zeta| + 2\varepsilon'\}, \alpha, \beta \in N^n$. Therefore by use of the standard characteristic curve method we can find the solution

$\vartheta(t,z,\zeta)$ of (3.35) which satisfies that there are $C_\vartheta > 0, \tau_\vartheta > 0, \varepsilon_\vartheta > 0$ and $\rho_0 > 0$ such that

(3.36)

$$|\vartheta^{(\alpha)}_{(\beta)}(t,z,\zeta)| \le C_\vartheta \rho \varepsilon_\vartheta^{-|\alpha+\beta|}(< Rez > + < Re\zeta >) < Rez >^{-|\beta|}< Re\zeta >^{-|\alpha|} |\alpha+\beta|!$$

for $z,\zeta \in \Gamma_{2\tau_\vartheta,2\varepsilon_\vartheta}, \alpha,\beta \in N^n, |\rho| \le \rho_0$ and $|t| \le T$. Now we shall construct a solution of (3.32) of form below,

$$E(t) = I_{\phi(t)}(x,D)e(t,x,D), \tag{3.37}$$

where $\phi(t) = x\xi - i\vartheta(t,x,\xi)$. Noting that $\partial_t E(t) = I_{\phi(t)}\partial_t e(t) + (\partial_t\vartheta)_{\phi(t)}e(t)$, we get from (3.32),

$$\partial_t e(t) = I^{-1}_{\phi(t)}(\rho p_0(x,D)I_{\phi(t)} - \partial_t\vartheta_{\phi(t)})e(t), |t| \le T, e(0) = I. \tag{3.38}$$

It follows from Theorem 3.7 that we have

(3.39)

$$\sigma(I^{-1}_{\phi(t)}\rho p_0 I_{\phi(t)})(x,\xi) = \rho(p_0(x+i\nabla_\xi\vartheta(t,\Phi',\xi),\xi-i\nabla_x\vartheta(t,\Phi',\xi))+\tilde{p}(t,x,\xi)+r_1(t,x,\xi)),$$

where $\tilde{p} \in AS(1,g)$ and $r_1 \in A_{-\varepsilon_0,-\varepsilon_0}$. On the other hand from Lemma 3.4 we get

$$\sigma(I^{-1}_{\phi(t)}\partial_t\vartheta_{\phi(t)}(x,\xi) = \partial_t\vartheta(t,x+i\nabla_\xi\vartheta(t,\Phi',\xi),\xi)+\vartheta_1(t,x,\xi)+r_2(t,x,\xi), \tag{3.40}$$

where $\vartheta_1 \in AS(|\rho|,g)$ and $r_2 \in A_{-\varepsilon_0,-\varepsilon_0}$ uniformly in $t \in [-T,T]$. Therefore taking account of (3.35) we obtain from (3.38)-(3.40),

$$\partial_t e(t,x,D) = (\rho p(t,x,D) + r(t,x,D))e(t,x,D), |t| \le T, e(0,x,D) = I, \tag{3.41}$$

where $p \in AS(1,g), r \in A_{-\varepsilon_0,-\varepsilon_0}$ uniformly in $t \in [-T,T]$. Here we note that $p(t,x,\xi)$ satisfies that there are $C_p > 0, \tau_p > 0, \varepsilon_p > 0$ such that

$$|p^{(\alpha)}_{(\beta)}(t,x,\zeta)| \le C_p\varepsilon_p^{-|\alpha+\beta|} < x >^{-|\beta|}< Re\zeta >^{-|\alpha|} |\alpha+\beta|! \tag{3.42}$$

for $x \in R^n, \zeta \in \Gamma_{\tau_p,\varepsilon_p}, \alpha,\beta \in N^n, |\rho| \le \rho_0$, and $|t| \le T$.
We construct an astmptotic solution of (3.41) as follows,

$$e(t,x,D) = \sum_{j=0}^{\infty} E_j(t,x,D).$$

Then the symbols $E_j(t,x,\zeta)$ satisfy

$$(\partial_t - \rho p(t,x,\zeta))E_0(t,x,\zeta) = 0, |t| \le T, E(0,x,\zeta) = 1 \tag{3.43}$$

$$(\partial_t - \rho p(t,x,\zeta))E_j(t,x,\zeta) = p_j(t,x,\zeta), |t| \le T, E_j(0,\zeta) = 0, (j \ge 1), \tag{3.44}$$

for $x,\zeta \in \Gamma_{\tau',\varepsilon'}, \alpha,\beta \in N$, where

$$p_j(t,x,\zeta) = \sum_{k=0}^{j-1}\sum_{|\alpha|+k=j} \alpha!^{-1}\rho p^{(\alpha)}(t,x,\zeta)E_{k(\alpha)}(t,x,\zeta), (j \ge 1). \tag{3.45}$$

Solving (3.43)-(3.44), we get

$$E_0(t,x,\zeta) = e^{\int_0^t \rho p(s,x,\zeta)ds}, \tag{3.46}$$

$$E_j(t,x,\zeta) = \int_0^t p_j(\tau,x,\zeta)E_0(\tau,x,\zeta)^{-1}d\tau E_0(t,x,\zeta), (j \geq 1). \tag{3.47}$$

We put $e_0(t,x,\zeta) = 1$ *and for* $j \geq 1$

$$e_j(t,x,\zeta) = E_j(t,x,\zeta)E_0(t,x,\zeta)^{-1},$$

$$q_j(t,x,\zeta) = p_j(t,x,\zeta)E_0(t,x,\zeta)^{-1}.$$

Denote

$$\omega_\alpha(t,x,\zeta) = e^{-\int_0^t \rho p(s,x,\zeta)ds} D_x^\alpha e^{\int_0^t \rho p(s,x,\zeta)ds}. \tag{3.48}$$

Then noting that p satisfies (3.42), we can prove by induction on α *that there are* $A_i(i=1,2) > 0$ *such that*

$$|D_x^\beta \omega_\alpha(t,x,\zeta)| \leq A_1^{|\alpha+\beta|} < x >^{-|\alpha+\beta|} \sum_{k=1}^{|\alpha|} (A_2|\rho t|)^k (|\alpha+\beta|+1-k)!, \tag{3.49}$$

for $t, \rho \in R, x \in R^n, \zeta \in \Gamma_{\tau_p,\varepsilon_p}, \alpha(\neq 0), \beta \in N^n$. *Taking account that*

$$D_x^\alpha E_k = \sum_{\alpha' \leq \alpha} \binom{\alpha}{\alpha'} e_{k(\alpha')}\omega_{\alpha-\alpha'}E_0,$$

holds, we get from (3.45),(3.48),

$$q_j(t,x,\zeta) = \sum_{k=0}^{j-1} \sum_{|\alpha|+k=j} \alpha!^{-1} \rho p^{(\alpha)}(\dot{x},\zeta) \sum_{\alpha' \leq \alpha} \binom{\alpha}{\alpha'} e_{k(\alpha')}\omega_{\alpha-\alpha'}, \tag{3.50}$$

and consequently we have from (3.47)

$$e_j(t,x,\zeta) = \int_0^t q_j(\tau,x,\zeta)d\tau. \tag{3.51}$$

Now we shall prove by induction on j that there are $c_i(i=1,2,3) > 0$ *such that*

$$|D_x^\beta e_j(t,x,\zeta)| \leq c_1^j c_2^{|\beta|} < x >^{-j-|\beta|} < Re\zeta >^{-j} (j+1)!^{-1} \tag{3.52}$$

$$\times \sum_{\ell=0}^{2j} (c_3|t\rho|)^\ell (|\beta|+2j+1-\ell)!,$$

for $|t| \le T, |\rho| \le \rho_0, x \in R^n, \zeta \in \Gamma_{\tau_p,\varepsilon_p}$ *and* $\beta \in N^n, j \ge 1$. *To simplify the notation we prove (3.52) in the case of* $t, \rho \ge 0$. *For* $j = 1$ *we have from (3.47)*

$$q_1(t,x,\zeta) = \sum_{|\alpha|=1} \rho p^{(\alpha)}\omega_{(\alpha)} = \sum_{|\alpha|=1} \rho^2 t p^{(\alpha)} \int_0^t p_{(\alpha)}(s,x,\zeta)ds.$$

Hence we get from (2.37),

$$(3.53) \qquad |D_x^\beta q_1| \le \sum_{|\alpha|=1} t\rho^2 \sum_{\beta' \le \beta} \binom{\beta}{\beta'} |p^{(\alpha)}_{(\beta')} p_{(\alpha+\beta-\beta')}$$

$$\le C_p^2(2\rho\varepsilon_p^{-1})^2 t(2\varepsilon_p^{-1})^{|\beta|}(2+|\beta|)! < x >^{-1-|\beta|} < Re\zeta >^{-1},$$

where we used the inequality,

$$\sum_{\beta' \le \beta} \binom{\beta}{\beta'} (1+|\beta'|!(1+|\beta-\beta'|)! \le 2^{2+|\beta|}(2+|\beta|)!.$$

Integrating (3.53) with respect to t we get (3.52) for $j = 1$, *if we choose* $c_1 \ge C_p^2, c_2 \ge 2\varepsilon_p^{-1}, c_3 \ge 2\varepsilon_p^{-1}$. *Assume (3.52) is valid for* $j \ge 1$. *¿From (3.50) we have*

$$D_x^\beta q_j = \sum \binom{\alpha}{\alpha'}\binom{\beta}{\beta'}\binom{\beta-\beta'}{\beta''} \alpha!^{-1} \rho p^{(\alpha)}_{(\beta')} e_{k(\alpha'+\beta'')}\omega_{\alpha-\alpha'(\beta-\beta'-\beta'')},$$

where the summation ranges over $k, \alpha, \alpha', \beta', \beta''$ *such that* $1 \le |\alpha| + k = j, \alpha' \le \alpha, \beta' \le \beta$, $\beta'' \le \beta - \beta'$. *Hence we get from the assumption of induction and (3.42),(3.49),*

$$(3.54) \qquad |D_x^\beta q_j| \le \sum_{|\alpha|+k=j} \frac{\rho n^{|\alpha|}}{|\alpha|!} \sum_{\alpha' \le \alpha} \binom{|\alpha|}{|\alpha'|} \sum_{\beta' \le \beta, \beta'' \le \beta-\beta'} \binom{\beta}{\beta'}\binom{\beta-\beta'}{\beta''}$$

$$\times C_p \varepsilon_p^{-|\alpha+\beta'|} < x >^{-k-|\alpha'+\beta|} < Re\zeta >^{-|\alpha|-k} |\alpha+\beta'|! c_1^k c_2^{|\alpha'+\beta''|}(k+1)!^{-1}$$

$$\times \sum_{\ell=0}^{2k} (c_3 t\rho)^\ell (|\alpha'+\beta''|+2k+1-\ell)! \frac{A_1^{|\alpha-\alpha'+\beta-\beta'-\beta''|}}{< x >^{|\alpha-\alpha'+\beta-\beta'-\beta''|}}$$

$$\times \sum_{h=1}^{|\alpha-\alpha'|} (A_2\rho t)^h (|\alpha-\alpha'+\beta-\beta'-\beta''|+1-h)!$$

$$\le \sum_{k=0}^{j-1} \frac{\rho n^{j-k}}{(j-k)!(k+1)!} < x >^{-j-|\beta|} < Re\zeta >^{-j} \sum_{|\alpha'|=0}^{j-k} \binom{j-k}{|\alpha'|} \sum_{\beta',\beta''} \binom{\beta}{\beta'}\binom{\beta-\beta'}{\beta''} C_p \varepsilon_p^{-(j-k+|\beta'|)}$$

$$\times (1+j-k+|\beta'|)! c_1^k c_2^{j-k+|\beta''|} A_1^{j-k+|\beta-\beta'-\beta''|} \sum_{s=1}^{2k+|\alpha|} (c_3\rho t)^s$$

$$\times \sum_{h+\ell=s,h\le j-k} (A_2c_3^{-1})^h(|\alpha'+\beta''|+2k+1-\ell)!(j-k+|\beta-\beta'-\beta''|-|\alpha'|+1-h)!$$

$$\le C_p\rho c_1^j c_2^{|\beta|} < x >^{-j-|\beta|}< Re\zeta >^{-j} (j+1)!^{-1}$$

$$\times \sum (c_3\rho t)^s \binom{j+1}{k+1}\binom{\beta}{\beta'}\binom{\beta-\beta'}{\beta''}(nA_1c_2\varepsilon_p^{-1}c_1^{-1})^{j-k}$$

$$\times(\varepsilon_p^{-1}c_2^{-1})^{|\beta'|}(A_1c_2^{-1})^{|\beta-\beta'-\beta''|}(j-k-|\alpha'|)!(A_2c_3^{-1})^\ell$$

$$\times(|\alpha'+\beta''|+2k+1-\ell)!(1+j-k+|\beta'|+|\beta-\beta'-\beta''|+1-h)!$$

where the last summation ranges over $k, s, |\alpha'|, \beta', \beta'', h, \ell$ *such that* $0 \le k \le j-1, 1 \le s \le j+k, 0 \le |\alpha'| \le j-k, h+\ell = s, \ell \le 2k, \beta'' \le \beta-\beta', \beta' \le \beta$. *We can see easily that the following inequality holds,*

$$(j-k-|\alpha'|+|\beta-\beta'-\beta''|+1-h)! \le 2^{j-k+|\beta-\beta'-\beta''|}(j-k-|\alpha'|+|\beta-\beta'-\beta''|-h)!.$$

Hence we have

$$\sum_{\beta''\le\beta-\beta'} \binom{\beta-\beta'}{\beta''}(A_1c_2^{-1})^{|\beta-\beta'-\beta''|}(|\alpha'+\beta''|+2k+1-\ell)! \tag{3.55}$$

$$\times(j-k-|\alpha'|+|\beta-\beta'-\beta''|+1-h)!$$

$$\le 2^{j-k}\sum_{|\beta''|=0}^{|\beta-\beta'|}(2A_1c_2^{-1})^{|\beta-\beta'-\beta''|}(j+k+1+|\beta-\beta'|-\ell-h)!$$

$$\le 2^{j-k}(1-2A_1c_2^{-1})^{-1}(j+k+1+|\beta-\beta'|-\ell-h)!$$

if $2A_1c_2^{-1} < 1$, *here we used*

$$\binom{\beta-\beta'}{\beta''} \le \binom{|\beta-\beta'|}{|\beta''|}\binom{j+k+1-\ell-h}{|\alpha'|+2k+1-\ell} \le \binom{j+k+|\beta-\beta'|-\ell-h}{|\alpha'+\beta''|+2k+1-\ell}$$

for $\ell+h \le j+k, |\alpha'| \le j-k$ *and* $\ell \le 2k$. *Moreover taking account of the following inequality,*

$$\sum_{h=0}^{|\alpha-\alpha'|}(A_2c_3^{-1})^{s-h} \le \sum_{h=0}^{j-k}(A_2c_3^{-1})^{s-h} \le (A_2c_3^{-1})^s(2c_3A_2^{-1})^{j-k},$$

if $A_2c_3^{-1} \le 1$, *we obtain from (3.54), (3.55)*

$$|D_x^\beta q_j| \le \rho C_p c_1^j c_2^{|\beta|} < x >^{-j-|\beta|}< Re\zeta >^{-j} \tag{3.56}$$

$$\times(j+1)^{-1}(1-2A_1c_2^{-1})^{-1}\sum_{k=0}^{j-1}\sum_{s=1}^{j+k}(c_3\rho t)^s(A_2c_3^{-1})^s$$

$$\times \sum_{|\beta'|=0}^{|\beta|} \binom{j+1}{k+1} \binom{|\beta|}{|\beta'|} (4c_3 A_2^{-1} A_1 c_2 \varepsilon_p^{-1} c_1^{-1})^{j-k} (\varepsilon_p^{-1} c_2^{-1})^{|\beta'|}$$

$$\times (j-k+|\beta'|)!(j+k+1+|\beta-\beta'|-s)!.$$

Noting that $(j-k+|\beta'|)! \le 2^{j-k+|\beta'|}(j-k-1+|\beta'|)$ *for* $k \le j-1$ *and*

$$\binom{|\beta|}{|\beta'|}(j-k-1+|\beta'|)!(j+k+1+|\beta-\beta'|-s)!(2j+|\beta|-s)!^{-1}$$

$$= \binom{|\beta|}{|\beta'|}\binom{2j+|\beta|-s}{j-k+|\beta|}^{-1} \le \binom{2j-s}{j-k}^{-1},$$

we get from (3.56) if $2c_p c_2^{-1} < 1$,

$$|D_x^\beta q_j| \le \rho C_p c_1^j c_2^{|\beta|} < x >^{-j-|\beta|} < Re\zeta >^{-j} (j+1)!^{-1} \tag{3.57}$$

$$\times (1-2A_1 c_2^{-1})^{-1}(1-2\varepsilon_p^{-1}c_2^{-1})^{-1} \sum_{k=0}^{j-1}\sum_{s=1}^{2j-1} (c_3\rho t)^s \binom{j+1}{k+1}$$

$$\times \binom{2j-s}{j-k}^{-1} (c_3^{-1}A_2)^s (8c_3 A_2^{-1} c_2 \varepsilon_p^{-1} c_1^{-1})^{j-k} (2j+|\beta|-s)!.$$

For $s \le j-1$ *we have*

$$\sum_{k=0}^{j-1} \binom{j+1}{k+1}\binom{2j-s}{j-k}^{-1} (c_3^{-1}A_2)^s (8c_3 A_2^{-1} c_2 \varepsilon_p^{-1} c_1^{-1})^{j-k}$$

$$\le \sum_{k=0}^{j-1} \binom{j+1}{k+1} (c_3^{-1}A_2)^s (8c_3 A_2^{-1} c_2 \varepsilon_p^{-1} c_1^{-1})^{j-k} \binom{j+1}{j-k}^{-1}$$

$$= \sum_{k=0}^{j-1} (c_3^{-1}A_2)^s (8c_3 A_2^{-1} c_2 \varepsilon_p^{-1} c_1^{-1})^{j-k} \le (1-8c_3 A_2^{-1} c_2 \varepsilon_p^{-1} c_1^{-1})^{-1} \le 2,$$

if $8c_3 A_2^{-1} c_2 \varepsilon_p^{-1} c_1^{-1} < 1/2$ *and* $c_3^{-1}A_2 \le 1$, *Besides, for* $s \ge j$ *we have*

$$\sum_{k=0}^{j-1} \binom{j+1}{k+1}\binom{2j-s}{j-k}^{-1} (c_3^{-1}A_2)^s (8c_3 A_2^{-1} c_2 \varepsilon_p^{-1} c_1^{-1})^{j-k}$$

$$\le \sum_{k=0}^{j-1} \binom{j+1}{k+1} (c_3^{-1}A_2)^j (8c_3 A_2^{-1} c_2 \varepsilon_p^{-1} c_1^{-1})^{j-k} \le (c_3^{-1}A_2 + 8c_3 A_2^{-1} c_2 \varepsilon_p^{-1} c_1^{-1})^{j+1} \le 1,$$

if $c_3^{-1}A_2 + 8c_3 A_2^{-1} c_2 \varepsilon^{-1} \varepsilon_p^{-1} c_1^{-1} \le 1$. *Thus we get from (2.52)*

$$|D_x^\beta q_j| \le 2\rho C_p (1-A_1 c_2^{-1})^{-1}(1-2\varepsilon_p^{-1}c_2^{-1})^{-1} c_1^j c_2^{|\beta|} (j+1)!^{-1}$$

$$\times <x>^{-j-|\beta|}<Re\zeta>^{-j}\sum_{s=1}^{2j-1}(c_3\rho t)^s(2j+|\beta|-s)!.$$

Integrating $|q_{j(\beta)}|$ with respect to t, we obtain (3.52) from (3.51), if we choose positive constants c_1, c_2, c_3 satisfying the above restrictions and $2c_3^{-1}(1-A_1c_2^{-1})^{-1}(1-2\varepsilon_p^{-1}c_2^{-1}) \leq 1$. Next we shall construct the exact solution $E(t)$ of the equation of (3.41). It follows from (3.52) that we can see easily $e_j(t,x,\xi)$ satisfying

(3.58)
$$|e^{(\alpha)}_{j(\beta)}(t,x,\xi)| \leq c_4^{1+j+|\alpha+\beta|}(<x><\xi>)^{-j}j!e^{c_4|\rho t|}<x>^{-|\beta|}<\xi>^{-|\alpha|}|\alpha+\beta|!,$$

for $|t| \leq T, |\rho| \leq \rho_0, x,\xi \in R^n$, where c_4 is a positive constant independent of j. Therefore applying Lemma 2.14 to $\{e_j\}$ we can construct $d(t,x,\xi)$ sstisfying

$$(3.59)\quad |\partial_\xi^\alpha D_x^\beta(d(t,x,\xi)-\sum_{j=0}^{N}e_j(t,x,\xi))| \leq c_5^{1+N+|\alpha+\beta|}(<x><\xi>)^{-N}N!$$
$$\times e^{c_4|t\rho|}<x>^{-|\beta|}<\xi>^{-|\alpha|}|\alpha+\beta|!,$$

for $|t| \leq T, |\rho| \leq \rho_0, x,\xi \in R^n$ and for any non negative integer N. Putting $\tilde{e}(t,x,\xi) = d(t,x,\xi)e^{\rho\int_0^t p(s,x,\xi)ds}$, we obtain the following relation,

$$(\partial_t - \rho p(x,D))\tilde{e}(t,x,D) = R(t,x,D), |t| < T, \tilde{e}(0,x,D) = I,$$

where $R(t,x,\xi)$ satisfies from (3.43)-(3.45) and (3.59)

$$(3.60)\qquad |R^{(\alpha)}_{(\beta)}(t,x,\xi)| \leq c_7^{1+|\alpha+\beta|}e^{-\varepsilon_0/2(<x>+<\xi>)}|\alpha+\beta|!,$$

if $|t\rho| \leq \varepsilon_0/2c_4$. Now we can construct the exact solution of (3.41) as follows,

$$(3.61)\qquad e(t,x,D) = \tilde{e}(t,x,D) + \int_0^t \tilde{e}(t-\tau)W(\tau,x,D)d\tau,$$

where $W(t,x,D)$ satisfies

$$(3.62)\qquad W(t,x,D) = -\int_0^t R(t-\tau,x,D)W(\tau,x,D)d\tau.$$

We can solve the above integral equationas follows

$$(3.63)\qquad W(t,x,D) = \sum_{j=1}^{\infty}W_j(t,x,D),$$

where

$$W_1(t,x,D) = -R(t,x,D),$$

$$(3.64)\qquad W_j(t,x,D) = \int_0^t W_1(t-\tau,x,D)W_{j-1}(\tau,x,D)d\tau, (j \geq 2)$$

$$= \int_0^t \int_0^{\tau_1} \cdots \int_0^{\tau_{j-1}} (-1)^j R(t-\tau_1)R(\tau_1-\tau_2)\cdots R(\tau_{j-1}-\tau_j)R(\tau_j)d\tau_1\cdots d\tau_j$$

Since we can regard $R(t,x,\xi)$ as an element of $A_{0,0}$ from (3.60), we can apply Lemma 2.12 to the right side in (3.64). Then $W_j(t,x,\xi)$ satisfies the following estimate from (2.33) of Lemma 2.12 and by j-times integrations,

$$(3.65) \qquad |W_{j(\beta)}^{(\alpha)}(t,x,\xi)| \le c_7^{1+|\alpha+\beta|}(C|t|)^{j-1}(j-1)!^{-1}|\alpha+\beta|!,$$

for any $j \ge 1, |t\rho| \le \epsilon_0/c_4, x,\xi \in R^n, \alpha,\beta \in N^n$. Therefore $W(t,x,\xi)$ given by (3.63) converges in $A_{0,0}$. On the other hand, since $W(t)$ satisfies (3.62) and $R(t,x,\xi)$ belongs to $A_{-\epsilon_0,-\epsilon_0}$, it follows from (i) of Proposition 2.3 that $W(t,x,\xi)$ also belongs to $A_{-\epsilon_0,-\epsilon_0}$. Therefore we can see from (3.61) that $e(t,x,\xi)$ satisfies (3.34) modifying ϵ_0 if necessary. Thus we completed the proof of Proposition 3.8. Q.E.D.

1.5 CRITERION TO L^2–WELL POSED CAUCHY PROBLEM

For $T > 0$ let consider the following Cauchy problem,

$$(4.1) \qquad \partial_t u(t,x) - ia(x,D)u(t,x) - b(t,x,D)u(t,x) = f(t,x),$$

$$(4.2) \qquad u(0,x) = u_0(x),$$

for $(t,x) \in (0,T)\times R^n$. We assume that the principal symbol $a(x,\xi) \in S^2_{1,0}$ has a real value and $b(t,x,\xi)$ is in $C^0([0,T];S^1_{1,0})$. Moreover we suppose that there are $C \in R, K > 0$ and $\theta(x,\xi) \in C^\infty(R^{2n})$ such that

$$(4.3) \qquad Reb(t,x,\xi) - 1/2\sum_{j=1}^{n} a_{x_j\xi_j}(x,\xi) \le C,$$

for $\xi \in R^n$ and $x \in R^n$ with $|x| \ge K$, and $\theta(x,\xi)$ satisfies

$$(4.4) \qquad H_a\theta(x,\xi) \ge c_0|\xi| - c_1, (c_0 > 0, c_1 \in R)$$

for $x,\xi \in R^n$ and

$$(4.5) \qquad |\theta_{(\beta)}^{(\alpha)}(x,\xi)| \le C_{\alpha\beta} <x><\xi>^{-|\alpha|}$$

for $x,\xi \in R^n$ and $\alpha,\beta \in N^n$. Then we can prove the following theorem by use of the same method as that of [3] and [7].

Theorem 4.1. *Assume that the above conditions (4.3)-(4.5) are valid. For any $u_0 \in L^2$ and $f \in C^0([0,T];L^2)$ there exists a unique solution $u \in C^0([0,T];L^2)\cap C^1([0,T];H^{-2})$ of the Cauchy problem (4.1)-(4.2).*

1.6 PROOF OF THEOREM

We shall prove (8) for $t > 0$. We may assume that $a(x,\xi)$ satisfying (5) is non negative. We can prove it as a same way otherwise For $p_0(x,\xi) = (a(x,\xi)+1)^{1/2} \in AS(<\xi>, g)$, where $a(x,\xi)$ satisfies (5) and $\rho > 0$, let us denote by $E(t)$ the fundamental solution of the Cauchy problem for $\rho p_0(x,D)$, that is,

$$\frac{d}{dt}E(t) = \rho p_0(x,D)E(t), E(0) = I. \tag{5.1}$$

Then it follows from Proposition 3.8 that $E(t) = I_{\phi(t)}(x,D)e(t,x,D)$,where $\phi(t) = x\xi - i\vartheta(t,x,\xi)$ satisfies (3.35) becomes a Fourier integral operator which commute with $p_0(x,D)$. Therefore noting that from Lemma 3.1 we can see $a(x,D) = p_0(x,D)^2 + p_1(x,D) + r_0(x,D) - 1$, where $p_1(x,\xi) \in AS(<\xi><x>^{-1}, g)$ and $r_0 \in A_{-\varepsilon_0,-\varepsilon_0}$, we get from Theorem 3.7, Lemma 3.1 and Proposition 2.3,

$$\begin{aligned} E(t)a(x,D)E(t)^{-1} &= p(x,D)^2 + E(t)(p_1(x,D) + r_0(x,D) - 1)E(t)^{-1} \\ &= a(x,D) + a_1(t,x,D) + r_1(x,D) \end{aligned} \tag{5.2}$$

where $a_1(t,z,\zeta) \in AS(<\xi><x>^{-1}, g)$ for $|t| \le T, |\rho| \le \rho_0$ and $r_1 \in A_{-\varepsilon_0+c\rho_0 T,-\varepsilon_0}$. Let u be satisfied with (1)-(2). Put $v(t,x) = E(t)u(t,x)$. Then v satiesfies the following Cauchy problem from (5.1)-(5.2),

$$\frac{\partial}{\partial t}v(t,x) = (ia(x,D) + c(t,x,D))v(t,x) + g(t,x), \tag{5.3}$$

$$v(0,x) = u_0(x), \tag{5.4}$$

where $g = E(t)f$ and

$$c(t,x,D) = \rho p(x,D) + a_1(t,x,D) + r_1(t,x,D) + E(t)b(t,xD)E(t)^{-1}$$

$$= \rho p(x,D) + b(t,x,D) + b_1(t,x,D) + r_2(t,x,D), \tag{5.5}$$

where $b_1(x,\xi) \in AS(< \xi > < x >^{-1}, g), r_2(t,z,\zeta) \in A_{-\varepsilon_0+c\rho_0 T,-\varepsilon_0}$from Theorem 3.7.. Oncemore we change the unknown function v to w as follows,

$$w(t,x) = I_\phi(x,D)v(t,x), \tag{5.6}$$

where $\phi = x\xi - i\epsilon\theta(x,\xi)$. It follows from Lemma 3.5 that if $|\epsilon|$ is sufficiently small, we have the inverse $I_\phi(x,D)^{-1}$. Therefore we get the following Cauchy problem of w from (5.3)-(5.4),

$$\frac{\partial}{\partial t}w(t,x) = I_\phi(ia(x,D) + c(t,x,D))I_\phi(x,D)^{-1}w(t,x), \tag{5.7}$$

(5.8) $$w(0,x) = I_\phi(x,D)u_0(x).$$

Since $\theta(x,\xi)$ belongs to $AS(<x>,g)$, it follows from (3.12) that $\nabla_\xi\theta(x,\Phi(x,\xi)) \in S(<x><\xi>^{-1},g), \nabla_x\theta(x,\Phi(x,\xi)) \in S(1,g)$ and $\Phi(x,\xi)-\xi \in S(1,g)$, we have from (3.27) in Theorem 3.7 and Proposition 2.3

(5.9)

$$\sigma(I_\phi a I_\phi^{-1})(x,\xi) = a(x - i\epsilon\nabla_\xi\theta(x,\Phi),\xi + i\epsilon\nabla_x\theta(x,\Phi)) + a_1(x,D) + r_3(x,\xi),$$

$$= a(x - i\epsilon\nabla_\xi\theta(x,\xi),\xi + i\epsilon\nabla_x\theta(x,\xi)) + a_1'(x,\xi) + r_3(x,D)$$
$$= a(x,\xi) + i\epsilon H_a\theta(x,\xi) + a_1''(x,D) + r_3(x,\xi),$$

where $a_1, a_1' \in S(<\xi><x>^{-1},g), a_1'' \in S(<x>^{-1}<\xi> + |\varepsilon|^2<\xi>,g)$ and $r_3 \in A_{-\varepsilon_0+c\rho_0T,-\varepsilon_0+c|\varepsilon|}$ for some $c>0$(independent of ε,ρ_0). Here we choose ρ_0,ε such that $c\rho_0T<\varepsilon_0, c|\varepsilon|<\varepsilon_0$. and so r_3 belongs to $S(1,g)$. Thus we obtain the equation of w from (5.7)-(5.9),

(5.10)

$$\frac{\partial w}{\partial t} = (ia(x,D)+p(x,D)+b(t,x,D)-\epsilon(H_a\theta)(x,D))w+r_4(t,x,D))w(t,x)+I_\phi g,$$

(5.11) $$w(0) = I_\phi(x,D)u_0(x),$$

where $r_4 \in S(<\xi><x>^{-1} + |\varepsilon|^2<\xi>,g)$. Moreover taking account of the assumptions (6) and (7) in the introduction we can choose conviniently ϵ and ρsuch that we have

$$\rho p(x,\xi) + Reb(t,x,\xi) - \epsilon H_a\theta(x,\xi) + Rer_4(t,x,\xi) \le 0,$$

for $x,\xi \in R^n$ with $|x| \ge K$,where $K>0$ is sufficiently large. Therefore we can solve the Cauchy problem (5.10)-(5.11) by use of Proposition 4.1, if $w(0) = I_\phi u_0$ belongs to L^2, and cosequently we get the solution $u = E(t)^{-1}I_\phi(x,D)^{-1}w(t,x) = e(t,x,D)^{-1}I_{\phi(t)}(x,D)^{-1}I_\phi(x,D)^{-1}w$, which satisfies (8) by from Lemma 1.2. In fact taking account of the assumptions in the introduction we can see that $Re\vartheta(t,x,\xi) \ge 2c_0\rho t|\xi|$ and $|\theta(x,\xi)| \le C<x>/2$ and consequently that u belongs to $H_{c_0t\rho,-C|\epsilon|}$ and $c_0t\rho>0$. This completes the proof of Theorem.

References

[1] *De Bouard A. Hayashi N. & Kato K.* Regularizing effect for the (generalized) Korteweg-de Vrie equations and nonlinear Schrödinger equations, *Ann. Inst. Henri Poincaré Analyse nonlinear vol. 12 pp. 673-725 (1995).*

[2] *Boutet de Monvel L. & Krée P.* Pseudo-differential operators and Gevrey classes, *Ann. Inst. Fourier Grenoble vol.17 pp. 295-323 (1967).*

[3] *Doi S.* Remarks on the Cauchy problem for Schrödinger type equations, *Comm. P.D.E. vol. 21 pp. 163-178 (1996).*

[4] *Hayashi N. & Saitoh S.* Analyticity and smoothing effect for Schrödinger equation, *Ann. Inst. Henri Poincaré Math. vol 52 pp. 163-173 (1990).*

[5] *Hayashi S., Nakamitsu K. & Tsutsumi M.* On solutions of the initial value problem for the nonlinear Schrödinger equations in one space dimension, *MathŻvol. 192 pp. 637-650 (1986).*

[6] *Jensen A.* Commutator method and a Smoothing property of the Schrödinger evolution group, *MathŻ. vol. 191 pp. 53-59 (1986).*

[7] *Kajitani K.* The Cauchy problem for Schrödinger type equations with variable coefficients , *to appear in Jour. Math. Soc. Japan (1997).*

[8] *Kajitani K.* Analytically smoothing effect for Schrödinger equations, *Proceedings of the International Conference on Dynamical Systems & Differential Equations in Southwest Missouri State University (1996).*

[9] *Kajitani K. & Baba A.* The Cauchy problem for Schrödinger type equations , *Bull. Sci. math. vol. 119 pp. 459-473 (1995)*

[10] *Kato K. & Taniguti K.* Gevrey regularizing effect for nonlinear Schrödinger equations, *Osaka J. Math. vol. 33 pp. 863-880 (1996).*

[11] *Kato T.& Yajima K.* Some examples of smoothing operators and the associated smoothing effect, *Rev. Math. Phys. vol.1 pp. 481-496 (1989).*

[12] *Kumanogo H.* Pseudo-Differential Operators, *MIT Press (1981).*

[13] *Matsumoto W.* Ultradifferentiable classes and Pseudo-Differential Operators, *in Japanese (1982).*

[14] *Wakabayashi S.* Classical microlocal analysis in the space of hyperfunctions, *preprint (1997).*

INITIAL BOUNDARY VALUE PROBLEM FOR THE VISCOUS INCOMPRESSIBLE FLOWS

Hisako Kato

Graduate School of Mathematics,
Kyushu University, Ropponmatsu,
Fukuoka 810, Japan

Abstract: The present paper is concerned with a new model for the viscous incompressible fluids. We propose that the motion of the viscous incompressible fluids is governed by two type equations, namely the Navier-Stokes equations at times when the velocity gradient is below a given constant and the 'Ladyzhenskaia equations' at times when the velocity gradient is above the constant.

1.1 INTRODUCTION AND SUMMARY

The present paper is concerned with a new model for the viscous incompressible fluids. We propose that the motion of the viscous incompressible fluids is governed by two type equations, namely the Navier-Stokes equations at times when the velocity gradient is below a given constant and the 'Ladyzhenskaia equations' at times when the velocity gradient is above the constant. For the new model we will prove the global (in time) existence and uniqueness of strong solutions under no smallness assumption on initial data.

R.P. Gilbert et al.(eds.), Direct and Inverse Problems of Mathematical Physics, 221–232.

The initial boundary value problem for the Navier-Stokes equations is formulated as follows:

$$\frac{\partial u}{\partial t} - \nu_0 \Delta u + u \cdot \nabla u = -\nabla p \qquad (x \in \Omega,\ t > 0), \tag{1.1}$$

$$\text{div } u = 0 \qquad (x \in \Omega, \quad t > 0), \tag{1.2}$$

with the conditions

$$u = 0 \qquad (x \in \partial\Omega, \quad t > 0), \tag{1.3}$$

$$u(x,0) = a(x) \qquad (x \in \Omega), \tag{1.4}$$

where Ω is a bounded domain in R^3 with smooth boundary $\partial\Omega$, $\nu_0 > 0$ is the coefficient of kinematic viscosity, and the initial velocity $a = (a^1(x), a^2(x), a^3(x))$ is given.

The vector function $u = (u^1(x,t), u^2(x,t), u^3(x,t))$ and scalar function $p = p(x,t)$ represent, respectively, the unknown velocity and unknown pressure. We use the notation

$$u \cdot \nabla v = \sum_{i=1}^{3} u^i \frac{\partial v}{\partial x_i}, \qquad \text{div } u = \sum_{i=1}^{3} \frac{\partial u^i}{\partial x_i}$$

for vector functions u and v.

In this paper we will prove the global (in time) existence of strong solutions satisfying equations (1.1)–(1.4) for t when $\nabla u(t)$ is small, and satisfying the 'Ladyzhenskaia model' instead of (1.1) for t when $\nabla u(t)$ is large, and further satisfying equations (1.1)–(1.4) for all large t. Here, the Ladyzhenskaia model is as follows:

$$\frac{\partial u}{\partial t} - (\alpha + \beta \parallel \nabla u \parallel^2)\Delta u + u \cdot \nabla u = -\nabla p \qquad (x \in \Omega, \quad t > 0) \tag{1.5}$$

with positive constants α and β. (See [8](pp.193-201), [9], [10].) Du and Gunzburger studied on the Ladyzhenskaia model in [2].

To this end we introduce the following equation:

$$\frac{\partial u}{\partial t} - \frac{\nu_0}{2}(1 + \nu(\parallel \nabla u \parallel^2)\ )\Delta u + u \cdot \nabla u = -\nabla p \qquad (x \in \Omega, \quad t > 0) \tag{1.6}$$

with a function

$$\nu(t) = \frac{1}{K} \max\{t, K\}$$

for any $K > 0$. Then the solutions u of (1.6) with conditions (1.2)–(1.4) satisfy equation (1.1) for t when $||\nabla u(t)||^2 \leq K$, and satisfy the Ladyzhenskaia model for t when $||\nabla u(t)||^2 > K$. Here, we note the function $\nu(t)$ satisfies the remarkable inequality

$$K|\nu(t) - \nu(s)| \leq |t - s|.$$

By the inequality the desired convergence for 'approximate solutions' will be obtained.

First, we will show the global existence of strong solutions u of equation (1.6) with (1.2)–(1.4) under no smallness assumption on initial data and show that the solution is unique. Next, we will prove that there exists a positive number T independent of K such that the solution u satisfies equations (1.1)–(1.4) in $[T, \infty)$. Thus, equation (1.6) gives a regularization of (1.1) and moreover implies an improvement on the Ladyzhenskaia model.

Finally, we will determine the value of K such that $\| \nabla u(t) \|$ is nonincreasing in $[0, \infty)$. In addition we will consider values of K in connection with the viscosity, namely we will show that the global strong solution u of (1.6) with (1.2)–(1.4) satisfies (1.1) for t when $\|\nabla u(t)\| \leq \nu_0$, and satisfies Ladyzhenskaia model (1.5) with $\alpha = \nu_0/2$ and $\beta = 1/(2\nu_0)$ for t when $\|\nabla u(t)\| > \nu_0$. Here, $1/\nu_0$ can be considered as the Reynolds number.

For the Navier-Stokes initial boundary value problem in a bounded domain, up to the present the local (in time) existence of a regular solution, and the global existence of a regular solution in the case where the initial velocity $a(x)$ is small have been proved. (See Fujita and T.Kato [3], Giga and Miyakawa [5], Kiselev and Ladyzhenskaia [7], Ladyzhenskaia [8], Lions [11], Masuda [12], Serrin [13], Shinbrot [14], Sohr and von Wahl [15], Temam [16], Wahl [17], and papers cited therein.)

To state our theorems precisely we first introduce some basic function spaces and notion. We define

$$C_{0,\sigma}^{\infty} \equiv \{\varphi \in C_0^{\infty}(\Omega) \ ; \ \operatorname{div} \varphi = 0 \ \}.$$

In addition, we define H_σ as the closure of $C_{0,\sigma}^{\infty}$ in $L_2(\Omega)$, and $H_{0,\sigma}^1$ as the closure of $C_{0,\sigma}^{\infty}$ in $H^1(\Omega)$. Throughout this paper, $L_2(\Omega)$ represents the Hilbert space equipped with the inner product

$$(u, v) = \sum_{i=1}^{3} \int_{\Omega} u^i \ v^i \ dx.$$

We denote the $L_2(\Omega)$-norm by $\| \cdot \|$. $H^m(\Omega)$ is the Sobolev space of vector-valued functions in $L_2(\Omega)$ together with their derivatives up to order m. $H_0^m(\Omega)$ is the completion of the set $C_0^{\infty}(\Omega)$ in $H^m(\Omega)$. Further, for a Hilbert space H, $L_p((0,T); H)$ $(1 \leq p \leq \infty)$ denotes the set of H-valued measurable functions f in $(0, T)$ such that

$$\| f \|_{L_p((0,T);H)} = \left(\int_0^T \| f(t) \|_H^p \ dt\right)^{1/p} < \infty \quad (1 \leq p < \infty),$$

$$\| f \|_{L_\infty((0,T);H)} = \operatorname*{ess\ sup}_{0<t<T} \| f(t) \|_H < \infty.$$

$W^{m,p}((0,T);H)$ denotes the set of functions which are in $L_p((0,T);H)$ together with their t-derivatives up to order m.

Let P be the orthogonal projection from $L_2(\Omega)$ onto H_σ. By the Stokes operator A, we denote the Friedrichs extension of the symmetric operator $-P\Delta$ in H_σ with domain $D(A) = H^2(\Omega) \cap H_0^1(\Omega) \cap H_\sigma$. Then, equation (1.6) with conditions (1.2)–(1.4) is formulated as an equation with the initial condition in the Hilbert space H_σ:

$$\frac{du}{dt} + \frac{\nu_0}{2}(1 + \nu(\| \nabla u \|^2)\ )Au + Pu \cdot \nabla u = 0 \qquad (t > 0), \tag{1.7}$$

$$u(0) = a, \tag{1.8}$$

by the orthogonal decomposition ([4]):

$$L_2(\Omega) = H_\sigma \oplus \ \{\nabla p \ ; \ p \in H^1(\Omega)\}$$

We next recall the estimate of the nonlinear term of (1.1) (Giga and Miyakawa [5],p.270):

$$\| Pv \cdot \nabla v \| \leq c_1 \| A^{1/4} v \| \| Av \| \quad \text{for any } v \in D(A) \tag{1.9}$$

with an absolute constant c_1, which gives

$$\|Pv \cdot \nabla v\| \leq c_1 \| A^{-1/4} \| \| A^{1/2} v \| \| Av \| \quad \text{for any } v \in D(A). \tag{1.10}$$

Here, A^α represents the fractional power of the Stokes operator A.

We now state our result.

Theorem 1.1.1 (*Unique existence.*) *Let* $a \in H^1_{0,\sigma} \cap H^2(\Omega)$. *Then there exists a unique strong solution* $u = u^{(K)}(x,t)$ *in* $[0,\infty)$ *of equations* (1.7) – (1.8) *satisfying*

$$u \in L_\infty((0,T);H_\sigma) \ \cap \ L_\infty((0,T);H^1_{0,\sigma}) \ \cap \ L_\infty((0,T);D(A)),$$
$$\textit{and} \quad u \in W^{1,\infty}((0,T);H_\sigma) \ \cap \ W^{1,2}((0,T);H^1_{0,\sigma})$$

for any $T > 0$. *In addition,* $u(t)$ *is strongly continuous in (0,T) as* $H^1_{0,\sigma}$*-valued functions and its derivative* $u_t(t)$ *is weakly continuous in* $(0,T)$ *as* H_σ*-valued functions.*

Theorem 1.1.2 (*Regularization.*) *Let* $u = u^{(K)}(x,t)$ *be the solution in Theorem* 1.1.

(a) Set

$$\Lambda_1 = \{t \in (0,\infty);\ \ \| \nabla u(t) \|^2 \leq K\},$$
$$\Lambda_2 = \{t \in (0,\infty);\ \ \| \nabla u(t) \|^2 > K\}.$$

Then the solution u satisfies the following equations:

$$\frac{du}{dt} + \nu_0 Au + Pu\cdot\nabla u = 0 \quad in \quad \Lambda_1, \tag{1.11}$$
$$\frac{du}{dt} + \frac{\nu_0}{2}(1 + \frac{1}{K} \| \nabla u \|^2)Au + Pu\cdot\nabla u = 0 \quad in \quad \Lambda_2. \tag{1.12}$$

(b) Suppose that $K \geq (\nu_0/2c_0)^2$ with $c_0 = c_1\|A^{-1/4}\|$. Then there exists a positive number T_K bounded above with a constant independent of K :

$$T_K \leq \frac{4c_0^2\|a\|^2}{\nu_0^3}$$

such that the solution $u = u^{(K)}$ satisfies (1.11) *for all $t \in [T_K,\infty)$. (The constant c_1 here is defined in* (1.9).)

Concerning Theorems 1.1 and 1.2 we remark the following.

Remark 1. For an initial velocity satisfying $\| \nabla a \| < \nu_0/2c_0$, it is easy to see the global existence of strong solutions of (1.1)–(1.4).

Remark 2. There exists a subsequence of $u = u^{(K)}(x,t)$ converging to a Hopf weak solution of the Navier-Stokes equations (1.1)–(1.4) as $K \to \infty$.

Remark 3. If global strong solutions v of the Navier-Stokes equations (1.1)–(1.4) exist, then the solution $u = u^K$ satisfying (1.7) and (1.8) is equal to the solution v for any $K \geq K_0 \equiv \sup_{0\leq t<\infty} \| \nabla v(t) \|^2$.

We consider the values of K such that $\|\nabla u(t)\|$ is nonincreasing in $[0,\infty)$. In addition, it is interesting to consider values of K in connection with the viscosity ν_0. Really we obtain the following theorem.

Theorem 1.1.3 *Let $u = u^{(K)}(x,t)$ be the global solution in Theorem* 1.1.

(a) For $K =$ the critical value $(\nu_0/2c_0)^2$, it follows $\| \nabla u(t) \|$ is nonincreasing in $[0,\infty)$.

(b) For $K = \nu_0^2$ with $\nu_0 \geq 1$ it follows that

$$\frac{du}{dt} + \nu_0 Au + Pu\cdot\nabla u = 0, \quad in \quad \|\nabla u(t)\| \leq \nu_0, \tag{1.13}$$
$$\frac{du}{dt} + \frac{1}{2}(\nu_0 + \frac{1}{\nu_0}\|\nabla u\|^2)Au + Pu\cdot\nabla u = 0, \quad in \quad \|\nabla u(t)\| > \nu_0. \tag{1.14}$$

Furthermore, for $K = \sqrt{\nu_0}$ with $\nu_0 \in (0,1)$ it follows that

$$\frac{du}{dt} + \nu_0 Au + Pu \cdot \nabla u = 0, \quad \text{in} \quad \|\nabla u(t)\|^2 \leq \sqrt{\nu_0}, \tag{1.15}$$

$$\frac{du}{dt} + \frac{1}{2}(\nu_0 + \sqrt{\nu_0}\|\nabla u\|^2)Au + Pu \cdot \nabla u = 0, \tag{1.16}$$
$$\text{in} \quad \|\nabla u(t)\|^2 > \sqrt{\nu_0}.$$

Remark 4. In Theorem 1.3 it easily follows that (a) is also valid for $K < (\nu_0/2c_0)^2$ by considering Remark 1 mentioned above.

1.2 PROOF OF THEOREMS 1.1–1.3

We see the existence of approximate solutions of equations (1.7) and (1.8) for an initial velocity $a \in H^1_{0,\sigma} \cap H^2(\Omega)$ by applying the Galerkin procedure. In fact, let w_i $(i = 1, 2, \cdots)$ be the completely orthonormal system in H_σ consisting of the eigenfunctions of the Stokes operator A, and for each n define an approximate solution u_n as

$$u_n(t) = \sum_{i=1}^{n} c_{in}(t) w_i, \tag{2.1}$$

satisfying

$$(u_{nt} + \frac{\nu_0}{2}[1 + \nu(\| \nabla u_n \|^2)]Au_n + Pu_n \cdot \nabla u_n \ , \ w_i) = 0, \tag{2.2}$$
$$(i = 1, 2, \cdots, n),$$

$$u_n(0) = \sum_{i=1}^{n} c_{in}(0) w_i \qquad \text{with} \quad c_{in}(0) = (a, w_i). \tag{2.3}$$

Here, equations (2.2) form a system of ordinary differential equations for the functions $c_{in}(t)$ $(i = 1, 2, \cdots, n)$. Since the completely orthonormal system w_i $(i = 1, 2, \cdots)$ in H_σ is a complete system in $H^1_{0,\sigma}$, $u_n(0)$ converges to the initial velocity a in $H^1_{0,\sigma}$.

Thus, we have the following lemmas.

Lemma 1.2.1 *Let an initial velocity a be in $H^1_{0,\sigma} \cap H^2(\Omega)$. Then there exist approximate solutions $u_n = \sum_{i=1}^n c_{in}(t) w_i$ of equations (1.7) and (1.8) satisfying*

$$\| u_n(t) \|_{H^2(\Omega)} + \| u_{nt}(t) \| \leq C, \quad n = 1, 2, \cdots, \quad t > 0,$$
$$\int_0^t \| \nabla u_{nt} \|^2 \, dt \leq C, \quad n = 1, 2, \cdots, \quad t > 0,$$

where the constant C is independent of n and t.

Lemma 1.2.2 *Let u_n $(n = 1, 2, \cdots)$ be the approximate solutions in Lemma 2.1. Then there is a subsequence $u_{n'}$ of u_n such that*

$$\begin{aligned} &u_{n'} \to u \quad \text{in} \quad L_\infty((0,T); H^1_{0,\sigma}) \quad \text{strongly}, \\ &u_{n'} \to u \quad \text{in} \quad L_\infty((0,T); D(A)) \quad \text{weakly*}, \\ &u_{n't} \to u_t \quad \text{in} \quad L_2((0,T); H_\sigma) \quad \text{strongly}, \\ &u_{n't}(t) \to u_t(t) \quad \text{in} \quad H_\sigma \quad \text{weakly and uniformly in} \quad t \in [0,T], \end{aligned}$$

as $n' \to \infty$ and for any $T > 0$.

Lemma 1.2.3 *Let u be the limit function in Lemma 2.2. Then*

(1) $\| u(t) \|^2$ *is differentiable in $t \in (0, \infty)$, and*

$$\frac{d}{dt} \| u \|^2 = 2(u,\ u_t). \tag{2.4}$$

(2) $\| \nabla u(t) \|^2$ *is absolutely continuous, and*

$$\| \nabla u(t) \|^2 - \| \nabla a \|^2 = 2 \int_0^t (\nabla u,\ \nabla u_t) dt. \tag{2.5}$$

We now prove Theorems 1.1–1.3.

Proof of Theorem 1.1.

First, we find u satisfies equations (1.7)–(1.8) and the regularity properties by Lemmas 2.1–2.3. Next, we see the uniqueness. Indeed, it is sufficient to prove the uniqueness for $\nu_0 = 2$. Let u and v be solutions of (1.7)–(1.8) with $\nu_0 = 2$ and $w = u - v$. Then we get

$$\begin{aligned} &(w_t + Aw + \nu(\| \nabla u \|^2) Au - \nu(\| \nabla v \|^2) Av,\ w) \\ &\quad + (Pu \cdot \nabla w + Pw \cdot \nabla v,\ w) = 0, \end{aligned}$$

so that

$$\begin{aligned} &\frac{1}{2}\frac{d}{dt} \| w \|^2 + \| \nabla w \|^2 + \nu(\| \nabla u \|^2)(Aw,\ w) \\ &\leq C \,|\, \nu(\| \nabla u \|^2) - \nu(\| \nabla v \|^2) \,|\ \ |\, (Av, w) \,| \\ &\qquad + C \| w \|^{3/2} \ \| Av \| \ \| \nabla w \|^{1/2} \\ &\leq C' \,|\ \| \nabla u \|^2 - \| \nabla v \|^2 |\ \ \| w \| + C' \| w \|^{3/2}\ \ \| \nabla w \|^{1/2}, \end{aligned} \tag{2.6}$$

where we have used Lemmas 2.1–2.3. Hence, we see by the Young inequality

$$\begin{aligned} \frac{1}{2}\frac{d}{dt} \| w \|^2 + \| \nabla w \|^2 \ &\leq C'' \ \ \| \nabla w \| \ \ \| w \| + C'' \| w \|^{3/2} \ \ \| \nabla w \|^{1/2} \\ &\leq C''' \ \ \| w \|^2 + \frac{1}{2} \| \nabla w \|^2, \end{aligned}$$

which implies $w(t) \equiv 0$. Hence the proof of the uniqueness, and therefore Theorem 1.1 is complete.

Proof of Theorem 1.2.

We first note that (a) in Theorem 1.2 is a direct result of Theorem 1.1. We next show (b). Let $u = u^K(x,t)$ be the solution obtained in Theorem 1.1, and u_n be the approximate solutions. Then we get by (1.7)–(1.8) the energy inequality

$$\begin{aligned} &\| u(t) \|^2 + \nu_0 \int_0^t \| \nabla u \|^2 \, dt \\ &+ \nu_0 \int_0^t \nu(\| \nabla u \|^2) \| \nabla u \|^2 \, dt \leq \| a \|^2 \end{aligned} \tag{2.7}$$

for any t, and therefore

$$\int_0^\infty \| \nabla u \|^2 \, dt \leq \frac{1}{\nu_0} \| a \|^2 < \infty.$$

Hence, there exists $T'_K > 0$ such that

$$\| \nabla u(T'_K) \| < \frac{\nu_0}{2c_0}. \tag{2.8}$$

¿From this, for the approximate solutions u_n, it follows that there exists a positive integer n_0 such that $\| \nabla u_n(T'_K) \| < \nu_0/2c_0$ for any $n \geq n_0$. Since $\| \nabla u_n(t) \|$ is a continuous function, there exists $\delta = \delta(n) > 0$ such that

$$\| \nabla u_n(t) \| < \frac{\nu_0}{2c_0} \quad \text{for any } t \in [T'_K, T'_K + \delta).$$

Set

$$T^* = \sup\{T; \ \| \nabla u_n(t) \| \leq \frac{\nu_0}{2c_0} \quad \text{for any } t \in [T'_K, T) \ \}.$$

Then we have $T^* = \infty$.

In fact, if T^* were finite it would follow that

$$\| \nabla u_n(T^*) \| = \frac{\nu_0}{2c_0}, \tag{2.9}$$

and

$$\| \nabla u_n(t) \| \leq \frac{\nu_0}{2c_0} \quad \text{for any } t \in [T'_K, T^*). \tag{2.10}$$

On the other hand, considering (2.2) we see

$$\begin{aligned} &\frac{1}{2}\frac{d}{dt} \| \nabla u_n \|^2 + \frac{\nu_0}{2}[1 + \nu(\| \nabla u_n \|^2)] \| Au_n \|^2 \\ &= -(u_n \cdot \nabla u_n, \ Au_n). \end{aligned} \tag{2.11}$$

Therefore, for such a value $t = T^*$ we have the following estimate of the right-hand side of (2.11):

$$\begin{aligned} |(u_n(t)\cdot\nabla u_n(t),\ Au_n(t))| &\le c_1\|A^{-1/4}\|\,\|\nabla u_n(t)\|\ \ \|Au_n(t)\|^2 \\ &\le \frac{\nu_0}{2}\|Au_n(t)\|^2, \end{aligned} \tag{2.12}$$

where we have used (1.10),(2.9) and $c_0 = c_1\|A^{-1/4}\|$, so that

$$\frac{d}{dt}\|\nabla u_n\|^2 < 0 \quad \text{at } t = T^*.$$

This gives $\|\nabla u_n\| < \nu_0/2c_0$ in some neighborhood of $t = T^*$. This is a contradiction.

Thus, we get $\|\nabla u_n(t)\|^2 \le (\nu_0/2c_0)^2$ in $[T'_K,\infty)$, which implies $\|\nabla u(t)\|^2 \le (\nu_0/2c_0)^2$ in $[T'_K,\infty)$. Hence, for any $K \ge (\nu_0/2c_0)^2$ we have $\nu(\|\nabla u(t)\|^2) = 1$ for any $t \in [T'_K,\infty)$.

Furthermore, set

$$T_K = \inf\{T'_K;\ \|\nabla u(t)\| \le \frac{\nu_0}{2c_0} \quad \text{for any } t \in [T'_K,\infty)\}.$$

Then, we get

$$\begin{aligned} &\|\nabla u(t)\| \le \frac{\nu_0}{2c_0} \quad \text{for any } t \in [T_K,\infty), \\ &\|\nabla u(t)\| \ge \frac{\nu_0}{2c_0} \quad \text{for any } t \in [0,T_K), \end{aligned} \tag{2.13}$$

so that

$$\nu_0\int_0^{T_K}\left(\frac{\nu_0}{2c_0}\right)^2 dt \le \nu_0\int_0^{T_K}\|\nabla u\|^2 dt \le \|a\|^2$$

from (2.7). This gives

$$T_K \le \frac{4c_0^2\|a\|^2}{\nu_0^3}.$$

Hence the proof of (b), and therefore Theorem 1.2 is complete.

Proof of Theorem 1.3.

We first note that (b) in Theorem 1.3 is a direct result of Theorem 1.1. We next show (a). Take $(\nu_0/2c_0)^2$ as K. Then, from (2.11) and (1.10) it follows that

$$\begin{aligned} &\frac{1}{2}\frac{d}{dt}\|\nabla u_n\|^2 + \frac{\nu_0}{2}[1+\nu(\|\nabla u_n\|^2)]\,\|Au_n\|^2 \\ &\le c_0\|\nabla u_n\|\ \ \|Au_n\|^2, \end{aligned}$$

so that

$$\frac{1}{2}\frac{d}{dt} \| \nabla u_n \|^2 + \frac{\nu_0}{2} \| Au_n \|^2 \tag{2.14}$$
$$= \| Au_n \|^2 (c_0 \| \nabla u_n \| - \frac{\nu_0}{2}\nu(\| \nabla u_n \|^2)).$$

Here, set

$$\alpha = c_0 \| \nabla u_n \| - \frac{\nu_0}{2}\nu(\| \nabla u_n \|^2).$$

Then,

$$\alpha = \begin{cases} c_0 \| \nabla u_n(t) \| (1 - \frac{2c_0}{\nu_0} \| \nabla u_n(t) \|), & \text{in } \{\| \nabla u_n \| > \nu_0/2c_0\} \\ c_0 \| \nabla u_n \| - \frac{\nu_0}{2}, & \text{in } \{\| \nabla u_n \| \leq \nu_0/2c_0\}, \end{cases}$$

so that $\alpha(t) \leq 0$ in $[0, \infty)$. Consequently, $\| \nabla u_n(t) \|^2$ is nonincreasing in $[0, \infty)$, so that $\| \nabla u(t) \|$ is also nonincreasing. Hence the proof of (a), and therefore Theorem 1.3 is complete.

1.3 APPENDIX

For the stationary Navier-Stokes problem:

$$-\nu_0 \Delta u + u \cdot \nabla u = f(x) - \nabla p \qquad (x \in \Omega), \tag{3.1}$$
$$\text{div } u = 0 \qquad (x \in \Omega), \tag{3.2}$$

with the condition

$$u = 0 \qquad (x \in \partial\Omega), \tag{3.3}$$

we obtain the following theorem.

Theorem 1.3.1 *Let $f = f(x) \in H_\sigma(\Omega)$. Then, there exists a unique solution $u = u(x) \in H^1_{0,\sigma} \cap D(A)$ satisfying our equation*

$$\frac{\nu_0}{2}[1 + \nu(\|\nabla u\|^2)]Au + Pu \cdot \nabla u = f \tag{3.4}$$

with $K = (\frac{\nu_0}{2c_0})^2$.

Therefore, we can find a value K such that for any $f \in H_\sigma$ there exists a unique solution u satisfying either (a): the stationary Navier-Stokes equations $\nu_0 Au + Pu \cdot \nabla u = f$ with $\|\nabla u\|^2 \leq K$ and (3.2)–(3.3), or (b): equations (3.4) $\frac{\nu_0}{2}[1 + \frac{1}{K}\|\nabla u\|^2]Au + Pu \cdot \nabla u = f$ with $\|\nabla u\|^2 > K$ and (3.2)–(3.3).

References

[1] Courant R. and D. Hilbert. (1968). *Methoden der Mathematishen Physik II*, Springer, Berlin, Heidelberg, New York.

[2] Du Q. and M. Gunzburger. (1991). *Analysis of a Ladyzhenskaya model for incompressible viscous flow*, J. Math. Anal. Appl., Vol. 155, (pages 21-45).

[3] Fujita H. and T. Kato. (1964). *On the Navier-Stokes initial value problem 1*, Arch. Rational Mech. Anal., Vol. 16, (pages 269-315).

[4] Fujiwara D. and H. Morimoto. (1977). *An L_r-theorem of the Helmholtz decomposition of vector fields*, J. Fac. Sci. Univ. Tokyo, Vol. 24, (pages 685-700).

[5] Giga Y. and T. Miyakawa. (1985). *Solutions in L_r of the Navier-Stokes initial value problem*, Arch. Rational Mech. Anal., Vol. 89, (pages 267-281).

[6] Kinderlehrer D. and G. Stampacchia. (1980). *An Introduction to Variational Inequalities and Their Applications*, Academic Press, New York, London, Toronto Sydney, San Francisco.

[7] Kiselev, A. A. and O. A. Ladyzhenskaia. (1957). *On the existence and uniqueness of the solution of the nonstationary problem for a viscous incompressible fluid*, Izv. Akad. Nauk SSSR, Ser. Mat., Vol. 21, (pages 655-680).

[8] Ladyzhenskaia, O. A. (1969). *The Mathematical Theory of Viscous Incompressible Flow*, Gorden and Breach, Revised English edn., New York, London.

[9] Ladyzhenskaia, O. A. (1967). *New equations for the description of motion of viscous incompressible fluids and solvability in the large of boundary value problems for them*, Proc. Steklov Inst. Math., Vol. 102, (pages 95-118), Trudy Mat. Inst. Steklov, Vol. 102, (pages 85-104).

[10] Ladyzhenskaia, O. A. (1983). *Limit states for modified Navier-Stokes equations in three dimensional space*, J. Soviet Math., Vol. 21, (pages 345-356).

[11] Lions, J. L. (1969). *Quelque Methodes de Résolution des Problèmes aux Limites Non Linéaires*, Dunod, Paris.

[12] Masuda, K. (1984). *Weak solutions of the Navier-Stokes equations* Tohoku Math. J., Vol. 36, (pages 623-646).

[13] Serrin, J. (1963). *The initial value problem for the Navier-Stokes equations*, in Nonlinear Problem'(R. E. Langer ed), The University of Wisconsin Press, Madison.

[14] Shinbrot, M. (1973). *Lectures on Fluid Mechanics*, Gordon-Breach, New York, London, Paris.

[15] Sohr, H. and W. von Wahl. (1984). *On the singular set and the uniqueness of weak solutions of the Navier-Stokes equations*, Manuscripta Math., Vol. 49), (pages 27-59).

[16] Temam, R. (1984). *Navier-Stokes Equations*, North-Holland, Amsterdam, New York, Oxford.

[17] von Wahl, W. (1985). *The Equations of Navier-Stokes and Abstract Parabolic Equations*, Vieweg, Braunschweig-Wiesbaden.

INITIAL-BOUNDARY VALUE PROBLEMS FOR AN EQUATION OF INTERNAL WAVES IN A STRATIFIED FLUID

P. A. Krutitskii

Department of Mathematics
Faculty of Physics
Moscow State University
Moscow 119899, Russia
e-mail: krutitsk@math.phys.msu.su

In the plane $x = (x_1, x_2) \in R^2$ we consider the internal or external multiply connected domain D bounded by closed curves $\Gamma \in C^{2,0}$. The following PDE of composite type

$$(1) \qquad \frac{\partial^4 u}{\partial t^2 \partial x_1^2} + \frac{\partial^4 u}{\partial t^2 \partial x_2^2} + \omega_1^2 \frac{\partial^2 u}{\partial x_1^2} + \omega_2^2 \frac{\partial^2 u}{\partial x_2^2} = 0; \quad \omega_1,\ \omega_2 \geq 0,$$

describes internal waves in the ocean [1]. The potential theory has been constructed for eq.(1) recently. Some applications of potentials to solving problems are presented in [1–8]. In particular, explicit solutions of some problems in canonical domains were obtained in [1–4], [9-10]. In the present note we study the solvability of the

R.P. Gilbert et al.(eds.), Direct and Inverse Problems of Mathematical Physics, 233–236.

initial-boundary value problems with either Dirichlet or Neumann boundary condition in arbitrary domains with the help of the potential technique and the boundary integral equation method [5–8]. Boundary value problems for equations of composite type in multiply connected domains were not treated before.

Definition 1. A function $u(t,x)$ defined on $[0,\infty)\times D$ belongs to the smoothness class G if

1) $u, u_t \in C^0\left([0,\infty)\times D\right)$,

2) at each $t\geq 0$ there exists a limit of $u(t,x)$ along the normal to the boundary Γ,

3) and, in addition,

$$\frac{\partial^k}{\partial t^k}\frac{\partial^p}{\partial x_j^k}u \in C^0\left((0,\infty)\times D\right), \quad k,p=0,1,2; \;\; j=1,2. \tag{2}$$

Let $u(t,x)$ be a sufficiently smooth function for $t\geq 0,\ x\in D$. Assuming that n is the normal at the point $x(s)\in\Gamma$, we define the differential operator $N_{t,x}$ at points $\bar{x}\in D$ by the relationship

$$N_{t,x}u(t,\bar{x}) = \frac{\partial^2}{\partial t^2}\frac{\partial}{\partial n}u(t,\bar{x}) + \omega_1^2\cos(n,x_1)u_{x_1}(t,\bar{x}) + \omega_2^2\cos(n,x_2)u_{x_2}(t,\bar{x}),$$

where $\cos(n,x_j)$ is the cosine of the angle between the normal n and the axis $Ox_j,\ j=1,2$.

By $N_{t,x}u(t,x)$ we denote the limiting value of $N_{t,x}u(t,\bar{x})$ as $\bar{x}$ approaches $x(s)\in\Gamma$ along the normal n, if the limit exists uniformly for all $x(s)\in\Gamma$.

Definition 2. A function $u(t,x)$ defined on $[0,\infty)\times D$ belongs to the smoothness class G_1 if

1) condition (2) holds,

2) $u, u_t \in C^0([0,\infty)\times \overline{D}),\ \ \nabla u, \nabla u_t \in C^0([0,\infty)\times D)$,

3) and the expression $N_{t,x}u(t,x)$ exists at each point of Γ in the sense of the uniform limit for all $x(s)\in\Gamma$.

Problem **N** (Neumann). To find $u(t,x)$ of class G_1 satisfying eq.(1) in $(0,\infty)\times D$, satisfying the initial conditions

$$u(0,x) = u_t(0,x) = 0, \tag{3}$$

and the boundary conditions

$$N_{t,x}u|_\Gamma = f(t,x)|_\Gamma \in C^0([0,\infty); C^0(\Gamma)).$$

By $C^k([0,\infty);H)$ we denote the class of abstract functions $\varphi(t)$, which are k times continuously differentiable in t for $t \geq 0$ and which take values in the Banach space H. Besides, we put

$$C_0^k([0,\infty);H) = \{\varphi(t) : \varphi(t) \in C^k([0,\infty);H),\ \varphi(0) = \varphi'(0) = \ldots = \varphi^{(k-1)}(0) = 0\}.$$

Problem **D** (Dirichlet). To find $u(t,x)$ of class G satisfying eq.(1) in $(0,\infty) \times D$, satisfying initial conditions (3) and the boundary condition

$$u|_\Gamma = f(t,x)|_\Gamma \in C_0^2([0,\infty);C^0(\Gamma)).$$

All conditions of the problems must be satisfied in the classical sense. In the case of the external domain D the following conditions at infinity must be included in the formulation of the problems **N** and **D**:

$$|\partial_t^k u| \leq q_1(t), \quad |\partial_t^k u_{x_j}| \leq q_2(t)|x|^{-2}, \quad k = 0,1,2; \;\; j = 1,2;$$

where $|x| = \sqrt{x_1^2 + x_2^2} \to \infty$ and $q_1(t),\ q_2(t) \in C^0[0,\infty)$.

On the basis of energy equalities [1, 6–8] we prove the following theorem.

Theorem 1. **(a).** *A necessary condition for the solvability of the problem* **N** *is*

$$\int_\Gamma f(t,x)dl_x = 0, \quad t \geq 0. \tag{4}$$

If a solution of the problem **N** *exists, then it is determined up to an arbitrary additive function of time* $c(t) \in C_0^2[0,\infty)$.

(b). *There is no more, than one solution of the problem* **D** *in the class* G_1.

By means of potentials the problems are reduced to the Fredholm integral equations at the boundary [6–8]. Studying the solvability of these equations in the appropriate Banach spaces, we obtain the solvability theorem for the problems.

Theorem 2. **(a).** *If the condition (4) holds, then the classical solution of the problem* **N** *exists and it is defined up to an arbitrary function of class* $C_0^2[0,\infty)$.

(b). *A classical solution of the problem* **D** *exists. Moreover, if* $\Gamma \in C^{2,\lambda}$, $f(t,x) \in C_0^2([0,\infty);C^{1,\lambda}(\Gamma))$, *where the Hölder exponent* $\lambda \in (0,1]$, *then this solution belongs to the class* G_1, *and according to Theorem 1(b) it is unique.*

The theorems holds for both internal and external domain D. The problems **N** and **D** can be easily computed by means of finding numerical solutions of Fredholm boundary integral equations derived for these problems. The Fredholm equations have invertible integral operators, and so their numerical solutions can be obtained by standard codes, i.e. by means of discretization and inversion of the matrix.

The research was supported by the RFBR grant 96-01-01411.

References.

1. Krutitskii P.A. *Math.Meth.Appl.Sci.*, 1995, v.18, pp.897-926.

2. Krutitskii P.A. *Russian Acad.Sci.Dokl.Math.*, 1993, v.46, No.1, p.63-69, and p.118-125.

3. Krutitskii P.A. *ZAMM*, 1996, v.76, special issues, No.2, p.581-582.

4. Krutitskii P.A. *Applicable Analysis*, 1996, v.61, p.209-217.

5. Krutitskii P.A. *Russian Math. Surveys*, 1995, v.50, No.4, p.740.

6. Krutitskii P.A. *Comput. Maths. Math. Phys.*, 1996, v.36, No.1, p.113-123.; and 1997, v.37, No.1, p.113-123.

7. Krutitskii P.A. *Mathematical Notes*, 1996, v.60, No.1, p.29-41.

8. Krutitskii P.A. *Differential Equations*, 1996, v.32, No.10, p.1383-1392.

9. Krutitskii P.A. *J. Math. Kyoto Univ.* , 1997, v.37, No.2, p.343-365.

10. Krutitskii P.A. *Appl. Math. Letters*, 1997, v.10, No.2, p.117-122.

Homogenization of the System Equations of High Frequency Nonlinear Acoustics

Evgueny A. LAPSHIN and Gregory P. PANASENKO

Abstract - High frequency nonlinear acoustics equations proposed in [1] are considered. The acoustic characteristics of the medium rapidly oscillate. This model describes the propagation of pulses of sound shocks generated by supersonic planes, blast waves in atmosphere and ocean, continuous radiation of sound sources. An asymptotic solutions of high frequency nonlinear acoustics equations is constructed (when the characteristic size of inhomogeneity is small with respect to the height of the layer where the problem is posed).

1. Formulation of the problem and the main results.

Consider the following system of equations proposed in [1]

$$(\nabla\psi_\varepsilon)^2 = n^2(z,\frac{z}{\varepsilon}),\ z \in (0,z_0),\ x \in I\!R \tag{1}$$

$$\begin{aligned}\nabla\psi_\varepsilon\cdot\nabla p_\varepsilon + \tfrac{1}{2}\Delta\psi_\varepsilon p_\varepsilon - \tfrac{1}{2}p_\varepsilon\nabla\psi_\varepsilon\cdot\nabla \ln\rho(z,\tfrac{z}{\varepsilon}) \quad - \\ -E(z,\tfrac{z}{\varepsilon})p_\varepsilon\tfrac{\partial p_\varepsilon}{\partial\tau} = 0,\ z\in(0,z_0),\ x\in I\!R,\ \tau\in I\!R,\end{aligned} \tag{2}$$

where $\nabla = (\frac{\partial}{\partial x},\frac{\partial}{\partial z}), \Delta = \frac{\partial^2}{\partial x^2}+\frac{\partial^2}{\partial z^2}$.

Here the functions $\psi_\varepsilon(x,z)$ and $p_\varepsilon(\tau,x,z)$ are unknown and the functions $n(z,\frac{z}{\varepsilon})$, $\rho(z,\frac{z}{\varepsilon})$, $E(z,\frac{z}{\varepsilon})$ are given. The last three functions are the characteristics of some stratified medium which depends on a small heterogeneity parameter $\varepsilon << 1$.

In applications the most important characteristic is the pressure $\mathcal{P}_\varepsilon$ which is a function of three arguments : x,z (space coordinates) and t (time), i.e $\mathcal{P}_\varepsilon(x,z,t) = p_\varepsilon(x,z,t-\psi_\varepsilon/c_0)$, where the pair $p_\varepsilon,\psi_\varepsilon$ is the solution of the system (1), (2), c_0 is the positive constant. In this connection we complete the system of equations (1) , (2) by the third equation, relating τ and t :

$$\tau = t - \psi_\varepsilon/c_0. \tag{3}$$

Equip the equations (1) and (2) with the following boundary conditions for $z=0$:

$$\psi_\varepsilon|_{z=0} = \psi_0(x), \tag{4}$$

$$p_\varepsilon|_{z=0} = p_0(x,\tau), \tag{5}$$

where ψ_0, p_0 are given functions.

The system of equations of high frequency non linear acoustics (1) - (3) was introduced in [1] where it was derived from the non linear wave equation for an acoustic pressure p :

R.P. Gilbert et al.(eds.), Direct and Inverse Problems of Mathematical Physics, 237–250.

$$\frac{1}{c^2\rho}\frac{\partial^2 p}{\partial t^2} - \frac{e}{c^4\rho^2}\frac{\partial^2 p^2}{\partial t^2} = div(\frac{1}{\rho}gradp);$$

here $c^2 = c_0^2/n^2$, ρ, $e = Ec^3\rho/n$ are given functions of a point (x, z). The coefficient e of the non linear term $\frac{\partial^2 p^2}{\partial t^2}$ is called non linearity of the medium. The derivation of the equations (1) - (3) from the equation (6) is made in [1] under the assumptions of the high frequency of oscillations and of the smalless of the Mach number $M = p/c^2\rho$. The equation (1) is an eikonal equation with the refraction coefficient n, and the equation (2) is a non linear transfer equation. It is shown in [1] that the accounting of non linearity of the model is important in some situations. The parameter ε is a characteristic size of an inhomogenity when the caracteristic size in the direction z is taken as a unit (i.e z_0 is the value of a magnitude 1). In the case of periodic in z media we take ε equal to the period. This parameter ε is assumed to be small ($\varepsilon << 1$). The aim of this paper is to study an asymptotic behaviour of the solution of the problem (1) - (5) as $\varepsilon \to 0$.

Other models of high frequency wave propagation in non-homogeneous media were considered in [2-4].

We replace the equation (1) by the equation

$$\frac{\partial \psi_\varepsilon}{\partial z} = \sqrt{n^2(z, \frac{z}{\varepsilon}) - (\frac{\partial \psi_\varepsilon}{\partial x})^2}, \; z \in (0, z_0), \; x \in I\!R \tag{6}$$

Thus we consider the solution with non-negative derivative (we shall prove that it exists for sufficiently small z_0).

The assumptions on the data are

i) $\psi_0 \in C^3(I\!R)$, $p_0 \in C^1(I\!R^2)$, $n, \rho, E \in C^1(I\!R_+^2)$,

ii) there exist two constants $\kappa_1, \kappa_2 > 0$ independent of ε such that $\kappa_1 \le n(z,\xi), \rho(z,\xi), E(z,\xi) \le \kappa_2$;

iii) $sup_{\eta \in I\!R}\ (\psi_0'(\eta))^2 < inf_{(z,\xi)\in I\!R_+^2}\, n^2(z,\xi)$.

We need now some definitons to formulate the fourth assumption.

Definition 1. *Let F be a function of three variables $z \in [0, \bar{z}_0]$, $\xi \in I\!R_+$ and $a \in [-a_0, a_0]$. One says that F satifies the weak property of good mixture on $[0, \bar{z}_0] \times I\!R_+ \times [-a_0, a_0]$ of order $\beta(\varepsilon)$ if and only if*

1) for any $z \in [0, \bar{z}_0], a \in [-a_0, a_0]$ there exists a limit

$$\lim_{T\to+\infty} \frac{1}{T}\int_0^T F(z,\xi,a)d\xi$$

denoted by $< F > (z, a)$ or $< F(z, ., a) >$, and

2) for any $z \in [0, \bar{z}_0]$, $a \in [-a_0, a_0]$ the estimate holds true :

$$|\int_0^z F(z', \frac{z'}{\varepsilon}, a)dz' - \int_0^z < F > (z', a)dz'| \le C\beta(\varepsilon),$$

where C is a positive constant independent of ε and β is a function such that $\lim_{\varepsilon\to+0} \beta(\varepsilon) = 0$.

Definition 2. *One says that F satisfies the strong property of good mixture on* $[0,\bar{z}_0]\times \mathbb{R}_+ \times [-a_0,a_0]$ *of order* $\beta(\varepsilon)$ *if and only if F satisfies the weak property of good mixture and*

3) for any differentiable function b with values $b(z)\in[-a_0,a_0]$ *and the derivative* $|b'(z)|$ *bounded by a constant* b_0 *, i.e* $|b'(z)|\le b_0$, *and for all* $z\in[0,\bar{z}_0]$

$$|\int_0^z F(z',\frac{z'}{\varepsilon},b(z'))dz' - \int_0^z <F>(z',b(z'))dz'|\le C\beta(\varepsilon),$$

where C is a positive constant independent of ε *and* β *is a function , such that*

$$\lim_{\varepsilon\to+0}\beta(\varepsilon)=0.$$

Definition 3. *One sais that F is formally differentiable in z in average if there exist* $<F>(z,0)$; $<\frac{\partial F}{\partial z}>(z,a)$, $\frac{\partial}{\partial z}(<F>(z,a))$ *and* $\frac{\partial}{\partial z}<F>(z,a)=<\frac{\partial F}{\partial z}>(z,a)$.

In the same way we can define formal differentiability in a (in average).

Remark 1. *Let* $F,\frac{\partial F}{\partial z}\in C([0,\bar{z}_0]\times \mathbb{R}\times[-a_0,a_0])$ *and* $F(z,\xi,a)$ *be 1-periodic in* ξ. *Then for all* $z\in[0,\bar{z}_0]$ *and* $a\in[-a_0,a_0]$

$$<F>(z,a)=\int_0^1 F(z,\xi,a)d\xi$$

and F satisfies the weak property of good mixture of order ε. *If moreover* $\frac{\partial F}{\partial a}\in C([0,\bar{z}_0]\times\mathbb{R}\times[-a_0,a_0])$ *then F satisfies the strong property of good mixture of order* ε *and F is formally differentiable in a and in z in average.*

The proof is based on the approximation of the interval of integration in $\int_0^z F(z',\frac{z'}{\varepsilon},a)dz'$ by an integer number of periods and the application on each period of the mean value theorem.

The fourth assumption on the data is :

iv) the weak property of good mixture of the functions

$(n^2(z,\xi)-a^2)^{-1/2}$, $n^2(z,\xi)(n^2(z,\xi)-a^2)^{-3/2}$ and $\sqrt{\rho(z,\xi)}E(z,\xi)(n^2(z,\xi)-a^2)^{-3/4}$ of orders $\beta(\varepsilon)$, $\delta(\varepsilon)$ and $\mu(\varepsilon)$ respectively, and

the strong property of good mixture of order $\gamma(\varepsilon)$ for $(n^2(z,\xi)-a^2)^{1/2}$ (with $b_0=\sup_{n\in\mathbb{R}}|\psi_0'(\eta)|$), and

the formal differentiability in average of these four functions with respect to z and a ; here β,δ,μ,γ tend to zero as $\varepsilon\to0$, $z\in[0,z_1]$, $\xi\in\mathbb{R}_+$, $a\in[-\sup_{\eta\in\mathbb{R}}|\psi_0'(\eta)|,\sup_{\eta\in\mathbb{R}}|\psi_0'(\eta)|]$, z_1 does not depend on ε.

Let $q\in(0,1)$ independent of ε,

$$z_2=\frac{q(\underline{n}^2-\bar{a}_0^2)^{3/2}}{\bar{a}_1\bar{n}^2}, \tag{7}$$

where $\bar{a}_1=\sup_{\eta\in\mathbb{R}}|\psi_0''(\eta)|$, $\bar{a}_0=\sup_{\eta\in\mathbb{R}}|\psi_0'(\eta)|$, $\underline{n}=\inf_{z,\xi\in\mathbb{R}}n(z,\xi)$, $\bar{n}=\sup_{z,\xi\in\mathbb{R}_+}n(z,\xi)$.

Consider the following equation with unknown z_3 :

$$q = z_3\bar{p}_1 \frac{\bar{E}\sqrt{\bar{\rho}}}{(\underline{n}^2 - \bar{a}_0^2)^{3/4}}\, exp\, \{\frac{z_3\bar{n}^2\bar{a}_1}{2(\underline{n}^2 - \bar{a}_0^2)^{3/2}(1-q)}\}, \tag{8}$$

where $\bar{p}_1 = \sup_{\eta,\tau\in I\!R_+} |\frac{\partial p_0(\eta,\tau)}{\partial\tau}|,\ \ \bar{E} = \sup_{z,\xi\in I\!R_+} E(z,\xi),\ \ \bar{\rho} = \sup_{z,\xi\in I\!R_+} \rho(z,\xi)$

Lemma 1 . *Let* $z_4 = min\ (z_1, z_2, z_3)$. *Then for* $z_0 = z_4$ *there exist a solution of the problem (1) - (5)* $(\psi_\varepsilon, p_\varepsilon, \tau)$ *with* $\psi_\varepsilon \in C^2(I\!R \times [0, z_4])$, $p_\varepsilon \in C^1(I\!R \times [0, z_4] \times I\!R)$, $\tau \in C^2(I\!R \times [0, z_4] \times I\!R)$.

Now we are going to construct the asymptotic solution of the problem (1) - (5) and to discuss the estimates. The following five-steps algorithm is proposed. Each step does not depend on the small parameter.

Step 1. For each $z \in [0, z_4]$, $x \in I\!R$, solve the functional equation for $b(x, z)$

$$b = a_0(x - b\int_0^z < \frac{1}{\sqrt{n^2(z',.) - b^2}} > dz'), \tag{9}$$

where $a_0(\eta) = \psi_0'(\eta)$.

Lemma 2. *For all* $z \in [0, z_4]$, $x \in I\!R$ *there exist a unique solution* b *of the equation (9) such that*

$|b| \le \bar{a}_0$; *b is differentiable .*

Step 2. Define the function $\bar{\psi}$,

$$\bar{\psi}(x,z) = \psi_0(z) + \int_0^z < \sqrt{n^2(z,'.) - b^2(x,z')} > dz', \tag{10}$$

It is a solution of the homogenized eikonal equation

$$\frac{\partial\bar{\psi}}{\partial z} =< \sqrt{n^2 - (\frac{\partial\bar{\psi}}{\partial x})^2} >,\ \ z \in (0, z_4) \tag{11}$$

with the "initial" condition

$$\bar{\psi}|_{z=0} = \psi_0(x) \tag{12}$$

Step 3. Introduce new space variable $\bar{\eta} = \bar{\eta}(x,z)$ "along the beam" by the equation :

$$x = \bar{\eta} + a_0(\bar{\eta})\int_0^z < \frac{1}{\sqrt{n^2(z',.) - a_0^2(\bar{\eta})}} > dz' \tag{13}$$

Lemma 3. *For all* $z \in [0, z_4]$, $x \in I\!R$ *there exist a unique solution* $\bar{\eta}$ *such that* $|x - \bar{\eta}| \le \bar{a}_0 z_4 \frac{1}{\sqrt{\underline{n}^2 - \bar{a}_0^2}}$; $\bar{\eta}$ *is differentiable;*

$$\bar{\psi}(x(z,\bar{\eta}), z) = \psi_0(\bar{\eta}) + \int_0^z < \frac{n^2(z',.)}{\sqrt{n^2(z',.) - a_0^2(\bar{\eta})}} > dz', \tag{14}$$

where $\bar{\psi}$ *is defined by (10).*

Equations (13), (14) give a parametric form of the solution of homogenized eikonal problem (11), (12).

Step 4. For each $z \in [0, z_4]$, $\bar{\eta} \in I\!R$, $\xi \in I\!R_+$, $\tau \in I\!R$ define $\bar{p} = \bar{p}(\bar{\eta}, z, \xi, \tau)$ as a solution of the equation

$$\bar{p} = \bar{S}(\bar{\eta}, z, \xi)\, p_0(\bar{\eta}, \tau + $$
$$+ \bar{p}\bar{S}^{-1}(\bar{\eta}, z, \xi) \int_0^z < \frac{E(z', .)}{\sqrt{n^2(z', .) - a_0^2(\bar{\eta})}} > \bar{S}(\bar{\eta}, z', \xi) dz'), \tag{15}$$

where

$$\bar{S}(\bar{\eta}, z, \xi) =$$
$$= \frac{\sqrt{\rho(z, \xi)}}{(n^2(z, \xi) - a_0^2(\bar{\eta}))^{1/4}} exp\{-1/2 \int_0^z < \frac{n^2(z', .)}{(n^2(z', .) - a_0^2(\bar{\eta}))^{3/2}} > \times$$
$$\times a_0'(\bar{\eta}) \bar{R}(\bar{\eta}, z') dz'\}, \tag{16}$$

$$\bar{R}(\bar{\eta}, z) = (1 + a_0'(\bar{\eta}) \int_0^z < \frac{n^2(z', .)}{(n^2(z', .) - a_0^2(\bar{\eta}))^{3/2}} > dz')^{-1}, \tag{17}$$

$a_0(\bar{\eta}) = \psi_0'(\bar{\eta})$.

Lemma 4. *For all $z \in [0, z_4]$, $x \in I\!R$, $\xi \in I\!R_+$, $\tau \in I\!R$ there exist a unique solution $\bar{p}$ of the equation (15), such that*

$$|\bar{p}| \le \bar{p}_0 \frac{\sqrt{\bar{\rho}}}{(\underline{n}^2 - \bar{a_0}^2)^{1/4}} exp\{\frac{z_4 \bar{n}^2 \bar{a}_1}{2(\underline{n}^2 - \bar{a}_0^2)^{3/2}(1-q)}\},$$

where $\bar{p}_0 = \sup_{\eta, \tau \in I\!R} |p_0(\eta, \tau)|$; $\bar{p}$ is differentiable in $\bar{\eta}$ and in τ and these partial derivatives are bounded (uniformly with respect to all variables).

Step 5. Define

$$\bar{\tau} = t - \bar{\psi}(x, z)/c_0. \tag{18}$$

Theorem . *For $z_0 = z_4$ there exist such a solution of the problem (1) - (5) $(\psi_\epsilon p_\epsilon, \tau)$ that the estimates hold true :*

$$\sup_{x \in I\!R, z \in [0, z_4]} |\psi_\varepsilon(x, z) - \bar{\psi}(x, z)| \le C_1(\beta(\varepsilon) + \gamma(\varepsilon)) \tag{19}$$

and

$$\sup_{x \in I\!R, z \in [0, z_4], t \in I\!R_+} |p_\varepsilon(x, z, \tau)|_{\tau = t - \psi_\varepsilon(x,z)/c_0} - \bar{p}(\bar{\eta}(x, z), z, \frac{z}{\varepsilon}, \bar{\tau})|_{\bar{\tau} = t - \bar{\psi}(x,z)/c_0}| \le$$
$$\le C_2(\beta(\varepsilon) + \gamma(\varepsilon) + \delta(\varepsilon) + \mu(\varepsilon)), \tag{20}$$

where C_1, C_2 are positive constants independent of ε; β, γ, δ, μ are the functions of the assumption (iv), tending to zero as $\varepsilon \to 0$. The functions

$\bar{\psi} = \bar{\psi}(x,z)$; $\bar{p} = \bar{p}(\bar{\eta}, z, \xi, \tau)$, $\bar{\tau} = t - \bar{\psi}(x,z)/c_0$ and $\bar{\eta} = \bar{\eta}(x,z)$ do not depend on ε and are defined by the relations (9), (10), (13) , (15) - (18) . In the case of periodic in ξ coefficients the estimates (19) , (20) have right hand sides $C_1\varepsilon$ and $C_2\varepsilon$ respectively.

The main steps of proofs of lemmas 1-4 and of the theorem are given in sections 2-4.

Remark 2 .*The algorithm of construction of an asymptotic solution of the problem (1) - (5) can be generalized for the piecewise - constant (in ξ) coefficients with the interface conditons of continuity of the solution $(\psi_\varepsilon, p_\varepsilon, \tau)$ of the problem (1) - (5) . In the case of piecewise - smooth coefficients there exist such data η, p and E that the interval of existance of the solution $[0, z_4]$ is of order of ε and therefore the solution "exploses" very fastly in z. However the physical sense of equations (1) - (5) is not clear because the model (1) - (5) was never used for description of the process of reflection of waves from the interface of different media.*

Remark 3 .*One of the main questions of homogenization theory is the question of existence of the "effective medium" described by a homogenized problem independent of rapid variable $\xi = \frac{z}{\varepsilon}$. The solution of this "effective medium" described by a homogenized problem should be close to the solution of the original problem describing the heterogeneous (real) medium. In the present article the asymptotic solution was constructed and it essentially depends on $\frac{z}{\varepsilon}$. More precisely the function $\bar{p}(\bar{\eta}, z, \frac{z}{\varepsilon}, \bar{\tau})$ depends on $\frac{z}{\varepsilon}$. Thus in this sens the solution of the initial problem (1) - (5) cannot be approximated by any "effective medium" . Nevertheless the eikonal equation (1) can be homogenized (separately): equation (11) describes the "effective medium" from the point of view of the propagation of a sound beam. It can be presented in a form*

$$(\nabla\bar{\psi})^2 = \hat{n}^2,$$

where $\hat{n}^2 =< \sqrt{n^2 - (\frac{\partial\bar{\psi}}{\partial x})^2} >^2 + (\frac{\partial\bar{\psi}}{\partial x})^2$, i. e. the refraction coefficient of the "effective" medium depends on angle of a beam in each point. It is a new quality of "effective medium" with respect to homogeneous one.

Remark 4.*Linear and quasi-linear partial differential equations of first order were homogenized in [5,6].*

The main results of the present article were formulated in [7].

2. Eikonal Equation.

Consider the problem (1), (4) with the unknown function ψ_ε and the problem (11) , (12) with the unknown function $\bar{\psi}$ Our aim is to prove the existence theorems for the solutions of these equations and to prove the estimate (19) for the difference $\psi_\varepsilon - \bar{\psi}$.

1. Proof of existence of the solution of the problem (1),(4).

As announced we consider the solution of (1) , (4) with a positive derivative $\frac{\partial\psi_\varepsilon}{\partial z}$, i.e (4), (6) . Deriving formally (4) and (6) in x and changing

$$a_\varepsilon = \frac{\partial \psi_\varepsilon}{\partial x}$$

we obtain the quasi linear equation for a_ε

$$\frac{\partial a_\varepsilon}{\partial z} = -\frac{a_\varepsilon}{\sqrt{n^2(z, \frac{z}{\varepsilon}) - a_\varepsilon^2}} \frac{\partial a_\varepsilon}{\partial x}, \; z \in (0, z_0), \; x \in I\!R, \tag{21}$$

$$a_\varepsilon|_{z=0} = a_0(x), \tag{22}$$

where $a_0(x) = \psi'(x)$.

This equation can be deduced by method of characteristics to the functional equation

$$a_\varepsilon = a_0(x - a_\varepsilon \int_0^z \frac{dz'}{\sqrt{n^2(z', \frac{z'}{\varepsilon}) - a_\varepsilon^2(x, z)}}), \tag{23}$$

Consider now this equation. It can be derived from (7) that the operator (23) is contraction operator in the ball $\{|a_\varepsilon| \leq \bar{a}_0\}$ with the contraction rate q.

Therefore the solution in this ball exists and it is unique . It is differentiable by the theorem on implicit function. Differentiating (23) in x and in z and expressing $\frac{\partial a_\varepsilon}{\partial x}$ and $\frac{\partial a_\varepsilon}{\partial z}$ we obtain the relation (21). The relation (22) can be verified directly.

These relations (21) , (22) imply

$$a_\varepsilon = \int_0^z \frac{\partial}{\partial x} \sqrt{n^2(z', \frac{z'}{\varepsilon}) - a_\varepsilon^2(x, z')dz'} + a_0(x). \tag{24}$$

From the other hand the function ψ_ε defined by the relation

$$\psi_\varepsilon(x, z) = \psi_0(x) + \int_0^z \sqrt{n^2(z', \frac{z'}{\varepsilon}) - a_\varepsilon^2(x, z)dz'}, \tag{25}$$

is differentiable and $\frac{\partial \psi_\varepsilon}{\partial x}$ satisfies the same equation (24) :

$$\frac{\partial \psi_\varepsilon}{\partial x} = \int_0^z \frac{\partial}{\partial x} \sqrt{n^2(z', \frac{z'}{\varepsilon}) - a_\varepsilon^2(x, z')dz'} + a_0(x) \tag{26}$$

Therefore $a_\varepsilon = \frac{\partial \psi_\varepsilon}{\partial x}$.

Differentiating (25) in z and substituting $a_\varepsilon = \frac{\partial \psi_\varepsilon}{\partial x}$ we obtain (6).

Thus the existence of the solution of the problem (4) , (6) is proved for $z \in [0, z_4]$.

Since $a_\varepsilon \in C^1(I\!R \times [0, z_4])$, $a_\varepsilon = \frac{\partial \psi_\varepsilon}{\partial x}$ and $\frac{\partial \psi_\varepsilon}{\partial z} = \sqrt{n^2(z, \frac{z}{\varepsilon}) - a_\varepsilon^2(x, z)}$ we obtain that $\psi_\varepsilon \in C^2(I\!R \times [0, z_4])$.

2 . Proof of the lemma 2.

In a same way we obtain the existence and uniqueness of the solution b of the equation (9), its derivability and the relations (11) , (12) for the function $\bar{\psi}$ defined by (10) , i.e lemma 2 is proved analogously.

3. Proof of the estimate (19).

1) First we shall prove that the difference of the solutions of the problems (9) and (23) satisfies the estimate

$$\sup_{x\in I\!R, z\in[0,z_4]} |a_\varepsilon(x,z)-b(x,z)| \le C_3\beta(\varepsilon) \tag{27}$$

where C_3 is a constant independent of ε. Indeed it follows from (9) and (23) that

$$|a_\varepsilon(x,z)-b(x,z)| \le$$

$$\le \overline{a}_1 |a_\varepsilon(x,z)\int_0^z \frac{dz'}{\sqrt{n^2(z',\frac{z'}{\varepsilon})-a_\varepsilon^2(x,z)}} - b(x,z)\int_0^z \langle \frac{1}{\sqrt{n^2(z',.)-b^2(x,z)}}\rangle dz'| \le$$

$$\le \overline{a}_1\{|a_\varepsilon(x,z)\int_0^z \frac{dz'}{\sqrt{n^2(z',\frac{z'}{\varepsilon})-a_\varepsilon^2(x,z)}} - b(x,z)\int_0^z \frac{dz'}{\sqrt{n^2(z',\frac{z'}{\varepsilon})-b^2(x,z)}}|+$$

$$+|b(x,z)\{\int_0^z \frac{dz'}{\sqrt{n^2(z'\frac{z'}{\varepsilon})-b^2(x,z)}} - \int_0^z \langle \frac{1}{\sqrt{n^2(z',.)-b^2(x,z)}}\rangle dz'\}|.$$

Applying the finite difference formula we estimate the first modulus by

$$\overline{s}|a_\varepsilon(x,z)-b(x,z)|,$$

with

$$\overline{s} = \sup_{|u|\le\overline{a}_0,\ x\in I\!R,\ 0\le z\le z_4} |\frac{d}{du}\int_0^z \frac{u\,dz'}{\sqrt{n^2(z',\frac{z'}{\varepsilon})-u^2}}|$$

and the second modulus we estimate by $\sqrt{\overline{a}_0}\beta(\varepsilon)$. Taking in consideration the relation $\overline{a}_1\overline{s}\le q$ we obtain the final estimate

$$|a_\varepsilon(x,z)-b(x,z)| \le \frac{\overline{a}_1\sqrt{\overline{a}_0}\beta(\varepsilon)}{1-q}.$$

2) In a same way we estimate

$$|\psi_\varepsilon(x,z)-\overline{\psi}(x,z)| \le$$

$$\le |\int_0^z (\sqrt{n^2(z',\frac{z'}{\varepsilon})-a_\varepsilon^2(x,z')} - \sqrt{n^2(z',\frac{z'}{\varepsilon})-b^2(x,z')})dz'|+$$

$$+\left|\int_0^z \sqrt{n^2(z',\frac{z'}{\varepsilon})-b^2(x,z')}-\langle\sqrt{n^2(z',\xi)-b^2(x,z')}\,\rangle dz'\right|\leq$$

$$\leq C_4(\beta(\varepsilon)+\gamma(\varepsilon)),$$

where C_4 does not depend on ε . Thus the estimate (19) is proved .

If $n(z,\xi)$ is periodic in ξ then we can obtain the estimate :

$$|\psi_\varepsilon(x,z)-\overline{\psi}(x,z)|\leq C_5\varepsilon,$$

where C_5 is a constant.

3. Parametrization of the eikonal equation along the beam.

Introduce new space variables "along the beam" $\eta_\varepsilon = \eta_\varepsilon(x,z)$ and $\overline{\eta} = \overline{\eta}(x,z)$, associated respectively to the eikonal problem (4) , (6) and to the homogenized eikonal problem (11), (12)

$$x=\eta_\varepsilon+a_0(\eta_\varepsilon)\int_0^z \frac{1}{\sqrt{n^2(z',\frac{z'}{\varepsilon})-a_0^2(\eta_\varepsilon)}}dz', \tag{28}$$

$$x=\overline{\eta}+a_0(\overline{\eta})\int_0^z \langle\frac{1}{\sqrt{n^2(z',.)-a_0^2(\overline{\eta})}}\rangle\, dz'. \tag{29}$$

Lemma 7. *For all $z\in[0,z_4]$, $x\in I\!R$ there exist a unique solution η_ε of (28) and a unique solution $\overline{\eta}$ of (29) such that $|x-\eta_\varepsilon|, |x-\overline{\eta}|\leq \overline{a}_0 z_4\frac{1}{\sqrt{\underline{n}^2-\overline{a}_0^2}}$, $\eta_\varepsilon,\overline{\eta}$ are differentiable ;*

$$\psi_\varepsilon(x(z,\eta_\varepsilon),z)=\psi_0(\eta_\varepsilon)+\int_0^z \frac{n^2(z',\frac{z'}{\varepsilon})}{\sqrt{n^2(z',\frac{z'}{\varepsilon})-a_0^2(\eta_\varepsilon)}}\, dz', \tag{30}$$

$$\overline{\psi}(x(z,\overline{\eta}),z)=\psi_0(\overline{\eta})+\int_0^z \langle\frac{n^2(z',.)}{\sqrt{n^2(z',.)-a_0^2(\overline{\eta})}}\rangle dz', \tag{31}$$

where ψ_ε and $\overline{\psi}$ are defined by (25) and by (10) respectively.

Proof.

The existence and uniqueness of the solutions of the equations (28) and (29) follows from the fixed point theorem. Indeed, consider the equation (28). Changing the unknown function η_ε by $\mu_\varepsilon=\eta_\varepsilon-x$ we obtain the equation

$$\mu_\varepsilon=\Phi(\mu_\varepsilon,x,z)$$

with the contraction operator Φ :

$$\Phi(\mu_\varepsilon, x, z) = -a_0(\mu_\varepsilon + x) \int_0^z \frac{dz'}{\sqrt{n^2(z', \frac{z'}{\varepsilon}) - a_0^2(\mu_\varepsilon + x))}},$$

$$|\Phi| \leq \overline{a}_0 z_4 \frac{1}{\sqrt{\underline{n}^2 - \overline{a}_0^2}}$$

Thus the solution μ_ε exists and it is inique in the ball $|\mu_\varepsilon| \leq \overline{a}_0 z_4 \frac{1}{\sqrt{\underline{n}^2 - \overline{a}_0^2}}$.

In the same way the existance and uniqueness are proved for $\overline{\eta}$.

Derivability of η_ε follows form the fact that $|\frac{\partial \Phi}{\partial \mu_\varepsilon}| \leq q < 1$ for all $x \in I\!R$ and $z \in [0, z_4]$ and from the derivability of Φ in x and z) . For the proof of derivability of $\overline{\eta}$ we use the condition of formal differentiability in average of the function $< \frac{1}{\sqrt{n^2(z,.) - a^2}} >$ with respect to z and a .

Changing the variables x, z according to (28), (29) in the equations (23) , (9) we check that the functions $a_0(\eta_\varepsilon)$ and $a_0(\overline{\eta})$ are their unique solutions , i.e.

$$a_\varepsilon(x(\eta_\varepsilon, z), z) = a_0(\eta_\varepsilon), \quad b(x(\overline{\eta}, z), z) = a_0(\overline{\eta}). \tag{32}$$

The relations (30) and (31) are checked by differentiation in z , where the equalities

$$\frac{\partial \psi_\varepsilon}{\partial x}(x, z)|_{x=x(\eta_\varepsilon, z)} = a_0(\eta_\varepsilon), \quad \frac{\partial \overline{\psi}}{\partial x}(x, z)|_{x=x(\overline{\eta}, z)} = a_0(\overline{\eta}) \tag{33}$$

are taken into account (as well as the expressions of $\frac{\partial \eta_\varepsilon(x,z)}{\partial z}$ and $\frac{\partial \overline{\eta}(x,z)}{\partial z}$ obtained from (28) , (29) by differentiation in z) .

Thus the solutions of the problems (23) and (9) can be parametrized in a form (28) , (30) and (29) , (31) respectively .

Lemma 5 is proved . Lemma 3 is its past. Analogously to the proof of (19) we obtain:

Lemma 6. *The estimate holds true :*

$$\sup_{x \in I\!R, z \in [0, z_4]} |\eta_\varepsilon(x, z) - \overline{\eta}(x, z)| \leq C_6 \beta(\varepsilon),$$

where C_6 is a constant independent of the small parameter .

4. Non linear transfer equation.

Consider equation (2) . Introducing the space variable "along the beam" $\eta_\varepsilon(x, z)$ we calculate :

$$\frac{\partial \psi_\varepsilon}{\partial x} = a_0(\eta_\varepsilon), \quad \frac{\partial^2 \psi_\varepsilon}{\partial x^2} = a_0'(\eta_\varepsilon) R_\varepsilon(\eta_\varepsilon, z),$$

where

$$R_\varepsilon(\eta_\varepsilon, z) = (1 + a_0'(\eta_\varepsilon) \int_0^z \frac{n^2(z', \frac{z'}{\varepsilon}) dz'}{(n^2(z', \frac{z'}{\varepsilon}) - a_0^2(\eta_\varepsilon))^{3/2}})^{-1};$$

In the same way we express $\frac{\partial \psi_\varepsilon}{\partial z}$ and $\Delta \psi_\varepsilon$ and represent the equation (2) in a form

$$\frac{\partial \hat{p}_\varepsilon}{\partial z} + \frac{\hat{p}_\varepsilon}{2} S_\varepsilon(\eta_\varepsilon, z) = \frac{E(z, \frac{z}{\varepsilon}) \hat{p}_\varepsilon \frac{\partial \hat{p}_\varepsilon}{\partial \tau}}{\sqrt{n^2(z, \frac{z}{\varepsilon}) - a_0^2(\eta_\varepsilon)}}, \tag{35}$$

where

$$S_\varepsilon(\eta_\varepsilon, z) = \frac{\partial}{\partial z} \, ln \, \{ \frac{\sqrt{n^2(z, \frac{z}{\varepsilon}) - a_0^2(\eta_\varepsilon)}}{\rho(z, \frac{z}{\varepsilon})} \} +$$

$$+ \frac{\partial}{\partial z} \int_0^z \frac{n^2(z', \frac{z'}{\varepsilon})}{(n^2(z', \frac{z'}{\varepsilon}) - a_0^2(\eta_\varepsilon))^{3/2}} a_0'(\eta_\varepsilon) R_\varepsilon(\eta_\varepsilon, z') \, dz'; \tag{36}$$

here z and η_ε are independent variables.

The unknown function $\hat{p}_\varepsilon(\eta_\varepsilon, z, \tau) = p_\varepsilon(x(\eta_\varepsilon, z), z, \tau)$ satisfies the initial condition

$$\hat{p}_\varepsilon|_{z=0} = p_0(\eta_\varepsilon, \tau), \tag{37}$$

because for $z = 0$ $x(\eta_\varepsilon, 0) = \eta_\varepsilon$.

Applying the method of characteristics we reduce (36) , (37) to a functional equation

$$\hat{p}_\varepsilon = \Phi_\varepsilon(\eta, z, \tau + \hat{p}_\varepsilon \varphi_\varepsilon(\eta_\varepsilon, z)), \tag{38}$$

where

$$\Phi_\varepsilon(\eta_\varepsilon, z, \tau) = \tilde{S}_\varepsilon(\eta_\varepsilon, z) p_0(\eta_\varepsilon, \tau),$$

$$\tilde{S}_\varepsilon(\eta_\varepsilon, z) = e^{-\frac{1}{2} \int_0^z S_\varepsilon(\eta_\varepsilon, z') dz'},$$

$$\varphi_\varepsilon(\eta_\varepsilon, z) = \tilde{S}_\varepsilon^{-1}(\eta_\varepsilon, z) \int_0^z \frac{E(z', \frac{z'}{\varepsilon})}{\sqrt{n^2(z', \frac{z'}{\varepsilon}) - a_0^2(\eta_\varepsilon)}} \tilde{S}_\varepsilon(\eta_\varepsilon, z') dz'.$$

Lemma 7. *For all* $z \in [0, z_4]$, $x \in I\!R$, $\tau \in I\!R$ *there exist a unique solution* $\hat{p}_\varepsilon$ *of the equation (45) , such that* $|\hat{p}_\varepsilon| \leq p_s$ *(from lemma 6),* $\hat{p}_\varepsilon$ *is differentiable in* η_ε *and* τ *and these partial derivatives are uniformly bounded.*

Proof.

The existence and uniqueness of the soultion of the equation (45) follows from the fixed point theorem with a contraction rate q :

$$|\frac{\partial \Phi_\varepsilon}{\partial \hat{p}_\varepsilon}| = |\frac{\partial p_0}{\partial \tau}(\eta_\varepsilon, \tau + \hat{p}_\varepsilon \varphi_\varepsilon(\eta_\varepsilon, z)) \int_0^z \frac{E(z', \frac{z'}{\varepsilon}) S_\varepsilon(\eta_\varepsilon, z')}{\sqrt{n^2(z', \frac{z'}{\varepsilon}) - a_0^2(\eta_\varepsilon)}} dz'| \leq$$

$$\leq \bar{p}_1 z_4 \frac{\overline{E}\sqrt{\overline{\rho}}}{(\underline{n}^2 - \overline{a}_0^2)^{3/4}} exp\{\frac{z_4 \overline{n}^2 \overline{a}_1}{2(\underline{n}^2 - \overline{a}_0^2)^{3/2}(1-q)}\} \leq q.$$

On the other hand $|\Phi_\varepsilon| \leq p_s$.

The differentiability of the solution follows from the theorem on differentiability of implicit function.

The partial derivatives of $\Phi_\varepsilon(\eta_\varepsilon, z, \tau + \hat{p}_\varepsilon \varphi_\varepsilon(\eta_\varepsilon, z))$ in η_ε and τ are uniformly bounded by a constant C_7 independent of $\eta_\varepsilon, z, \tau, \hat{p}_\varepsilon, \varepsilon$. Therefore $\frac{\partial \hat{p}_\varepsilon}{\partial \eta_\varepsilon}$ and $\frac{\partial \hat{p}_\varepsilon}{\partial \tau}$ are also uniformly bounded.

The following lemma can be derived by simple differentiating of (38).

Lemma 8. *The solution $\hat{p}_\varepsilon$ of the problem (35) ($|\hat{p}_\varepsilon| \leq p_s$) is the solution of the problem (36) , (37).*

Corollaly 1. *The function $\hat{p}_\varepsilon(\eta_\varepsilon(x, z), z, \tau)$ satisfies the relations (2) , (5).*

Thus we obtain the proposition of lemma 1. This proposition follows from (35) - (42).

The proof of Lemma 4 is the same as the proof of Lemma 7.

Now we use the assumptions of good mixture to obtain some estimatesand complete the proof of the theorem. Let us mention the main steps of this reasoning.

1. The following estimates hold true :

for all $\eta \in I\!R$, $z \in [0, z_4]$

$$|R_\varepsilon(\eta, z) - \overline{R}(\eta, z)| \leq C_8 \, \delta(\varepsilon),$$

$$|\tilde{S}_\varepsilon(\eta, z) - \overline{S}(\eta, z, \frac{z}{\varepsilon})| \leq C_9 \, \delta(\varepsilon),$$

$$|\tilde{S}_\varepsilon^{-1}(\eta, z) - \overline{S}^{-1}(\eta, z, \frac{z}{\varepsilon})| \leq C_{10} \, \delta(\varepsilon),$$

$$|\varphi_\varepsilon(\eta, z) - \overline{\varphi}(\eta, z, \frac{z}{\varepsilon})| \leq C_{11} \, (\delta(\varepsilon) + \mu(\varepsilon)),$$

where

$$\overline{\varphi}(\eta, z, \xi) = \overline{S}^{-1}(\eta, z, \xi) \int_0^z \langle \frac{E(z', .)\sqrt{\rho(z', .)}}{n^2(z', .) - a_0^2(\eta))^{3/4}} \rangle \overline{\sigma}(\eta, z') dz'.$$

2. For all $\eta \in I\!R,\ z \in [0, z_4],\ \tau \in I\!R,\ p \in \{|p| \leq p_s\}$ the estimate holds true

$$|\Phi_\varepsilon(\eta, z, \tau + p\varphi_\varepsilon(\eta, z)) - \overline{\Phi}(\eta, z, \frac{z}{\varepsilon}, \tau + p\overline{\varphi}(\eta, z, \frac{z}{\varepsilon}))| \leq$$

$$\leq C_{12}\ (\delta(\varepsilon) + \mu(\varepsilon)),$$

where $\overline{\Phi}(\eta, z, \xi, \tau) = \overline{S}(\eta, z, \xi)p_0(\eta, \tau)$.

3. Let $\hat{p}_\varepsilon(\eta, z, \tau)$ be the solution of equation

$$\hat{p}_\varepsilon = \Phi_\varepsilon(\eta, z, \tau + \hat{p}_\varepsilon\varphi_\varepsilon(\eta, z))$$

and $\overline{p}(\eta, z, \xi, \tau)$ be the solution of equation

$$\overline{p} = \overline{\Phi}(\eta, z, \xi, \tau + \overline{p}\overline{\psi}(\eta, z, \xi)).$$

Then for all $\eta \in I\!R,\ z \in [0, z_4],\ \tau \in I\!R$

$$|\hat{p}_\varepsilon(\eta, z, \tau) - \overline{p}(\eta, z, \frac{z}{\varepsilon}, \tau)| \leq C_{13}\ (\delta(\varepsilon) + \mu(\varepsilon)),$$

and for all $x \in I\!R,\ z \in [0, z_4],\ \tau \in I\!R$

$$|\hat{p}_\varepsilon(\eta_\varepsilon(x, z); z, \tau) - \overline{p}(\overline{\eta}(x, z), z, \frac{z}{\varepsilon}, \tau)| \leq C_{14}\ (\delta(\varepsilon) + \mu(\varepsilon) + \beta(\varepsilon)).$$

Here the constants $C_8, C_9, C_{10}, C_{11}, C_{12}, C_{13}, C_{14}$ do not depend on ε.

4. For all $x \in I\!R,\ z \in [0, z_4],\ t \in I\!R_+$

$$|p_\varepsilon(x, z, \tau)|_{\tau = t - \overline{\psi}(x,z)/c_0} - \overline{p}(\overline{\eta}(x, z), z, \frac{z}{\varepsilon}, \overline{\tau})|_{\overline{\tau} = t - \overline{\psi}(x,z)/c_0}| \leq$$

$$\leq C_2\ (\beta(\varepsilon) + \gamma(\varepsilon) + \delta(\varepsilon) + \mu(\varepsilon)),$$

where C_2 is the constant of the theorem.

The proof of these steps is based on the same ideas as the proof of the estimate (19).

References

[1] O.V. RUDENKO , A.K. SUKHORUKOVA and A.P. SUKHORUKOV, Equations of high frequency nonlinear acoustics of heterogeneous media *Acoustic Journal* , 40, No 2 , 1994 (in Russian)

[2] N.S. BAKHVALOV and G.P PANASENKO, *Homogenization : Averaging Processes in Periodic Media* Kluwer Ac. Publ., Dordrecht/London/ Boston, 1989.

[3] A.L. PIATNITSKY, Refraction problem for a stratified medium, *Math. USSR Sbornik,* 115, No 3, 1981.

[4] V. BERDICHEVSKY and V. SUTYRIN, The dynamics of periodic structures , *Soviet Phys. Doklady,* 28 , No 3, 1983, pp. 239-241.

[5] Y. AMIRAT, K. HAMDACHE, A.ZIANI , Homogénisation d'un modèle d'écoulements miscibles en milieu poreux, *Asymptotic Analysis,* 3, 1990, pp. 77-89.

[6] A. BOURGEAT, A. MIKELIC , Homogenization of two-phase immiscible flows in a one-dimensional porous medium, *Asymptotic Analysis,* 9, 1994, pp. 359-380.

[7] E.A.LAPSHIN, G.P.PANASENKO, Homogenization of the equations of high frequency nonlinear acoustics, *C.R.Acad.Sci.Paris,* 325, serie 1, 1997, pp. 931-936.

E.LAPSHIN: Math. - Mech. Department,
Moscow State University,
Vorobievy Gory, 119899, Moscow, Russia.
G.PANASENKO: CNRS UMR 5585,
Equipe d'Analyse Numérique,
Université Jean Monnet, 23, rue P. Michelon,
42023 Saint-Etienne Cedex, France.
The work was done during the stay of E.Lapshin
in the University Jean Monnet.

IDENTIFICATION OF A REFLECTION BOUNDARY COEFFICIENT IN AN ACOUSTIC WAVE EQUATION BY OPTIMAL CONTROL TECHNIQUES

SUZANNE LENHART[1,2]
VLADIMIR PROTOPOPESCU[2]
JIONGMIN YONG[3]

ABSTRACT: We apply optimal control techniques to find approximate solutions to an inverse problem for the acoustic wave equation. The inverse problem (assumed here to have a solution) is to determine the boundary refection coefficient from partial measurements of the acoustic signal. The sought reflection coefficient is treated as a control and the goal - quantified by an objective functional - is to drive the model solution close to the experimental data by adjusting this coefficient. The problem is solved by finding the optimal control that minimizes the objective functional. Then by driving the "cost of the control" to zero one proves that the sequence of optimal controls represents a converging sequence of estimates for the solution of the inverse problem. Compared to classical regularization methods (e.g. Tikhonov coupled with optimization schemes), our approach yields: (i) a systematic procedure to solve inverse problems of identification type and (ii) an explicit expression for the approximations of the solution.

1. Introduction

Over the last two decades, parameter identification, i.e. reconstruction of model properties from observed data has become one of the most active and work intensive areas of applied mathematics. Traditionally, identification techniques have been associated to model (retro)fitting and validation, reverse engineering, and signal detection and interpretation. With the advent of high power computers, new standards of performance have become the

[1] University of Tennessee, Mathematics Department, Knoxville, TN 37996-1300
[2] Oak Ridge National Laboratory, Computer Science and Mathematics Division, Oak Ridge, TN 37831-6364
[3] Fudan University, Department of Mathematics, Shanghai, 200433 China

R.P. Gilbert et al.(eds.), Direct and Inverse Problems of Mathematical Physics, 251–266.

required norm in modern fields such as communication systems, computer networks, astrophysics, bioengineering, or sophisticated military command and control systems. These standards, especially in domains where on-line responses are crucial, make it necessary to have a much better understanding of the quantitative models involved and warrant the continuing interest in the development of new robust and rigorous identification methods.

From a formal mathematical standpoint, the parameter identification problem is an inverse problem that consists of two separate albeit related subproblems, namely: (a) the identifiability problem and (b) the parameter estimation problem. To understand more precisely the nature of these inverse problems we first describe briefly the direct problem. In the direct problem, a physical system is described (modeled) by the state function, u, which satisfies the abstract operator state equation with data F:

$$A(u(\xi); F(\xi)) = 0. \tag{1.1}$$

The state of the system and the data may be scalar or vector. The independent variables of the problem, ξ - which for evolution problems contain the time, t - take values in the domain $\Omega \subset \mathbb{R}^n$, with sufficiently regular boundary, $\partial\Omega$. The operator A may be rather general, including nonlinear equations of higher order, ordinary differential equations, hybrid systems, etc.

If the system (1.1) accurately describes a realistic physical situation, the direct problem is well-posed, i.e. the data F (parameters, functions, coefficients, sources, initial and boundary values, etc.) determine uniquely a regular solution u [4,14]. The inverse problem is usually ill-posed in the sense that existence, uniqueness, and/or regularity cannot be expected in general [1,5,8].

The identifiability problem consists in studying the well-posedness of the parameter identification problem. Identifiability is loosely defined as the injectivity of the mapping between the sought parameter and ther output within the used model that is supposed to be completely and accurately known. The identifiability problem consists in determining whether one can uniquely recover (a part of) F from the model solution, u. The parameter estimation problem consists in finding an estimated value of the unknown parameter from the data within the admissible set that is consistent with identifiability. In the following, we shall assume that the parameters are uniquely identifiable. The remaining problem is then related to regularity, explicit characterization, and approximation of the parameters.

More precisely, we shall tackle the following identification problem corresponding to the direct problem (1.1): "Given partial observations, $B\tilde{u}$, of the true solution $\tilde{u}$, in a subdomain Ω' of the phase space, $\Omega' \subset \Omega$, and a known part, ϕ of the data F, we seek to determine the unknown part of the data, f."

The standard methods of solving parameter estimation problems are based on Tikhonov's regularization. In this approach, one constructs - starting from actual observations, $B\tilde{u}$, - a cost functional:

$$J_\beta(f) = \frac{1}{2}\|Bu - B\tilde{u}\|^2 + \frac{\beta}{2}\|f\|^2, \quad \beta > 0 \tag{1.2}$$

where u is the solution of (1.1) for the data $F = (\phi, f)$. The exact form of the cost functional and the types of norms involved depend on the concrete problem that has to be solved. Tikhonov's approach [28] and its variants seek to minimize the functional $J_\beta(f)$ - for a fixed β - over the set of unknown data. It assures - in principle - that the observation Bu of the model solution, $u(f)$, obtained with parameter f, will approach the actual observation, $B\tilde{u}$. The functions f_β which achieve the absolute minimization of $J_\beta(f)$ represent the approximate solution of the parameter identification problem. The disadvantages of this approach are:

- when β is very small, the problem is unstable;
- when β is very large, the solution is not accurate;
- there is no systematic procedure for finding the absolute minimum;
- there are no systematic means to evaluate the approximations.

To eliminate most of the disadvantages above, we propose a different approach to the inverse problem of identification, based on optimal control for operator equations as developed by J.-L. Lions [17 20].

Our idea is to consider a family of functionals (1.2) for $\beta \geq 0$. For each β strictly positive one considers the unknown data, f, as a control which belongs to a certain bounded set, U; the control has to be adjusted in order to minimize the functional $J_\beta(f)$. The minimum of the cost functional over f is attained at the optimal control, $f = f_\beta$:

$$J_\beta(f_\beta) = \inf_{f \in U} J_\beta(f).$$

Letting the sequence of β tend toward zero, one can verify that the sequence f_β converges in an appropriate sense to an element of the control set, $f^* \in U$. This element represents the (unique) solution of the parameter identification problem.

In this paper, we shall specifically apply the formalism to a hyperbolic (acoustic wave) equation. The paper is structured as follows. In Section 3, we shall briefly review the existence of a weak solution to the state problem stated in Section 2, and of an optimal control (for a more detailed analysis see Ref 12). In Section 4, we derive the necessary conditions that an optimal control and its corresponding state must satisfy. Uniqueness of the optimal control for sufficiently short times is proved in Section 5. Section 6 illustrates the application of this optimal control problem to determining an estimate of the reflection coefficient from partial measurements of the solution.

2. Statement of the Problem

Given a bounded domain $D \subset \mathbb{R}^2$ with C^1 boundary, define the spatial domain

$$\Omega = \{(x, y, z) | (x, y) \in D, \ u(x, y) < z < 0\},$$

where $u : D \to (-\infty, 0)$ is a C^2 function. Assume the region Ω contains a certain medium (like water in a section of the ocean) with a known velocity tensor E. Let $K > 0$ be a finite constant and define the control set

$$U_K = \{\sigma \in L^\infty(D) | 0 \leq \sigma(x, y) \leq K\}.$$

Given a control $\sigma \in U_K$, we consider the solution $w = w(\sigma)$ of the acoustic wave equation:

$$
\begin{aligned}
(2.1) \qquad & w_{tt} - \nabla(E\nabla w) = f \quad \text{in } Q \\
& w = 0 \quad \text{on } \Sigma \times (0,T), \quad \text{sides of spatial domain} \\
& \frac{\partial w}{\partial \nu} = 0 \quad \text{on } D \times \{z = 0\} \times (0,T), \quad \text{top of spatial domain} \\
& \frac{\partial w}{\partial \nu} + \sigma w = 0 \quad \text{on } \Gamma \times (0,T), \quad \text{bottom of spatial domain} \\
& w = g_1, \; w_t = g_2 \quad \text{on } \Omega \times \{0\},
\end{aligned}
$$

where

$$
\begin{aligned}
\Gamma &= \{(x, y, u(x,y)) | (x,y) \in D\} \\
\Sigma &= \{(x,y,z) | (x,y) \in \partial D, \; u(x,y) < z < 0\} \\
Q &= \Omega \times (0,T)
\end{aligned}
$$

and

$$
\frac{\partial w}{\partial \nu} = E\nabla w \cdot \eta \quad \text{with } \eta, \text{ the outward unit normal vector at the boundary.}
$$

The objective functional $J_\beta(\sigma)$ is defined by

$$
(2.2) \qquad J_\beta(\sigma) = \frac{1}{2}\int_{G\times(0,T)} (w-h)^2 dx\,dy\,dz\,dt + \frac{\beta}{2}\int_\Gamma (\sigma(x,y))^2 ds
$$

where $G \subset \Omega$ with positive measure. The first term in $J_\beta(\sigma)$ drives w close to the target h on $G \times (0,T)$ and the second term is the cost of the control. We seek to characterize σ_β such that

$$
J_\beta(\sigma_\beta) = \inf_{\sigma \in U_K} J_\beta(\sigma).
$$

After this characterization has been completed, we let the parameter β go to zero to approximate a solution to an inverse problem. The inverse problem is to identify σ from observations h of a solution w on $G \times (0,T)$, resulting from a signal source f. Assuming that this inverse problem has a unique solution, then for β small, the optimal control determined from J_β, σ_β, will approximate it reasonably well.

This bilinear optimal control problem is new for wave equations. See [3] for a controllability result for wave equation using a velocity damped control term. See [14] for a similar bilinear boundary optimal control problem in the parabolic case.

The approach of using optimal control techniques with adjoint equations to approximate solutions to inverse problems of identification type is different from traditional approaches which couple Tikhonov regularization (with a functional like ours) with an optimization algorithm [1]. Our approach has the advantage of an explicit characterization of the approximation σ_β. In a series of recent papers [21,22,29], Puel and Yamamoto obtain uniqueness, stability, and reconstruction results for the inverse problem of identifying the source in a wave equation. Their work is based on exact controllability results and the Hilbert Uniqueness Method [18,19,20] that apply to *linear* equations only. Moreover, the identification is realized only for special types of observed data.

The other approaches which use control theory for solving inverse problems are limited to particular cases. Indeed, Russell [7, 23-26] developed the control theory for the wave equation, including many controllability and stability results. Our framework is applicable to more general equations. Lasiecka and Triggiani [10,11] further extended control theory for wave equations, in particular developing better trace estimates and the Riccati equation framework for linear-quadratic problems. Our control problem cannot be treated within this scheme since it is bilinear - σw - in the boundary condition. Liang [15] analyzes a simpler bilinear optimal control problem for a wave equation with the control in the state equation itself.

Finally, ideas more closely related to the present approach have been developed by Tikhonov [28], Chavent [4], James and Sepulveda [9] and by the authors [12,13].

Tikhonov introduces the regularization to stabilize an optimal control problem where the objective functional does not depend explicitly on the control [28].

Chavent [4] has investigated the stability of applying optimal control techniques to identification problems and conditions to guarantee the identifiability of the parameters. Barbu and Pavel [2] recently solved an optimal control problem, which approximates the inverse problem of identifying the acoustic impedance function in a one dimensional wave equation.

James and Sepulveda [9] solve the parameter identification problem by treating it as a constrained optimization problem. They formally tackle the latter by the Lagrangian method and solve it numerically without establishing any rigorous result related to its solvability.

The results contained in this paper further the application of our general formalism [13] to the parameter identification problem for the acoustic wave equation [12].

3. Existence of an Optimal Control

To define the solution space for the state problem (2.1), let

$$V = \{v \in H^1(\Omega) | v = 0 \text{ on } \Sigma\}$$

with norm

$$\|v\|_V = \left(\int_\Omega |\nabla v|^2 dx\, dy\, dz\right)^{\frac{1}{2}}.$$

Note that this norm on V is equivalent to the usual H^1 norm due to zero boundary conditions on Σ and Poincaré's inequality. We make the following assumptions:

(3.1) $E \in C(\overline{\Omega}; \mathbf{R}^{3\times 3})$ and $E(x,y,z) \geq \delta I$ for all $(x,y,z) \in \overline{\Omega}$, for some $\delta > 0$

(3.2) $f, f_t \in L^2(Q)$

(3.3) $G \subset \Omega$ with positive Lebesgue measure

(3.4) $h \in L^2(G \times (0,T))$

(3.5) $u \in C^2(D), \quad u(x,y) < 0$

(3.6) $g_1 \in H^2(\Omega) \cap V, \quad g_2 \in H^1(\Omega)$.

We introduce the bilinear form:

$$B[w,v;t] = \int_\Omega E\nabla w \cdot \nabla v(x,y,z,t)\,dx\,dy\,dz + \int_\Gamma \sigma\, w\, v(x,y,u(x,y),t)\,ds$$

for $w, v \in L^2(0,T;V)$, $0 \leq t \leq T$.

Definition: A function $w \in L^2(0,T;V)$ with $w_t \in L^2(0,T;L^2(\Omega))$ and $w_{tt} \in L^2(0,T;V')$ is a weak solution of (2.1) if

(i) $\langle w_{tt}, v\rangle + B[w,v;t] = \int_\Omega f(t)\,v\,dx\,dy\,dz$ for each $v \in V$ and for a.e. $0 \leq t \leq T$,

(ii) at $t = 0$, $w = g_1$, $w_t = g_2$,

where $\langle \cdot, \cdot \rangle$ denotes the V', V duality pairing.

Note that we also have $w \in C([0,T]; L^2(\Omega))$ and $w_t \in C([0,T]; V')$ (see [3]), so (ii) makes sense.

The weak solution to the state problem (2.1) is constructed by applying Galerkin's method of finite dimensional approximations [6]. Unlike alternative methods used to prove existence here (e.g. semigroups), the Galerkin method clearly displays the regularity and the dependence on σ. The detailed derivation of these results is contained in [12].

Theorem 3.1. *For $\sigma \in U_K$, the state problem has a unique weak solution w that satisfies the following regularity estimate:*

$$\sup_{0\leq t\leq T} (\|w(t)\|_V + \|w_t(t)\|_{L^2(\Omega)}) + \|w_{tt}\|_{L^2(0,T;V')} \leq C(\|g_1\|_V + \|g_2\|_{L^2(\Omega)} + \|f\|_{L^2(Q)}).$$

Sketch of the Proof. One considers a sequence of approximations w_n to the solution w

(3.7) $$w_m(t) = w_m(x,y,z,t) = \sum_{k=1}^{m} d_k^m(t)\psi_k, m \in \mathbf{N}$$

$$\{\psi_k\}_{k\in\mathbf{N}} \text{ is a basis for } V$$

and

$$\{\psi_k\}_{k\in\mathbf{N}} \text{ is an orthonormal basis of } L^2(\Omega),$$

where $\{\psi_k\}$ are eigenvectors (corresponding to the eigenvalue λ_k) of the following Sturm-Liouville problem:

$$\begin{aligned} &\frac{\partial\psi}{\partial\nu} + \sigma\psi = 0 \quad \text{on } \Gamma \\ &\psi = 0 \quad \text{on } \Sigma \\ &\frac{\partial\psi}{\partial\nu} = 0 \quad \text{on } D \times \{z = 0\}. \end{aligned}$$

The functions $d_k^m(t)$ satisfy

$$d_k^{m\prime\prime}(t) + \sum_{\ell=1}^{m} B[w_\ell, w_k; t] d_\ell^m(t) = \int_\Omega f(t) w_k \, d\Omega \tag{3.8}$$

and

$$d_k^m(0) = (g_1, \psi_k)_{L^2}, \quad d_k^{m\prime}(0) = (g_2, \psi_k)_{L^2}, \quad k = 1, \ldots, m. \tag{3.9}$$

Using the orthogonality of $\{\psi_k\}$ in $L^2(\Omega)$ and the definition of w_m, one shows that w_m approximately satisfies the definition of the weak solution for $v = \psi_k$ and thus for $v = w_m$. One can then derive a priori estimates for w_m from which, by passing to the limit, one obtains convergences on subsequences. In turn, these convergences are sufficient to show that the limit w of the sequence w_m is the unique weak solution of the equation (2.1). □

Additional regularity on the state solution is needed to derive the necessary conditions that an optimal control and corresponding state must satisfy. In fact, one can prove [12]:

Proposition 3.2. *The weak solution of (2.1) has additional regularity:*

$$\sup_{0\le t\le T} \left[\int_\Omega (w_{tt})^2(x,y,z,t) d\Omega + \int_\Omega |\nabla w_t|^2(x,y,z,t) d\Omega + \int_\Gamma \sigma (w_t)^2(x,y,u(x,y),t) ds \right] \le C$$

with C depending only on $\|f\|_{L^2}$, $\|f_t\|_{L^2}$, $\|g_1\|_{H^2}$ and $\|g_2\|_{H^1}$.

Remarks: 1. We do not get the $L^2(0,T;H^2(\Omega))$ estimate on w since we have a Robin boundary condition on the bottom with only L^∞ regularity on the coefficient σ. See [6] for this estimate for Dirichlet boundary conditions.

2. Note that we also obtain

$$\sup_{0\le t\le T} \int_\Gamma w_t^2(x,y,u(x,y),t) ds \le C$$

from the bound on

$$\int_{\Omega} |\nabla w_t|^2(x,y,z,t)d\Omega,$$

and Poincaré's Inequality, since

$$\begin{aligned}\int_{\Gamma}(w_t)^2(t)ds &\le C\int_{\Omega}|\nabla w_t|^2(t)d\Omega + \int_{\Omega}(w_t)^2(t)d\Omega\\ &\le C_1\int_{\Omega}|\nabla w_t|^2(t)d\Omega\end{aligned}$$

and $w_t = 0$ on $\Sigma \times \{t\}$.

Based on the existence and regularity results contained in Theorem 3.1 we can now prove the existence of an optimal control.

Theorem 3.3: *There exists a control σ_β in U_K that minimizes the functional $J_\beta(\sigma)$ over U_K.*

Proof. Let $\{\sigma^n\}$ be a minimizing sequence, i.e.

$$\lim_{n\to\infty} J_\beta(\sigma^n) = \inf_{\sigma\in U_K} J_\beta(\sigma).$$

Let $w^n = w(\sigma^n)$ be the corresponding state solution. By estimates in Theorem 3.1,

$$\|w^n\|_{L^2(0,T,V)} + \|w_t^n\|_{L^2(0,T;L^2(\Omega))} + \|w_{tt}^n\|_{L^2(0,T,V')} \le C_1$$

where the constant C_1 is independent of n. Then there exists $\sigma_\beta \in U_K$, $w_\beta \in L^2(0,T;V)$ such that, on a subsequence,

$$\begin{aligned}w^n &\rightharpoonup w_\beta \text{ weakly in } L^2(0,T;V)\\ w_{tt}^n &\rightharpoonup (w_\beta)_{tt} \text{ weakly}^* \text{ in } L^2(0,T;V')\end{aligned}$$

and

$$\sigma^n \rightharpoonup \sigma_\beta \text{ weakly on } L^2(\Gamma).$$

Using a compactness result [27],

$$w^n \to w_\beta \text{ strongly in } L^2(0,T,H^{\frac{1}{2}+\epsilon}_{\{0\}}(\Omega))$$

where the subscript $\{0\}$ means zero on ∂D, and $0 < \varepsilon < \frac{1}{2}$. By trace results,

$$w^n \to w_\beta \text{ strongly in } L^2(0,T;\partial\Omega),$$

which is needed for convergence of the term

$$\int_{\Gamma\times(0,T)} \sigma^n w^n \phi ds$$

in the weak form of solution w^n. Thus we can pass to the limit and conclude that $w_\beta = w(\sigma_\beta)$.

The objective functional $J_\beta(\sigma)$ is lower semi-continuous with respect to L^2 weak convergence and hence the minimum of $J_\beta(\sigma)$ is attained by σ_β. □

4. Necessary Conditions

To derive necessary conditions that an optimal pair σ_β, $w_\beta = w(\sigma_\beta)$ must satisfy, we differentiate the map

$$\sigma \rightarrow J_\beta(\sigma)$$

with respect to σ. Since $J_\beta(\sigma)$ also depends on w, we first differentiate the map

$$\sigma \rightarrow w(\sigma).$$

We give the two theorems from [12] deriving the necessary conditions and briefly summarize the proofs.

Theorem 4.1. *The map*

$$\sigma \in U_K \rightarrow w(\sigma) \in L^2(0,T;V)$$

is differentiable in the following sense:

$$\frac{w(\sigma+\varepsilon\ell) - w(\sigma)}{\varepsilon} \rightharpoonup \xi \quad \text{in } L^2(0,T;V)$$

as $\varepsilon \rightarrow 0$, *for* $\sigma + \varepsilon\ell$, $\sigma \in U_K$, $\ell \in L^\infty(D)$. *Furthermore,* ξ *is a weak solution of*

$$\begin{aligned} \xi_{tt} - \nabla(E\nabla\xi) &= 0 \quad \text{in } Q \\ \xi &= 0 \quad \text{on } \Sigma\times(0,T) \\ \frac{\partial\xi}{\partial\nu} &= 0 \quad \text{on } D\times\{z=0\}\times(0,T) \\ \frac{\partial\xi}{\partial\nu} + \sigma\xi &= -\ell w \quad \text{on } \Gamma\times(0,T) \\ \xi = \xi_t &= 0 \quad \text{on } \Omega\times\{0\} \end{aligned} \tag{4.1}$$

where $w = w(\sigma)$.

Proof: Using the notation of $w = w(\sigma)$, $w^\varepsilon = w(\sigma + \varepsilon\ell)$ and noting the added regularity of w and w^ε from Proposition 3.2, we choose $\left(\frac{w^\varepsilon - w}{\varepsilon}\right)_t$ as test function in the weak solution form of the $\frac{w^\varepsilon - w}{\varepsilon}$ equation. On $Q_s = \Omega \times (0,s)$, $\Gamma_s = \Gamma \times (0,s)$, $0 < s \leq T$, we obtain

$$\int_{Q_s} \left(\frac{w^\varepsilon - w}{\varepsilon}\right)_{tt} \left(\frac{w^\varepsilon - w}{\varepsilon}\right)_t d\Omega dt + \int_{Q_s} E\nabla\left(\frac{w^\varepsilon - w}{\varepsilon}\right) \cdot \nabla\left(\left(\frac{w^\varepsilon - w}{\varepsilon}\right)_t\right) d\Omega dt$$
$$+ \int_{\Gamma_s} \sigma\left(\frac{w^\varepsilon - w}{\varepsilon}\right)\left(\frac{w^\varepsilon - w}{\varepsilon}\right)_t d\Gamma dt = -\int_{\Gamma_s} \ell w^\varepsilon \left(\frac{w^\varepsilon - w}{\varepsilon}\right)_t d\Gamma dt.$$

Integrating by parts and using Cauchy-Schwarz one obtains

$$\int_\Gamma \left(\left(\frac{w^\varepsilon - w}{\varepsilon}\right)_t\right)^2 (s) d\Omega + \int_\Omega \left|\nabla\left(\frac{w^\varepsilon - w}{\varepsilon}\right)\right|^2 (s) d\Omega + \int_\Gamma \sigma\left(\frac{w^\varepsilon - w}{\varepsilon}\right)^2 (s) d\Omega$$
$$\leq C\int_{\Gamma_s} (w_t^\varepsilon)^2 d\Gamma dt + C_\delta \int_\Gamma (w^\varepsilon)^2 (s) d\Gamma + \int_{\Gamma_s} \left(\frac{w^\varepsilon - w}{\varepsilon}\right)^2 d\Gamma + \delta \int_\Gamma \left(\frac{w^\varepsilon - w}{\varepsilon}\right)^2 (s) d\Gamma.$$

Using trace estimates and Poincaré's Inequality on the right hand side, we have

$$\int_\Omega \left(\left(\frac{w^\varepsilon - w}{\varepsilon}\right)_t\right)^2 (s) d\Omega + (1 - \delta C)\int_\Omega \left|\nabla\left(\frac{w^\varepsilon - w}{\varepsilon}\right)\right|^2 (s) d\Omega + \int_\Gamma \sigma\left(\frac{w^\varepsilon - w}{\varepsilon}\right)^2 (s) d\Gamma$$
$$\leq C + \int_0^s \int_\Omega \left|\nabla\left(\frac{w^\varepsilon - w}{\varepsilon}\right)\right|^2 (t) d\Omega dt.$$

Finally, choosing δ small so that $1 - \delta C > 0$ and applying Gronwall's Inequality [6] gives

$$\sup_{0 \leq d \leq T} \int_\Omega \left(\left(\frac{w^\varepsilon - w}{\varepsilon}\right)_t\right)^2 (s) d\Omega + \int_\Omega \left|\nabla\left(\frac{w^\varepsilon - w}{\varepsilon}\right)\right|^2 (s) d\Omega + \int_\Gamma \sigma\left(\frac{w^\varepsilon - w}{\varepsilon}\right)^2 (s) d\Gamma \leq C_1.$$

As in the proof of Theorem 3.1, this estimate implies

$$\left\|\frac{w^\varepsilon - w}{\varepsilon}\right\|_{L^2(0,T;V')} \leq C$$

and on a subsequence, as $\varepsilon \to 0$,

$$\frac{w^\varepsilon - w}{\varepsilon} \rightharpoonup \xi \quad \text{in } L^2(0,T;V)$$
$$\left(\frac{w^\varepsilon - w}{\varepsilon}\right)_{tt} \rightharpoonup \xi_{tt} \quad \text{in } L^2(0,T;V')$$
$$\left(\frac{w^\varepsilon - w}{\varepsilon}\right)_t \rightharpoonup \xi_t \quad \text{in } L^2(0,T;L^2(\Omega))$$
$$\frac{w^\varepsilon - w}{\varepsilon} \rightharpoonup \xi \quad \text{in } L^2(0,T;L^2(\Gamma)).$$

Also ξ satisfies (4.1) in the weak sense:

$$\langle \xi_{tt}, v\rangle + B[\xi, v; t] = -\int_\Gamma \ell w v(t)d\Gamma, \quad v \in V, \text{ a.e. } t. \quad \square$$

Next, we state the necessary conditions:

Theorem 4.2. *Given an optimal control σ_β and a corresponding state $w_\beta = w(\sigma_\beta)$, there exists an adjoint solution p in $L^2(0,T;V)$, solving in a weak sense the adjoint system:*

$$\begin{aligned}
p_{tt} - \nabla(E\nabla p) &= (w_\beta - h)\chi_G \quad \text{in } Q \\
p &= 0 \quad \text{on } \Sigma \times (0,T) \\
\frac{\partial p}{\partial \nu} &= 0 \quad \text{on } D \times \{z=0\} \times (0,T) \\
\frac{\partial p}{\partial \nu} + \sigma_\beta p &= 0 \quad \text{on } \Gamma \times (0,T) \\
p = p_t &= 0 \quad \text{on } \Omega \times \{T\}.
\end{aligned} \tag{4.3}$$

Furthermore,

$$\sigma_\beta(x,y) = \min\left\{ \left(\frac{1}{\beta}\int_0^T w_\beta\, p(x,y,u(x,y),t)dt\right)^+, K\right\}.$$

Proof: Since $(w_\beta - h)\chi_G \in L^2(Q)$, we can obtain the existence and uniqueness of the weak solution p by arguments like in Theorem 3.1.

Suppose $\sigma_\beta + \varepsilon\ell \in U_K$ for ε small. Consider the directional derivative of J with respect to σ at σ_β in the direction ℓ:

$$\begin{aligned}
0 &\le \lim_{\varepsilon\to 0+} \frac{J_\beta(\sigma_\beta + \varepsilon\ell) - J_\beta(\sigma_\beta)}{\varepsilon} \\
&= \int_{G\times(0,T)} \xi(w_\beta - h)dGdt + \beta\int_\Gamma \sigma_\beta \ell d\Gamma \\
&= \int_0^T \langle p_{tt}, \xi\rangle dt + \int_0^T B[p,\xi;t]dt + \beta\int_\Gamma \sigma_\beta\, \ell\, d\Gamma \\
&= -\int_{\Gamma\times(0,T)} \ell\, w_\beta\, p\, d\Gamma\, dt + \beta\int_\Gamma \sigma_\beta\, \ell\, d\Gamma \\
&= \int_\Gamma \ell(x,y)\left(-\int_0^T w_\beta p(x,y,u(x,y),t)dt + \beta\sigma_\beta(x,y)\right) d\Gamma
\end{aligned}$$

by using the adjoint equation (4.3) and then using the equation (4.1) satisfied by ξ. Standard control arguments give

$$\sigma_\beta(x,y) = \min\left\{ \left(\frac{1}{\beta}\int_0^T w_\beta p(x,y,u(x,y),t)dt\right)^+, K\right\}. \qquad \square$$

5. Uniqueness of the Optimal Control

Uniqueness was not addressed in [12]; we now prove the following result:

Theorem 5.1. *For T sufficiently small, the solution of the optimality system (OS) is unique:*

$$\begin{aligned}
w_{tt} - \nabla(E\nabla w) &= f && \text{in } Q\\
p_{tt} - \nabla(E\nabla p) &= (w-h)\chi_G && \text{in } Q\\
w = p &= 0 && \text{on } \Sigma\times(0,T)\\
\frac{\partial w}{\partial \nu} = \frac{\partial p}{\partial \nu} &= 0 && \text{on } D\times\{z=0\}\times(0,T)\\
w = g_1,\quad w_t &= g_2 && \text{on } \Omega\times\{0\}\\
p = p_t &= 0 && \text{on } \Omega\times\{T\}\\
\frac{\partial w}{\partial \nu} + \min\left\{\frac{1}{\beta}\left(\int_0^T wp(x,y,u(x,y),t)dt\right)^+, K\right\} w &= 0 && \text{on } \Gamma\times(0,T)\\
\frac{\partial p}{\partial \nu} + \min\left\{\frac{1}{\beta}\left(\int_0^T wp(x,y,u(x,y),t)dt\right)^+, K\right\} p &= 0 && \text{on } \Gamma\times(0,T).
\end{aligned}$$

Proof. Suppose w, p and $\bar w, \bar p$ are solutions of (OS) system. Let $\hat w = w - \bar w$, $\hat p = p - \bar p$. Consider the $\hat w$ equation on $\Omega\times(0,t), 0 < t \le T$, and obtain

$$\int_\Omega \left[(\hat w_t(t))^2 + |\nabla \hat w(t)|^2\right] d\Omega + \int_\Gamma \sigma \hat w(t)^2 d\Gamma = -2\int_0^t\int_\Gamma (\sigma - \bar\sigma)\bar w \hat w_t d\Gamma d\tau \tag{5.1}$$

where

$$\sigma(x,y) = \min\left\{\frac{1}{\beta}\left(\int_0^T wp(x,y,u(x,y),t)dt\right)^+, K\right\}$$

and

$$\bar\sigma(x,y) = \min\left\{\frac{1}{\beta}\left(\int_0^T \bar w\bar p(x,y,u(x,y),t)dt\right)^+, K\right\}$$

Similarly on $\Omega \times (t,T), 0 < t \leq T$, in the adjoint case, we obtain

$$\int_\Omega \left[(\hat{p}_t(t))^2 + |\nabla \hat{p}(t)|^2\right] d\Omega + \int_\Gamma \sigma(\hat{p}(t))^2 d\Gamma = -2\int_0^T \int_G \hat{w}\hat{p}_t dG d\tau + 2\int_t^T \int_\Gamma (\sigma - \bar{\sigma})\bar{p}\hat{p}_t d\Gamma d\tau. \tag{5.2}$$

Integration by parts gives

$$\int_t^T \int_\Gamma (\sigma - \bar{\sigma})\bar{p}\hat{p}_t d\Gamma d\tau = -\int_t^T \int_\Gamma (\sigma - \bar{\sigma})\bar{p}_t \hat{p} d\Gamma d\tau - \int_\Gamma (\sigma - \bar{\sigma})\bar{p}\hat{p}(t) d\Gamma. \tag{5.3}$$

Continuing to estimate, we have

$$\begin{aligned}
&\int_t^T \int_\Gamma |\sigma - \bar{\sigma}||\bar{p}_t||\hat{p}| d\Gamma d\tau \\
&\leq \frac{1}{2}\int_t^T \int_\Gamma (\sigma - \bar{\sigma})^2 d\Gamma d\tau + \frac{1}{2}\int_t^T \int_\Gamma |\bar{p}_t|^2 |\hat{p}|^2 d\Gamma d\tau \\
&\leq C\int_t^T \int_\Gamma \int_0^T \bar{w}^2\hat{p}^2 + p^2\hat{w}^2 d\tau_1 d\Gamma d\tau + \frac{1}{2}\int_t^T \left(\int_\Gamma (\bar{p}_t)^4 d\Gamma\right)^{\frac{1}{2}} \left(\int_\Gamma \hat{p}^4 d\Gamma\right)^{\frac{1}{2}} d\tau \\
&\leq C\int_t^T \int_0^T \left(\int_\Gamma \bar{w}^4(\tau_1) d\Gamma\right)^{\frac{1}{2}} \left(\int_\Gamma \hat{p}^4(\tau_1) d\Gamma\right)^{\frac{1}{2}} d\tau_1 d\tau \\
&\quad + \int_t^T \int_0^T \left(\int_\Gamma p^4(\tau_1) d\Gamma\right)^{\frac{1}{2}} \left(\int_\Gamma \hat{w}^4(\tau_1) d\Gamma\right)^{\frac{1}{2}} d\tau_1 d\tau + C_2 \int_t^T \left(\int_\Gamma \hat{p}^4 ds\right)^{\frac{1}{2}} d\tau
\end{aligned}$$

where we used $\bar{p}$ satisfies the additional regularity like in Proposition 3.2 and

$$\bar{p}_t(t) \in H^1(\Omega) \hookrightarrow L^4(\Gamma).$$

If a similar estimate on $\bar{w}, p$ is used, one obtains

$$\int_t^T |\sigma - \bar{\sigma}||\bar{p}_t||\hat{p}| d\Gamma d\tau \leq CT \sup_{0 \leq t \leq T} \left(\|\nabla \hat{p}(t)\|^2_{L^2(\Omega)} + \|\nabla \hat{w}(t)\|_{L^2(\Omega)}\right).$$

For $\epsilon > 0$, estimate the other type of boundary term in (5.3)

$$\begin{aligned}
&\int_\Gamma |\sigma - \bar{\sigma}||\bar{p}||\hat{p}|(t) d\Gamma \\
&\leq C_\epsilon \int_\Gamma (\sigma - \bar{\sigma})^2 d\Gamma + \epsilon \int_\Gamma |\bar{p}|^2 |\hat{p}|^2(t) d\Gamma \\
&\leq \hat{C}_\epsilon T \sup_{0 \leq t \leq T} \left[\|\nabla \hat{p}(t)\|^2_{L^2(\Omega)} + \|\nabla \hat{w}(t)\|^2_{L^2(\Omega)}\right] + \epsilon \|\nabla \hat{p}(t)\|^2_{L^2(\Omega)},
\end{aligned}$$

using similar imbeddings involving $L^4(\Gamma)$.

Taking the supremum over t in the adjoint estimate, gives

$$\sup_{0\le t\le T}\left(\|\hat{p}_t(t)\|^2_{L^2(\Omega)}+\|\nabla\hat{p}(t)\|^2_{L^2(\Omega)}+\int_\Gamma \sigma\hat{p}^2(t)d\Gamma\right)$$
$$\le \bar{C}_\epsilon T\sup_{0\le t\le T}\left(\|\nabla\hat{p}(t)\|^2_{L^2(\Omega)}+\|\nabla\hat{w}(t)\|^2_{L^2(\Omega)}+\|\hat{w}_t(t)\|^2_{L^2(\Omega)}+\|\hat{p}_t(t)\|^2_{L^2(\Omega)}\right)$$
$$+\epsilon\sup_{0\le t\le T}\left(\|\nabla\hat{p}(t)\|^2_{L^2(\Omega)}+\|\nabla\hat{w}(t)\|^2_{L^2(\Omega)}\right).$$

We have a similar estimate for $\hat{w}$.

Adding the supremum estimates for $\hat{p},\hat{w}$ and taking ϵ small enough and then T small so that

$$2(\bar{C}_\epsilon T+\epsilon)\ <\ 1,$$

we obtain

$$\sup_{0\le t\le T}\left(\|\hat{p}_t(t)\|^2_{L^2(\Omega)}+\|\nabla\hat{p}(t)\|^2_{L^2(\Omega)}++\|\hat{w}_t(t)\|^2_{L^2(\Omega)}+\|\nabla\hat{w}(t)\|^2_{L^2(\Omega)}\right)\ \le\ 0$$

which gives uniqueness. □

6. Approximate Solution of the Parameter Identification Problem

Having solved the optimal control problem, we now use a sequence of optimal controls to approximate the solution of the paramter identification problem. We view h as an observation of the solution w on $G\times(0,T)$ and seek to identify a reflection coefficient σ_0 such that the solution $w=w(\sigma_0)$ satisfies

$$w|_{G\times(0,T)}=h.$$

Assuming the existence of a solution of this inverse problem, we will approximate σ_0 by the optimal control σ_β as β gets small.

Theorem 6.1. *Suppose the inverse problem has a solution, i.e. there exists $\sigma^*\in U_K$ such that $w^*=w(\sigma^*)$ satisfies $w^*=h$ a.e. on $G\times(0,T)$. Then there exists $0\le\sigma_0\le K$ such that on a subsequence of $\beta\to 0$,*

$$\sigma_\beta\rightharpoonup\sigma_0\quad in\ L^2(\Gamma)$$
$$w_\beta=w(\sigma_\beta)\rightharpoonup w_0\quad in\ L^2(0,T;V)$$

and

$$w_0=h\quad a.e.\ on\ G\times(0,T).$$

Proof. Notice that $\sigma^* \in U_K$ and let σ_β be an optimal control that minimizes $J_\beta(\sigma)$. Then we have

$$J_\beta(\sigma_\beta) \leq J_\beta(\sigma^*)$$

and

$$\frac{1}{2}\int_{G\times(0,T)}(w_\beta - h)^2 dG\,dt + \frac{\beta}{2}\int_\Gamma \sigma_\beta^2 d\Gamma \leq \frac{\beta}{2}\int_\Gamma (\sigma^*)^2 d\Gamma. \tag{6.1}$$

Using the above a priori estimates on w_β, which are independent of β, for a (sub)sequence $\beta \to 0$, we have

$$\begin{aligned} w_\beta &\rightharpoonup w_0 \quad \text{in } L^2(0,T;V) \\ (w_\beta)_{tt} &\rightharpoonup (w_0)_{tt} \quad \text{in } L^2(0,T;V') \\ (w_\beta)_t &\rightharpoonup (w_0)_t \quad \text{in } L^2(0,T;L^2(\Omega)) \\ \sigma^\beta &\rightharpoonup \sigma_0 \quad \text{in } L^2(\Gamma) \end{aligned}$$

and

$$w_0 = w(\sigma_0).$$

Letting $\beta \to 0$ in (6.1), we also obtain

$$\int_{G\times(0,T)}(w_0 - h)^2 dG\,dt = 0.$$

Thus w_0 has the desired property, i.e., $w = h$ a.e. on $G \times (0,T)$, and σ_0 is the coefficient that is "identified". If the solution of the inverse problem, σ^*, is unique, then $\sigma_0 = \sigma^*$. □

7. Discussion

Optimal control techniques applied to an acoustic wave equation provide a systematic method to approximate a boundary reflection coefficient, σ, that yields a solution consistent with the actual observations. By the same procedure, we also have an explicit characterization of the approximations σ_β.

In the previous discussion we did not take into account causality effects that may occur due to the finite speed of propagation of signals in hyperbolic systems. Indeed, if we suppose that: (i) at $u = u_t = 0$ at $t = 0$, (ii) the source is localized, and (iii) the receptors are far from the source, then there will be a time interval during which both the measured signal and the model solution are zero at the receptor location. Thus, for that time interval (whose length will depend on the signal speed and the actual geometric configuration), the first term in the functional (1.3) is zero, which means that the minimum of J is realized for $\sigma = 0$, independently of the value of the actual reflection coefficient at the boundary.

In order to obtain meaningful results, the time interval should be large enough to allow signals reflected by the boundary to actually reach the receptors and overcome the spurious threshold effects. Of course, this may hamper the uniqueness result which, as noted, is valid only for short times.

Another source of meaningless results is related to model inadequacies and/or the presence of noise. Indeed, suppose the model does not predict any (reflected) signal at the receptor, i.e. $u = 0$, while the actual measured signal, h, is different from zero (or vice-versa). Again the minimizer is given by the (most likely wrong) result $\sigma = 0$, independently of its actual value or functional form. This spurious result occurs because our method assumes the model is completely and correctly known. The knowledge of the model is paramount to this parameter identification method and any model imperfection will be reflected - sometimes in the form of severe anomalies - in the identified parameters.

Finally, we expect the quality of the recovery to increase with the amount and quality of the available information. In other words, the larger G is the better the results are expected to become. However this relationship cannot be expressed by a simple monotonic dependence.

Acknowledgments

S.L. and V.P. acknowledge partial support from DOE's Office of BES under contact No. AC05-96OR22464 with Lockheed Martin Energy Research Corporation. J.Y. is supported in part by the NNSF of China, the Chinese State Education Commission Science Foundation, and the Trans-Century Training Programme Foundation for Talents of the State Education Commission of China.

References

1. H.T. Banks and K. Kunisch, *Estimation Techniques for Distributed Parameter Systems*, Birkhäuser, Boston, 1989.
2. V. Barbu and N. H. Pavel, *Determining the acoustic impedance in the 1-D wave equation via an optimal control problem*, SIAM J. Optimal Control **35** (1997), 1544-1556.
3. B.P. Belinskiy, J.P. Dauer, C. Martin, and M.A. Shubov, *On controllability of an oscillating continuum with a viscous damping*, preprint.
4. J. Chavent, *On parameter idenitifiability*, Proceedings of the 7th IFAC Symposium on Identification and System Parameter Estimations, Pergamon Press, York, 1985, pp. 531–36.
5. J. D. Craig and C. Brown, *Inverse Problems in Astronomy*, Adam Hilger Ltd., Bristol and Boston, 1986.
6. L.C. Evans, *Partial Differential Equations* Vol. 3B (1993), Berkeley Math Lecture Notes, Berkeley.
7. K.D. Graham and D.L. Russell, *Boundary value control of the wave equation in a spherical region*, SIAM Journal on Control **13** (1975), 174–196.
8. C.W. Groetsch, *Inverse Problems in the Mathematical Sciences*, Vieweg, Braunschweig, Wiesbaden, 1993.
9. F. James and M. Sepulveda, *Parameter identification for a model of chromatographic column*, Inverse Problems 10 (1994), 367-385.

NUMERICAL SOLUTIONS TO ACOUSTIC SCATTERING IN SHALLOW OCEANS BY PERIODIC WAVELETS[1]

Wei Lin and Xianbiao Wang

Department of Mathematics
Zhongshan University
Guangzhou, 510275
People's Republic of China

Abstract: In this paper the approximate solution to the problem of a time-harmonic acoustic wave scattering from a obeject with a sound soft surface in a shallow ocean is investigated by means of wavelets. We reduce the problem into a boundary integral equation in which the kernel function is an infinite series. The Daubechies orthonormal wavelet basis is periodized and its corresponding properties are discussed. The kernel function first is truncated approximately and then is approximated via periodic wavelet. Error estimates are obtained and convergence discussions are given. Finally some numerical examples are presented.

1.1 INTRODUCTION: UNDERWATER ACOUSTIC SCATTERING PROBLEMS

We consider scattering problems which describe the scattering of acoustic waves from a cylindrical object with a sound soft boundary in a shallow ocean. We

[1]THIS RESEARCH WAS SUPPORTED IN PART BY THE NATIONAL NATURAL SCIENCE FOUNDATION OF CHINA & THE GUANGDONG NATURAL SCIENCE FOUNDATION

R.P. Gilbert et al.(eds.), Direct and Inverse Problems of Mathematical Physics, 267–279.

may model it as a boundary value problem in a waveguide. Assume $R_d^2 = \{(x_1, x_2) : x_1 \in R, 0 \leq x_2 \leq d\}$ be a region which represents the shallow ocean with ocean depth d. Consider an obeject Ω imbedded in R_d^2 and assume Ω to be a bounded,simply connected domain with a C^2 boundary $\partial\Omega$. Suppose the obeject has a sound soft boundary $\partial\Omega$. By setting $\Omega_e = R_d^2\backslash\bar{\Omega}$, the direct scattering problem in a waveguide may be formulated as a Dirichlet boundary value problem for the scattering of time-harmonic acoustic waves in:Ω_e[10]:

$$\Delta u + k^2 u = 0 \quad in \quad \Omega_e \tag{1.1}$$

$$u = 0 \qquad x_2 = 0 \tag{1.2}$$

$$\frac{\partial u}{\partial x_2} = 0 \qquad x_2 = d \tag{1.3}$$

$$u = -u^i \qquad \text{on} \quad \partial\Omega \tag{1.4}$$

$$\lim_{|x|\to\infty} \left(\frac{\partial u_n^s}{\partial r} - ika_n u_n^s \right) = 0, \quad r := |x| \tag{1.5}$$

where $u = u^i + u^s, u^i$ is the incident wave and u^s is the scattered wave. The u_n^s is the $n-$th normal propagating mode of scattered wave u^s, and

$$a_n = \left[1 - \frac{(2n-1)^2\pi^2}{4k^2d^2} \right]^{1/2} \tag{1.6}$$

$$k = k_0 \cdot n(x_1, x_2), \quad k_0 = \frac{\omega}{c_0}, \quad n(x_1, x_2) = \frac{c_2}{c(x_1, x_2)} \tag{1.7}$$

k —wave numher, w—angular frequency of sound, c_0—coustant referece sound speed, $c(x_1, x_2)$—range and depth-dependent speed of sound. In this paper we assume that $k \neq (2n-1)\pi/(2d)$.

The direct and inverse scattering problems in shallow ocean are meaningful in applications. For example the inverse problems: from the sampling data to determine the shape of Ω or to determine the index of refraction $n(x_1, x_2)$[2][7][9]. In order to solve the inverse problems, we should study the direct scattering problems and their numerical solution. There are various methods contributing to them: the parabolic approximation [5][8], the transmutation theory[6], the boundary integral equation methods[10][12] etc. The bondary integral equation (BIE) method usually uses the fundamental solution and represents the solution as a layer potential. The BIE method can reduce the problem to a problem in a lower dimensional space, but leads to an integral equation on both the baundary $\partial\Omega$ and the two unbounded boundaries. In order to avoid the integral equation on the two unbounded boundaries, instead of applying fundamental solution Xu and Yan use the Green function of the Helmholtz equation in R_d^2:[6][10]

$$\begin{aligned} G(x,y) &= G(x_1, x_2; y_1, y_2) \\ &= \textstyle\sum_{n=1}^{\infty} \frac{i}{\pi k a_n} \phi_n(x_2)\phi_n(y_2) e^{ika_n|x_1-y_1|} \end{aligned} \tag{1.8}$$

which filfuls the boundary conditions (1.2) (1.3) and the radiation condition (1.5), where

$$\phi_n(x_2) = \sin\left[k\left(1-a_n^2\right)^{1/2} x_2\right] \tag{1.9}$$

By $\nu_x = (\nu_1, \nu_2)$ denote the outward normal vector at the point $x = (x_1, x_2)$. By introducing a double layer potential

$$u(x) = \int\limits_{\partial\Omega} \partial G(x,y)\partial\nu_y \psi(y)dy, \quad \text{for} \quad x \in \Omega_e \tag{1.10}$$

and by means of the Green formula, Xu and Yan reduce the problem into solving the BIE

$$\psi(x) + 2\int\limits_{\partial\Omega} \frac{\partial G}{\partial\nu_y}(x,y)\psi(y)d\sigma_y = -2u^i(x), \quad \text{for} \quad x \in \partial\Omega \tag{1.11}$$

Since the kernel function in (1.11) is given in an infinite series,a truncation method for evaluation of the kernel of (1.11) is given in [10]. Then they use a good quadrature method for solving this equation.Their approach has a good accuracy and involves a fine CPU Time.

As we know, wavelets have a series of fine properties such as: vanishing moments, compact support, smoothness and the fast algorithms available for their constructions , they are applied to solve differential equations and integral equations. In particalar, the method based on orthonormal families of Daubechies wavelets has been effectively used for solving the propagation problem of time harmonic acoustic wave in a finite ocean[5]. The main task of this paper is to utilize the periodic wavelets to investigate the boundary integral equation (1.11). As Xu and Yan pointed out, it is a challenge to develop an efficieut numerical method for solving (1.11) where its kernel is given in an infinite series. In order to preserve a certain accuacy and minimize arithmetic operations, we have to truncate this infinite series for its evaluation. Here we provide another approach of the truncation which bases on separating the Green function $G(x,y)$ into two parts:

$$G(x,y) = G_0(x,y) + R(x,y)$$

Where $G_0(x,y)$ corresponds to the case $k = 0$. It is obvious that $G_0(x,y)$ is the corresponding Green function of Laplacian $\Delta u = 0$ and yields a singularity. However we will show that $G_0(x,y)$ has the representation of closed form. We will preserve $R(x,y)$ in the form of infinite series, but it already has not singularity. Above approach, which divides $G(x,y)$ into two terms — one with singularity has finite close form and another with infinite series has been regular, will give us advantage for computation.

This paper is arranged as follows: In the section 2, we investigate with a great care the truncting of the kernel in (1.8). Periodic wavelet and its properties will be presented in the section 3. Error estimates are obtained in section 4. Some numerical results and convergence discussion are given in the last section.

1.2 TRUNCATION TO THE GREEN FUNCTION

According to [6] [10], the Green function of Helmholtz equation (1.1) to (1.3) can be written as (1.8). For the simplicity, we may assume $d = \pi$, and then

$$G(x,y) = \frac{1}{\pi}\sum_{n=1}^{\infty} S_n(x_2)S_n(y_2)\frac{1}{\alpha_n}e^{-\alpha_n|x_1-y_1|} \tag{2.1}$$

where $S_n(x) = \sin\left(n-\frac{1}{2}\right)x, \alpha_n = \left[\left(n-\frac{1}{2}\right)^2 - k^2\right]^{\frac{1}{2}}$.

Obviously, conducting the numerical solution of equation (1.11) involves inevitably the evaluation of the kernel $\frac{\partial G(x,y)}{\partial \nu_y}$. Since $G(x,y)$ is given only as a sum of the infinite series, this numerical evalution can only be done approximately. In this section we shall provide a truncating to the kernel $\frac{\partial G(x,y)}{\partial \nu_y}$.

For this, we first present some lemmas and theorems. In the case when $k = 0, d = \pi$, we have

$$G_0(x,y) = \frac{1}{\pi}\sum_{n=1}^{\infty} \frac{1}{n-1/2}S_n(x_2)S_n(y_2)e^{-(n-1/2)|x_1-y_1|} \tag{2.2}$$

¿From 1.448.4 in Ref [13], and after some computation we get

$$G_0(x,y) = -\frac{1}{4\pi}\left[F(x_1-y_1, x_2-y_2) - F(x_1-y_1, x_2+y_2)\right] \tag{2.3}$$

where

$$F(s,t) = \log\frac{\cosh\frac{s}{2} - \cos\frac{t}{2}}{\cosh\frac{s}{2} + \cos\frac{t}{2}}.$$

Lemma 2.1 The following identities hold:

$$\frac{1}{\pi}\sum_{n=1}^{\infty}\frac{1}{n-1/2}S_n(x_2)S_n(y_2)e^{-(n-1/2)|x_1-y_1|}$$

$$= \frac{1}{4\pi}\ln\frac{\left(\cosh\frac{x_1-y_1}{2} + \cos\frac{x_2-y_2}{2}\right)\left(\cosh\frac{x_1-y_1}{2} - \cos\frac{x_2+y_2}{2}\right)}{\left(\cosh\frac{x_1-y_1}{2} - \cos\frac{x_2-y_2}{2}\right)\left(\cosh\frac{x_1-y_1}{2} + \cos\frac{x_2+y_2}{2}\right)} \triangleq G_{12} \tag{2.4}$$

$$\frac{1}{\pi}\sum_{n=1}^{\infty} S_n(x_2)S_n(y_2)e^{-(n-1/2)|x_1-y_1|}$$

$$= \frac{1}{4\pi}\left[\frac{\sinh\frac{|x_1-y_1|}{2}\cos\frac{x_2+y_2}{2}}{\sinh^2\frac{x_1-y_1}{2} + \sin^2\frac{x_2+y_2}{2}} - \frac{\sinh\frac{|x_1-y_1|}{2}\cos\frac{x_2-y_2}{2}}{\sinh^2\frac{x_1-y_1}{2} + \sin^2\frac{x_2-y_2}{2}}\right] \triangleq G_{11} \tag{2.5}$$

$$\frac{1}{\pi}\sum_{n=1}^{\infty}\frac{1}{n-1/2}S_n(x_2)C_n(y_2)e^{-(n-1/2)|x_1-y_1|}$$
$$=\frac{1}{2\pi}\left[\arctan\frac{\sin\frac{x_2-y_2}{2}}{\sinh\frac{|x_1-y_1|}{2}}+\arctan\frac{\sin\frac{x_2+y_2}{2}}{\sinh\frac{|x_1-y_1|}{2}}\right]\ stackrel\Delta{=}G_{22} \qquad (2.6)$$

$$\frac{1}{\pi}\sum_{n=1}^{\infty}S_n(x_2)C_n(y_2)e^{-(n-1/2)|x_1-y_1|}$$
$$=\frac{1}{4\pi}\left[\frac{\cosh\frac{|x_1-y_1|}{2}\sin\frac{x_2-y_2}{2}}{\sinh^2\frac{x_1-y_1}{2}+\sin^2\frac{x_2-y_2}{2}}+\frac{\cosh\frac{|x_1-y_1|}{2}\sin\frac{x_2+y_2}{2}}{\sinh^2\frac{x_1-y_1}{2}+\sin^2\frac{x_2+y_2}{2}}\right]\triangleq G_{21} \qquad (2.7)$$

where $C_n(x)=\cos\left(n-\frac{1}{2}\right)x$.

Now we consider $\frac{\partial G}{\partial y_1}$. Setting $G_1:=\frac{\partial G}{\partial y_1}$, from direct computation we have

$$G_1=\frac{\partial G}{\partial y_1}=\frac{1}{\pi}\sum_{n=1}^{\infty}S_n(x_2)S_n(y_2)e^{-\alpha_n|x_1-y_1|}\cdot\mathrm{sgn}(x_1-y_1) \qquad (2.8)$$

It is easy to see

$$\alpha_n=\left[\left(n-\frac{1}{2}\right)^2-k^2\right]^{\frac{1}{2}}\sim\left(n-\frac{1}{2}\right),\text{as}\quad n\to\infty$$
$$e^{-\alpha_n|x_1-y_1|}=e^{-\left(n-\frac{1}{2}\right)|x_1-y_1|}\cdot e^{\beta_n|x_1-y_1|} \qquad (2.9)$$

where

$$\begin{aligned}\beta_n &= \left(n-\tfrac{1}{2}\right)-\alpha_n\\ &= \frac{k^2}{\left(n-\frac{1}{2}\right)+\sqrt{\left(n-\frac{1}{2}\right)-k^2}}\sim\frac{k^2}{2n-1}\quad \text{as}\quad n\to\infty\end{aligned}$$

Obviously, $e^{-\alpha_n|x_1-y_1|}$ has the same convergence rate with $e^{-\left(n-\frac{1}{2}\right)|x_1-y_1|}$. We only discuss $e^{\beta_n|x_1-y_1|}$. To this end, we rewrite $e^{\beta_n|x_1-y_1|}$ into the following form:

$$e^{\beta_n|x_1-y_1|}=\left(e^{\beta_n|x_1-y_1|}-\frac{k^2}{2n-1}|x_1-y_1|-1\right)+\frac{k^2}{2n-1}|x_1-y_1|+1 \quad (2.10)$$

¿From (2.8)(2.9)(2.10) and (2.4),(2.5),we get

$$\begin{aligned}G_1 &= \mathrm{sgn}(x_1-y_1)\cdot G_{11}+\tfrac{k^2}{2}(x_1-y_1)\cdot G_{12}\\ &+\tfrac{1}{\pi}\sum_{n=1}^{\infty}S_n(x_2)S_n(y_2)e^{-\left(n-\frac{1}{2}\right)|x_1-y_1|}\cdot\mathrm{sgn}(x_1-y_1)\\ &\cdot\left(e^{\beta_n|x_1-y_1|}-\tfrac{k^2}{2n-1}|x_1-y_1|-1\right)\end{aligned}$$

Setting

$$\begin{aligned} G_1^p &= \mathrm{sgn}(x_1 - y_1)\cdot G_{11} + \tfrac{k^2}{2}(x_1 - y_1)\cdot G_{12} \\ &+\tfrac{1}{\pi}\sum_{n=1}^{p} S_n(x_2)S_n(y_2)e^{-(n-\frac{1}{2})|x_1-y_1|}\cdot \mathrm{sgn}(x_1-y_1) \\ &\cdot\left(e^{\beta_n|x_1-y_1|} - \tfrac{k^2}{2n-1}|x_1-y_1| - 1\right) \end{aligned}$$

We approximate $G_1 := \frac{\partial G}{\partial y_1}$ by G_1^p, then we have the following error estimate theorem:

Theorem 1.2.1 *If* $p > \frac{\sqrt{k^4\cdot d_\Omega + 4k^2 + 1}}{2}$, *then*

$$|G_1(x,y) - G_1^p(x,y)| \le \frac{C_1 k^4}{p^2 - k^2}, x, y, \in \partial\Omega$$

where $d_\Omega = \max_{x,y\in\partial\Omega} |x_1 - y_1|^2, C_1$ *is constant independent of* p, k, x, y.

Before the proof of our theorem,we prove

Lemma 1.2.2 *For* $0 \le t < 1$, *we have*

$$\left|e^t - 1 - t\right| < t^2$$

In fact, as ,it is easy to see:

$$0 \le e^t - 1 - t = \sum_{n=2}^{\infty}\frac{t^n}{n!} \le t^2\sum_{n=2}^{\infty}\frac{1}{n!} \le t^2.$$

Now we return to prove Theorem 2.2. As $n > p > \frac{\sqrt{k^4\cdot d_\Omega + 4k^2 + 1}}{2}$, *we have*

$$0 \le \beta_n|x_1 - y_1| < 1.$$

By means of Lemma 2.3, it follows:

$$\begin{aligned} &\left|e^{\beta_n|x_1-y_1|} - \tfrac{k^2}{2n-1}|x_1-y_1| - 1\right| \\ &= \left|e^{\beta_n|x_1-y_1|} - \beta_n|x_1-y_1| - 1 + \beta_n|x_1-y_1| - \tfrac{k^2}{2n-1}|x_1-y_1|\right| \\ &\le (\beta_n|x_1-y_1|)^2 + \left|\beta_n - \tfrac{k^2}{2n-1}\right|\cdot|x_1-y_1| \\ &= \beta_n^2(x_1-y_1)^2 + \left|\frac{k^2}{n-1/2+\sqrt{(n-1/2)^2-k^2}} - \frac{k^2}{2n-1}\right|\cdot|x_1-y_1| \\ &= \beta_n^2(x_1-y_1)^2 + \left|\frac{k^2(n-1/2-\sqrt{(n-1/2)^2-k^2})}{((n-1/2)+\sqrt{(n-1/2)^2-k^2})(2n-1)}(x_1-y_1)\right| \\ &= \beta_n^2(x_1-y_1)^2 + \beta_n^2\cdot\tfrac{1}{2n-1}\cdot|x_1-y_1| \le \beta_n^2(|x_1-y_1|^2 + |x_1-y_1|) \\ &\le \left[\frac{k^2}{2\sqrt{(n-1/2)^2-k^2}}\right]^2\cdot(|x_1-y_1|^2 + |x_1-y_1|) \\ &\le \tfrac{k^4}{(2p+1)^2-4k^2}(|x_1-y_1|^2 + |x_1-y_1|) \\ &\le \tfrac{k^4}{p^2-k^2}\cdot\tfrac{|x_1-y_1|^2+|x_1-y_1|}{4} \end{aligned}$$

Noticing $|S_n| \leq 1$,we have
$|G_1 - G_1^p|$

$$= \left| \frac{1}{\pi} \sum_{n=p+1}^{\infty} S_n(x_2)S_n(y_2)e^{-(n-\frac{1}{2})|x_1-y_1|}\mathrm{sgn}(x_1-y_1)\left(e^{\beta_n|x_1-y_1|} - \frac{k^2}{2n-1}|x_1-y_1| - 1\right)\right|$$

$$\leq \left| \frac{1}{\pi} \sum_{n=p+1}^{\infty} e^{-(n-\frac{1}{2})|x_1-y_1|} \cdot \frac{k^2}{p^2-k^2}(|x_1-y_1|^2 + |x_1-y_1|)/4 \right|$$

$$\leq \left| \frac{k^4}{4\pi} \cdot \frac{|x_1-y_1|^2+|x_1-y_1|}{p^2-k^2} \cdot \frac{e^{-(p-1/2)|x_1-y_1|}}{1-e^{-|x_1-y_1|}} \right|$$

$$\leq \left| \frac{k^4(|x_1-y_1|+1)e^{-(p-1/2)|x_1-y_1|}}{4\pi(p^2-k^2)} \right|$$

$$\leq \left| \frac{k^4(|x_1-y_1|+1)e^{-|x_1-y_1|}}{4\pi(p^2-k^2)} \right| (p \geq 2)$$

$$\leq \left| \frac{k^4}{4\pi(p^2-k^2)} \right|.$$

The proof of Theorem 2.2 is completed.

Similarly, we truncate $G_2 := \frac{\partial G}{\partial y_2}$ by

$$G_2^p = G_{21} + \frac{k^2}{2}|x_1-y_1|G_{22} + \frac{1}{\pi}\sum_{n=1}^{p} S_n(x_2)C_n(y_2)e^{-(n-1/2)|x_1-y_1|} \cdot \left(\frac{n-1/2}{\alpha_n}e^{\beta_n|x_1-y_1|} - \frac{k^2}{2n-1}|x_1-y_1| - 1\right),$$

thus we have the same error estimate as theorem 2.2:

Theorem 1.2.3 *If $p > \frac{\sqrt{k^4 \cdot d_\Omega + 4k^2} + 1}{2}$, then*

$$|G_2(x,y) - G_2^p(x,y)| \leq \frac{C_2 k^4}{p^2-k^2}, \quad x, y, \in \partial\Omega$$

where $d_\Omega = \max_{x,y\in\partial\Omega} |x_1-y_1|^2, C_2$ is a constant independent of p, k, x, y.

Finally,we use $G_y^p := (G_1^p, G_2^p) \cdot \nu_y$ to approximate the kernel function $\frac{\partial G}{\partial \nu_y}$. Combining theorem 2.3 with theorem 2.4, it is not difficult to get the following estimation.

Theorem 2.5

If $p > \frac{\sqrt{k^4 \cdot d_\Omega + 4k^2} + 1}{2}$, then

$$\left|\frac{\partial G}{\partial \nu_y} - G_y^p\right| \leq \frac{C_3 k^4}{p^2-k^2}, \quad x, y, \in \partial\Omega$$

where $d_\Omega = \max_{x,y\in\partial\Omega} |x_1-y_1|^2, C_3$ is a constant independent of p, k, x, y.

Remark: It can be obversed from theorem 2.5 that the order of convergence in our trunction is not less than 2.

1.3 PERIODIC WAVELET AND ITS PROPERTIES

We choose the scaling function and the wavelet function from Daubechies compactly supported wavelets. They possess the following properties:

(i) $\text{supp}\phi = [1-N, N]$
$\text{supp}\psi = [1-N, N]$
(ii) $\int_{-\infty}^{+\infty} \psi(x)x^m dx = 0, m = 0, \cdots, N-1$

$$\text{(iii)} \quad \begin{cases} \phi(x) = \sqrt{2} \sum\limits_{k=1-N}^{N} h_k \phi(2x-k) \\ \psi(x) = \sqrt{2} \sum\limits_{k=1-N}^{N} g_k \phi(2x-k) \quad \text{where} \quad g_k = (-1)^k h_{1-k} \end{cases}$$

By periodizing we can obtain :

$$\widetilde{\phi}_{jk} := \sum_{l \in Z} \phi_{jk}(x+l) = \sum_{l \in Z} 2^{j/2} \phi(2^j x + 2^j l - k)$$

$$\widetilde{\psi}_{jk} := \sum_{l \in Z} \psi_{jk}(x+l) = \sum_{l \in Z} 2^{j/2} \psi(2^j x + 2^j l - k)$$

Define

$$\widetilde{V}_j = \text{span}\{\widetilde{\phi}_{jk}; k = \bar{0}, 1, \cdots, 2^j - 1\}$$

and

$$\widetilde{W}_j = \text{span}\{\widetilde{\psi}_{jk}; k = \bar{0}, 1, \cdots, 2^j - 1\}$$

then it can be proved that (see [4])

$$V_{j+1} = V_j \bigoplus W_j \quad \text{and} \quad V_j \bigoplus \sum_{l=j}^{\infty} W_l = T \bigcap L^2[0,1]$$

where $T = \{f : f(x) = f(x+1)\}$.
In other word, we have

Theorem 3.1
$\{\widetilde{\phi}_{00}(x), \widetilde{\psi}_{jk}(x); k = 0, \cdots, 2^j-1, j \geq 0\}$ form an orthnormal base for $T \bigcap L^2[0,1]$.

Denoting $c_{jk} =< f, \widetilde{\psi}_{jk} >= \int_0^1 f(x)\widetilde{\psi}_{jk}(x)dx$, we can characterize the function space $C^\alpha[0,1]$ by wavelet coeffcients c_{jk}. In fact, we have

Theorem 3.2
If $f \in T$ and $c_{jk} =< f, \widetilde{\psi}_{jk} >$, then $f \in C^\alpha[0,1]$ if and only if

$$|c_{jk}| \leq C_4 2^{-(\frac{1}{2}+\alpha)j}.$$

For simpilcity, we introduce new notations for the set $\{\widetilde{\phi}_{00}(x), \widetilde{\psi}_{lk}(x) : 0 \leq k \leq 2^l-1, 0 \leq l \leq j-1\}$ and denote them by $\{\phi_k^j(x); \quad k \in E_j := \{0, 1, \cdots, 2^j-1\}\}$,

where

$$\begin{cases} \phi_0^j = \widetilde{\phi}_{00} \\ \phi_{2^l+k}^j = \widetilde{\psi}_{l,k}, 0 \le k \le 2^l - 1, 0 \le l \le j-1. \end{cases}$$

Define project operators as follows:

$$P_j f(x) = \sum_{k=0}^{2^j-1} < f, \phi_k^j > \phi_k^j(x),$$

$$Q_j f(x) = \sum_{k=0}^{2^j-1} < f, \psi_k^j > \psi_k^j(x),$$

Theorem 3.3
If $f \in T \bigcap C^\alpha[0,1]$, then

$$|f(x) - P_j f(x)| \le C_4 2^{-\alpha j}$$

where C_4 is a constant independent of j.
Similarly, we can define project operator as follows:

$$\bar{P}_j f(x,y) = \sum_{k,r=0}^{2^j-1} < f(x,y), \phi_k^j(x)\phi_r^j(y) > \phi_k^j(x)\phi_r^j(y).$$

Theorem 3.4
If $f \in T \bigcap C^\alpha[0,1]^2$, then

$$|f(x,y) - \bar{P}_j f(x,y)| \le C_5 2^{-\alpha j}$$

where C_5 is a constant independent of j.

1.4 NUMERICAL SOLUTION TO EQUATION (1.7)

Suppose $\partial\Omega$ satisfying:

$$\gamma(s) = (\gamma_1(s), \gamma_2(s)), \quad s \in [0,1],$$

and

$$\begin{array}{ll} \text{(i)} & \text{for all } s \in [0,1], \gamma'(s)| \neq 0, \\ \text{(ii)} & \gamma(s+1) = \gamma(s), \\ \text{(iii)} & \gamma \in C^2 \end{array}$$

Denotes

$$\omega(s) = \psi(\gamma(s)),$$

$$g(s) = -2u^i(\gamma(s)),$$

$$K(s,\sigma) = 2\frac{\partial G}{\partial \nu_y}(\gamma(s), \gamma(\sigma))|\gamma'(\sigma)|,$$

$$K^p(s,\sigma) = 2G_y^p(\gamma(s),\gamma(\sigma))|\gamma'(\sigma)|,$$

Equation (1.11) can be reduced into

$$\omega(s) + \int_0^1 K(s,\sigma)\omega(\sigma)d\sigma = g(s), s \in [0,1] \tag{4.1}$$

Thus we only need to consider its approximate equation:

$$\omega_j^p(s) + \int_0^1 K_j^p(s,\sigma)\omega_j^p(\sigma)d\sigma = g(s), s \in [0,1] \tag{4.2}$$

where $K_j^p(s,\sigma) = \bar{P}_j K^p(s,\sigma)$.
Setting

$$B\omega = g - \int_0^1 K(s,\sigma)\omega(\sigma)d\sigma$$

$$B^p\omega = g - \int_0^1 K^p(s,\sigma)\omega(\sigma)d\sigma$$

$$B_j^p\omega = g - \int_0^1 K_j^p(s,\sigma)\omega(\sigma)d\sigma$$

Hence

$$\omega = B\omega, \;\; \omega_j^p = B_j^p\omega_j^p,$$

$$(I - B_j^p)(\omega - \omega_j^p) = \omega - \omega_j^p - B_j^p\omega + B_j^p\omega_j^p = \omega - B_j^p\omega$$

So we have

$$\omega - \omega_j^p = (I - B_j^p)^{-1}(\omega - B_j^p\omega)$$

$$\begin{aligned} ||\omega - \omega_j^p|| \;\; &\le ||(I - B_j^p)^{-1}|| \cdot ||\omega - B_j^p\omega|| \\ &= ||(I - B_j^p)^{-1}|| \cdot ||B\omega - B^p\omega + B^p\omega - B_j^p\omega|| \end{aligned}$$

Considerring that $||(I - B_j^p)^{-1}|| \le C$(see [11]), we have

$$\begin{aligned} ||\omega - \omega_j|| \;\; &\le C[||B\omega - B^p\omega|| + ||B^p\omega - B_j^p\omega||] \\ &= C[||A\omega - A^p\omega|| + ||A^p\omega - A_j^p\omega||] \\ &\le C[||K - K^p|| \cdot ||\omega|| + ||K^p - K_j^p|| \cdot ||\omega||] \\ &= C[||K - K^p|| + ||K^p - K_j^p||] \cdot ||\omega|| \\ ||\omega - \omega_j^p|| \;\; &\le C\left[\tfrac{k^4}{p^2-k^2} + 2^{-\alpha j}\right] \cdot ||\omega|| \end{aligned}$$

Summarizing above discussion,we have proved that

Theorem 4.1
If $\omega \in C^\alpha[0,1], p > \frac{\sqrt{k^4 \cdot d_\Omega + 4k^2 + 1}}{2}$, then

$$||\omega - \omega_j^p|| \le C\left[\frac{k^4}{p^2 - k^2} + 2^{-\alpha j}\right] \cdot ||\omega||$$

where C is a constant independent of j, p.

1.5 NUMERICAL EXAMPLES

We consider an object Ω centered at $(0, z_0)$ in the finite depth with $d = \pi$. The boundary of Ω is given by the ellipse

$$\partial\Omega = \left\{(x_1, x_2) : \frac{x_1^2}{\beta_1^2} + \frac{(x_2 - z_0)^2}{\beta_2^2} = 1\right\}$$

The incident wave $u^i(x)$ is given for test by

$$u^i(x) = -G(x_1, x_2; 0, z_0)$$

so the solution of the boundary value problem is

$$u(x) = G(x_1, x_2; 0, z_0)$$

In the following processing we truncate 30 terms of Green function as the exact solution.
All calculations were performed in PC 486/DX66.The potential points $x = (1.01, \pi/2), (2, \pi/2), (4, \pi/2)$ are chosen so that the distances between x and Ω are 0.01, 1 and 3 respectively . In Tables 1,2,3 and 4 we take $N = 2^j$, where j is the corresponding scales in section 3. The errors between the exact solution and the approximation solution are reported in Tables.

Table 1. $p = 10$ $z_0 = \frac{\pi}{2}$ $(\beta_1, \beta_2) = (1, 1)$ $k = 2$

N	$\lvert e_h(1.01, \pi/2)\rvert$	$\lvert e_h(2, \pi/2)\rvert$	$\lvert e_h(4, \pi/2)\rvert$
4	5.69822e-1	2.0085e-1	4.57e-1
8	1.47479e-1	5.19832e-2	1.18279e-1
16	2.71262e-2	1.30862e-2	2.97754e-2
32	8.9654e-3	3.16011e-3	7.19029e-3
64	1.8332e-3	6.46165e-4	9.47024e-4
128	3.74845e-4	1.32125e-4	1.00628e-4
256	3.7547e-5	1.32302e-5	1.17527e-5

Table 2. $p = 10$ $z_0 = \frac{\pi}{2}$ $(\beta_1, \beta_2) = (1, 1)$ $k = 5$

N	$\lvert e_h(1.01, \pi/2)\rvert$	$\lvert e_h(2, \pi/2)\rvert$	$\lvert e_h(4, \pi/2)\rvert$
16	4.32145e-1	1.10841e-2	8.10654e-3
32	1.04356e-1	2.67663e-2	1.9576e-3
64	2.13383e-2	5.47306e-3	4.00281e-4
128	4.36315e-3	1.11911e-3	8.18477e-5
256	4.36889e-4	1.1206e-4	8.19572-6

Table 3. $p = 10 \;\; z_0 = \frac{\pi}{2} \;\; (\beta_1, \beta_2) = (1, 1) \;\; k = 10$

N	$\|e_h(1.01, \pi/2)\|$	$\|e_h(2, \pi/2)\|$	$\|e_h(4, \pi/2)\|$
64	6.7778e-1	2.86041e-2	1.5327e-2
128	1.38589e-1	5.84884e-3	3.134e-3
256	1.38775e-2	5.85666e-4	3.13819e-4

Table 4. $p = 20 \;\; z_0 = \frac{\pi}{2} \;\; (\beta_1, \beta_2) = (1, 1) \;\; k = 2$

N	$\|e_h(1.01, \pi/2)\|$	$\|e_h(2, \pi/2)\|$	$\|e_h(4, \pi/2)\|$
4	1.90123e-2	1.00012e-2	1.56e-2
8	4.92069e-3	1.8376e-3	2.01877e-3
16	1.15578e-3	4.31616e-4	4.74169e-4
32	2.60411e-4	9.72484e-5	1.06836e-4
64	5.32476e-5	1.9885e-6	2.18454e-5
128	1.06638e-5	3.9823e-6	4.37493e-6
256	2.09166e-6	7.81113e-7	8.58124e-7

Table 5. $p = 20 \;\; z_0 = \frac{\pi}{2} \;\; (\beta_1, \beta_2) = (1, 1) \;\; k = 5$

N	$\|e_h(1.01, \pi/2)\|$	$\|e_h(2, \pi/2)\|$	$\|e_h(4, \pi/2)\|$
16	1.23021e-1	4.01289e-3	3.78015e-3
32	2.77182e-2	9.04155e-4	8.51715e-4
64	5.66769e-3	1.84878e-4	1.74155e-4
128	1.13505e-3	3.7025e-5	3.48776e-5
256	2.22636e-4	7.2623e-6	6.8411e-6

Table 6. $p = 20 \;\; z_0 = \frac{\pi}{2} \;\; (\beta_1, \beta_2) = (1, 1) \;\; k = 10$

N	$\|e_h(1.01, \pi/2)\|$	$\|e_h(2, \pi/2)\|$	$\|e_h(4, \pi/2)\|$
64	1.76583e-1	1.69078e-2	1.21864e-2
128	3.53638e-2	3.38608e-3	2.44054e-3
256	6.93647e-3	6.64167e-4	4.78702e-4

References

[1] Colton, D. and R. Kress. (1983). *Integral Equation Methods in Scattering Theory*, John Wiley, New York.

[2] Colton, D. and P. Monk. (1988). *The inverse scattering problem for time-harmonic acoustic waves in an inhomogeneous medium*, Mech.Appl. Math., Vol. 41.

[3] Daubechies, I. (1988). *Orthonormal bases of compactly supported wavelets*, Comm. Pure and Applied Math., Vol. XLI.

[4] Daubechies, I. *Ten Lectures on Wavelets*, CBMS Lecture Notes nr. 61, SIAM, Philadephia.

[5] Gilbert, R. P. and Lin Wei. (1993). *Wavelet solutions for time harmonic acoustic waves in a finite ocean*, J.Comput. Acoust., Vol. 1(1), (pages 31-60).

[6] Gilbert, R. P. and Yongzhi Xu. (1992). *Acoustic waves and far-field patterns in two dimensional oceans with porous-elastic seabeds*, Result in Mathematics, Vol.22, (pages 685-700).

[7] Lin, W., Yongzhi Xu and Yuqiu Zhao. (1996). *Normal modes analysis for sounds waves in an ocean with an ice cap and a perfectly reflecting bottom*, Applicable Analysis, Vol. 63, (pages 167-182).

[8] Tappert, R. D. (1977). *The parabolic approximation method*, Wave Propagation and Underwater Acoustics, (eds. J.B. Keller and J.S. Papadakis), Springer-Verlag, Berlin, Chap. V, (pages 224-287).

[9] Xu, Yongzhi. (1991). *An injective far-field pattern operator and inverse scattering problem in a finite depth ocean*, Proc.Edinburgh Mathematical Society, Vol. 34, (pages 295-311).

[10] Xu, Y. and Y. Yan. (1992). *Boundary integral equation method for source localization with a continuous wave sonar*, J. Acoust. Soc. AM., Vol. 92(2), (pages 995-1002).

[11] Wang, X. B. and W. Lin (1998). *ID-wavelet method for Hammerstein integral equations*, to appear in JCM.

[12] Yan, Y. (1991). *A Fast Boundary Integral Equation Method for the Two Dimensional Helmholtz Equation.*

[13] Grodshteyn, I. S. and I. M. Ryzhik. (1980). *Table of Integrals, Series and Products*, Academic Press Inc.

SOLUTION OF THE ROBIN AND DIRICHLET PROBLEM FOR THE LAPLACE EQUATION

Dagmar Medková, Praha

Mathematical Institute of Czech Academy of Sciences
Žitná 25
115 67 Praha 1
Czech Republic
e-mail:medkova@math.cas.cz

Abstract: Suppose that $G \subset R^m$ $(m \geq 2)$ is an open set with a non-void compact boundary ∂G such that $\partial G = \partial(\mathrm{cl}\ G)$, where cl G is the closure of G. Fix a nonnegative element λ of $\mathcal{C}'(\partial G)$ (=the Banach space of all finite signed Borel measures with support in ∂G with the total variation as a norm) and suppose that the single layer potential $\mathcal{U}\lambda$ is bounded and continuous on ∂G. (In R^2 it means that $\lambda = 0$. If $G \subset R^m, (m > 2), \partial G$ is locally Lipschitz, $\lambda = f\mathcal{H}$, $\mathcal{H}$ is the surface measure on the boundary of G, f is a nonnegative bounded measurable function, then $\mathcal{U}\lambda$ is bounded and continuous.)Here

$$\mathcal{U}\nu(x) = \int_{R^m} h_x(y)\, d\nu(y),$$

where $\nu \in \mathcal{C}'(\partial G)$,

$$h_x(y) = \begin{cases} (m-2)^{-1}A^{-1}|x-y|^{2-m}, & m > 2, \\ A^{-1}\log|x-y|^{-1}, & m = 2, \end{cases}$$

A is the area of the unit sphere in R^m.

R.P. Gilbert et al.(eds.), Direct and Inverse Problems of Mathematical Physics, 281–290.

If G has a smooth boundary,$u \in C^1(\text{cl } G)$ is a harmonic function on G and

$$\frac{\partial u}{\partial n} + fu = g \text{ on } \partial G$$

where $f, g \in C(\partial G)$(= the space of all bounded continuous functions on ∂G equipped with the maximum norm) and n is the exterior unit normal of G then for $\phi \in \mathcal{D}$ (=the space of all compactly supported infinitely differentiable functions in R^m)

$$\int_{\partial G} \phi g \, d\mathcal{H}_{m-1} = \int_G \nabla \phi \cdot \nabla u \, d\mathcal{H}_m + \int_{\partial G} \phi f u \, d\mathcal{H}_{m-1}. \tag{1}$$

Here $\mathcal{H}_k$ is the k-dimensional Hausdorff measure normalized such that $\mathcal{H}_k$ is the Lebesgue measure in R^k. If we denote by $\mathcal{H}$ the restriction of $\mathcal{H}_{m-1}$ on ∂G and by $N^G u$ the distribution

$$\langle \phi, N^G u \rangle = \int_G \nabla \phi \cdot \nabla u \, d\mathcal{H}_m \tag{2}$$

then (1) has a form

$$N^G u + fu\mathcal{H} = g\mathcal{H}. \tag{3}$$

Here $N^G u$ is a characterization in the sense of distributions of the normal derivative of u.

Let now G be general. The formula (3) motivates our definition of the solution of the Robin problem for the Laplace equation

$$\Delta u = 0 \quad \text{in } G, \tag{4}$$

$$N^G u + u\lambda = \mu,$$

where $\mu \in C'(\partial G)$.

We introduce in R^m the fine topology, i. e. the weakest topology in which all superharmonic functions in R^m are continuous. This topology is stronger than ordinary topology.

If u is a harmonic function on G such that

$$\int_H |\nabla u| \, d\mathcal{H}_m < \infty \tag{5}$$

for all bounded open subsets H of G we define the weak normal derivative $N^G u$ of u as a distribution

$$\langle \varphi, N^G u \rangle = \int_G \nabla \varphi \cdot \nabla u \, d\mathcal{H}_m$$

for $\varphi \in \mathcal{D}$.

Let $\mu \in C'(\partial G)$.Now we formulate the Robin problem for the Laplace equation (4) as follows: Find a function $u \in L^1(\lambda)$ on cl G,the closure of G, harmonic on G and fine continuous in λ-a. a. points of ∂G for which ∇u is integrable over all bounded open subsets of G and $N^G u + u\lambda = \mu$.

The single layer potential $\mathcal{U}\nu$, where $\nu \in C'(\partial G)$, has all these properties and if we look for a solution of the Robin problem in the form of the single layer potential we obtain the equation

$$N^G \mathcal{U}\nu + (\mathcal{U}\nu)\lambda = \mu.$$

It was shown by J. Král for $\lambda = 0$ (see [8]) and independently by Burago, Maz'ya (see [2]) and by I. Netuka ([21] for a general λ that $N^G\mathcal{U}\nu + (\mathcal{U}\nu)\lambda \in C'(\partial G)$ for each $\nu \in C'(\partial G)$ if and only if $V^G < \infty$,where

$$V^G = \sup_{x \in \partial G} v^G(x),$$

$$v^G(x) = \sup\{\int_G \nabla\, \phi \cdot \nabla\, h_x\; d\mathcal{H}_m; \phi \in \mathcal{D}, |\phi| \leq 1, spt\; \phi \subset R^m - \{x\}\}.$$

There are more geometrical characterizations of $v^G(x)$ which ensure $V^G < \infty$ for G convex or for G with $\partial G \subset \cup_{i=1}^k L_i$,where L_i are $(m-1)$-dimensional Ljapunov surfaces (i. e. of class $C^{1+\alpha}$). Denote

$$\partial_e G = \{x \in R^m; \bar{d}_G(x) > 0, \bar{d}_{R^m - G}(x) > 0\}$$

the essential boundary of G where

$$\bar{d}_M(x) = \limsup_{r \to 0+} \frac{\mathcal{H}_m(M \cap \mathcal{U}(x;r))}{\mathcal{H}_m(\mathcal{U}(x;r))}$$

is the upper density of M at x, $\mathcal{U}(x;r)$ is the ball with the centre x and the radius r. Then

$$v^G(x) = \frac{1}{A} \int_{\partial \mathcal{U}(0;1)} n(\theta, x)\; d\mathcal{H}_{m-1}(\theta),$$

where $n(\theta, x)$ is the number of all points of $\partial_e G \cap \{x + t\theta; t > 0\}$ (see [7]). It means that $v^G(x)$ is the total angle under which is G visible from the point x. (For example if $G_1, ..., G_k$ are convex sets and $\partial G \subset \cup \partial G_j$ then $V^G \leq k$.) This expression is a modification of the similar expression in [9]. Let us recall another characterization of $v^G(x)$ using a notion of an interior normal in Federer's sense.

If $z \in R^m$ and θ is a unit vector such that the symmetric difference of G and the half-space $\{x \in R^m; (x - z) \cdot \theta > 0\}$ has m-dimensional density zero at z then $n^G(z) = \theta$ is termed the interior normal of G at z in Federer's sense. (The symmetric difference of B and C is equal to $(B - C) \cup (C - B)$.) If there is no interior normal of G at z in this sense, we denote by $n^G(z)$ the zero vector in R^m. The set $\{y \in R^m; |n^G(y)| > 0\}$ is called the reduced boundary of G and will be denoted by $\widehat{\partial} G$. Clearly $\widehat{\partial} G \subset \partial_e G$.

If $\mathcal{H}_{m-1}(\partial_e G)$ the perimeter of G is finite then $\mathcal{H}_{m-1}(\partial_e G - \widehat{\partial} G) = 0$ and

$$v^G(x) = \int_{\widehat{\partial} G} |n^G(y) \cdot \nabla h_x(y)|\; d\mathcal{H}_{m-1}(y)$$

for each $x \in R^m$.

If G has a piecewisise-$C^{1+\alpha}$ boundary than $V^G < \infty$. But there is a domain G with C^1 boundary and $V^G = \infty$ (see [19]). On the other hand there is a domain G with $V^G < \infty$ and $\kappa_m(\partial G) > 0$. So open sets with a locally Lipschitz boundary and open sets with $V^G < \infty$ are incomparable.

Suppose now that $V^G < \infty$.Then the operator

$$\tau : \nu \mapsto N^G(\mathcal{U}\nu) + (\mathcal{U}\nu)\lambda$$

is a bounded linear operator on $\mathcal{C}'(\partial G)$ and

$$\tau\nu(M) = \int_{\partial G \cap M} \mathcal{U}\nu \, d\lambda + \int_{\partial G \cap M} d_G(x) \, d\nu(x) -$$

$$- \int_{\partial G} \int_{\partial G \cap M} n^G(y) \cdot \nabla h_x(y) \, d\mathcal{H}_{m-1}(y) \, d\nu(x).$$

Denote by $\mathcal{H}$ the restriction of $\mathcal{H}_{m-1}$ on $\widehat{\partial} G$. Then $\mathcal{H}(\partial G) < \infty$. If $\lambda = f\mathcal{H}, \nu = h\mathcal{H} \in \mathcal{C}'(\partial G)$ then

$$\tau(h\mathcal{H}) = (Th)\mathcal{H}$$

where

$$Th(x) = \frac{1}{2}h(x) - \int_{\partial G} n^G(x) \cdot \nabla h_y(x) h(y) \, d\mathcal{H}(y)$$
$$+ f(x)\mathcal{U}(h\mathcal{H})(x).$$

The Robin problem $N^G(\mathcal{U}\nu) + (\mathcal{U}\nu)\lambda = \mu$ leads to the equation

$$\tau\nu = \mu.$$

If we want to solve this equation we need the operator τ to be a Fredholm operator with null index. It is true for example if the Fredholm radius of $(\tau - \frac{1}{2}I)$ is greater than 2. The condition that the Fredholm radius of $(\tau - \frac{1}{2}I)$ is greater than 2 does not depend on λ and has a local character. It is well-known that this condition is fulfilled for sets with a smooth boundary (of class $C^{1+\alpha}$) (see [10]) and for convex sets (see [24]). J. Radon proved this condition for open set with a piecewise smooth boundary without cusps in the plane (see [29],[30]). R. S. Angell, R. E. Kleinman, J. Král and W. L. Wendland proved that rectangular domains (i. e. formed from rectangular parallelepipeds) in R^3 have this property (see [1], [12]). A. Rathsfeld showed in [31], [32] that polyhedral cones in R^3 have this property. (By a polyhedral cone in R^3 we mean an open set Ω which boundary is locally a hypersurface (i. e. every point of $\partial\Omega$ has a neighbourhood in $\partial\Omega$ which is homeomorphic to R^2) and $\partial\Omega$ is formed by a finite number of plane angles. By a polyhedral open set with bounded boundary in R^3 we mean an open set Ω which boundary is locally a hypersurface and $\partial\Omega$ is formed by a finite number of polygons.) N. V. Grachev and V. G. Maz'ya obtained independently analogical result for polyhedral open sets with bounded boundary in R^3 (see [6]). (Remark that there is a polyhedral set in R^3 which has not a locally Lipschitz boundary.) This condition is fullfiled for $G \subset R^3$ with "piecewise-smooth" boundary i. e. such that for each $x \in \partial G$ there are $r(x) > 0$, a domain D_x which is polyhedral or convex or a complement of a convex domain or an open set with a smooth boundary and a diffeomorphism $\psi_x : \mathcal{U}(x; r(x)) \to R^3$ of class $C^{1+\alpha}$,$\alpha > 0$, such that $\psi_x(G \cap \mathcal{U}(x : r(x))) = D_x \cap \psi_x(\mathcal{U}(x; r(x)))$ (see [16]). V. G. Maz'ya and N. V. Grachev proved this condition for several types of sets with "piecewise-smooth" boundary in general Euclidean space (see [3],[4]).

Theorem 1.0.1 *Theorem on the Robin problem. ([17],[18]) Let the Fredholm radius of* $(\tau - \frac{1}{2}I)$ *is greater than 2,* $\mu \in C'(\partial G)$. *Then there is a harmonic function u on G, which is a solution of the Robin problem*

$$N^G u + u\lambda = \mu, \tag{6}$$

if and only if $\mu \in C'_0(\partial G)$ (= *the space of such* $\nu \in C'(\partial G)$ *that* $\nu(\partial H) = 0$ *for each bounded component H of cl G for which* $\lambda(\partial H) = 0$). *If* $\mu \in C'_0(\partial G)$ *then there is a unique* $\nu \in C'_0(\partial G)$ *such that*

$$\tau\nu = \mu \tag{7}$$

and for such ν *the single layer potential* $\mathcal{U}\nu$ *is a solution of (6). If*

$$\beta > \frac{1}{2}(V^G + 1 + \sup_{x \in \partial G} U\lambda(x))$$

then

$$\nu = \sum_{n=0}^{\infty} (\frac{\beta I - \tau}{\beta})^n \frac{\mu}{\beta} \tag{8}$$

and there are $q \in (0,1), C \in < 1, \infty)$ *such that*

$$||(\frac{\tau - \beta I}{\beta})^n \mu|| \le Cq^n ||\mu|| \text{ for } \mu \in C'_0(\partial G).$$

If $\lambda = 0$ *then*

$$\nu = \mu + \sum_{n=0}^{\infty} (2\tau - I)^n (2\tau)\mu \tag{9}$$

and there are $q \in (0,1), C \in < 1, \infty)$ *such that*

$$||(2\tau - I)^n (2\tau)\mu|| \le Cq^n ||\mu|| \text{ for } \mu \in C'_0(\partial G).$$

If $\lambda = 0$ *and* $R^m - G$ *is unbounded and connected then*

$$\nu = \sum_{n=0}^{\infty} (I - 2\tau)^n (2\mu) \tag{10}$$

and there are $q \in (0,1), C \in <1,\infty)$ *such that*

$$||(I-2\tau)^n\mu|| \le Cq^n||\mu|| \text{ for } \mu \in C'_0(\partial G).$$

If $\lambda = 0$ *then*

$$\sum_{n=0}^{\infty}(I-2\tau)^n \tag{11}$$

converges on $C_0'(\partial G)$ *if and only if* $R^m - G$ *is unbounded and connected.*

Remark 1. In the theorem we need estimate

$$\sup_{x\in\partial G} U\lambda(x)).$$

If $\lambda = f\mathcal{H}$, where $|f| \le c$ and we can estimate V^G we can use the estimate

$$\sup_{x\in\partial G} U\lambda(x)) \le c(V^G + \frac{1}{2})2^{m-2}(m+2)^m(\text{diam } \partial G),$$

where diam $B = \sup\{|x-y|; x,y \in B\}$.

Remark 2. Let the Fredholm radius of $(\tau - \frac{1}{2}I)$ is greater than 2. Then $\mathcal{H}_{m-1}(\partial G) < \infty$ and $\mathcal{H}$ is the restriction of $\mathcal{H}_{m-1}$ on ∂G. If $\lambda = f\mathcal{H}$, $\mu = g\mathcal{H}$, $\mu \in C'_0(\partial G)$ then there is $h \in L^1(\mathcal{H})$ such that for $\nu = h\mathcal{H}$ we have $\nu \in C'_0(\partial G)$, $\tau\nu = \mu$ and the single layer potential $\mathcal{U}\nu$ is a solution of (6). If

$$\beta > \frac{1}{2}(V^G + 1 + \sup_{x\in\partial G} U\lambda(x))$$

then

$$h = \sum_{n=0}^{\infty}(\frac{\beta I - T}{\beta})^n\frac{g}{\beta}$$

and there are $q \in (0,1), C \in <1,\infty)$ such that

$$||(\frac{T-\beta I}{\beta})^n g|| \le Cq^n||g|| \text{ for } g \in L^1(\mathcal{H}), g\mathcal{H} \in C'_0(\partial G).$$

If $f = 0$ then

$$h = g + \sum_{n=0}^{\infty}(2T-I)^n(2T)g$$

and there are $q \in (0,1), C \in <1,\infty)$ such that

$$||(2T-I)^n(2T)g|| \le Cq^n||g|| \text{ for } g \in L^1(\mathcal{H}), g\mathcal{H} \in C'_0(\partial G).$$

If $f = 0$ and $R^m - G$ is unbounded and connected then

$$h = \sum_{n=0}^{\infty}(I-2T)^n(2g)$$

and there are $q \in (0,1), C \in< 1, \infty)$ such that

$$||(I-2T)^n g|| \leq Cq^n ||g|| \text{ for } g \in L^1(\mathcal{H}), g\mathcal{H} \in C'_0(\partial G).$$

Remark 3. C. Neumann (see [25]-[27]) had the idea to solve the Neumann problem for the Laplace equation with the boundary condition g in the form of the single layer potential where the corresponding density g is given by the series

$$h = \sum_{n=0}^{\infty} (I-2T)^n (2g).$$

Unfortunately, the series $\sum_{n=0}^{\infty}(I-2T)^n$ converges for no open set G. (Similarly, the series $\sum_{n=0}^{\infty}(I-2\tau)^n$ does not converge.) It was shown by J. Král and I. Netuka (see [11]) that for $\tilde{\tau}$ the restriction of τ on the space $C'_0(\partial G)$ the series $\sum_{n=0}^{\infty}(I-2\tilde{\tau})^n$ converges if the set G is convex. This series converges if and only if the set $R^m - G$ is unbounded and connected. For a general open set we can solve the equation (7) by the series (9) (if the right side of (7) is in $C'_0(\partial G)$). But we cannot use the series (9) for the calculation of the solution of the equation (7) for Robin problem, because for each open set G there is such measure λ and a real measure $\mu \in C'_0(\partial G)$) that the series (9) does not converge. We express the solution of the equation (7) by the series (8). The parameter β in the series (8) does depend on λ because for each positive β and each open set G there is such measure λ and a real measure $\mu \in C'_0(\partial G)$) that the series (8) does not converge.

This result has an interesting consequence for the Dirichlet problem for the Laplace equation

$$\Delta u = 0 \text{ in } G, \qquad (12)$$
$$u = g \text{ on } \partial G,$$

where $g \in C(\partial G)$ is a continuous function on the boundary of G. If we look for the solution of the Dirichlet problem (12) in the form of the double layer potential with a continuous density on the boundary of G

$$Wf(x) = \int_{\partial G} f(y) n^G(y) \cdot \nabla h_x(y) \, d\mathcal{H}_{m-1}(y)$$

we obtain the integral operator

$$Df(x) = (1 - d_G(x)) f(x) + \int_{\partial G} f(y) n^G(y) \cdot \nabla h_x(y) \, d\mathcal{H}_{m-1}(y).$$

on $C(\partial G)$. The adjoint operator of D is the operator corresponding to the Neumann problem for the Laplace operator on the complementary domain to G (see [9]). We obtain as a consequence of the theorem for the Neumann problem the following result:

Theorem 1.0.2 *Theorem on the Dirichlet problem. ([17]) Let $V^G < \infty$, the Fredholm radius of $(\tau - \frac{1}{2}I)$ be greater than 2. If the set $R^m - G$ is unbounded and connected and $g \in C(\partial G)$ then the double layer potential*

$$Wf(x) = \int_{\partial G} f(y) n^G(y) \cdot \nabla h_x(y) \, d\mathcal{H}_{m-1}(y)$$

is a classical solution of the Dirichlet problem for the Laplace equation with the boundary condition g, where

$$f = g + \sum_{j=0}^{\infty} (2D - I)^j 2Dg.$$

Moreover, there are constants $C > 1, q \in (0,1)$ such that

$$||(2D - I)^j 2Dg|| \leq Cq^j ||g||,$$

$$\sup_{x \in G} |W[(2D - I)^j 2Dg](x)| \leq Cq^j ||g||$$

for each $g \in C(\partial G)$ and a natural number j.

References

[1] Angell, T. S., R. E. Kleinman, and J. Král. (1988). *Layer potentials on boundaries with corners and edges*, Cas. pěst. mat., Vol. 113, (pages 387-402).

[2] Yu., D. Burago, V. G. Maz' ya. (1969). *Potential theory and function theory for irregular regions*, Seminars in mathematics V. A. Steklov Mathematical Institute, Leningrad.

[3] Grachev, N. V. and V. G. Maz'ya. (1986). *On the Fredholm radius for operators of the double layer potential type on piecewise smooth boundaries*, Vest. Leningrad. Univ., Vol. 19(4), (pages 60–64).

[4] Grachev, N. V. and V. G. Maz'ya. *Estimates for kernels of the inverse operators of the integral equations of elasticity on surfaces with conic points*, Report LiTH-MAT-R-91-06, Linköping Univ., Sweden.

[5] Grachev, N. V. and V. G. Maz'ya. *Invertibility of boundary integral operators of elasticity on surfaces with conic points*, Report LiTH-MAT-R-91-07, Linköping Univ., Sweden.

[6] Grachev, N. V. and V. G. Maz'ya. *Solvability of a boundary integral equation on a polyhedron*, Report LiTH-MAT-R-91-50, Linköping Univ., Sweden.

[7] Chlebík , M. (1988). *Tricomi potentials*, Thesis, Mathematical Institute of the Czechoslovak Academy of Sciences Praha (in Slovak).

[8] Král, J. (1964). *On double-layer potential in multidimensional space*, Dokl. Akad. Nauk SSSR, Vol. 159.

[9] Král, J. (1980). *Integral Operators in Potential Theory*, Lecture Notes in Mathematics 823, Springer-Verlag, Berlin.

[10] Král, J. (1966). *The Fredholm method in potential theory*, Trans. Amer. Math. Soc., Vol. 125, (pages 511-547).

[11] Král, J. and I. Netuka. (1977). *Contractivity of C. Neumann's operator in potential theory*, Journal of the Mathematical Analysis and its Applications, Vol. 61, (pages 607-619).

[12] Král, J. and W. L. Wendland. (1986). *Some examples concerning applicability of the Fredholm-Radon method in potential theory*, Aplikace Matematiky, Vol. 31, (pages 239-308).

[13] Kress, R. and G. F. Roach. (1976). *On the convergence of successive approximations for an integral equation in a Green's function approach to the Dirichlet problem*, Journal of Mathematical Analysis and Applications, Vol. 55, (pages 102-111).

[14] Maz'ya, V. G. (1988). *Boundary integral equations*, Sovremennyje problemy matematiki, fundamental'nyje napravlenija, Vol. 27, Viniti, Moskva (Russian).

[15] Maz'ya, V. and A. Solov'ev. (1993). *On the boundary integral equation of the Neumann problem in a domain with a peak*, Amer. Math. Soc. Transl., Vol. 155, (pages 101-127).

[16] Medková, D. (in print). *The third boundary value problem in potential theory for domains with a piecewise smooth boundary*, Czech. Math. J..

[17] Medková, D. (in print). *Solution of the Neumann problem for the Laplace equation*, Czech. Math. J..

[18] Medková, D. (1997). *Solution of the Robin problem for the Laplace equation*, preprint No.120, Academy of Sciences of the Czech republic.

[19] Netuka, I. (1971). *Smooth surfaces with infinite cyclic variation*, Čas. pěst. mat., Vol. 96.

[20] Netuka, I. (1971). *The Robin problem in potential theory*, Comment. Math. Univ. Carolinae, Vol. 12, (pages 205-211).

[21] Netuka, I. (1972). *Generalized Robin problem in potential theory*, Czech. Math. J., Vol. 22(97), (pages 312-324).

[22] Netuka, I. (1972). *An operator connected with the third boundary value problem in potential theory*, Czech Math. J., Vol. 22(97), (pages 462-489).

[23] Netuka, I. (1972). *The third boundary value problem in potential theory*, Czech. Math. J., Vol. 2(97), (pages 554-580).

[24] Netuka, I. (1975). *Fredholm radius of a potential theoretic operator for convex sets*, Čas. pěst. mat., Vol. 100, (pages 374-383).

[25] Neumann, C. (1877). *Untersuchungen über das logarithmische und Newtonsche Potential*, Teubner Verlag, Leipzig.

[26] Neumann, C. (1870). *Zur Theorie des logarithmischen und des Newtonschen Potentials*, Berichte über die Verhandlungen der Königlich Sachsischen Gesellschaft der Wissenschaften zu Leipzig, Vol. 22, (pages 49–56, 264–321).

[27] Neumann, C. (1888). *Über die Methode des arithmetischen Mittels*, Hirzel, Leipzig, 1887 (erste Abhandlung), 1888 (zweite Abhandlung).

[28] Plemelj, J. (1911). *Potentialtheoretische Untersuchungen*, B. G. Teubner, Leipzig.

[29] Radon, J. (1919). *Über Randwertaufgaben beim logarithmischen Potential*, Sitzber. Akad. Wiss. Wien, Vol. 128, (pages 1123–1167).

[30] Radon, J. (1987). *Über Randwertaufgaben beim logarithmischen Potential*, Collected Works, Vol. 1, Birkhäuser, Vienna.

[31] Rathsfeld, A. (1992). *The invertibility of the double layer potential in the space of continuous functions defined on a polyhedron. The panel method*, Applicable Analysis, Vol. 45, (pages 1-4, 135-177).

[32] Rathsfeld, A. (1995). *The invertibility of the double layer potential in the space of continuous functions defined on a polyhedron. The panel method*, Erratum. Applicable Analysis, Vol. 56, (pages 109-115).

EXISTENCE AND DECAY OF SOLUTIONS OF SOME NONLINEAR DEGENERATE PARABOLIC EQUATIONS

Tokumori Nanbu

Toyama Medical and Pharmaceutical University
2630,Sugitani, Toyama, 930-01, Japan

Abstract: We study the existence and the decay estimates of solutions of the initial-boundary value problem for some nonlinear degenerate parabolic equations

$$u_t = \Delta(|u|^{m-1}u) + b\cdot\nabla(B(u)) - q(t)A(u)$$

where $u = u(x,t)$ is a scalar function of the spatial variable $x \in \Omega$ and time $t > 0$, $b \in R^N$ ($b \neq O$) and Ω is a regular unbounded domain in R^N.

1.1 INTRODUCTION

We consider the existence and the decay estimates of solutions of the following initial-boundary value problem $(P)_0$:

$$(1.1)\qquad u_t = \Delta(|u|^{m-1}u) + b\cdot\nabla(B(u)) - q(t)A(u) \quad \text{in} \quad Q = \Omega\times R^+,$$

$$(1.2)\qquad u(x,t) = 0 \quad \text{on} \quad \Gamma = \partial\Omega\times R^+,$$

$$(1.3)\qquad u(x,0) = u_0(x) \quad \text{in} \quad \Omega.$$

Here Ω is a regular unbounded domain in $R^N (N > 1)$ with the smooth boundary $\partial\Omega$, $R^+ = (0, +\infty)$, Δ denotes the N-dimensional Laplace operator, $m > 1$,

R.P. Gilbert et al.(eds.), Direct and Inverse Problems of Mathematical Physics, 291–300.

$b \in R^N$ ($b = (b_1, b_2, \cdot, b_N) \neq O$), and functions $B(u)$, $q(t), A(u)$ and $u_0(x)$ satisfy some hypotheses which will be given later.
The special case of (1.1) is the 'porous media equation 'with convection and absorption terms

$$(1.4) \qquad u_t = \Delta(|u|^{m-1}u) + \sum_{k=1}^{N} b_k \frac{\partial |u|^{n-1}u)}{\partial x_k} - \lambda |u|^{\nu-1}u$$

where b_k , $m(> 1), n(\geq 1)$, $\nu(\geq 1)$ and $\lambda(\geq 0)$ are constants. When Ω is bounded, for the equation (1.4) under the conditions (1.2) and (1.3) several authors ([1],[2],[3],[5]. [7],[9] etc.) have studied the decay estimates in the norm $\|u(t)\|_{L^p(\Omega)}$ $(1 \leq p \leq \infty)$ of solutions.

M.Escobedo and E.Zuazua ([6]) have considered the large time behavior of solutions of the initial-value problem for the convection-diffusion equation

$$(1.5) \qquad u_t - \Delta u = a \cdot \nabla(|u|^{q-1}u) \quad in \quad (0, \infty) \times R^N$$

with $a(\in R^N)$ is a constant vector and $q \geq 1 + \frac{1}{N}$.

In this paper, we shall prove the existence of solution and the decay-estimates of solution as $t \to \infty$ of the problem (1.1)-(1.3) when Ω is unbounded.

1.2 ASSUMPTIONS AND THEOREMS

Throughout this paper, the notations of function spaces and their norms are as usual and we use the following notations: $R = (-\infty, +\infty)$, R^+=$(0, +\infty)$, $Q_T = \Omega \times (0, T]$, $Q = \Omega \times (0, +\infty)$, $\Gamma_T = \partial\Omega \times (0, T]$, $\Gamma = \partial\Omega \times (0, +\infty)$ and $\nabla u = (u_{x_1}, u_{x_2}, ..., u_{x_n})$.

Definition 1. A *solution* u of the problem (1.1) - (1.3) on $[0,T]$ is a function with the following properties :

$$(i) \qquad u \in C([0, T] : L^2_{loc}(\Omega)) \cap L^\infty(Q_T),$$

$$(ii) \qquad \int_\Omega u(t)\varphi(t)dx - \int_0^t \int_\Omega [u\varphi_t + |u|^{m-1}u\Delta\varphi - \sum_{k=1}^{N} b_k B(u)\varphi_{x_k}]dxd\tau$$

$$= \int_\Omega u_0(x)\varphi(x, 0)dx - \int_0^t \int_\Omega q(t)A(u)\varphi dxd\tau$$

for any $t \in (0, T]$ and $\varphi \in C^2(\overline{Q_T})$ such that $\varphi = 0$ on $\partial\Omega \times [0, T]$.

Definition 2. A *solution* u of the problem (1.1)-(1.3) on $[0, +\infty)$ means a solution on *each* $[0, T]$.

For the functions $B(u)$ we assume

(H.B.1) $$\begin{cases} B(u) \in C^2(R), & b(u) \equiv \frac{\partial B(u)}{\partial u}, \\ B(0) = b(0) = 0, & b(u) > 0 \quad (u \neq 0). \end{cases}$$

For the function $q(t)$ we assume

(H.q.1) $$q(t) \in C^1([0,\infty)), q(t) \geq 0 \text{ and } q'(t) \leq 0 \ (t \in [0,\infty)).$$

For the function $A(u)$ we impose

(H.A.1) $$\begin{cases} A(u) \in C^1(R), & a(u) \equiv \frac{\partial A(u)}{\partial u}, \\ A(0) = a(0) = 0, & a(u) > 0 \quad (u \neq 0). \end{cases}$$

(H.u_0) $|u_0|^{m-1}u_0 \in W_0^{1,2}(\Omega) \cap L^\infty(\Omega)$ and $u_0(x)$ has a bounded support.

We now state our main results.

Theorem 1. *Suppose that (H.B.1), (H.q.1),(H.A.1) and (H.u_0) are satisfied. Then the problem (1.1) - (1.3) has a solution on $[0,+\infty)$ such that*

(i) $$\|\nabla(|u|^{m-1}u)(t))\|_{L^2(\Omega)} \leq c_0(u_0) \quad \text{on} \quad [0,+\infty),$$

(ii) $$\|(|u|^{m-1}u)_t\|^2_{L^2(0,T:L^2(\Omega))} \leq (1 + mM^{m-1})c_0(u_0).$$

Here $M = \sup_{x\in\Omega} |u_0(x)|$, and $c_0(u_0)$ is a constant depending only on the data.

Theorem 2. *Suppose that the assumptions of Theorem 1 are satisfied. Let u be a solution of the Problem (1.1) - (1.3) on $[0,+\infty)$. Then we have :*

(i) $$\|u(\cdot,t)\|_{L^{p+1}(\Omega)} \leq (a_1 t + a_2)^{-\gamma_1} \qquad (t \in R^+)\ (p \geq m)$$

and

(ii) $$|u(x,t)|_{L^\infty(\Omega)} \leq C(a_3 t + a_4)^{-\gamma_2} \quad (t \in R^+)$$

where $\gamma_1 = \frac{p}{p+1}\frac{N}{N(m-1)+2}$, $\gamma_2 = \frac{m}{m+1}\frac{N}{N(m-1)+2}$, and a_1,a_2, a_3 and a_4 and C are constants depending only on $m, p, u_0(x)$.

Furthermore we set the following assumptions on $q(t)$ and $A(u)$:

(H.q.2) $$\delta_0(t) \equiv \delta_0(1+t)^{-\epsilon} \leq q(t), \quad (\epsilon \geq 0).$$

Here δ_0 is a positive constant.
There exist positive constants μ_A and $\nu(\geq 1)$ such that

(H.A.2) $$\mu_A|u|^{\nu+1} \leq A(u)u \quad (u \in R).$$

Theorem 3. *Suppose that the assumptions of Theorem 2, (H.q.2) and (H.A.2) are satisfied. Let u be a solution of the Problem (1.1) - (1.3) on* $[0,+\infty)$. *Then if* $\nu > 1$,

$$(i-1) \qquad \|u(\cdot,t)\|_{L^{m+1}(\Omega)} \leq \min\left\{(a_1t+a_2)^{-\gamma_2}, \omega_\nu(t)\right\},$$

where

$$\omega_\nu(t) = \left\{\left(\int_\Omega |u_0|^{m+1}dx\right)^{\frac{-(\nu-1)}{m}} + \frac{\nu-1}{m}\mu_A\|u_0\|_1^{-\frac{\nu-1}{m}} \int_0^t \delta_0(s)ds\right\}^{\frac{-m}{\nu-1}}$$

and if $\nu = 1$,

$$(i-2) \qquad \|u(\cdot,t)\|_{L^{m+1}(\Omega)} \leq \min\left\{(a_1t+a_2)^{-\gamma_2}, \omega_1(t)\right\}.$$

where

$$\omega_1(t) = \left\{\left(\int_\Omega |u_0|^{m+1}dx\right) \times exp(-\mu_A \int_0^t \delta_0(s)ds)\right\}.$$

Here γ_2 *is the constant in Theorem 2 (i). And*

$$(ii) \qquad |u(x,t)|_{L^\infty(\Omega)} \leq \omega_\nu(t) \quad (t \in R^+)$$

where $\omega_\nu(t)$ *is defined in (i-1) or (i-2)*.

We remark that $\omega_\nu(t)$ satisfies $\lim_{t\to\infty}\omega_\nu(t) = 0$.

1.3 PROOF OF THEOREM 1

Proof of Theorem 1. Suppose that the support of $u_0(x)$ is contained in a ball B_ρ in R^N. For the sake of brevity, we assume that $u_0(x)$ is of class $C_0^\infty(B_\rho \cap \Omega)$. Set $\Phi(u) = |u|^{m-1}u$, $\Omega_n \equiv \Omega \cap B_n$, $Q_n \equiv \Omega_n \times R^+$, $\Gamma_n = \partial\Omega_n \times R^+$ and $Q_{n,T} \equiv \Omega_n \times [0,T]$. Assume that $\rho < n < \infty$.

Let us consider the following regularized problem:

$$(3.1)_n \qquad u_t = \Delta(\frac{1}{n}u + \Phi(u)) + b \cdot \nabla(B(u)) - q(t)A(u) \quad \text{in } Q_n,$$

$$(3.2)_n \qquad u(x,t) = 0 \quad \text{on } \Gamma_n,$$

$$(3.3)_n \qquad u(x,0) = u_0(x) \quad \text{in } \Omega_n.$$

It is easy to show that $M_+ = \|u_0\|_\infty$ ($M_- = -\|u_0\|_\infty$) is a supersolution (subsolution) of the problem $(3.1)_n - (3.3)_n$. By the standard theory, for each $T \in R^+$ this regularized problem $(3.1)_n - (3.3)_n$ has a unique classical solution $u_n(x,t) \in C_0^{2,1}(\overline{Q_{n,T}})$ ([8],Ch.V). We can easily show that

$$(3.4) \qquad |u_n(x,t)| \leq M \quad \text{on} \quad \bar{Q_n} = \bar{\Omega_n} \times [0,+\infty).$$

Here $M = \sup_{x\in\Omega}|u_0(x)|$.

Set $\Phi_n(s)=\frac{1}{n}s + \Phi(s)$. From now on we denote u_n by u.
Multiplying (3.1) by $\int_0^u (b(s))^2 ds$ and integrating it over Ω_n, we have

$$\text{(3.5)} \qquad \int_{\Omega_n} \left\{ \nabla_x \Phi_n(u) \nabla_x \left(\int_0^u (b(s))^2 ds \right) \right\} dx$$

$$\leq -\frac{d}{dt} \left\{ \int_{\Omega_n} \left[\int_0^u \left(\int_0^s (b(\tau))^2 d\tau \right) ds \right] dx \right\}.$$

Multiplying (3.1) by $(\Phi_n(u))_t$ and integrating it over Ω_n, we have

$$\int_{\Omega_n} \Phi_n'(u)(u_t)^2 dx + (1/2)\frac{d}{dt} \left(\int_{\Omega_n} |\nabla_x \Phi_n(u)|^2 dx \right)$$

$$\leq \frac{d}{dt} \left\{ \int_{\Omega_n} \left[-q(t) \left(\int_0^u \Phi_n'(s) A(s) ds \right) \right] dx \right\}$$

$$+\frac{1}{2} \int_{\Omega_n} \Phi_n'(u)(u_t)^2 dx + \|\mathbf{b}\|^2 \left\{ \sum_{k=1}^N \left[\int_{\Omega_n} (b(u))^2 (\frac{\partial u}{\partial x_k})^2 \Phi_n'(u) dx \right] \right\}.$$

Here $\|\mathbf{b}\|^2 = \left\{ \sum_{k=1}^N b_k^2 \right\}$. Hence we have

$$\text{(3.6)} \qquad \int_{\Omega_n} \Phi_n'(u)(u_t)^2 dx + \frac{d}{dt} \left(\int_{\Omega} |\nabla_x \Phi_n(u)|^2 dx \right)$$

$$+\frac{d}{dt} \left\{ 2 \int_{\Omega_n} \left[q(t) \left(\int_0^u \Phi_n'(s) A(s) ds \right) \right] dx \right\}$$

$$\leq 2\|\mathbf{b}\|^2 \int_{\Omega_n} \{ \nabla_x \left(\int_0^u (b(s))^2 ds \right) \nabla_x (\Phi_n(u)) \} dx.$$

From (3.5) and (3.6) it follows that

$$\text{(3.7)} \qquad \int_{\Omega_n} \Phi_n'(u)(u_t)^2 dx + \frac{d}{dt} \left(\int_{\Omega_n} |\nabla_x \Phi_n(u)|^2 dx \right)$$

$$+\frac{d}{dt} \left\{ 2 \int_{\Omega_n} \left[q(t) \left(\int_0^u \Phi_n'(s) A(s) ds \right) \right] dx \right\}$$

$$\leq -2\|\mathbf{b}\|^2 \frac{d}{dt} \{ \int_{\Omega_n} \left[\int_0^u \left(\int_0^s (b(\tau))^2 d\tau \right) ds \right] dx \}.$$

Integrating (3.7) in t on $[0, T]$, we obtain

$$\int_0^T \int_{\Omega_n} \Phi_n'(u)(u_t)^2 dx dt + \int_{\Omega_n} |\nabla_x \Phi_n(u(x, T))|^2 dx$$

$$+2q(T) \int_{\Omega_n} \left[\left(\int_0^{u(x,T)} \Phi_n'(s) A(s) ds \right) \right] dx$$

$$+2\|\mathbf{b}\|^2\int_{\Omega_n}\left[\int_0^{u(x,T)}(\int_0^s(b(\tau))^2d\tau)ds\right]dx$$

$$\leq\int_{\Omega}|\nabla_x\Phi_n(u_0(x))|^2dx+2q(0)\int_{\Omega}\left(\int_0^{u_0(x)}\Phi_n^{'}(s)A(s)ds\right)dx$$

$$+2\|\mathbf{b}\|^2\int_{\Omega}\left[\int_0^{u_0(x)}\left(\int_0^s\Big(b(\tau)\Big)^2d\tau\right)ds\right]dx$$

$\equiv C(u_0)$.

Here $C(u_0)$ is a constant depending on only u_0 and data. Thus we obtain

$$\int_0^T\int_{\Omega_n}\Phi_n^{'}(u)(u_t)^2dxdt+\int_{\Omega_n}|\nabla_x\Phi_n(u(x,T))|^2dx \tag{3.8}$$

$$+\|\mathbf{b}\|^2\int_{\Omega_n}\left[\int_0^{u(x,T)}(\int_0^s(b(\tau))^2d\tau)ds\right]dx\leq C(u_0).$$

Furthermore we have

$$\|(\Phi(u_n(t))_t\|^2_{L^2(0,T:L^2(\Omega_n))}\leq(1+mM^{m-1})\int_0^T\int_{\Omega_n}\Phi_n^{'}(u)(u_t)^2dxdt.$$

Hence from (3.8) we have

$$\|\nabla_x\Phi(u_n(t))\|_{L^2(\Omega_n)}\leq C(u_0)\quad(t>0), \tag{3.9}$$

$$\|\mathbf{b}\|^2\int_{\Omega_n}\left[\int_0^{u_n(x,t)}(\int_0^s(b(\tau))^2d\tau)ds\right]dx\leq C(u_0)\quad(t>0), \tag{3.10}$$

and

$$\|(\Phi(u_n(t))_t\|^2_{L^2(0,T:L^2(\Omega_n))}\leq(1+mM^{m-1})C(u_0). \tag{3.11}$$

Here the constant $C(u_0)$ depends only on u_0 and data, but not on n and T. We set $v_n(x,t)=\Phi(u_n(x,t))$. By Ascoli-Arzela's theorem, we can show that there exist a function $v(x,t)$ and a subsequence $\{v_{k_n}(x,t)\}$ of $\{v_n(x,t)\}$ such that

$$v(x,t)\in C([0,T]:L^2_{loc}(\Omega)),$$

and

$$v_{k_n}(x,t)\to v(x,t)\quad\text{as}\quad k_n\to\infty\quad\text{in}\quad C([0,T]:L^2_{loc}(\Omega)).$$

Put $u_n(x,t)=u_{k_n}(x,t)=\Phi^{-1}(v_{k_n}(x,t))$ and $u(x,t)=\Phi^{-1}(v(x,t))$.

We can easily verify that $u(x,t)$ is a solution of the problem (1.1)-(1.3) on $[0,+\infty)$.

Remark. We remark that if $u_0(x)$ is a given non-negative function, then the solution $u(x,t)$ of the problem (1.1) - (1.3) is non-negative.

4. Proof of Theorems 2 and 3

Proof of Theorem 2.

Proof of (i) Let $u_n(x,t)$ be a solution of the regularized problem $(3.1)_n - (3.3)_n$. We denote $u_n(x,t)$ by $u(x,t)$.

Multiplying Eq.$(3.1)_n$ by $sgn(u(t,x))$ and integrating in Ω_n we have

$$\int_{\Omega_n} |u|dx = \|u(t)\|_1 \le \|u_0\|_1 \quad t \in [0,T]. \tag{4.1}$$

Multiply $(3.1)_n$ by $|u|^{p-1}u \quad (m \le p)$ and integrate it in Ω_n. Then we have

$$\int_{\Omega_n} |u|^{p-1}uu_t dx$$

$$\le - \int_{\Omega_n} \{\nabla_x \Phi_n(u) \nabla_x \left(|u|^{p-1}u\right)\} dx - \delta_0(t) \int_{\Omega_n} A(u)|u|^{p-1}udx.$$

Then we obtain

$$\frac{d}{dt}\{\frac{1}{p+1}\int_{\Omega_n} |u|^{p-1}u^2 dx\} \tag{4.2}$$

$$\le - \frac{4p}{(p+m)^2}(\mu_0) \int_{\Omega_n} |\nabla_x(|u|^{(p+m-2)/2}u)|^2 dx - \delta_0(t) \int_{\Omega_n} A(u)|u|^{p-1}udx.$$

We remark that $v \equiv |u|^{(p+m-2)/2}u \in W_0^{1,2}(\Omega_n)$.

When $N \ge 3$,by Sobolev's lemma ([4],[12]) we have

$$(\|u\|_{\frac{N(p+m)}{N-2}})^{p+m} \le C_N \|\nabla(|u|^{(p+m-2)/2}u)\|_2^2. \tag{4.3}$$

By Schwarz's inequality, (4.1) and (4.3), we have

$$\|u_0\|_1^{-\frac{N(m-1)+2p+2}{Np}} \times \|u\|_{p+1}^{\frac{(p+1)(pN+N(m-1)+2)}{pN}} \le \|u\|_{\frac{N(p+m)}{N-2}}^{p+m} \tag{4.4}$$

$$\le C_N \|\nabla(|u|^{(p+m-2)/2}u)\|_2^2.$$

From (4.2) and (4.4) it follows that

$$\frac{d}{dt}\{\int_{\Omega} |u|^{p+1} dx\} \tag{4.5}$$

$$\leq -\frac{4mp(p+1)}{(p+m)^2}(\mu_0)\frac{1}{C_N}\|u_0\|_1^{-\frac{N(m-1)+2p+2}{Np}} \times (\|u\|_{p+1}^{p+1})^{\frac{(pN+N(m-1)+2)}{pN}}.$$

When N=2, by Gagliardo's inequality([4],[12]) we have

$$\|u\|_{\frac{p+m}{2}q}^{\frac{p+m}{2}} \leq A\|u\|_{\frac{p+m}{2}r}^{\frac{p+m}{2}\frac{r}{q}} \|\nabla(|u|^{(p+m-2)/2}u)\|_2^{1-\frac{r}{q}} \tag{4.6}$$

where $1 \leq r < q < \infty$.

By Schwarz's inequality and (4.1), we have

$$\|u\|_{p+1}^{p+1} \leq \|u_0\|_1^{\frac{(p+m)q-2p-2}{(p+m)q-2}} \times \|u\|^{\frac{\frac{(p+m)}{2}\frac{2q}{q-r}\frac{p(q-r)}{p+m)q}}{2}}. \tag{4.7}$$

Set $r = \frac{2(p+1)}{(p+m)}$ and $q = 2r$. Combining (4.6) and (4.7) we obtain

$$(\|u\|_{p+1}^{p+1})^{\frac{(p+m)}{p}} \|u_0\|_1^{-\frac{(p+m)}{p}} \leq A^4 \|\nabla(|u|^{\frac{p+m}{2}-1}u)\|_2^2. \tag{4.8}$$

Then from (4.2) and (4.8) we have

$$\frac{d}{dt}\{\int_\Omega |u|^{p+1}dx\} \tag{4.9}$$

$$\leq -\frac{4mp(p+1)}{(p+m)^2}(\mu_0)A^{-4} \times \|u_0\|_1^{-\frac{(p+m)}{p}} \times (\|u\|_{p+1}^{p+1})^{\frac{(p+m)}{p}}.$$

From (4.5) and (4.9) inequality (i) follows.

Proof of (ii).

We follows Alikakos([1]). Using the estimate (i) with p=m, we can prove (ii) (See [1],[9],[10]).

Proof of Theorem 3.

Proof of (i). By (H.A.3),(H.q.3) and (4.2) we have

$$\frac{d}{dt}\{\frac{1}{p+1}\int_{\Omega_n} |u|^{p-1}u^2dx\} \tag{4.10}$$

$$\leq -\frac{4p}{(p+m)^2}(\mu_0)\int_{\Omega_n} |\nabla_x(|u|^{(p+m-2)/2}u)|^2dx$$

$$-\delta_0(t)\mu_A\int_{\Omega_n} |u|^{p+\nu}dx.$$

When $\nu > 1$, by Schwarz's inequality we have

$$\int_{\Omega_n} |u|^{p+1}dx \leq (\int_{\Omega_n} |u|dx)^{\frac{\nu-1}{\nu+p-1}}(\int_{\Omega_n} |u|^{\nu+p}dx)^{\frac{p}{\nu+p-1}}. \tag{4.11}$$

Putting $p = m$ in (4.10) and (4.11), we obtain

$$\frac{d}{dt}\left(\int_{\Omega_n} |u|^{m+1} dx\right) \tag{4.12}$$

$$\leq -\frac{m+1}{m}(\mu_0)\frac{1}{C_N}\|u_0\|_1^{-\frac{N(m-1)+2m+2}{Nm}}\left(\|u\|_{m+1}^{m+1}\right)^{\frac{(mN+N(m-1)+2)}{mN}}$$

$$-\delta_0(t)\mu_A\|u_0\|_1^{-\frac{\nu-1}{m}}\left(\|u\|_{m+1}^{m+1}\right)^{\frac{\nu+m-1}{m}}.$$

Combining (4.12) and Theorem 2 (i), we have

$$\|u(\cdot\, t)\|_{L^{m+1}(\Omega)} \leq \min\left\{(a_1 t + a_2)^{-\gamma_2}, \omega_\nu(t)\right\} \tag{4.13}$$

where

$$\omega_\nu(t) = \left\{\left(\int_\Omega |u_0|^{m+1} dx\right)^{\frac{-(\nu-1)}{m}} + \frac{\nu-1}{m}\mu_A\|u_0\|_1^{-\frac{\nu-1}{m}}\int_0^t \delta_0(s)ds\right\}^{\frac{-m}{\nu-1}}.$$

When $\nu = 1$, we have

$$\|u(\cdot\, t)\|_{L^{m+1}(\Omega)} \leq \min\left\{(a_1 t + a_2)^{-\gamma_2}, \omega_1(t)\right\} \tag{4.14}$$

where

$$\omega_1(t) = \left\{\left(\int_\Omega |u_0|^{m+1} dx\right) \times exp\left(-\mu_A \int_0^t \delta_0(s)ds\right)\right\}.$$

Here γ_2 is the constant in Theorem 2 (i).

Combining (i) and Theorem 2(ii) we can prove

$$|u(x,t)|_{L^\infty(\Omega)} \leq \omega_\nu(t) \quad (t \in R^+) \tag{4.15}$$

where $\omega_\nu(t)$ is defined in (4.13) and (4.14) .

References

[1] Alikakos, N.D. (1979). *L^p bounds of solutions of reaction-diffusion equations*, Comm. Partial Diff. Eq., Vol.4, (pages 827–868).

[2] Aronson, D.G. and Peletier, L.A. (1981). *Large time behaviour of solutions of the porous medium equation in bounded domains*, Jour. Diff. Eq., Vol. 39, (pages 378–412).

[3] Bertsch, M., Nanbu, T. and Peletier, L.A. (1982). *Decay of solutions of a degenerate nonlinear diffusion equation*, Nonlinear Analysis T. M. A., Vol.6, (pages 539–554).

[4] DiBenedetto, E. (1991). *Degenerate Parabolic Equations*, Springer-Verlag.

[5] Ebihara, Y. and Nanbu, T. (1980). *Global classical solutions to* $u_t - \Delta(u^{2m+1}) + \lambda u = f$, Jour. Diff. Eq., Vol. 38, (pages 260–277).

[6] Escobedo, M. and Zuazua,E. (1991). *Large Time Behavior for Convection Diffusion Equations in* R^N, Jour. Func.Anal., Vol. 100, (pages 119–161).

[7] Kalashinikov,A.S. (1987). *Some problems of the qualitative theory of non-linear degenerate second-order parabolic equations*, Russian Math. Surveys, Vol. 42, (pages 169–222).

[8] Ladyzhenskaya, O. A. Solonikov, V. A. and Ura'lceva, N. N. (1968). *Linear and quasilinear equations of parabolic type*, (Translation M. M. 23, Amer. Math. Soc.).

[9] Nakao, M. (1985). *Existence and decay of global solutions of some nonlinear degenerate parabolic equations*, Jour. Math. Anal. Appl., Vol. 109, (pages 118–129).

[10] Nanbu, T. (1984). *Some degenerate nonlinear parabolic equations*, Math. Rep. Coll. Gen. Ed. Kyushu Univ., Vol. 14, (pages 91–110).

ON REGULARITY RESULTS FOR VARIATIONAL-HEMIVARIATIONAL INEQUALITIES

Z. Naniewicz

Warsaw University
Institute of Applied Mathematics and Mechanics
Banach Str. 2
02-097 Warsaw, Poland
E-mail: znaniew@mimuw.edu.pl

P.D. Panagiotopoulos

Aristotle University of Thessaloniki
Department of Civil Engineering
Institute of Steel Structures
54006 Thessaloniki, Greece
E-mail: pdpana@heron.civil.auth.gr

R.P. Gilbert et al.(eds.), Direct and Inverse Problems of Mathematical Physics, 301–322.
© 2000 Kluwer Academic Publishers.

Abstract: The aim of the present paper is to investigate the regularity of the nonlinear term which results from the nonconvex part of the energy in variational-hemivariational inequalities. This term expresses the virtual work of the nonmonotone multivalued stress-strain or reaction-displacement law which gives rise to the variational-hemivariational inequality under consideration.

1.1 INTRODUCTION

The works of the second author of the present paper initiated the theory of hemivariational inequalities in the early eighties. The main reason for the creation of this theory is the need for the derivation of variational expressions for mechanical problems involving nonconvex, nonsmooth energy functions. For informations concerning the theory of hemivariational inequalities cf. [18][13]. Until now a basic open problem was the problem of the regularity of the nonlinear term $j^0(u,v)$ appearing in every hemivariational inequality. This is a problem with concrete mechanical meaning, because it gives a characterization of the nature of the reactions related to the nonsmooth nonlinearity of the hemivariational inequality. Let us consider first a simple hemivariational inequality studied in [16][17], in order to explain better the arising question. Let V be a real Hilbert space such that

$$V \subset [L^2(\Omega)]^n \subset V^\star.$$

Here $V^\star$ denotes the dual space of V, Ω is an open bounded subset of R^n, and the injections are continuous and dense. We denote in this Section by $(\cdot,\cdot)$ the $[L^2(\Omega)]^n$ inner product and the duality pairing, by $||\cdot||$ the norm of V and by $|\cdot|_2$ the $[L^2(\Omega)]^n$-norm. We recall that the form $(\cdot,\cdot)$ extends uniquely from $V \times L^2[(\Omega)]^n$ to $V \times V^\star$. Moreover let $L : V \to L^2(\Omega)$, $Lu = \hat{u}$, $\hat{u}(x) \in R$ be a linear continuous mapping. We assume that $l \in V^\star$, that

$$L : V \to L^2(\Omega) \qquad \text{is compact}$$

and that

$$\tilde{V} = \{v \in V : \hat{v} \in L^\infty(\Omega)\} \qquad \text{is dense in } V \text{ for the } V\text{-norm.}$$

Moreover let $\tilde{V}$ have a Galerkin base. It is also assumed that $a(\cdot,\cdot): V \times V \to R$ is a bilinear symmetric continuous form, which is coercive, i.e. there exists $c > 0$ constant such that

$$a(v,v) \geq c||v||^2 \qquad \forall v \in V.$$

We denote by $j: R \to R$, a locally Lipschitz function defined in the following way: let $\beta \in L^\infty_{\text{loc}}(R)$ and consider the functions $\bar{\beta}_\mu$ and $\bar{\bar{\beta}}_\mu$ defined by

$$\bar{\beta}_\mu(\xi) = \operatorname*{essinf}_{|\xi_1-\xi|\leq\mu} \beta(\xi_1) \text{ and } \bar{\bar{\beta}}_\mu(\xi) = \operatorname*{esssup}_{|\xi_1-\xi|\leq\mu} \beta(\xi_1).$$

They are decreasing and increasing functions of μ, respectively; therefore the limits for $\mu \to 0_+$ exist. We denote them by $\bar{\beta}(\xi)$ and $\bar{\bar{\beta}}(\xi)$ respectively; the multivalued function $\tilde{\beta}$ is defined by

$$\tilde{\beta}(\xi) = [\bar{\beta}(\xi), \bar{\bar{\beta}}(\xi)].$$

If $\beta(\xi_{\pm 0})$ exists for every $\xi \in R$, then a locally Lipschitz function $j: R \to R$ can be determined up to an additive constant such that

$$\tilde{\beta}(\xi) = \partial j(\xi).$$

Now we formulate the following coercive hemivariational inequality (problem P^C): Find $u \in V$ such that

$$a(u, v-u) + \int_\Omega j^0(\hat{u}, \hat{v} - \hat{u}) d\Omega \geq (l, v-u) \qquad \forall v \in V.$$

An element $u \in V$ is said to be a solution of P^C if there exists $\chi \in L^1(\Omega)$ with $L^*\chi \in V^*$ (L^* denotes the transpose operator of L) such that [13][16]

$$a(u,v) + (L^*\chi, v) = (l, v) \quad \forall v \in V$$

and

$$\chi(x) \in \partial j(u(x)) \quad \text{a.e. on } \Omega,$$

and where

$$(L^*\chi, v) = \int_\Omega \chi L v d\Omega = \int_\Omega \chi \hat{v} d\Omega \quad \text{if } v \in \tilde{V}.$$

Therefore one can give the following definition. An element $u \in V$ is said to be a solution of P^C, if there exists $\chi \in L^1(\Omega)$ such that (Problem P_1^C)

$$a(u,v) + \int_\Omega \chi \hat{v} d\Omega = (l, v) \quad \forall v \in \tilde{V}$$

and

$$\chi(x) \in \partial j(u(x)) \quad \text{a.e. on } \Omega,$$

hold. Obviously due to the density assumption the two above definitions are equivalent. Indeed one can show that the second problem has a solution [13][16].

It arises naturally the question concerning the regularity of the product χv for $v \in V$ (and not for $v \in \tilde{V}$). Since v is a generalized displacement and χ is a generalized force for the corresponding mechanical problem, one expects that the product, i.e. the work, $v\chi$ should be an L^1-function at least, and that the problem P^C and P_1^C are completely equivalent. This result is derived in the present paper, using some natural growth assumptions analogous to the ones of [15].

We consider further a general variational-hemivariational inequality related to a multidimensional superpotential law, and with respect to this inequality we prove the aforementined result.

1.2 FORMULATION OF THE PROBLEM

Let $V = H^1(\Omega; R^N)$, $N \geq 1$, be a vector valued Sobolev space compactly imbedded into $L^p(\Omega; R^N)$, $p > 1$, where Ω is a bounded domain in R^m, $m \geq 1$, with sufficiently smooth boundary $\partial\Omega$. We write $\|\cdot\|_V$ and $\|\cdot\|_{L^p(\Omega;R^N)}$ for the norms in V and $L^p(\Omega; R^N)$, respectively. For the pairing over $V^\star \times V$ the symbol $\langle\cdot,\cdot\rangle_V$ will be used. $V^\star$ denotes the dual of V.

Let $A : V \to V^\star$ be a bounded operator, i.e. an operator which maps bounded sets into bounded sets, and we assume in addition that A is pseudo-monotone. Let us recall, following [3], the notion of the pseudo-monotonicity: a multivalued mapping $T : V \to 2^{V^\star}$ is said to be pseudo-monotone, if the following conditions hold:

(i) the set Tu is nonempty, bounded, closed and convex for all $u \in V$;

(ii) if $u_n \to u$ strongly in V, $u_n^\star \in Tu_n$, and $u_n^\star \to u^\star$ weakly in $V^\star$, then $u^\star \in Tu$;

(iii) if $u_n \to u$ weakly in V and $u_n^\star \in Tu_n$ is such that $\limsup\langle u_n^\star, u_n - u\rangle_V \leq 0$, then to each $v \in V$ there exists $u^\star(v) \in Tu$ with the property that

$$\liminf\langle u_n^\star, u_n - v\rangle_V \geq \liminf\langle u^\star(v), u - v\rangle_V.$$

Further, let $j : R^N \to R$ be a locally Lipschitz function from R^N into R. Throughout the paper we assume j that satisfies the following unilateral growth restriction [12]:

$$j^0(\xi; \eta - \xi) \leq \alpha(r)(1 + |\xi|^\sigma) \quad \forall\ \xi, \eta \in R^N,\ |\eta| \leq r,\ r \geq 0. \tag{1}$$

Here $1 \leq \sigma < p$, $\alpha : R^+ \to R^+$ is assumed to be a nondecreasing function from R^+ into R^+, and $j^0(\cdot,\cdot)$ denotes the generalized Clarke differential [4], i.e.

$$j^0(\xi; \eta) = \limsup_{\substack{h \to 0 \\ \lambda \to 0_+}} \frac{j(\xi + h + \lambda\eta) - j(\xi + h)}{\lambda}.$$

Let $\Phi : V \to R \cup \{+\infty\}$ be a convex, lower semicontinuous and proper function, $g \in V^\star$ be an element of $V^\star$. We study the following variational-hemivariational inequality problem:

Problem (P). Find $u \in V$ such that

$$\langle Au - g, v - u\rangle_V + \Phi(v) - \Phi(u) + \int_\Omega j^0(u; v - u)d\Omega \geq 0 \quad \forall\, v \in V, \tag{2}$$

where the integral above is assumed to be $+\infty$, whenever $j^0(u; v - u) \notin L^1(\Omega)$.

We are going to establish a stronger result, namely, to show the existence of $u \in V$, $\chi \in L^1(\Omega; R^N)$ and $u^\star \in V^\star$ fulfilling the relations

$$\langle Au - g, v - u\rangle_V + \langle u^\star, v - u\rangle_V + \int_\Omega \chi \cdot (v - u)d\Omega = 0 \ \forall v \in V \cap L^\infty(\Omega; R^N) \tag{3}$$

$$u^\star \in \partial\Phi(u), \quad \chi \in \partial j(u) \text{ a.e. in } \Omega, \quad \chi \cdot u \in L^1(\Omega). \tag{4}$$

Here the dot " $\cdot$ " denotes the inner product in R^N. We can derive the variational-hemivariational inequality (2) from (3) and (4).

However it should be noted for the present that the unilateral growth restriction (1) does not guarantee the finite integrability of $j^0(u; v-u)$ in Ω for each $v \in V$, even for $u \in V$ with $j(u) \in L^1(\Omega)$.

1.3 HEMIVARIATIONAL INEQUALITIES RELATED TO MULTIVALUED MAPPINGS

In this Section we prove the following result.

Theorem 3.1. Let $T : V \to 2^{V^\star}$ be a pseudo-monotone, bounded and coercive mapping, i.e.

$$\langle u^\star, u\rangle_V \geq c(||u||_V)||u||_V, \quad u^\star \in Tu,$$

for some function $c : R^+ \to R$ with $c(r) \to \infty$ as $r \to \infty$. Suppose that the unilateral growth condition (1) holds together with

$$j^0(\xi; -\xi) \leq k(1 + |\xi|) \quad \forall\, \xi \in R^N, \tag{5}$$

k being a constant. Then for a given $g \in V^\star$ there exist $u \in V$, $u_0^\star \in V^\star$ and $\chi \in L^1(\Omega; R^N)$ such that

$$\langle u_0^\star - g, v - u\rangle_V + \int_\Omega \chi \cdot (v-u) d\Omega = 0 \ \ \forall v \in V \cap L^\infty(\Omega; R^N) \tag{6}$$

$$u_0^\star \in T(u), \quad \chi \in \partial j(u) \text{ a.e. in } \Omega, \quad \chi \cdot u \in L^1(\Omega). \tag{7}$$

The proof of this theorem will follow from a sequence of lemmas below. We start with the finite dimensional approximation related to problem (6)–(7).

Let Λ be the family of all finite dimensional subspaces F of $V \cap L^\infty(\Omega; R^N)$, ordered by inclusion. Denote by $i_F : F \to V$ the inclusion mapping of F into V and by $i_F^\star : V^\star \to F^\star$ the dual projection mapping of $V^\star$ into $F^\star$, $F^\star$ being the dual of F. The pairing over $F^\star \times F$ will be denoted by $\langle \cdot, \cdot \rangle_F$. Set $T_F := i_F^\star T i_F$ and $g_F := i_F^\star g$.

For any $F \in \Lambda$ we formulate the finite dimensional problem.

Problem (P_F). Find $u_F \in F$ and $\chi_F \in L^1(\Omega; R^N)$ for which there exists $u_F^\star \in T_F(u_F)$ such that

$$\begin{gathered} \langle u_F^\star - g, v\rangle_V + \int_\Omega \chi_F \cdot v d\Omega = 0 \quad \forall\, v \in F \\ \chi_F \in \partial j(u_F) \quad \text{a.e. in } \Omega. \end{gathered} \tag{8}$$

Proposition 3.2. Let $T : V \to 2^{V^\star}$ be a pseudo-monotone, bounded and coercive mapping from V into $2^{V^\star}$. Suppose that (1) and (5) hold. Then for any $F \in \Lambda$ the (P_F) problem has at least one solution. Moreover, $\{u_F\}_{F\in\Lambda}$ is bounded in V independently of F and the corresponding set $\{\chi_F\}_{F\in\Lambda}$ is weakly precompact in $L^1(\Omega; R^N)$.

Proof. The proof of Proposition 3.2 will be divided into a sequence of steps. Step 1. Recall first that j has been assumed to be locally Lipschitz. Thus for each $v_F \in F$ a constant C_F can be determined such that

$$|j^0(v_F, w)| \leq C_F|w| \leq C_F||w||_{L^\infty(\Omega)}$$

for any $w \in L^\infty(\Omega; R^N)$. This fact allows us to define for each $v_F \in F$ a mapping $\Gamma_F : F \to 2^{L^1(\Omega;R^N)}$ by means of the following formula

$$\Gamma_F(v_F) = \{\psi \in L^1(\Omega; R^N) : \int\limits_\Omega \psi \cdot w d\Omega \leq \int\limits_\Omega j^0(v_F, w) d\Omega \ \ \forall w \in L^\infty(\Omega; R^N)\}.$$

Notice that if $\psi \in \Gamma_F(v_F)$, then $\psi \in \partial j(v_F)$ a.e. in Ω. It can be easily deduced that for each $v_F \in F$, $\Gamma_F(v_F)$ is a nonempty, convex and weakly compact subset of $L^1(\Omega; R^N)$ (for details see [13]).

Let $\tau_F : L^1(\Omega; R^N) \to F^\star$ be an operator which assigns to any $\psi \in L^1(\Omega; R^N)$ the element $\tau_F\psi \in F^\star$ defined by

$$\langle \tau_F\psi, v\rangle_F := \int\limits_\Omega \psi \cdot v d\Omega \quad \text{for any } v \in F. \tag{9}$$

Now let us consider a mapping $B_F : F \to 2^{F^\star}$ given by formula

$$B_F(v_F) := \tau_F\Gamma_F(v_F) \quad \text{for } v_F \in F. \tag{10}$$

The following properties of B_F can be proved.

Lemma 3.2. B_F is an upper semicontinuous mapping from F into $2^{F^\star}$ with nonempty, bounded, closed and convex values.

Proof. Let us note that τ_F defined by (9) is linear and continuous from the weak topology on $L^1(\Omega; R^N)$ to the (unique) topology on $F^\star$. Thus taking into account the already established properties of $\Gamma_F(v_F)$, $v_F \in F$, it is easily seen that $B_F(v_F)$ is a nonempty bounded, closed, convex subset of $F^\star$ for each $v_F \in F$. Thus it remains to show that B_F is upper semicontinuous from F into $2^{F^\star}$. For this purpose let us suppose that a sequence $\{v_{F_n}\} \subset F$ converges to v_F in F, and that the corresponding sequence $\{\tau_F\psi_n\} \subset F^\star$ with $\psi_n \in \Gamma_F(v_{F_n})$, converges to some $\psi^\star \in F^\star$. We have to prove that there exists $\psi \in \Gamma_F(v_F)$ such that $\psi^\star = \tau_F\psi$.

First we notice that the convergence of $\{v_{F_n}\}$ in F implies the existence of a finite upper bound for $\{||v_{F_n}||_{L^\infty(\Omega;R^N)}\}$, say $\hat{C}$, i.e.,

$$||v_{F_n}||_{L^\infty(\Omega;R^N)} \leq \hat{C} \quad \text{for } n = 1, 2, \dots . \tag{11}$$

The Lipschitz constant of j on the corresponding ball $\{\xi \in R^N; |\xi| \leq \hat{C}\}$ will be denoted by C. Taking into account that for any measurable $\omega \subset \Omega$

$$\int_\Omega \psi_n \cdot w d\Omega \leq \int_\Omega j^0(v_{Fn}, w) d\Omega \quad \forall\, w \in L^\infty(\Omega; R^N), \tag{12}$$

we obtain the estimate

$$\int_\omega |\psi_n| d\Omega \leq \sqrt{N} \int_\omega C d\Omega, \quad \omega \subset \Omega, \tag{13}$$

which implies immediately that for any $\varepsilon > 0$ there exists a $\delta > 0$ such that

$$\int_\omega |\psi_n| d\Omega \leq \varepsilon \quad n = 1, 2, \ldots, \tag{14}$$

provided $\operatorname{mes} \omega < \delta$. Applying the Dunford-Pettis theorem we obtain the weak precompactness of $\{\psi_n\}$ in $L^1(\Omega; R^N)$. Accordingly, a subsequence of $\{\psi_n\}$ can be extracted (again denoted by the same notation) such that $\psi_n \to \psi$ weakly in $L^1(\Omega; R^N)$ for some $\psi \in L^1(\Omega; R^N)$. This fact implies that $\tau_F \psi_n \to \tau_F \psi = \psi^\star$ in $F^\star$. We claim that $\psi \in \Gamma_F(v_F)$. Indeed, taking into account the upper semicontinuity of the function

$$F \ni v_{F_n} \to \int_\Omega j^0(v_{F_n}, w) d\Omega, \quad \forall\, w \in L^\infty(\Omega; R^N), \tag{15}$$

we are led, by means of (12), directly to the relation

$$\int_\Omega \psi \cdot w d\Omega \leq \int_\Omega j^0(v_F, w) d\Omega \quad \forall\, w \in L^\infty(\Omega; R^N), \tag{16}$$

implying immediately that $\psi \in \Gamma_F(v_F)$. The proof is complete. □

Lemma 3.3. For every $F \in \Lambda$ the problem (P_F) has at least one solution.

Proof. In order to obtain the solvability of (P_F) it suffices to show that g_F belongs to the range of the multivalued mapping $T_F + B_F$. We shall show that, in fact, Range $(T_F + B_F) = F^\star$. Indeed, due to Lemma 3.2, for each $v \in F$, $T_F(v) + B_F(v)$ is a nonempty, bounded, closed, convex subset of $F^\star$. Further, the pseudo-monotonicity of T implies the upper semicontinuity of T_F from F into $F^\star$. Therefore, $T_F + B_F$ is upper semicontinuous from F into $2^{F^\star}$. Moreover, the coercivity of T and (5) yield

$$\langle v^\star, v\rangle_F + \int_\Omega \psi \cdot v d\Omega \geq \langle v^\star, v\rangle_V - \int_\Omega j^0(v, -v) d\Omega$$

$$\geq c(\|v\|_V)\|v\|_V - \int_\Omega k|v| d\Omega \geq c(\|v\|_V)\|v\|_V - k(\operatorname{mes}\Omega)^{\frac{p-1}{p}} \|v\|_{L^p(\Omega)}$$

$$\geq c(\|v\|_V)\|v\|_V - k\gamma(\operatorname{mes}\Omega)^{\frac{p-1}{p}} \|v\|_V, \; v \in F, v^\star \in T_F(v), \; \psi \in \Gamma_F(v), \tag{17}$$

where γ is a positive constant resulting from the imbedding $V \subset L^p(\Omega)$ which is compact. This means that $T_F + B_F$ is also coercive and consequently, by the well known result (cf. [3]), Range $(T_F + B_F) = F^\star$. Thus for $g \in V^\star$ there exist $u_F \in F$, $u_F^\star \in T_F(u_F)$ and $\chi_F \in \Gamma_F(u_F)$ such that $g_F = u_F^\star + \tau_F \chi_F$, from which (6)–(7) can be easily deduced. The proof is complete. □

Lemma 3.4. There exists a positive constant M not depending on $F \in \Lambda$ such that

$$||u_F||_V \leq M, \quad F \in \Lambda. \tag{18}$$

Proof. To show that the family $\{u_F\}$ is uniformly bounded we make use of the coercivity of T and (5). The estimations follow

$$\begin{aligned} ||g||_{V^\star}||u_F||_V &\geq \langle g, u_F\rangle_V = \langle g_F, u_F\rangle_F = \langle u_F^\star, u_F\rangle_F + \int_\Omega \chi_F \cdot u_F d\Omega \\ &\geq \langle u_F^\star, u_F\rangle_V - \int_\Omega j^0(u_F, -u_F) d\Omega \qquad (19) \\ &\geq c(||u_F||_V)||u_F||_V - \int_\Omega k|u_F| d\Omega \\ &\geq c(||u_F||_V)||u_F||_V - k(\operatorname{mes}\Omega)^{\frac{p-1}{p}}||u_F||_{L^p(\Omega)} \\ &\geq c(||u_F||_V)||u_F||_V - k\gamma(\operatorname{mes}\Omega)^{\frac{p-1}{p}}||u_F||_V, \end{aligned}$$

which, due to the behaviour of $c = c(\cdot)$ at infinity, lead easily to the assertion. □

The next lemma concerns the compactness property of the set $\{\chi_F : F \in \Lambda\}$ in $L^1(\Omega; R^N)$.

Lemma 3.5. Let for some $F \in \Lambda$ a pair $(u_F, \chi_F) \in F \times L^1(\Omega; R^N)$ be a solution of (P_F). Then the set $\{\chi_F \in L^1(\Omega; R^N) : (u_F, \chi_F)$ is a solution of (P_F) for some $u_F \in F$, $F \in \Lambda\}$ is weakly precompact in $L^1(\Omega; R^N)$.

Proof. According to the Dunford-Pettis theorem it suffices to show that for each $\varepsilon > 0$ a $\delta_\varepsilon > 0$ can be determined such that for any $\omega \subset \Omega$ with $\operatorname{mes}\omega < \delta_\varepsilon$,

$$\int_\omega |\chi_F| d\Omega < \varepsilon, \quad F \in \Lambda. \tag{20}$$

Fix $r > 0$ and let $\eta \in R^N$ be such that $|\eta| \leq r$. Then we have

$$\chi_F \cdot (\eta - u_F) \leq j^0(u_F, \eta - u_F)$$

from which, by virtue of (5) it results that

$$\chi_F \cdot \eta \leq \chi_F \cdot u_F + \alpha(r)(1 + |u_F|^\sigma)$$

a.e. in Ω. Let us set

$$\eta = \frac{r}{\sqrt{N}}(\operatorname{sgn}\chi_{F_1}, \ldots, \operatorname{sgn}\chi_{F_N}),$$

where χ_{F_i}, $i = 1, 2, \ldots, N$, are the components of χ_F and where

$$\operatorname{sgn} y = \begin{cases} 1 & \text{if } y > 0 \\ 0 & \text{if } y = 0 \\ -1 & \text{if } y < 0 \end{cases}.$$

It is not difficult to verify that $|\eta| \leq r$ for almost all $x \in \Omega$ and that

$$\chi_F \cdot \eta \geq \frac{r}{\sqrt{N}}|\chi_F|.$$

Therefore using (1) we are led to the estimate

$$\frac{r}{\sqrt{N}}|\chi_F)| \leq \chi_F \cdot u_F + \alpha(r)(1 + |u_F|^\sigma).$$

Integrating this inequality over $\omega \subset \Omega$ yields

$$\begin{aligned} \int_\omega |\chi_F| d\Omega \quad &\leq \quad \frac{\sqrt{N}}{r}\int_\omega \chi_F \cdot u_F d\Omega + \frac{\sqrt{N}}{r}\alpha(r) \operatorname{mes}\omega \\ &+ \quad \frac{\sqrt{N}}{r}\alpha(r)(\operatorname{mes}\omega)^{\frac{p-\sigma}{p}}||u_F||^\sigma_{L^p(\Omega)}. \end{aligned} \tag{21}$$

Thus, from (18) we obtain

$$\begin{aligned} \int_\omega |\chi_F| d\Omega \quad &\leq \quad \frac{\sqrt{N}}{r}\int_\omega \chi_F \cdot u_F d\Omega \\ &+ \quad \frac{\sqrt{N}}{r}\alpha(r) \operatorname{mes}\omega + \frac{\sqrt{N}}{r}\alpha(r)(\operatorname{mes}\omega)^{\frac{p-\sigma}{p}}\gamma^\sigma||u_F||^\sigma_V \\ &\leq \quad \frac{\sqrt{N}}{r}\int_\omega \chi_F \cdot u_F d\Omega \\ &+ \quad \frac{\sqrt{N}}{r}\alpha(r) \operatorname{mes}\omega + \frac{\sqrt{N}}{r}\alpha(r)(\operatorname{mes}\omega)^{\frac{p-\sigma}{p}}\gamma^\sigma M^\sigma. \end{aligned} \tag{22}$$

Now we will show that

$$\int_\omega \chi_F \cdot u_F d\Omega \leq C \tag{23}$$

for some positive constant C not depending on $\omega \subset \Omega$ and $F \in \Lambda$. Indeed, from (5) one can easily deduce that

$$\chi_F \cdot u_F + k(1 + |u_F|) \geq 0 \text{ a.e. in } \Omega.$$

It follows that

$$\int_\omega [\chi_F \cdot u_F + k(1+|u_F|)]d\Omega \le \int_\Omega [\chi_F \cdot u_F + k(1+|u_F|)]d\Omega,$$

and consequently that

$$\begin{aligned}
\int_\omega \chi_F \cdot u_F d\Omega &\le \int_\Omega \chi_F \cdot u_F d\Omega \\
&+ k \operatorname{mes}\Omega + k\gamma(\operatorname{mes}\Omega)^{\frac{p-1}{p}} ||u_F||_V \\
&\le \int_\Omega \chi_F \cdot u_F d\Omega \\
&+ k \operatorname{mes}\Omega + k\gamma(\operatorname{mes}\Omega)^{\frac{p-1}{p}} M. \qquad (24)
\end{aligned}$$

But T maps bounded sets into bounded sets. Therefore, by means of (5) and (18) we conclude that

$$\int_\Omega \chi_F \cdot u_F d\Omega = -\langle u_F^\star - g, u_F\rangle_V \le ||u_F^\star - g||_{V^\star} ||u_F||_V \le \hat{C}, \quad \hat{C} = \text{const.}$$

¿From the last two estimates (23) is easily obtained. Further, from (22) and (23), for $r > 0$ one gets

$$\int_\omega |\chi_F| d\Omega \le \frac{\sqrt{N}}{r} C + \frac{\sqrt{N}}{r}\alpha(r) \operatorname{mes}\omega + \frac{\sqrt{N}}{r}\alpha(r)(\operatorname{mes}\omega)^{\frac{p-\sigma}{p}} \gamma^\sigma M^\sigma. \qquad (25)$$

This estimation is important for the derivation of (20). Let $\varepsilon > 0$. Fix $r > 0$ with

$$\frac{\sqrt{N}}{r} C < \frac{\varepsilon}{2} \qquad (26)$$

and determine $\delta_\varepsilon > 0$ small enough such that

$$\frac{\sqrt{N}}{r}\alpha(r) \operatorname{mes}\omega + \frac{\sqrt{N}}{r}\alpha(r)(\operatorname{mes}\omega)^{\frac{p-\sigma}{p}} \gamma^\sigma M^\sigma \le \frac{\varepsilon}{2} \qquad (27)$$

whenever $\operatorname{mes}\omega < \delta_\varepsilon$. Then from (25), (26) and (27) it follows

$$\int_\omega |\chi_F| d\Omega \le \varepsilon \quad F \in \Lambda, \qquad (28)$$

for any $\omega \subset \Omega$ with $\operatorname{mes}\omega < \delta_\varepsilon$. Accordingly, the weak precompactness of $\{\chi_F : F \in \Lambda\}$ in $L^1(\Omega; R^N)$ is proved. □

The combination of the results of Lemmas 3.2 – 3.5 implies Proposition 3.2. Now we are in a position to complete the proof of Theorem 3.1.

Proof of Theorem 3.1.
Step 1. For $F \in \Lambda$ let

$$W_F = \bigcup_{\substack{F' \in \Lambda \\ F' \supset F}} \{(u_{F'}, \chi_{F'}) \in V \times L^1(\Omega; R^N) : (u_{F'}, \chi_{F'}) \text{ satisfies } (P_{F'})\}. \quad (29)$$

We use the symbol weakcl (W_F) to denote the closure of W_F in the weak topology of $V \times L^1(\Omega; R^N)$. Moreover, let

$$Z = \bigcup_{F \in \Lambda} \{\chi_F \in L^1(\Omega; R^N) : (u_F, \chi_F) \text{ satisfies } (P_F)\}.$$

Denoting by weakcl (Z) the closure of Z in the weak topology of $L^1(\Omega; R^N)$ we get

$$\text{weakcl}\,(W_F) \subset B_V(O, M) \times \text{weakcl}\,(Z) \quad \forall\, F \in \Lambda.$$

Since $B_V(O, M)$ is weakly compact in V and, by Lemma 3.5, weakcl (Z) is weakly compact in $L^1(\Omega; R^N)$, the family $\{\text{weakcl}(W_F) : F \in \Lambda\}$ is contained in the weakly compact set $B_V(O, M) \times \text{weakcl}(Z)$ in $V \times L^1(\Omega; R^N)$. Now let us notice that for any $F_1, \ldots, F_k \in \Lambda$, $k = 1, 2, \ldots$, we have the inclusion $W_{F_1} \cap \ldots \cap W_{F_k} \supset W_F$, with $F = F_1 + \ldots + F_k$, from which it follows by Lemma 3.4 that the family $\{\text{weakcl}(W_F) : F \in \Lambda\}$ has the finite intersection property. Thus the intersection

$$\bigcap_{F \in \Lambda} \text{weakcl}\,(W_F)$$

is not empty. Let (u, χ) be an element of this intersection.

Step 2. Let us fix $v \in V \cap L^\infty(\Omega; R^N)$ arbitrarily. We choose $F \in \Lambda$ such that $v \in F$. Thus there exists a sequence $\{(u_{F_n}, \chi_{F_n})\} \subset W_F$ (for the sake of simplicity this sequence will be denoted by (u_n, χ_n)), for which there exists $u_n^\star \in F^\star$ with the properties that

$$\begin{gathered} \langle u_n^\star - g, w - u_n \rangle_V + \int_\Omega \chi_n \cdot (w - u_n) d\Omega = 0 \quad \forall\, w \in F_n \\ u_n^\star \in T_{F_n}(u_n), \quad \chi_n \in \partial j(u_n) \end{gathered} \quad (30)$$

and

$$\begin{gathered} u_n \to u \quad \text{weakly in } V \\ \chi_n \to \chi \quad \text{weakly in } L^1(\Omega; R^N). \end{gathered} \quad (31)$$

Taking into account (30) and the fact that $v \in F \subset F_n$ we have

$$\langle u_n^\star - g, v \rangle_V + \int_\Omega \chi_n \cdot v d\Omega = 0, \quad n = 1, 2 \ldots . \quad (32)$$

Now we can pass to the limit in (32) for $n \to \infty$. The boundedness of $\{u_n^\star\}$ (T has been assumed to be bounded) allows us to conclude that for some $u_0^\star \in V^\star$,

$u_n^\star \to u_0^\star$ weakly in $V^\star$ (by passing to a subsequence, if necessary). Since $v \in V \cap L^\infty(\Omega; R^N)$ has been chosen arbitrarily, the equality

$$\langle u_0^\star - g, v\rangle_V + \int_\Omega \chi \cdot v d\Omega = 0 \tag{33}$$

is valid for any $v \in V \cap L^\infty(\Omega; R^N)$.

Step 2. We prove that the first claim in (25), i.e. $\chi \in \partial j(u)$, holds. Since V is compactly imbedded into $L^p(\Omega; R^N)$, from (31) we obtain (by passing to a subsequence, if necessary)

$$u_n \to u \quad \text{strongly in } L^p(\Omega; R^N). \tag{34}$$

This implies that for a subsequence of $\{u_n\}$ (again denoted by the same symbol) one gets

$$u_n \to u \quad \text{a.e. in } \Omega.$$

Thus Egoroff's theorem can be applied from which it follows that for any $\varepsilon > 0$ a subset $\omega \subset \Omega$ with $\operatorname{mes}\omega < \varepsilon$ can be determined such that

$$u_n \to u \quad \text{uniformly in } \Omega \setminus \omega$$

with $u \in L^\infty(\Omega \setminus \omega; R^N)$. Let $v \in L^\infty(\Omega \setminus \omega; R^N)$ be an arbitrary function. From the estimate

$$\int_{\Omega\setminus\omega} \chi_n \cdot v d\Omega \le \int_{\Omega\setminus\omega} j^0(u_n, v) d\Omega$$

combined with the weak convergence in $L^1(\Omega; R^N)$ of χ_n to χ, (34) and with the upper semicontinuity of

$$L^\infty(\Omega \setminus \omega; R^N) \ni w \longmapsto \int_{\Omega\setminus\omega} j^0(w, v) d\Omega$$

we obtain

$$\int_{\Omega\setminus\omega} \chi \cdot v d\Omega \le \int_{\Omega\setminus\omega} j^0(u, v) d\Omega \quad \forall\, v \in L^\infty(\Omega \setminus \omega; R^N).$$

But the last inequality implies that

$$\chi \in \partial j(u) \quad \text{a.e. in } \Omega \setminus \omega.$$

Since $\operatorname{mes}\omega < \varepsilon$ and ε was chosen arbitrarily,

$$\chi \in \partial j(u) \quad \text{a.e. in } \Omega, \tag{35}$$

as claimed.

Step 3. Now we will show that $\chi \cdot u \in L^1(\Omega)$. For this purpose we shall need the following truncation result for vector-valued Sobolev spaces.

Theorem 3.6 (Naniewicz, [15]). For each $v \in H^1(\Omega; R^N)$ there exists a sequence of functions $\{\varepsilon_n\} \subset L^\infty(\Omega)$ with $0 \leq \varepsilon_n \leq 1$ such that

$$\begin{aligned} &\{(1-\varepsilon_n)v\} \subset H^1(\Omega; R^N) \cap L^\infty(\Omega; R^N) \\ &(1-\varepsilon_n)v \to v \quad \text{strongly in } H^1(\Omega; R^N). \end{aligned} \tag{36}$$

According to the aforementioned theorem, for the $u \in V$ determined in Step 1 of the proof of Theorem 3.1 one can find a sequence $\{\varepsilon_k\} \in L^\infty(\Omega)$ with $0 \leq \varepsilon_k \leq 1$ such that $\hat{u}_k := (1-\varepsilon_k)u \in V \cap L^\infty(\Omega; R^N)$ and $\hat{u}_k \to u$ in V as $k \to \infty$. By the compactness argument we can assume without loss of generality that $\hat{u}_k \to u$ a.e. in Ω. Since we already know that $\chi \in \partial j(u)$, one can apply (5) to obtain

$$\chi \cdot (-u) \leq j^0(u; -u) \leq k(1+|u|).$$

Hence

$$\chi \cdot u \geq -k(1+|u|) \tag{37}$$

and consequently,

$$\chi \cdot \hat{u}_k = (1-\varepsilon_k)\chi \cdot u \geq -k(1+|u|). \tag{38}$$

This implies that the sequence $\{\chi \cdot \hat{u}_k\}$ is bounded from below by a function which is integrable in Ω. On the other hand, due to (33) we get

$$C \geq \langle -u_0^* + g, \hat{u}_k \rangle_V = \int_\Omega \chi \cdot \hat{u}_k d\Omega$$

for a positive constant C. Thus by Fatou's lemma $\chi \cdot u \in L^1(\Omega)$, as required.

Step 4. In this step we will obtain the estimate

$$\liminf \int_\Omega \chi_n \cdot u_n d\Omega \geq \int_\Omega \chi \cdot u d\Omega, \tag{39}$$

where a sequence (u_n, χ_n) satisfies (30) and (31). We can suppose here, again by the compactness argument, that $u_n \to u$ a.e. in Ω. Fix $v \in L^\infty(\Omega; R^N)$ arbitrarily. Since $\chi_n \in \partial j(u_n)$, one gets by (1)

$$\chi_n \cdot (v - u_n) \leq j^0(u_n; v - u_n) \leq \alpha(||v||_{L^\infty(\Omega;R^N)})(1 + |u_n|^\sigma).$$

Thus by Fatou's lemma and by the upper semicontinuity of $V \ni w \mapsto \int_\Omega j^0(u; w - u)d\Omega$, we obtain

$$\liminf \int_\Omega \chi_n \cdot u_n d\Omega \geq \int_\Omega \chi \cdot v d\Omega - \int_\Omega j^0(u; v-u) d\Omega \quad \forall\, v \in V \cap L^\infty(\Omega; R^N). \tag{40}$$

On substituting $v = \hat{u}_k$ (with $\hat{u}_k$ being described by the truncation argument of Theorem 3.6) into the right hand side of (40) one gets

$$\liminf_n \int_\Omega \chi_n \cdot u_n d\Omega \geq \liminf_k \int_\Omega \chi \cdot \hat{u}_k d\Omega - \limsup_k \int_\Omega j^0(u; \hat{u}_k - u) d\Omega, \quad (41)$$

which is valid for each number k. Further, it is easy to check that

$$j^0(u; \hat{u}_k - u) = \varepsilon_k j^0(u; -u) \leq \varepsilon_k k(1 + |u|) \leq k(1 + |u|). \quad (42)$$

Therefore due to the fact that $\hat{u}_k \to u$ a.e. in Ω we can apply Fatou's lemma to deduce

$$\limsup_k \int_\Omega j^0(u; \hat{u}_k - u) d\Omega \leq 0,$$

whereas from (38) we get

$$\liminf_k \int_\Omega \chi \cdot \hat{u}_k d\Omega \geq \int_\Omega \chi \cdot u d\Omega.$$

Finally, combining the last two inequalities with (41) yields (39), as desired.

Step 5. In this step we claim that

$$\langle u_0^\star - g, u \rangle_V + \int_\Omega \chi \cdot u d\Omega = 0. \quad (43)$$

Indeed, (33) yields

$$\langle u_0^\star - g, \hat{u}_k \rangle_V + \int_\Omega \chi \cdot \hat{u}_k d\Omega = 0.$$

Recall that $\chi \cdot u \in L^1(\Omega)$ and we have the following estimates

$$-k(1 + |u|) \leq \chi \cdot \hat{u}_k = (1 - \varepsilon_k)\chi \cdot u \leq |\chi \cdot u|.$$

Thus we can apply the dominated convergence theorem to deduce

$$\int_\Omega \chi \cdot \hat{u}_k d\Omega \to \int_\Omega \chi \cdot u d\Omega,$$

which means that (43) holds in the limit.

Step 6. Finally the pseudo-monotonicity of T will permit the derivation of (6)–(7). We first check that (30) (when substituting $w = 0$) and (43) yield

$$\langle u_n^\star, u_n - u \rangle_V = \langle g, u_n \rangle_V - \int_\Omega \chi_n \cdot u_n d\Omega - \langle u_n^\star, u \rangle_V + \langle u_0^\star - g, u_n \rangle_V + \int_\Omega \chi \cdot u_n d\Omega.$$

Hence by virtue of (39)

$$\limsup \langle u_n^\star, u_n - u\rangle_V \le -\liminf \int_\Omega \chi_n \cdot u_n d\Omega + \int_\Omega \chi \cdot u d\Omega \le 0.$$

Thus the pseudo-monotonicity of T allows us to conclude that $\langle u_n^\star, u_n\rangle_V \to \langle u_0^\star, u\rangle_V$ and $u_n^\star \to u_0^\star$ weakly in $V^\star$ as $n \to \infty$ and $u_0^\star \in T(u)$. Hence from (30) and (39) we get (6)–(7) immediately. The proof of Theorem 3.1 is complete. □

Corollary 3.7 Let us assume that all the hypotheses of Theorem 3.1 hold. Then there exists at least one $u \in V$ and $u_0^\star \in T(u)$ such as to satisfy the hemivariational inequality

$$\langle u_0^\star - g, v - u\rangle_V + \int_\Omega j^0(u; v-u) d\Omega \ge 0 \quad \forall\, v \in V, \tag{44}$$

where the integral above is assumed to take $+\infty$ as its value, whenever $j^0(u; v-u) \notin L^1(\Omega)$.

Proof. Let us first choose an arbitrary $v \in V \cap L^\infty(\Omega; R^N)$. ¿From (1) we have

$$\chi \cdot (v-u) \le j^0(u; v-u) \le \alpha(||v||_{L^\infty(\Omega;R^N)})(1+|u|^\sigma)$$

with $\chi\cdot(v-u) \in L^1(\Omega)$ and $\alpha(||v||_{L^\infty(\Omega;R^N)})(1+|u|^\sigma) \in L^1(\Omega)$. Hence $j^0(u; v-u)$ is finite integrable and (44) follows immediately from (6).

Now let us consider the case $j^0(u; v-u) \in L^1(\Omega)$ with $v \notin V \cap L^\infty(\Omega; R^N)$. According to Theorem 3.2 one can find a sequence $\hat{v}_k = (1-\varepsilon_k)v$ such that $\{\hat{v}_k\} \subset V \cap L^\infty(\Omega; R^N)$ and $\hat{v}_k \to v$ strongly in V. As we already know

$$\langle u_0^\star - g, \hat{v}_k - u\rangle_V + \int_\Omega j^0(u; \hat{v}_k - u) d\Omega \ge 0. \tag{45}$$

Thus in order to establish (44) it remains to show that

$$\limsup_k \int_\Omega j^0(u; \hat{v}_k - u) d\Omega \le \int_\Omega j^0(u; v-u) d\Omega. \tag{46}$$

For this purpose we note that $\hat{v}_k - u = (1-\varepsilon_k)(v-u) + \varepsilon_k(-u)$ and the convexity of $j^0(u;\cdot)$ yield

$$\begin{aligned} j^0(u; \hat{v}_k - u) &\le (1-\varepsilon_k) j^0(u; v-u) + \varepsilon_k j^0(u; -u) \\ &\le |j^0(u; v-u)| + k(1+|u|). \end{aligned}$$

Finally, (46) is obtained by Fatou's lemma. The proof of Corollary 3.7 is complete. □

1.4 VARIATIONAL-HEMIVARIATIONAL INEQUALITIES

It is our main purpose in the present section to show that the problem (P) has solutions. Recall that Φ is a convex, lower semicontinuous function $\Phi : V \to R \cup \{+\infty\}$. The strategy of our proof is the following: We first establish the existence of solutions for hemivariational inequalities involving the multivalued mapping $T_\lambda := A + (\partial\Phi)_\lambda$, $\lambda > 0$, by making use of Theorem 3.1. Here $(\partial\Phi)_\lambda$ is a modified form of the one involved in an approximation procedure due to Brézis-Crandall-Pazy [1]. Then we obtain our main result by letting $\lambda \to 0$.

Let $J : V \to 2^{V^*}$ be a normalized duality mapping of V into 2^{V^*}, i.e.

$$Jv = \{v^* \in V^* : \langle v^*, v\rangle_V = \|v^*\|_{V^*}\|v\|_V,\ \|v^*\|_{V^*} = \|v\|_V\}. \tag{47}$$

It is well known that J is a maximal monotone, bounded, coercive mapping and has the effective domain $D(J) = V$ [1]. By setting

$$(\partial\Phi)_\lambda(v) := (\partial\Phi^{-1} + \lambda J^{-1})^{-1}(v) = \{v^* \in V^* : \exists w \in D(\partial\Phi) \text{ such that } v^* \in \partial\Phi(w),\ \lambda v^* \in J(v-w)\}. \tag{48}$$

one obtains a bounded, maximal monotone mapping with the effective domain $D((\partial\Phi)_\lambda) = V$ [1]. Moreover, if $u^* \in \partial\Phi(u)$ and $u^*_\lambda \in (\partial\Phi)_\lambda(u)$, then $\|u^*_\lambda\|_{V^*} \le \|u^*\|_{V^*}$.

Lemma 4.1. Suppose that $\Phi : V \to R \cup \{+\infty\}$ is a convex, bounded from below and lower semicontinuous function with $\Phi(0) \in R$. Then the multivalued mapping $T_\lambda := A + (\partial\Phi)_\lambda$, $\lambda > 0$, is bounded, pseudo-monotone and coercive with the coercivity function not depending on $\lambda > 0$.

Proof. The sum of pseudo-monotone mappings is pseudo-monotone [3]. For the coercivity of T_λ it suffices to notice that for any $v^* \in (\partial\Phi)_\lambda(v)$ there exists $w \in V$ with $v^* \in (\partial\Phi)(w)$ and $\lambda v^* \in J(v-w)$. Hence

$$\Phi(0) - \Phi(w) \ge \langle v^*, -w\rangle_V = \langle v^*, -v\rangle_V + \frac{1}{\lambda}\|v-w\|_V^2 \ge \langle v^*, -v\rangle_V.$$

Accordingly, $\langle v^*, v\rangle_V \ge \Phi(w) - \Phi(0) \ge -R$, where $-R = \inf\{\Phi(w) - \Phi(0)\}$, and consequently

$$\langle Av + v^*, v\rangle_V \ge (c(\|v\|_V) - R)\|v\|_V,\quad v^* \in (\partial\Phi)_\lambda(v), \tag{49}$$

which implies the assertion. □

Definition. Let T be a mapping from V into 2^{V^*}, and α a function taken from (1). Then T will be called α-quasi-bounded if for each $M > 0$ there exist two constants $C(M) > 0$ and $K(M) > 0$ such that whenever $[v^*, v]$ lies in the Graph$(\partial\Phi)$ and

$$\langle v^*, v-w\rangle_V \le C(M)\big(\alpha(\|w\|_{L^\infty(\Omega;R^N)}) + \|w\|_V + 1\big) \quad \forall w \in V \cap L^\infty(\Omega; R^N),\ \ \|v\|_V \le M, \tag{50}$$

then

$$||v^{\star}||_{V^{\star}} \leq K(M).$$

Now we can formulate the following result.

Proposition 4.2. Let $A : V \to V^{\star}$ be a pseudo-monotone, bounded and coercive operator from V into $V^{\star}$, $\Phi : V \to R\cup\{+\infty\}$ a convex, lower semicontinuous, bounded from below function with $\Phi(0) \in R$. Moreover, assume that (1) and (5) hold and that $\partial\Phi$ is α-quasi-bounded. Then for any $\lambda > 0$ there exist $u_\lambda \in V$, $u_\lambda^{\star} \in V^{\star}$ and $\chi_\lambda \in L^{\infty}(\Omega; R^N)$ such that

$$\langle Au_\lambda - g, v - u_\lambda\rangle_V + \langle u_\lambda^{\star}, v - u_\lambda\rangle_V + \int_\Omega \chi_\lambda \cdot (v - u_\lambda) d\Omega = 0$$

$$\forall v \in V \cap L^{\infty}(\Omega; R^N) \tag{51}$$

$$u_\lambda^{\star} \in (\partial\Phi)_\lambda(u), \quad \chi_\lambda \in \partial j(u_\lambda) \text{ a.e. in } \Omega, \quad \chi_\lambda \cdot u_\lambda \in L^1(\Omega). \tag{52}$$

Furthermore, $\{u_\lambda\}_{\lambda>0}$ and $\{u_\lambda^{\star}\}_{\lambda>0}$ are bounded in V and $V^{\star}$, respectively, independently of λ and $\{\chi_\lambda\}_{\lambda>0}$ is weakly precompact in $L^1(\Omega; R^N)$.

Proof. The boundedness of $\{u_\lambda\}_{\lambda>0}$ in V follows from (5) and the fact that the coercivity function of $A + (\partial\Phi)_\lambda$ does not depend on λ (for details see the proof of Lemma 3.4). For the proof of precompactness of $\{\chi_\lambda\}_{\lambda>0}$ in $L^1(\Omega; R^N)$ it suffices to follow the lines of the proof of Lemma 3.5. The details will be omitted here. In order to get the boundedness of $\{u_\lambda^{\star}\}$ in $V^{\star}$ we shall use the α-quasi-boundedness property of $\partial\Phi(\cdot)$. Since $u_\lambda^{\star} \in (\partial\Phi)_\lambda(u_\lambda)$, by the definition of $(\partial\Phi)_\lambda(\cdot)$, there exists $w_\lambda \in V$ such that $u_\lambda^{\star} \in \partial\Phi(w_\lambda)$ and $\lambda u_\lambda^{\star} \in J(u_\lambda - w_\lambda)$. Thus from (51) it follows that whenever $||u_\lambda||_V \leq M$ we have

$$\begin{aligned}
&\langle u_\lambda^{\star}, w_\lambda - v\rangle_V + \frac{1}{\lambda}||u_\lambda - w_\lambda||_V^2 \\
&= \langle Au_\lambda - g, v - u_\lambda\rangle_V + \int_\Omega \chi_\lambda \cdot (v - u_\lambda) d\Omega \\
&\leq ||Au_\lambda - g||_{V^{\star}}(||u_\lambda||_V + ||v||_V) + \alpha\big(||v||_{L^{\infty}(\Omega;R^N)}\big) \int_\Omega (1 + |u_\lambda|^{\sigma}) d\Omega \\
&\leq K_1(M)(M + ||v||_V) + \alpha\big(||v||_{L^{\infty}(\Omega;R^N)}\big) K_2(M) \\
&\qquad \forall v \in V \cap L^{\infty}(\Omega; R^N),
\end{aligned} \tag{53}$$

Thus the α-quasi-boundedness of $\partial\Phi(\cdot)$ implies that $||u_\lambda^{\star}||_{V^{\star}} \leq K(M)$ for some constant $K(M) > 0$. The proof of Proposition 4.2 is complete. □

Remark 4.3. If Φ is lower bounded, i.e. $\Phi(v) \geq -C$ $\forall v \in V$, $C \in R$, then

$$||u_\lambda - w_\lambda||_V \to 0 \text{ as } \lambda \to 0. \tag{54}$$

Indeed, by substituting $v = 0$ into (53) and using the fact that $u_\lambda^\star \in \partial\Phi(w_\lambda)$ one gets

$$\frac{1}{\lambda}||u_\lambda - w_\lambda||_V^2 \leq K_1(M)M + \alpha(0)K_2(M) + \Phi(0) - \Phi(w_\lambda)$$
$$\leq K_1(M)M + \alpha(0)K_2(M) + \Phi(0) + C,$$

which proves the assertion.

Now we are ready to present our main result.

Theorem 4.4. Let $A : V \to V^\star$ be a pseudo-monotone, bounded and coercive operator from V into $V^\star$, $\Phi : V \to R \cup \{+\infty\}$ a convex, lower semicontinuous, bounded from below function with $\Phi(0) \in R$. Moreover, assume that (1) and (5) hold and that $\partial\Phi$ is α-quasi-bounded. Then there exist $u \in V$, $u^\star \in V^\star$ and $\chi \in L^\infty(\Omega; R^N)$ such that

$$\langle Au - g, v - u\rangle_V + \langle u^\star, v - u\rangle_V + \int_\Omega \chi \cdot (v - u)d\Omega = 0$$
$$\forall v \in V \cap L^\infty(\Omega; R^N) \qquad (55)$$

$$u^\star \in \partial\Phi(u), \quad \chi \in \partial j(u) \text{ a.e. in } \Omega, \quad \chi \cdot u \in L^1(\Omega). \qquad (56)$$

Proof. From Proposition 4.2 it follows that we are allowed to extract subsequences u_{λ_n}, $u_{\lambda_n}^\star$ and χ_{λ_n} such that

$$\langle Au_{\lambda_n} - g, v - u_{\lambda_n}\rangle V + \langle u_{\lambda_n}^\star, v - u_{\lambda_n}\rangle V + \int_\Omega \chi_{\lambda_n} \cdot (v - u_{\lambda_n})d\Omega = 0$$
$$\forall\, v \in V \cap L^\infty(\Omega; R^N), \qquad (57)$$

$$u_{\lambda_n}^\star \in (\partial\Phi)_\lambda(u_{\lambda_n}), \quad \chi_{\lambda_n} \in \partial j(u_{\lambda_n}), \quad \chi_{\lambda_n} \cdot u_{\lambda_n} \in L^1(\Omega) \qquad (58)$$

and

$$\left.\begin{aligned} u_{\lambda_n} &\to u \quad \text{weakly in } V \\ Au_{\lambda_n} &\to u_0^\star \quad \text{weakly in } V^\star \\ u_{\lambda_n}^\star &\to u^\star \quad \text{weakly in } V^\star \\ \chi_{\lambda_n} &\to \chi \quad \text{weakly in } L^1(\Omega; R^N) \\ ||u_{\lambda_n} - w_{\lambda_n}||_V &\to 0 \text{ as } n \to \infty \end{aligned}\right\} \qquad (59)$$

where $u_{\lambda_n}^\star \in \partial\Phi(w_{\lambda_n})$ and $\lambda_n u_{\lambda_n}^\star \in J(u_{\lambda_n} - w_{\lambda_n})$.

Further we follow the lines of the proof of Theorem 3.1. In order to deduce that

$$\chi \in \partial j(u) \qquad (60)$$

and

$$\chi \cdot u \in L^1(\Omega) \qquad (61)$$

the methods of Step 2 and Step 3 can be applied, respectively. For

$$\liminf \int_{\Omega} \chi_{\lambda_n} \cdot u_{\lambda_n} d\Omega \geq \int_{\Omega} \chi \cdot u d\Omega, \tag{62}$$

we apply the procedure of Step 4 while to get the relation

$$\langle u_0^\star - g, u \rangle_V + \langle u^\star, u \rangle_V + \int_{\Omega} \chi \cdot u d\Omega = 0. \tag{63}$$

it is enough to follow the reasoning of Step 5.

Further, from (57) when substituting $v = 0$, it follows

$$\langle Au_{\lambda_n} - g, -u_{\lambda_n} \rangle_V + \Phi(0) - \Phi(w_{\lambda_n}) - \frac{1}{\lambda} ||u_{\lambda_n} - w_{\lambda_n}||_V^2 + \int_{\Omega} \chi_{\lambda_n} \cdot (-u_{\lambda_n}) d\Omega \geq 0. \tag{64}$$

Hence

$$\langle Au_{\lambda_n} - g, -u_{\lambda_n} \rangle_V + \Phi(0) + \int_{\Omega} k(1 + |u_{\lambda_n}|) d\Omega \geq \Phi(w_{\lambda_n}).$$

Thus

$$\Phi(w_{\lambda_n}) \leq C$$

for some constant $C \in R$. Moreover, from (54) and (59) it follows that $w_{\lambda_n} \to u$ weakly in V. This means, by the lower semicontinuity of Φ, that $\Phi(u) \in R$, i.e. u belongs to the effective domain of Φ.

Our task now is to show that

$$\begin{aligned} &\limsup \langle Au_{\lambda_n}, u_{\lambda_n} - u \rangle_V \leq 0 \\ &\limsup \langle u_{\lambda_n}^\star, w_{\lambda_n} - u \rangle_V \leq 0, \end{aligned} \tag{65}$$

which will allow us to benefit from the monotone-type properties of A and $\partial\Phi$. For this purpose (63) and (64) will be used. They imply

$$\begin{aligned} &\langle Au_{\lambda_n}, u_{\lambda_n} - u \rangle_V + \langle u_{\lambda_n}^\star, w_{\lambda_n} - u \rangle_V \\ &= \langle -Au_{\lambda_n} + u_0^\star, u \rangle_V + \langle -u_{\lambda_n}^\star + u^\star, u \rangle_V - \frac{1}{\lambda_n} ||u_{\lambda_n} - w_{\lambda_n}||_V^2 \\ &\quad + \int_{\Omega} \chi \cdot u d\Omega - \int_{\Omega} \chi_{\lambda_n} \cdot u_{\lambda_n} d\Omega \end{aligned} \tag{66}$$

from which by virtue of (59) and (62) it follows

$$\limsup [\langle Au_{\lambda_n}, u_{\lambda_n} - u \rangle_V + \langle u_{\lambda_n}^\star, w_{\lambda_n} - u \rangle_V] \leq 0.$$

This in turn by maximal monotonicity of $\partial\Phi$ and pseudo-monotonicity of A easily implies (65). Hence

$$\left.\begin{aligned} Au_{\lambda_n} &\to u_0^\star = Au \quad \text{weakly in } V^\star \\ \langle Au_{\lambda_n}, u_{\lambda_n}\rangle_V &\to \langle Au, u\rangle_V \\ u_{\lambda_n}^\star &\to u^\star \in \partial\Phi(u) \quad \text{weakly in } V^\star \\ \langle u_{\lambda_n}^\star, w_{\lambda_n}\rangle_V &\to \langle u^\star, u\rangle_V \\ \int_\Omega \chi_{\lambda_n}\cdot u_{\lambda_n} d\Omega &\to \int_\Omega \chi\cdot u d\Omega. \end{aligned}\right\} \tag{67}$$

We may now pass to the limit in (57) as $n \to \infty$ to obtain the desired relations (55) and (56). This completes the proof of Theorem 4.4. □

Corollary 4.5 Let us posit all the hypotheses of Theorem 4.4. Then there exists at least one $u \in V$ such as to satisfy the variational-hemivariational inequality

$$\langle Au - g, v - u\rangle_V + \Phi(v) - \Phi(u) + \int_\Omega j^0(u; v-u) d\Omega \geq 0 \quad \forall\, v \in V, \tag{68}$$

where the integral above is assumed to take the value $+\infty$, whenever $j^0(u; v - u) \notin L^1(\Omega)$.

Proof. Let us first choose $v \in V \cap L^\infty(\Omega; R^N)$ arbitrarily. From (1) we have

$$\chi\cdot(v-u) \leq j^0(u; v-u) \leq \alpha(||v||_{L^\infty(\Omega;R^N)})(1 + |u|^\sigma)$$

with $\chi\cdot(v-u) \in L^1(\Omega)$ and $\alpha(||v||_{L^\infty(\Omega;R^N)})(1+|u|^\sigma) \in L^1(\Omega)$. Hence $j^0(u; v - u)$ is finite integrable and (68) follows immediately from the fact that $\langle u^\star, v - u\rangle_V \leq \Phi(v) - \Phi(u)\ \forall v \in V$.

Now let us consider the case $j^0(u; v - u) \in L^1(\Omega)$ with $v \notin V \cap L^\infty(\Omega; R^N)$. According to Theorem 3.2 one can find a sequence $\hat{v}_k = (1 - \varepsilon_k)v$ such that $\{\hat{v}_k\} \subset V \cap L^\infty(\Omega; R^N)$ and $\hat{v}_k \to v$ strongly in V. As we already know

$$\left\langle Au - g, \hat{v}_k - u\right\rangle_V + \left\langle u^\star, \hat{v}_k - u\right\rangle_V + \int_\Omega j^0(u; \hat{v}_k - u) d\Omega \geq 0,$$

so in order to establish (68) it remains to establish the validity of

$$\limsup_k \int_\Omega j^0(u; \hat{v}_k - u) d\Omega \leq \int_\Omega j^0(u; v - u) d\Omega,$$

because we already know that $\left\langle u^\star, \hat{v}_k - u\right\rangle_V \to \left\langle u^\star, v - u\right\rangle_V$. For this purpose let us notice that $\hat{v}_k - u = (1-\varepsilon_k)(v-u) + \varepsilon_k(-u)$ and the convexity of $j^0(u; \cdot)$ yields

$$\begin{aligned} j^0(u; \hat{v}_k - u) &\leq (1-\varepsilon_k) j^0(u; v-u) + \varepsilon_k j^0(u; -u) \\ &\leq |j^0(u; v-u)| + k(1 + |u|). \end{aligned}$$

Finally, by Fatou's lemma, (68) follows. The proof of Corollary 4.5 is complete.
□

References

[1] Brézis, H., M. G. Crandall, A. Pazy. (1970). *Perturbations of nonlinear maximal monotone sets in Banach spaces*, Comm. Pure Appl. Math., Vol. 23, (pages 123-144).

[2] Brézis, H. and F. E. Browder. (1982). *Some Properties of Higher Order Sobolev Spaces*, J. Math. Pures et Appl., Vol. 61, (pages 245-259).

[3] Browder, F. E. and P. Hess. (1972). *Nonlinear Mappings of Monotone Type in Banach Spaces*, J. Funct. Anal., Vol. 11, (pages 251-294).

[4] Clarke, F. H. (1983). *Optimization and Nonsmooth Analysis*, Wiley, New York.

[5] Ekeland, I. and R. Temam. (1976). *Convex Analysis and Variational Problems*, North Holland, Amsterdam and American Elsevier, New York.

[6] Hedberg, L I. (1978). *Two Approximation Problems in Function Spaces*, Ark. Mat., Vol. 16, (pages 51-81).

[7] Kinderlehrer, D. and G. Stampacchia. (1980). *An Introduction to Variational Inequalities and their Applications*, Academic Press, New York, San Francisco.

[8] Motreanu, D. and Z. Naniewicz. (1996). *Discontinuous Semilinear Problems in Vector - Valued Function Spaces*, Differential and Integral Equations, Vol. 9, (pages 581-598).

[9] Motreanu, D. and P. D. Panagiotopoulos. (1993). *Hysteresis: The Eigenvalue Problem for Hemivariational Inequalities. Models of Hysteresis* (ed. by A. Visintin) Pitman Research Notes in Mathematics, Longman, Harlow.

[10] Naniewicz, Z. (1992). *On the Pseudo-Monotonicity of Generalized Gradients of Nonconvex Functions*, Applicable Analysis, Vol. 47, (pages 151-172).

[11] Naniewicz, Z. (1994). *Hemivariational Inequality Approach to Constrained Problems for Star-Shaped Admissible Sets*, J. Opt. Theory Appl., Vol. 83, (pages 97-112).

[12] Naniewicz, Z. (1995). *Hemivariational Inequalities with Functionals which are not Locally Lipschitz*, Nonlin. Anal., Vol. 25, (pages 1307-1320).

[13] Naniewicz, Z. and P. D. Panagiotopoulos. (1995). *Mathematical Theory of Hemivariational Inequalities and Applications*, Marcel Dekker, Inc. New York.

[14] Naniewicz, Z. (1995). *On Variational Aspects of Some Nonconvex Nonsmooth Global Optimization Problem*, Journal of Global Optimization, Vol. 6, (pages 383-400).

[15] Naniewicz, Z. (1997). *Hemivariational inequalities as necessary conditions for optimality for a class of nonsmooth nonconvex functionals*, Nonlinear World, Vol. 4, (pages 117-133).

[16] Panagiotopoulos, P. D. (1985). *Inequality Problems in Mechanics and Applications. Convex and Nonconvex Energy Functions*, Birkhäuser Verlag, Basel, Boston. (Russian Translation MIR Publ., Moscow 1989).

[17] Panagiotopoulos, P. D. (1991). *Coercive and Semicoercive Hemivariational Inequalities*, Nonlin. Anal., Vol. 16, (pages 209-231).

[18] Panagiotopoulos, P. D. (1993). *Hemivariational Inequalities. Applications in Mechanics and Engineering*, Springer Verlag, Berlin.

[19] Panagiotopoulos, P. D. (1988). *Hemivariational Inequalities and their Applications*, In: Topics in Nonsmooth Mechanics (ed. by J.J.Moreau, P. D. Panagiotopoulos and G. Strang) Birkhäuser Verlag, Boston.

[20] Panagiotopoulos, P. D. (1988). *Nonconvex Superpotentials and Hemivariational Inequalities. Quasidifferentiability in Mechanics*, In: Nonsmooth Mechanics and Applications (ed. by J.J. Moreau and P.D. Panagiotopoulos) CISM Lect. Notes, Vol. 302, Springer Verlag, Wien, N.York.

[21] Panagiotopoulos, P. D. and J. Haslinger. (1989). *Optimal Control of Systems governed by Hemivariational Inequalities*, In: Math. Models for Phase Change Problems, (ed. by J.F.Rodrigues), Vol. 88, Birkhäuser Verlag, Basel Boston ISNM.

[22] Rauch, J. (1977). *Discontinuous Semilinear Differential Equations and Multiple Valued Maps*, Proc. Amer. Math. Soc., Vol. 64, (pages 277-282).

[23] Webb, J. R. L. (1980). *Boundary Value Problems for Strongly Nonlinear Elliptic Equations*, The Journal of the London Math. Soc., Vol. 21, (pages 123-132).

AN INVERSE PROBLEM IN ELASTODYNAMICS

Lizabeth V. Rachele

1395 Math Sciences Building
Purdue University
West Lafayette, IN 47907-1395

Abstract:
We show that the p-wave and s-wave speeds of an isotropic elastic object are determined in the interior by surface measurements, and that the density and elastic properties are determined to infinite order at the boundary. The material properties of the bounded, fully 3-dimensional object, that is, the density and elastic properties, are represented by the (nonconstant, leading) coefficients of the system of linear differential equations for elastodynamics. Surface measurements are modelled by the Dirichlet-to-Neumann map on a finite time interval.

The proof of these results makes use of high frequency asymptotic expansions, Hamilton-Jacobi theory, microlocal analysis, propagation of singularities results for systems of real principal type, and a result in integral geometry. Here we announce the results of [R I] and [R II].

We consider an inverse problem in elastodynamics. The physical setting for the problem is a bounded, 3-dimensional, isotropic elastic object with smooth boundary. The inverse problem can be formulated as the study of whether measurements made at the surface of the object determine the material properties (that is, the density and elastic properties) of the object. Surface measurements consist of the following pairs of surface data: all possible forces applied normal to the surface and the resulting displacements of the surface.

R.P. Gilbert et al.(eds.), Direct and Inverse Problems of Mathematical Physics, 323–332.

1.1 ELASTODYNAMICS

The elastic object is modelled by a bounded region $\Omega \subset \mathbb{R}^3$ with smooth boundary. The density and elastic properties of the object are represented by coefficients $\lambda(x), \mu(x), \rho(x)$ of the system P of linear equations for elastodynamics. Here we assume $\lambda, \mu, \rho \in C^\infty(\overline{\Omega})$, and $\rho(x) \geq \rho_0 > 0$, $\mu > 0$ on $\overline{\Omega}$.

The displacement of the object is modelled by solutions $u(x,t)$ of the initial-boundary-value problem (for $0 < T < \infty$)

$$\begin{cases} Pu = 0 & \text{in } \Omega \times (0,T) \\ u\big|_{\partial\Omega} = f & \text{for } t \in [0,T] \\ u\big|_{t=0} = 0, \quad (\partial_t u)\big|_{t=0} = 0 & \text{in } \Omega. \end{cases} \tag{1.1}$$

To guarantee the existence and uniqueness of solutions of this initial-boundary-value problem (1.1), we require that the Lamé parameters λ and μ satisfy the *strong convexity condition*

$$3\lambda + 2\mu > 0 \qquad \text{on} \quad \overline{\Omega}. \tag{1.2}$$

The operator P for elastodynamics is defined by

$$(Pu)_i = \rho \frac{\partial^2 u_i}{\partial t^2} - \Big[\sum_{j=1}^{n} \frac{\partial}{\partial x_i}(\lambda \frac{\partial u_j}{\partial x_j}) - \sum_{j=1}^{n} \frac{\partial}{\partial x_j}(\mu \frac{\partial u_i}{\partial x_j}) - \sum_{j=1}^{n} \frac{\partial}{\partial x_j}(\mu \frac{\partial u_j}{\partial x_i})\Big],$$

for $i = 1, 2, 3$. In vector notation P is written

$$Pu = \rho \partial_t^2 u - \nabla_x \cdot L(x,D)u$$

where

$$\begin{aligned} \nabla_x &= \left(\tfrac{\partial}{\partial x_1}, \tfrac{\partial}{\partial x_2}, \tfrac{\partial}{\partial x_3}\right), \\ \nabla_x \otimes u &= (\partial_{x_i} u_j)_{i,j=1..3} && \text{is the displacement gradient,} \\ \nabla_x u &= \tfrac{1}{2}[\partial_{x_i} u_j + \partial_{x_j} u_i] && \text{is the linearized strain,} \\ L(x,D)u &= \lambda(\nabla_x \cdot u)I + 2\mu(\nabla_x u) && \text{is the stress associated with } f. \end{aligned} \tag{1.3}$$

Surface measurements for this problem (which give the correspondence between any applied surface traction, i.e. force applied normal to the surface, and the resulting displacement of the surface) are modelled by the Dirichlet-to-Neumann map $\Lambda_{\lambda,\mu,\rho}$. The Dirichlet-to-Neumann map is defined by

$$\Lambda_{\lambda,\mu,\rho}\vec{f} = [L(x,D)u] \cdot \vec{\nu} \,\Big|_{\partial\Omega \times (0,T)} \tag{1.4}$$

where ν is the unit outer normal vector to the boundary $\partial\Omega$, and the displacement u solves the initial-boundary-value problem (1.1) with surface displacement f.

1.2 THE INVERSE PROBLEM

In general the inverse problem involves the study of the invertibility of the mapping

$$\lambda,\ \mu,\ \rho \quad \xrightarrow{\Lambda} \quad \Lambda_{\lambda,\mu,\rho},$$

in particular, the injectivity of the mapping Λ (that is, the unique determination of λ, μ, ρ by the Dirichlet-to-Neumann map $\Lambda_{\lambda,\mu,\rho}$) and the continuity of the inverse of Λ (that is, stability or the continuous dependence of λ, μ, ρ on the Dirichlet-to-Neumann map $\Lambda_{\lambda,\mu,\rho}$). An ultimate objective is the reconstruction of the material properties λ, μ, ρ of the medium given surface measurements in the form of the Dirichlet-to-Neumann map.

The main result announced here (see [R I] and [R II] for details of the proof) is the unique determination of the p-wave and s-wave speeds $c_p = \sqrt{\frac{\lambda+2\mu}{\rho}}$ and $c_s = \sqrt{\frac{\mu}{\rho}}$ in the interior Ω and the unique determination of all three coefficients λ, μ, ρ, to infinite order, at the boundary $\partial\Omega$. (Elastic waves here are composed of two basic types of waves: compressional waves, called p-waves, and shear waves, called s-waves. Compressional waves displace the medium in the same direction as the direction of propagation of the wave. In shear waves, on the other hand, the displacement of the medium is orthogonal to the direction of propagation of the wave.

Theorem 1.2.1 *Let $\Omega \subset \mathbb{R}^3$ be a bounded region with smooth boundary, $\lambda_j, \mu_j, \rho_j \in C^\infty(\overline{\Omega})$, and $\rho_j(x) \geq \rho_0 > 0$, $\mu_j > 0$, $3\lambda_j + 2\mu_j > 0$ on $\overline{\Omega}$ for $j = 1,2$. Suppose*

$$\Lambda_{\lambda_1,\mu_1,\rho_1} = \Lambda_{\lambda_2,\mu_2,\rho_2} \qquad \text{on } [0,T]$$

for some $0 < T < \infty$.
Boundary Determination: Then for $j = 0, 1, \ldots$

$$\partial_\nu^j \lambda_1 = \partial_\nu^j \lambda_2,\ \partial_\nu^j \mu_1 = \partial_\nu^j \mu_2,\ \partial_\nu^j \rho_1 = \partial_\nu^j \rho_2 \quad \text{at } \partial\Omega.$$

Uniqueness in the Interior: If, in addition, certain geometric conditions hold, in particular, if caustics do not arise; wave paths traverse Ω in finite time $< T$; there are unique wave paths between boundary points (i.e. distance-minimizing geodesic segments with respect to the metrics $\frac{1}{c_{p/s}^2}\delta_{ij}$); and wave paths do not graze $\partial\Omega$, then

$$\sqrt{\frac{\lambda_1 + 2\mu_1}{\rho_1}} = \sqrt{\frac{\lambda_2 + 2\mu_2}{\rho_2}} \qquad \text{and} \qquad \sqrt{\frac{\mu_1}{\rho_1}} = \sqrt{\frac{\mu_2}{\rho_2}} \qquad \text{in } \Omega.$$

1.3 RELATED WORK

Boundary determination by the Dirichlet-to-Neumann map of a coefficient and its higher derivatives has been an integral step in the proof of the related problem, the unique determination of the coefficient in the interior. In the case of the conductivity equation, $\mathrm{div}(\gamma\nabla u) = 0$, for example, Kohn and Vogelius [K-V], and then Sylvester and Uhlmann [S-U '88], prove boundary determination in the case of piecewise real analytic and then C^∞ conductivities γ. Then Sylvester and Uhlmann [S-U '87] and Nachman [Na] use the result at the boundary to show that the interior result holds for C^∞ conductivities in $\mathbb{R}^n$ when $n \geq 3$ and $n = 2$, respectively.

In the case of a hyperbolic operator, the wave operator $\partial_t^2 - \Delta_g$ associated with the Laplace-Beltrami operator Δ_g for a metric g, Sylvester and Uhlmann [Sy-U] show that the Dirichlet-to-Neumann map uniquely determines the metric (up to the pullback by a diffeomorphism of Ω that fixes $\partial\Omega$) to infinite order at the boundary $\partial\Omega$. In particular, they construct local asymptotic expansions $e^{i\varphi}\sum a_J$ of solutions u of this wave equation on $\mathbb{R}^n \times (0,T)$ with boundary values $f = u|_{\partial\Omega}$ that are particularly simple. (Here Ω is a bounded region in $\mathbb{R}^n$ with smooth boundary.) It is immediate in this case that $\partial_{\nu_g} u$ is determined by Λ_g at $\partial\Omega$. By collecting homogeneous terms in the expansion of $\partial_{\nu_g} u$, and using properties of the phase φ and the terms a_J of the amplitude, Sylvester and Uhlmann show that all higher derivatives of the phase φ are determined. The phase function solves an eikonal equation which involves the metric tensor g_{ij}, and so it follows that all higher derivatives of the metric are also determined.

The main feature of elastodynamics that leads to complications in the boundary determination that do not arise in the problem for $\partial_t^2 - \Delta_g$ is that the problem for elastodynamics is described in terms of a system of equations. It is due to this that there are two basic types of waves in the problem for elastodynamics, compressional waves (p-waves) and shear waves (s-waves). It follows that the asymptotic expansion, analogous to that in [Sy-U], of a solution u in elastodynamics consists of up to four types of waves, the inward and outward (or forward and backward), p and s-waves. Another consequence of the fact that the equation for elastodynamics is a system is that the amplitudes a_J in this case are vectors. In particular, the a_J are described not as (scalar) solutions of certain ordinary differential equations along wave paths, but as vectors that solve certain algebraic equations; when the a_J are written as combinations of certain basis vectors, the scalars involved can be described as solutions of ordinary differential equations, the *transport equations*, along wave paths. It is the complexity of the amplitudes and the two types of waves that lead to differences in the proof of the boundary determination for elastodynamics, especially in the inductive step for the boundary determination of the higher derivatives of the coefficients.

A global uniqueness problem which has some of the features of elastodynamics (see also Sacks and Symes [Sa-Sy], Bao and Symes [Ba-Sy], and Belishev and Kurylev [B-K]) is the problem solved by Rakesh and Symes in [Ra-Sy]. In [Ra-Sy] a disturbance that propagates through an object is modelled by a solution of the wave equation, $\left[\partial_t^2 - \Delta + q(x)\right] u = 0$ in $\Omega \times (0,T)$, for a bounded region Ω. Rakesh and Symes show that the integral of the potential, $q(x)$, over any path along which wave energy propagates, is determined by surface measurements. Knowing these integrals of the potential is enough to recover the potential itself inside Ω. In this model problem the paths along which energy propagates are straight lines, the differential equation is a scalar equation, and there is only one coefficient to describe, the potential $q(x)$.

In elastodynamics the paths along which wave energy propagates are not straight lines, the elasticity equation in the dynamic case is, in fact, a system of three equations in three unknowns, which, it seems, cannot be decoupled (without making simplifying assumptions), and there are three nonconstant leading coefficients to be determined: the density $\rho(x)$ and the Lamé parameters (or coefficients of elasticity), $\lambda(x)$ and $\mu(x)$. In addition, the nonhomogeneous medium is fully 3-dimensional (not layered).

In the problem for static linear elasticity (the isotropic case), Gen Nakamura and Gunther Uhlmann prove global uniqueness [N-U '94] again building on boundary determination [N-U '95] of the coefficients.

We do not reduce the dynamic problem to the static case by taking the Fourier transform in t; surface measurements are given on only a finite time interval $[0,T]$. Instead we apply techniques applicable to inverse problems for hyperbolic equations.

1.4 SKETCH OF THE PROOF

See [R I] and [R II] for details of the proof.

1.4.1 Representation of wave paths

Wave paths through Ω can be represented by geodesic segments between boundary points with respect to the metrics $g_{ij} = \frac{1}{c_{p/s}^2}\delta_{ij}$. These geodesic segments are characteristic curves $\big(t(s), x(s)\big)$ for P, so are projections of null bicharacteristic curves of P, $\big(t(s), x(s), \tau(s), \xi(s)\big)$, which lie in the cotangent bundle $T^*(\Omega \times \mathbb{R})$. Bicharacteristic curves of P form the flow out of an initial hypersurface $\{\ (t,x,\tau,\xi) \mid t = 0,\ \det p(t,x,\tau,\xi) = 0\ \}$ (for example) along the Hamilton vector fields $H_{q^{\pm}_{p/s}}$ of factors $q^{\pm}_{p/s} = \tau \mp c_{p/s}|\xi|$ of the determinant $\det p(t,x,\tau,\xi) = -\rho^3(\tau^2 - c_s^2|\xi|^2)^2(\tau^2 - c_p^2|\xi|^2)$ of the principal symbol $p(t,x,\tau\xi)$ of P.

It follows that wave paths can be described in terms of curves $\Big(t(s), x(s), \tau(s), \xi(s)\Big)$ which satisfy $\tau = \pm c_{p/s}|\xi|$ and have direction given by

$$\frac{dt}{ds} = 1, \quad \frac{dx}{ds} = \mp c_{p/s}\frac{\xi}{|\xi|}, \quad \frac{d\tau}{ds} = 0, \quad \frac{d\xi}{ds} = \pm(\nabla_x c_{p/s})|\xi|.$$

1.4.2 Uniqueness at $\partial\Omega$ of λ, μ, ρ and their normal derivatives of all orders

We have described wave paths in terms of the principal symbol of the operator P, which involves the coefficients λ, μ, and ρ. To show that the Dirichlet-to-Neumann map determines these coefficients at $\partial\Omega$, we observe that the Dirichlet-to-Neumann map is given in terms of solutions u of the initial-boundary-value problem (1.1). We write these solutions $u = u_p + u_s$ in terms of high frequency asymptotic expansions

$$u_{p/s} \sim e^{i\varphi^-_{p/s}} \sum_{J=0,-1,..} (a^-_{p/s})_J.$$

The condition that u solve the second-order system of differential equations for elastodynamics is reduced to

- **the eikonal equations**

$$(\partial_t \varphi_{p/s})^2 = c^2_{p/s} |\nabla_x \varphi_{p/s}|^2$$

  which are solved by the phase functions $\varphi^{\pm}_{p/s}(t, x, \sigma, \widetilde{\eta}')$ which are homogeneous of order 1 in the frequency $\omega = |(\sigma, \eta')|$ and have initial values given in terms of the parameters

$$\sigma \in \mathbb{R} \setminus 0 \;\text{ and }\; \widetilde{\eta}' \in \mathbb{R}^2 \setminus 0$$

  by

$$\varphi = t\sigma + \widetilde{x}' \cdot \widetilde{\eta}'$$

  in neighborhoods of points in $\partial\Omega$;

- **the transport equations**, which are first-order, linear differential equations along characteristics of the operator and are solved by scalars α that occur in the description of the amplitude; and

- **algebraic equations** which are solved by terms a_J of the amplitude.

In fact, the terms a_J of the amplitudes can be written as

$$a_J = \alpha_J N + \gamma_J M$$

where α_J and γ_J are scalars, and the vectors N have the direction of displacement of p/s-waves and the vectors M are orthogonal to the direction of displacement of p/s-waves.

Writing the Dirichlet-to-Neumann map in terms of normal and tangential derivatives at the boundary, we find, using only three choices of boundary data, that the three coefficients λ, μ, and ρ are determined at the boundary.

To show that the normal derivatives of λ, μ, and ρ are also determined at the boundary we observe that terms in the asymptotic expansion of the Dirichlet-to-Neumann map of lower order of homogeneity in the frequency ω involve terms a_J of the amplitudes for lower $J = 0, -1, \ldots$ We derive recurrence relations for the scalars α_J and γ_J, so that by an induction argument it follows that the scalars α_J and γ_J in fact depend only on higher derivatives of λ, μ, and ρ and on terms that are determined by the Dirichlet-to-Neumann map. By making a special choice of boundary data, we conclude that the higher derivatives of λ, μ, and ρ are determined by the Dirichlet-to-Neumann map at the boundary.

1.4.3 Uniqueness of the smooth extension of λ, μ, ρ outside Ω

It follows from the boundary determination of the normal derivatives (to all orders) of the density and Lamé parameters that the coefficients λ_j, μ_j, ρ_j, $(j = 1,2)$ can be extended to be smooth on $\mathbb{R}^3$ so that they are determined by the Dirichlet-to-Neumann map outside Ω. In particular, suppose $\Lambda_{\lambda_1,\mu_1,\rho_1} = \Lambda_{\lambda_2,\mu_2,\rho_2}$. We then extend λ_1, μ_1, ρ_1 to be smooth on $\mathbb{R}^3$ and extend λ_2, μ_2, ρ_2 to agree with λ_1, μ_1, ρ_1 outside Ω. It follows by the uniqueness of the normal derivatives of these coefficients (to all orders) that the extension of the second set of coefficients is in fact smooth.
Consequences:

- The operators $P_{\lambda_j,\mu_j,\rho_j}(t,x,D_{t,x})$ for elastodynamics corresponding to the two sets of coefficients are the same outside Ω.

- Certain solutions of the Cauchy problem on $\mathbb{R}^3$ for elastodynamics are determined outside Ω by the Dirichlet-to-Neumann map. This is an analogue of a result by Sylvester and Uhlmann [Sy-U] which they prove in the case of the wave equation for the Laplace-Beltrami operator for a metric. In fact, we show that $\Lambda_{\lambda,\mu,\rho}$ determines solutions of the following Cauchy problem outside Ω given fixed values at $\partial\Omega$ for $t \in (0,T)$ and given fixed initial data supported outside Ω.

  Theorem 1.4.1 *Suppose u_j $(j = 1,2)$ solve the following Cauchy problems on $\mathbb{R}^3 \times (0,T)$,*

  $$\begin{cases} P_j\, u_j &= 0 & on \mathbb{R}^3 \times (0,T) \\ u_j \big|_{t=0} &= \psi_0 & on \mathbb{R}^3 \\ (\partial_t u_j) \big|_{t=0} &= \psi_1 & on \mathbb{R}^3 \end{cases} \tag{4.5}$$

  where $(supp\ \psi_k) \cap \Omega = \emptyset$ for $k = 0,1$ and

  $$u_1 \big|_{\partial\Omega \times (0,T)} = u_2 \big|_{\partial\Omega \times (0,T)}.$$

  Then $\Lambda_{\lambda_1,\mu_1,\rho_1} = \Lambda_{\lambda_2,\mu_2,\rho_2}$ *implies*

  $$u_1(x,t) = u_2(x,t) \quad outside\ \Omega \quad for\ t \in (0,T).$$

- The wave front sets (the singular supports together with a sense of the direction of the singularity) of u_1 and u_2 agree outside Ω.

1.4.4 *Uniqueness of wave paths outside Ω and uniqueness of travel times through Ω*

We represent solutions of the Cauchy problem for elastodynamics on $\mathbb{R}^3 \times (0, T)$ in terms of Fourier integral operators. We then construct solutions with "minimal" wave front set by choosing initial data with wave front set restricted to a single point and direction (following Hörmander). It follows from propagation of singularities results for systems of real principal type [De] that the wave front set in this case is the union of the four (p/s-wave, forward/backward) bicharacteristic strips generated by the wave front set of the initial data. These four types of bicharacteristics can be distinguished: p-waves are faster than s-waves by the strong convexity condition (1.2), and a component of any bicharacteristic of P is either positive or negative depending on whether it is a forward or backward bicharacteristic.

It follows from the fact that the wave front set of these solutions is determined outside Ω by the Dirichlet-to-Neumann map that the bicharacteristics of P are also determined outside Ω.

The travel time of wave energy through the object corresponds to one of the components of these bicharacteristic curves. It follows that the elapsed travel time through the object is determined by the Dirichlet-to-Neumann map.

1.4.5 *Uniqueness in Ω of the wave speeds c_p and c_s*

Particle paths through the object correspond to geodesics with respect to the metric $g_{ij} = \frac{1}{c^2_{p/s}}\delta_{ij}$. In fact, travel times of wave energy along these paths is given by the geodesic distance. It follows from the uniqueness of the p and s-wave travel times through the object that the geodesic distances between boundary points of Ω are determined by the Dirichlet-to-Neumann map.

We now apply a result in integral geometry by Mukhometov and Romanov [M-R] and later Croke [C] to conclude that the metrics $g_{ij} = \frac{1}{c^2_{p/s}}\delta_{ij}$ are determined in Ω by the Dirichlet-to-Neumann map, given that certain geometric conditions hold. These conditions on the metrics g_{ij} are that the metrics be close to constant outside some large enough ball containing Ω, that wave paths do not graze the boundary of Ω, that there are no trapped geodesics, and that there exist unique distance-minimizing geodesic segments between boundary points.

We apply Croke's result to conclude that the wave speeds $c_{p/s}$ are determined in Ω.

1.5 OPEN PROBLEM: DETERMINATION OF THE DENSITY IN THE INTERIOR

We have shown that the two wave speeds c_p and c_s are determined in Ω. It remains an open problem to show that a third coefficient, the density ρ, say, is also determined in the interior.

1.6 ACKNOWLEDGEMENTS

The author would like to thank the organizers, R. Gilbert and Y. Xu, of the session on wave propagation for the opportunity to speak in their session. She would also like to express her appreciation to her advisor Gunther Uhlmann, for his direction and support on this project, and Antonio Sa Barreto for many helpful conversations.

References

[Ba-Sy] Bao, G. and W.W. Symes. *On the sensitivity of solutions of hyperbolic equations to the coefficients*, Inst. Math. Appl. preprint series, Vol. 1249, submitted to Comm. in PDE.

[B-K] Belishev, M. and Y. Kurylev. (1991). *Boundary control, wave field continuation and inverse problems for the wave equation*, Computers Math. Applic., Vol. 22(4/5), (pages 27-52).

[C] Croke, C. (1991). *Rigidity and the distance between boundary points*, Journal of Differential Geometry, Vol. 33(2), (pages 445–464).

[De] Dencker, N. (1982). *On the propagation of singularities for pseudodifferential operators of principal type*, Arkiv for Matematik, Vol. 20(1), (pages 23–60).

[D-L] G. Duvaut and J. L. Lions. *Inequalities in mechanics and physics*, Springer-Verlag, Berlin, 1976.

[Hö] Hörmander, L. (1985). *The Analysis of Linear Partial Differential Operators*, Vol. III, Springer-Verlag, Berlin, Heidelberg, New York, Tokyo.

[H-K-M] Hughes, T.J.R., T. Kato, and J. E. Marsden. (1977). *Well-posed, quasi-linear, second-order hyperbolic systems with applications to nonlinear elastodynamics and general relativity*, Arch. for Rat. Mech. Anal., Vol. 63(3), (pages 273–294).

[K-V] Kohn, R. and M. Vogelius. (1984). *Identification of an unknown conductivity by means of measurements at the boundary*, SIAM-AMS Proceedings, Vol. 14, (pages 113–123).

[Ku] Kupradze, V. D. (editor). (1979). *Three-dimensional problems of the mathematical theory of elasticity and thermoelasticity*, North-Holland, Amsterdam.

[Lu] Ludwig, D. (1960). *Exact and asymptotic solutions of the Cauchy problem*, Comm. Pure Appl. Math., Vol. 13, (pages 473–508).

[La] Lax, P.D. (1957). *Asymptotic solutions of oscillatory initial value problems*, Duke Math. J., Vol. 24, (pages 627–646).

[M-R] R. G. Mukhometov and V. G. Romanov. On the problem of finding an isotropic Riemannian metric in an n-dimensional space, *Dokl. Akad. Nauk. SSSR* **243** (1), 1978, 41–44.

[Na] A. Nachman. Global uniqueness for a two-dimensional inverse boundary value problem, *Ann. of Math. (2)* **143**(1), 1996, 71–96.

[N-U '93] Nakamura, G. and G. Uhlmann. (1993). *Identification of Lamé parameters by boundary measurements*, Amer. J. Math., Vol. 115(5), (pages 1161–1187).

[N-U '94] Nakamura, G. and G. Uhlmann. (1994). *Global uniqueness for an inverse boundary problem arising in elasticity*, Invent. Math., Vol. 118(3), (pages 457–474).

[N-U '95] Nakamura, G. and G. Uhlmann. (1995). *Inverse problems at the boundary for an elastic medium*, SIAM J. Math. Anal., Vol. 26, (pages 263–279).

[R I] Rachele, L. *Boundary determination for an inverse problem in elastodynamics*, Preprint: See http://www.math.purdue.edu/~lrachele.

[R II] Rachele, L. *An Inverse Problem in Elastodynamics: Determination of the wave speeds in the interior.* Preprint: See http://www.math.purdue.edu/~lrachele.

[Ra-Sy] Rakesh and W. W. Symes. (1988). *Uniqueness for an inverse problem for the wave equation.* Communications in Partial Differential Equations, Vol. 13(1), (pages 87–96).

[Sa-Sy] Sacks, P.E. and W.W. Symes. (1990). *The inverse problem for a fluid over a layered elastic half-space*, Inverse Problems, Vol. 6(6), (pages 1031-1054).

[S-U '87] Sylvester, J. and G. Uhlmann. (1987). *A global uniqueness theorem for an inverse boundary value problem*, Annals of Mathematics, Vol. 125, (pages 153–169).

[S-U '88] Sylvester, J. and G. Uhlmann. (1988). *Inverse boundary value problems at the boundary – continuous dependence.* Comm.of Pure and Applied Math., Vol. 41(2), (pages 197–219).

[Sy-U] Sylvester, J. and G. Uhlmann. (1991). *Inverse problems in anisotropic media*, Contemporary Mathematics, Vol. 122, (pages 105–117).

[U] Uhlmann, G. (1992). *Inverse boundary value problems and applications*, Methodes semi-classiques, Vol. 1 (Nantes, 1991), *Asterisque*, Vol. 207(6), (pages 153-211).

DENSENESS OF $C_0^\infty(R^N)$ IN THE GENERALIZED SOBOLEV SPACES $W^{M,P(X)}(R^N)$

Stefan Samko

1.Introduction

The spaces $L^{p(x)}(\Omega), \Omega \subseteq R^n$, with variable order $p(x)$ were studied recently. We refer to the pioneer work by I.I. Sharapudinov [6] and the later papers by O.Kováčik and J. Rákosník [2] and by the author [3]-[5]. In the paper [2] the Sobolev type spaces $W^{m,p(x)}(\Omega)$ were also studied. D.E.Edmunds and J. Rákosník [1] dealt with the problem of denseness of C^∞-functions in $W^{m,p(x)}(\Omega)$ and proved this denseness under some special monotonicity-type condition on $p(x)$. We prove that $C_0^\infty(R^n)$ is dense in $W^{m,p(x)}(R^n)$ without any monotonicity condition, requiring instead that $p(x)$ is somewhat better than just continuous - satisfies the Dini-Lipschitz condition. For this purpose we prove the boundedness of the convolution operators $\frac{1}{\epsilon^n}\mathcal{K}\left(\frac{x}{\epsilon}\right) * f$ in the space $L^{p(x)}$ uniform with respect to ϵ . This is the main result, the above mentioned denseness being its consequence, in fact.

In the one dimensional periodical case a similar result for the uniform boundedness in $L^{p(x)}$ of some family of operators K_ϵ, depending on ϵ, was proved by I.I.Sharapudinov [7].

2. Preliminaries

We refer to the papers [2]-[6] for basics of the spaces $L^{p(x)}$, but remind their definition and some important properties.

R.P. Gilbert et al.(eds.), Direct and Inverse Problems of Mathematical Physics, 333–342.

Let $p(x)$ be a measurable function on a domain $\Omega \subseteq R^n$ satisfying the condition $1 \leq p(x) \leq \infty$ and let

$$E_\infty = E_\infty(p) = \{x \in \Omega \ : p(x) = \infty\}.$$

We denote

$$P = \sup_{x \in \Omega \setminus E_\infty(p)} p(x) , \quad p_0 = \inf_{x \in \Omega} p(x).$$

where sup and inf stand for esssup and essinf, respectively. By $L^{p(x)}(\Omega)$ we denote the space of measurable functions $f(x)$ on Ω such that

$$I_p(f) := \int_{\Omega \setminus E_\infty} |f(x)|^{p(x)}\, dx < \infty \quad \textit{and} \quad f(x) \in L^\infty(E_\infty).$$

Let

$$\|f\|_{(p)} = \inf \left\{ \lambda > 0 : I_p\left(\frac{f}{\lambda}\right) \leq 1 \right\} . \tag{1}$$

In case of $P < \infty$ the space $L^{p(x)}$ is a Banach space with respect to the norm

$$\|f\|_p = \|f\|_{(p)} + \|f\|_{L^\infty(E_\infty)} \cdot \tag{2}$$

We emphasize that $\|f\|_p$ is finite for any $f(x) \in L^{p(x)}(\Omega)$ in the case $P = \infty$ as well, but $L^{p(x)}(\Omega)$ is not a linear space and $\|f\|_p$ is not a norm in this case.

We note the following properties of the space $L^{p(x)}(\Omega)$:

a) *the Hölder inequality* ([6],[2],[3]) :

$$\int_\Omega |f(x)\varphi(x)|\, dx \leq k\|f\|_p\|\varphi\|_q \tag{3),}$$

where $1 \leq p(x) \leq \infty$, $\frac{1}{p(x)} + \frac{1}{q(x)} \equiv 1$, $k = \sup_{x\in\Omega} \frac{1}{p(x)} + \sup_{x\in\Omega} \frac{1}{q(x)}$;

b) *inequalities between $I_p(f)$ and $\|f\|_{(p)}$* ([6],[2],[3]) :

$$\|f\|_{(p)}^P \leq I_p(f) \leq \|f\|_{(p)}^{p_0} , \quad \textit{if} \quad \|f\|_{(p)} \leq 1 , \tag{4}$$

$$\|f\|_{(p)}^{p_0} \leq I_p(f) \leq \|f\|_{(p)}^{P} , \quad \textit{if} \quad \|f\|_{(p)} \geq 1 , \tag{5}$$

the left-hand side inequality in (4) and the right-hand side one in (5) being trivial in the case $P = \infty$;

c) *estimates for the norm of the characteristic function of a set* ([3]) :

$$|E|^{\frac{1}{P}} \leq \|\chi_E\|_{(p)} \leq |E|^{\frac{1}{p_0}} , \quad \textit{if} \quad |E| \leq 1 , \ E \subseteq \Omega \setminus E_\infty(p), \tag{6}$$

the signs of the inequalities being opposite if $|E| \geq 1$; here $|E|$ is the Lebesgue measure of E ; as in (4)-(5), the corresponding inequalities are trivial in the case $P = \infty$;

d) *the embedding theorem* ([3]) : let $1 \leq r(x) \leq p(x) \leq P < \infty$ for $x \in \Omega$ and $|\Omega| < \infty$. Then $L^{p(x)} \subseteq L^{r(x)}$ and

$$\|f\|_r \leq (a_2 + (1 - a_1)|\Omega|)\|f\|_p \tag{7}$$

where $a_1 = \inf_\Omega \frac{r(x)}{p(x)}, a_2 = \sup_\Omega \frac{r(x)}{p(x)}$, see also [2] for this imbedding without the restriction $p(x) \leq P < \infty$, but with worse constants $a_2 = 1$ and $1 - a_1 = 1$

e) *denseness of step functions* ([3]): functions of the form $\sum_{k=1}^m c_k \chi_{\Omega_k}, \Omega_k \subset \Omega, |\Omega_k| < \infty$, with constant c_k, form a dense set in $L^{p(x)}(\Omega)$.

As in [4]-[5], we use the weak Lipschits condition (Dini-Lipschits condition):

$$|p(x) - p(y)| \leq \frac{A}{\log \frac{1}{|x-y|}} \quad , |x-y| \leq \frac{1}{2} . \tag{8}$$

Everywhere below we assume that $P < \infty$.

3. Statements of the main results

Let $\mathcal{K}(x)$ be a measurable function with support in the ball $B_R = B(0, R)$ of a radius $R < \infty$, and let

$$\mathcal{K}_\epsilon(x) = \frac{1}{\epsilon^n} \mathcal{K}\left(\frac{x}{\epsilon}\right) .$$

We consider the family of operators

$$K_\epsilon f = \int_\Omega \mathcal{K}_\epsilon(x-y) f(y) dy , \tag{9}$$

Ω being a bounded domain in R^n .

For the given domain Ω we define the larger domain

$$\Omega_R = \{x : dist(x, \Omega) \leq R\} \supseteq \Omega .$$

Let $p(x)$ be a function defined in Ω_R such that

$$1 \leq p(x) \leq P < \infty , \quad x \in \Omega_R . \tag{10}$$

Let also $\frac{1}{p(x)} + \frac{1}{q(x)} \equiv 1$ and

$$Q = \begin{cases} \sup_{x \in \Omega_R} q(x) = \frac{p_0}{p_0 - 1} , & \text{if } |E_1(p)| = 0 \\ \infty , & \text{if } |E_1(p)| > 0 \end{cases} \tag{11}$$

where $E_1(p) = \{x \in \Omega_R : p(x) = 1\}$.

Theorem 1. *Let $\mathcal{K}(x) \in L^Q(B_R)$ and let $p(x)$ satisfy (10) and (8) for all x and $y \in \Omega_R$. Then the operators K_ϵ are uniformly bounded from $L^{p(x)}(\Omega)$ into $L^{p(x)}(\Omega_R)$:*

$$\|K_\epsilon f\|_{L^{p(x)}(\Omega_R)} \leq c \|f\|_{L^{p(x)}(\Omega)} \tag{12}$$

where c does not depend on ϵ.

Theorem 2 . *Let $p(x)$ and $\mathcal{K}(x)$ satisfy the assumptions of Theorem 1 and*

$$\int_{B_R} \mathcal{K}(y) dy = 1 . \tag{13}$$

Then (9) is an identity approximation in $L^{p(x)}(\Omega)$:

$$\lim_{\epsilon\to 0} \|K_\epsilon f \;-\; f\|_{L^{p(x)}(\Omega_R)} \;=\; 0\;, \quad f(x) \in L^{p(x}(\Omega). \tag{14}$$

Let

$$f_\epsilon(x) \;=\; \frac{1}{\epsilon^n|B(0,1)|}\int_{y\in\Omega,|y-x|<\epsilon} f(y)dy \tag{15}$$

be the Steklov mean of the function $f(y)$.

Corollary 1. *Under the assumptions of Theorem 1 on* $p(x)$,

$$\lim_{\epsilon\to 0} \|f_\epsilon \;-\; f\|_{L^{p(x)}(\Omega)} \;=\; 0\;. \tag{16}$$

Remark 1. The statement (16) is an analogue of mean continuity property for $L^{p(x)}$-spaces, but with respect to the averaged "shift" operator (15). In the standard form, the mean continuity property $\lim_{h\to 0}\|f(x+h)-f(x)\|_p \;=\; 0$, generally speaking, is not valid for variable exponents $p(x)$ and, moreover, there exist functions $p(x)$ and $f(x) \in L^{p(x)}$ such that $f(x+h_k) \notin L^{p(x)}$ for some $h_k \to 0$, see [2], Example 2.9 and Theorem 2.10.

Corollary 2. *Let* $1 \le p(x) \le P < \infty, x \in R^n$, *and* $p(x)$ *satisfy the condition (8) in any ball in* R^n *(where A may depend on the ball) . Then* C_0^∞ *is dense in* $L^{p(x)}(R^n)$.

Remark 2. As it was shown in [2], $C_0^\infty(\Omega)$ is dense in $L^{p(x)}(\Omega), 1 \le p(x) \le P < \infty$, without requiring that $p(x)$ satisfies the condition (8).

Let $W^{m,p(x)} = W^{m,p(x)}(R^n)$ be the Sobolev type space of functions $f(x) \in L^{p(x)}(R^n)$ which have all the distributional derivatives $D^j f(x) \in L^{p(x)}(R^n), 0 \le |j| \le m$, and let

$$\|f\|_{W^{m,p(x)}} \;=\; \sum_{|j|\le m} \|D^j f\|_p\;.$$

Theorem 3. *Let* $p(x)$ *satisfy the assumptions of Theorem 3. Then* $C_0^\infty(R^n)$ *is dense in* $W^{m,p(x)}(R^n)$.

4. Proof of Theorem 1.

We assume that

$$\|f\|_p \le 1\;. \tag{17}$$

By (4)-(5) it suffices to show that

$$I_p\left(K_\epsilon f\right) \;=\; \int_{\Omega_R} |K_\epsilon f(x)|^{p(x)}dx \le c \tag{18}$$

with $c > 0$ not depending on ϵ . By the Hölder inequality (3) it is easy to show that $|K_\epsilon f(x)| \le c$ for all $x \in \Omega_R$ and $\epsilon \ge \epsilon^o (c = c(\epsilon^o)$ in this case). Therefore, it suffices to prove (18) for $0 < \epsilon \le \epsilon^o$ under some choice of ϵ^o.

Let

$$\Omega_R \;=\; \cup_{k=1}^N \omega_R^k$$

be any partition of Ω_R into small parts ω_R^k comparable with the given ϵ :

$$diam\ \omega_R^k \ \leq\ \epsilon\ , k = 1, 2, \cdots, N\ ;\ N = N(\epsilon).$$

We represent the integral in (18) as

$$I_p(K_\epsilon f) = \sum_{k=1}^{N} \int_{\omega_R^k} \left| \int_\Omega \mathcal{K}_\epsilon(x-y) f(y)\, dy \right|^{p(x)-p_k+p_k} dx \tag{19}$$

with

$$p_k \ = \ \inf_{x\in\Omega_R^k} p(x) \leq \inf_{x\in\omega_R^k} p(x) \tag{20}$$

where some larger portions $\Omega_R^k \supset \omega_R^k$ will be chosen later comparable with ϵ :

$$diam\ \Omega_R^k \leq m\epsilon\ , \quad m > 1\ . \tag{21}$$

We shall prove the uniform estimate

$$A_k(x,\epsilon)\ := \left| \int_\Omega \mathcal{K}_\epsilon(x-y) f(y)\, dy \right|^{p(x)-p_k} \leq c\ , \quad x \in \omega_R^k \tag{22}$$

where $c > 0$ does not depend on $x \in \omega_R^k$, k and $\epsilon \in (0, \epsilon^o)$ with some $\epsilon^o > 0$. To this end, we first obtain the estimate

$$A_k(x,\epsilon) \leq c_1\ \epsilon^{-n[p(x)-p_k]}\ , \quad x \in \Omega_R. \tag{23}$$

To get (23), we differ the cases $Q = \infty$ and $Q < \infty$.

Let $Q = \infty$. We have

$$A_k(x,\epsilon) \leq \left(\frac{M}{\epsilon^n} \int_\Omega \chi_{B(0,\epsilon R)}(y) |f(y)| dy \right)^{p(x)-p_k}$$

where $M = \sup_{B_R} |\mathcal{K}(x)|$. By the Hölder inequality (3) and the assumption (17) we obtain

$$A_k(x,\epsilon) \leq \left(\frac{Mk}{\epsilon^n} \|\chi_{B(0,\epsilon R)}\|_q \right)^{p(x)-p_k} . \tag{24}$$

According to (2) we have

$$\|\chi_{B(0,\epsilon R)}\|_q \ = \ \sup_{E_\infty(q)} \chi_{B(0,\epsilon R)}(x) \ + \ \|\chi_{B(0,\epsilon R)}\|_{(q)} = 1 + \|\chi_{B(0,\epsilon R)}\|_{(q)}\ .$$

In view of (6) we get

$$\|\chi_{B(0,\epsilon R)}\|_q \leq 1 + (\epsilon^n |B(0,R)|)^{\frac{1}{q_0}} \leq 2$$

under the asumption that

$$0 < \epsilon \leq |B(0,R)|^{-\frac{1}{n}}\ := \epsilon_1^o\ . \tag{25}$$

Then (24) provides the estimate (23) with $c_1 = (2kM)^{P-p_0}$ if $2kM \geq 1$ and $c_1 = 1$ otherwise.

Let $Q < \infty$. The estimate (23) is obtained in a similar way. Indeed, applying the Hölder inequality (3) again, we arrive at

$$A_k(x,\epsilon) \leq (k\|\mathcal{K}_\epsilon(x-y)\|_q)^{p(x)-p_k} \ .$$

By (4)-(5) we have

$$\|\mathcal{K}_\epsilon(x-y)\|_{(q)} = \frac{1}{\epsilon^n}\|\mathcal{K}\left(\frac{x-y}{\epsilon}\right)\|_{(q)} \leq \frac{1}{\epsilon^n}\left(\int_{\Omega\setminus E_\infty(q)}\left|\mathcal{K}\left(\frac{x-y}{\epsilon}\right)\right|^{q(y)} dy\right)^\theta$$

where $\theta = \frac{1}{Q}$ or $\theta = \frac{1}{q_o}$ depending on the fact whether the last integral in the parantheses is less or greater than 1, respectively. Hence,

$$\|\mathcal{K}_\epsilon(x-y)\|_{(q)} \leq \frac{1}{\epsilon^n}\left(\int_{|y|<R, x-\epsilon y\in\Omega\setminus E_\infty(q)} |\mathcal{K}(y)|^{q(x-\epsilon y)}\, dy\right)^\theta$$

$$\leq \frac{1}{\epsilon^n}\left[|B_R| + \int_{|y|<R, |\mathcal{K}(y)|\geq 1} |\mathcal{K}(y)|^Q\, dy\right]^\theta \leq \frac{1}{\epsilon^n}\left[|B_R| + \|\mathcal{K}\|_Q^Q\right]^\theta \leq c_2\epsilon^{-n} \tag{27}$$

where $c_2 = \max\{c_3^{\frac{1}{Q}}, c_3^{\frac{1}{q_o}}\}, c_3 = |B_R| + \|\mathcal{K}\|_Q^Q$.

Therefore, from (26) and (27) we obtain (23) in the case $Q < \infty$ as well, with $c_1 = (c_2k)^{P-p_o}$ if $c_2 > 1$ and $c_1 = 1$ otherwise.

The estimate (23) having been proved, we observe now that by (8)

$$p(x) - p_k = |p(x) - p(\xi_k)| \leq \frac{A}{\log\frac{1}{|x-\xi_k|}}$$

where $x \in \omega_R^k$, $\xi_k \in \Omega_R^k$. Evidently,

$$|x - \xi_k| \leq diam\Omega_R^k \leq m\epsilon$$

by (21). Therefore,

$$p(x) - p_k \leq \frac{A}{\log\frac{1}{m\epsilon}} \tag{28}$$

under the assumption that

$$0 < \epsilon \leq \frac{1}{2m} \ =: \ \epsilon_2^o \ . \tag{29}$$

Then from (23) and (28)

$$A_k(x,\epsilon) \ \leq \ c_1\epsilon^{-\frac{A}{\log\frac{1}{m\epsilon}}} \ , \ x \in \omega_R^k \ , \tag{30}$$

c_1 not depending on x and being given above. Then from (30)

$$A_k(x,\epsilon) \ \leq \ c_4 \ := \ c_1e^{2A}$$

for $x \in \omega_R^k$ and

$$0 < \epsilon \leq \epsilon_3^0 \ := \frac{1}{m^2}. \tag{31}$$

Therefore, we have the uniform estimate (22) with $c = c_1 e^{2A}$ and $0 < \epsilon \leq \epsilon^o$, $\epsilon^o = \min_{1\leq k\leq 3} \epsilon_k^0$, ϵ_k^o being given by (25), (29) and (31).

Using the estimate (22) we obtain from (19)

$$I_p(K_\epsilon f) \leq c\sum_{k=1}^{N} \int_{\omega_R^k} \left| \int_\Omega \mathcal{K}_\epsilon(x-y) f(y)\, dy \right|^{p_k} dx \,.$$

Here p_k are constants so that we may apply the usual Minkowsky inequality for integrals and obtain

$$I_p(K_\epsilon f) \leq c\sum_{k=1}^{N} \left\{ \int_{|y|<\epsilon R} |\mathcal{K}_\epsilon(y)|\, dy \left(\int_{\omega_R^k} |f(x-y)|^{p_k}\, dx \right)^{\frac{1}{p_k}} \right\}^{p_k}$$

$$= c\sum_{k=1}^{N} \left\{ \int_{|y|<R} |\mathcal{K}(y)|\, dy \left(\int_{x+\epsilon y\in\omega_R^k} |f(x)|^{p_k}\, dx \right)^{\frac{1}{p_k}} \right\}^{p_k} . \tag{32}$$

Obviously, the domain of integration in x in the last integral is embedded into the domain

$$\bigcup_{y\in B_{\epsilon R}} \{x : x+y \in \omega_R^k\} \tag{33}$$

which already does not depend on y . Now, we choose the sets Ω_R^k in (20), which were not determined until now, as the sets (33). Then, evidently, $\Omega_R^k \supset \omega_R^k$ and it is easily seen that

$$diam\ \Omega_R^k \leq (1+2R)\epsilon \tag{34}$$

so that the requirement (21) is satisfied with $m = 1+2R$.

From (32) we have

$$I_p(K_\epsilon f) \leq c\sum_{k=1}^{N} \left\{ \int_{|y|<R} |\mathcal{K}(y)|\, dy \right\}^{p_k} \int_{\Omega_R^k} |f(x)|^{p_k}\, dx$$

$$\leq c \left\{ \int_{|y|<R} |\mathcal{K}(y)|\, dy \right\}^{\theta} \sum_{k=1}^{N} \int_{\Omega_R^k\cap\Omega} |f(x)|^{p_k}\, dx$$

where $\theta = P$ if $\int_{|y|<R} |\mathcal{K}(y)| dy \leq 1$ and $\theta = p_o$ otherwise. In view of (34), the covering $\{\omega_k = \Omega_R^k \cap \Omega\}_{k=1}^N$ has a finite multiplicity (that is, each point $x \in \Omega$ belongs simultaneously not more than to a finite number n_o of the sets ω_k, $n_o \leq 1 + (1+2R)^n$ in this case). Therefore,

$$I_p(K_\epsilon f) \leq c_5 \int_\Omega |f(x)|^{\bar{p}(x)}\, dx \tag{35}$$

where

$$\tilde{p}(x) = \max_j p_j$$

the maximum being taken with respect to all the sets ω_j containing x . Evidently, $\tilde{p}(x) \leq p(x)$ for $x \in \Omega$. Then from (35) and (4)-(5) we obtain the estimate

$$I_p(K_\epsilon f) \leq c_5 \|f\|_{\tilde{p}}^{\theta_1}, \quad \theta_1 < P,$$

with $\theta_1 = \inf \tilde{p}(x)$ if $\|f\|_{\tilde{p}} \leq 1$ and $\theta_1 = \sup \tilde{p}(x)$ otherwise. Applying the imbedding theorem (7), we arrive at the final estimate

$$I_p(K_\epsilon f) \leq c_6 \|f\|_p^{\theta_1} \leq c_6 .$$

5. Proof of Theorem 2.

To prove (14), we use Theorem 1, which provides the uniform boundedness of the operators K_ϵ from $L^{p(x)}(\Omega)$ into $L^{p(x)}(\Omega_R)$. Then, by the Banach-Steinhaus theorem it suffices to verify that (14) holds for some dense set in $L^{p(x)}(\Omega)$, for example, for step functions, according to property e) of the spaces $L^{p(x)}(\Omega)$. So, it is sufficient to prove (14) for the characteristic function $\chi_E(x)$ of any bounded measurable set $E \subset \Omega$. We have

$$K_\epsilon(\chi_E) - \chi_E = \int_{B_R} \mathcal{K}(y) \left[\chi_E(x - \epsilon y) - \chi_E(x)\right] dy$$

by (13). Hence

$$\|K_\epsilon(\chi_E) - \chi_E\|_P \leq \int_{B_R} |\mathcal{K}(y)| \, \|\chi_E(\cdot - \epsilon y) - \chi_E(x)\|_P \, dy \to 0$$

as $\epsilon \to 0$ by the Lebesgue dominated convergence theorem and the P-mean continuity of functions in L^P with a constant P ($P = \sup_{x \in \Omega_R} p(x)$ in this case). Then, by (7), also

$$\|K_\epsilon(\chi_E) - \chi_E\|_p \to 0$$

with $p = p(x) \leq P < \infty$.□

6. Proof of Corollaries

To obtain Corollary 1 from Theorem 1, it suffices to choose $\mathcal{K}(y) = \frac{1}{|B(0,1)|}\chi_{B(0,1)}(y)$.

Proof of Corollary 2. Let $\chi_N(x) = \chi_{B(0,N)}(x)$. Then the functions $f^N(x) = \chi_N(x) f(x)$ have compact support and approximate $f(x) \in L^{p(x)}(R^n)$:

$$\|f - f^N\| \leq I_p^{\frac{1}{P}}(f - f^N) = \left(\int_{|x|>N} |f(x)|^{p(x)} dx \right)^{\frac{1}{P}} \to 0$$

as $N \to \infty$.

Therefore, we may consider $f(x)$ with a compact support in the ball B_N from the very beginning. To approximate $f(x)$ by C_0^∞, we use the identity approximation

$$f_\epsilon(x) = \int_{R^n} \mathcal{K}_\epsilon(x-t)f(t)dt = \int_{|y|<1} \mathcal{K}(y)f(x-\epsilon y)dy \qquad (36)$$

where $\mathcal{K}_\epsilon(x) = \frac{1}{\epsilon^n}\mathcal{K}\left(\frac{x}{\epsilon}\right)$ and $\mathcal{K}(y) \in C_0^\infty(R^n)$ with support in the ball B_1 and such that

$$\int_{|y|<1} \mathcal{K}(y)dy = 1 .$$

Then, evidently, $f_\epsilon(x) \in C_0^\infty(R^n)$ and has compact support because $f_\epsilon(x) \equiv 0$ if $|x| > N+\epsilon$. Therefore, for $\epsilon < 1$,

$$\|f_\epsilon - f\|_{L^{p(x)}(R^n)} = \|K_\epsilon f - f\|_{L^{p(x)}(B_{N+1})} \to 0$$

as $\epsilon \to 0$.

Proof of Theorem 3.

The proof follows from Theorem 2 and Corollary 2 in two steps.

1°. Let $f(x) \in W^{m,p(x)}(R^n)$ and let $\mu(r), 0 \le r \le \infty$, be a smooth step-function: $\mu(r) \equiv 1$ for $0 \le r \le 1, \mu(r) \equiv 0$ for $r \ge 2, \mu(r) \in C_0^\infty(R_+^1)$ and $0 \le \mu(r) \le 1$. Then

$$f^N(x) = \mu\left(\frac{|x|}{N}\right) f(x) \in W^{m,p(x)}(R^n) \qquad (37)$$

for every $N \in R_+^1$ and has compact support in B_{2N}.

The functions (37) approximate $f(x)$ in $W^{m,p(x)}(R^n)$. Indeed, denoting $\nu_N(x) = 1 - \mu\left(\frac{|x|}{N}\right)$, so that $\nu_N(x) \equiv 0$ for $|x| < N$, and using the Leibnitz formula for differentiation, we have

$$\|f - f^N\|_{W^{m,p(x)}} = \sum_{|j|\le m} \|D^j(\nu_N f)\|_p \le \sum_{|j|\le m}\sum_{0\le k\le j} c_k\|D^k(\nu_N)D^{j-k}f\|_p$$

$$\le \sum_{|j|\le m} \|\nu_N D^j f\|_p + c\sum_{|j|\le m}\sum_{0<k\le j} \|D^k(\nu_N)D^{j-k}f\|_p$$

$$\le \sum_{|j|\le m} \|\nu_N D^j f\|_p + c\sum_{|j|\le m}\sum_{0<k\le j} \frac{1}{N^{|k|}}\|D^{j-k}f\|_p \to 0 \qquad (38)$$

as $N \to 0$.

2. By the step 1° we may consider $f(x) \in W^{m,p(x)}$with compact support. Then we take $\mathcal{K}(y) \in C_0^\infty(R^n)$ with support in the ball B_1 and such that $\int_{|y|<1}\mathcal{K}(y)dy = 1$ and arrange the approximation (36). Then, evidently, $f_\epsilon \in C_0^\infty(R^n)$. Indeed, for any j we have

$$D^j f_\epsilon(x) = \frac{1}{\epsilon^{n+|j|}}\int_{|y|<1} (D^j\mathcal{K})\left(\frac{x-t}{\epsilon}\right) f(t)dt \in C^\infty(R^n)$$

and $f_\epsilon(x)$ has compact support because $f_\epsilon(x) \equiv 0$ if $|x| > 1 + \lambda$, where $\lambda = \sup_{x \in supp\, f} |x|$, *supp* standing for support of $f(x)$.

We have

$$\|f_\epsilon(x) - f\|_{W^{m,p(x)}} \leq \sum_{|j| \leq m} \|D^j f - K_\epsilon(D^j f)\|_{L^{p(x)}(R^n)}$$

$$= \sum_{|j| \leq m} \|D^j f - K_\epsilon(D^j f)\|_{L^{p(x)}(\Omega_1)}$$

where $\Omega_1 = \{x : dist(x, \Omega) \leq 1\}, \Omega = suppf(x)$. It suffices to apply Theorem 2.

References

[1] Edmunds, D.E. and J. Rákosník. (1992). *Density of smooth functions in* $W^{k,p(x)}(\Omega)$, Proc. R. Soc. Lond. A., Vol. 437, (pages 229-236).

[2] Kováčik, O. and J. Rákosník. (1991). *On spaces* $L^{p(x)}$ *and* $W^{k,p(x)}$, Czech. Math. J., Vol. 41(116), (pages 592-618).

[3] Samko, S.G. (1997). *Differentiation and integration of variable order and the spaces* $L^{p(x)}$, Proceed. of Intern. Conference "Operator Theory and Complex and Hypercomplex Analysis", 12-17 December 1994, Mexico City, Mexico, Contemp. Math..

[4] Samko S.G. (1997). *Convolution type operators in* $L^{p(x)}$*. Preliminary results*, Integr. Transf. Spec. Funct..

[5] Samko S.G. (1997). *Convolution and potential type operators in* $L^{p(x)}$, Integr. Transf. Spec. Funct..

[6] Sharapudinov, I.I. (1979). *On a topology of the space* $L^{p(t)}([0,1])$, Matem. Zametki, Vol. 26(4), (pages 613-632).

[7] Sharapudinov, I.I. (1996). *On uniform boundedness in* $L^p(p = p(x))$ *of some families of convolution operators (in Russian)*, Matem. Zametki, Vol. 59 (2), (pages 291-302).

SINGULARITIES OF THE REFLECTED AND REFRACTED RIEMANN FUNCTIONS OF ELASTIC WAVE PROPAGATION PROBLEMS IN STRATIFIED MEDIA

SENJO SHIMIZU

ABSTRACT. In this paper we shall study elastic mixed or initial-interface value problems and give an inner estimate of the location of singularities of the reflected and refracted Riemann functions by making use of the localization method.

1. Introduction

We consider elastic wave propagation problems in the following plane-stratified media $\mathbf{R}^3$ with the planar interface $x_3 = 0$:

$$(\lambda(x_3), \mu(x_3), \rho(x_3)) = \begin{cases} (\lambda_1, \mu_1, \rho_1) & \text{for} \quad x_3 < 0, \\ (\lambda_2, \mu_2, \rho_2) & \text{for} \quad x_3 > 0. \end{cases}$$

Here the constants λ_1, λ_2, μ_1, μ_2 are called the Lamé constants and the constants ρ_1, ρ_2 are densities. We shall denote the lower half-space $\mathbf{R}^3_-$ by *Medium I* and the upper half-space $\mathbf{R}^3_+$ by *Medium II*, respectively, as in Figure 1.

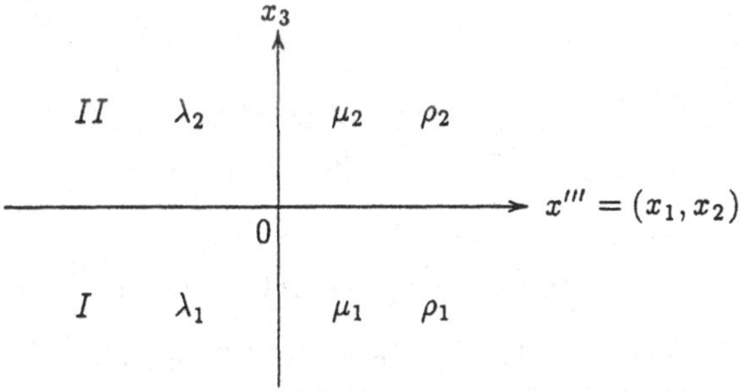

Figure 1 Stratified media I and II

We assume that

$$\lambda_i + \mu_i > 0, \quad \mu_i > 0, \quad \rho_i > 0, \quad i = 1, 2. \tag{1.1}$$

1991 *Mathematics Subject Classification.* Primary 35L67 ; Secondary 73C35 35E99.

This work was supported in part by Grants-in-Aid for Encouragement of Young Scientists (grant A-09740098) from the Ministry of Education, Science and Culture of Japan.

R.P. Gilbert et al.(eds.), Direct and Inverse Problems of Mathematical Physics, 343–362.

(1.1) is the natural assumption in practical situation. From the roots of the characteristic equations of $P^I(D)$ and $P^{II}(D)$ which are defined below 3×3 matrix valued hyperbolic partial diferential operators in Medium I and Medium II, respectively, we obtain two speeds correspond to Pressuer or Primary wave (P wave) and Share or Secondary wave (S wave) on each medium. c_{p_1} denotes the speed of P wave in Medium I and c_{s_1} denotes the speed of S wave in Medium I. c_{p_2} and c_{s_2} denote the speed of P and S wave in Medium II, respectively. They are given by

$$c_{p_i}^2 = \frac{\lambda_i + 2\mu_i}{\rho_i}, \quad c_{s_i}^2 = \frac{\mu_i}{\rho_i}, \quad i = 1, 2.$$

By assumption (1.1), the speed of P wave is greater than that of S wave in each medium. On account of this, these are six cases of the order relation of the speeds of $\{c_{p_1}, c_{s_1}. c_{p_2}, c_{s_2}\}$. Here we assume that

$$c_{s_1} < c_{p_1} \leq c_{s_2} < c_{p_2}. \tag{1.2}$$

It is the standard case (cf. [Sh, Section 3]). The other cases can be treated in a similar manner.

Let $x = (x_0, x_1, x_2, x_3) = (x', x_3) = (x_0, x'') = (x_0, x''', x_3)$ in $\mathbf{R}^4$. The variable x_0 will play a role of time, and $x'' = (x_1, x_2, x_3)$ will play that of space. ξ is a real dual variable of x and is equal to $(\xi_0, \xi_1, \xi_2, \xi_3) = (\xi', \xi_3) = (\xi_0, \xi'') = (\xi_0, \xi''', \xi_3)$ in $\mathbf{R}^4_\xi$. We use the differential symble $D_j = i^{-1}\partial/\partial x_j$ $(j = 0, 1, 2, 3)$, where $i = \sqrt{-1}$.

Let $u(x) = {}^t(u_1(x), u_2(x), u_3(x)) \in \mathbf{R}^3$ be the displacement vector at time x_0 and position x''. The propagation problems of elastic waves in the stratified media is formulated as mixed (initial-interface value) problem:

$$\begin{cases} P^I(D)u(x) = f(x), \quad x_0 > 0,\ x'' = (x_1, x_2, x_3) \in \mathbf{R}^3_-, \\ P^{II}(D)u(x) = f(x), \quad x_0 > 0,\ x'' = (x_1, x_2, x_3) \in \mathbf{R}^3_+, \\ u(x)|_{x_3=-0} = u(x)|_{x_3=+0}, \quad x_0 > 0,\ x''' \in \mathbf{R}^2, \\ B^I(D)u(x)|_{x_3=-0} = B^{II}(D)u(x)|_{x_3=+0}, \quad x_0 > 0,\ x''' \in \mathbf{R}^2, \\ D_0^k u(x)|_{x_0=0} = g_k(x''), \quad k = 0, 1, \quad x'' \in \mathbf{R}^3. \end{cases} \tag{1.3}$$

Here

$$P^I(D)u = -D_0^2 E u + \frac{\lambda_1 + \mu_1}{\rho_1}\nabla_{x''}(\nabla_{x''}\cdot u) + \frac{\mu_1}{\rho_1}\Delta_{x''}u,$$

is a 3×3 matrix valued second order hyperbolic differential operator with constant coefficients where E is a 3×3 identity matrix,

$$(B^I(D)u)_k = i\lambda_1(\nabla_{x''}\cdot u)\delta_{k3} + 2\mu_1\varepsilon_{k3}(u), \quad k = 1, 2, 3,$$

is the k-th component of symmetric stress tensors $B^I(D)u$ where

$$\varepsilon_{k3}(u) = i/2\,(D_3 u_k + D_k u_3), \quad k = 1, 2, 3,$$

are strain tensors. The $P^{II}(D)u$ and $B^{II}(D)u$ are defined by replacing λ_1, μ_1, ρ_1 by λ_2, μ_2, ρ_2, respectively.

If we put unit impulse Dirac's delta $\delta(x-y)$ at position $y = (0,\ y'')$ with $y_3 < 0$, that is, put in Medium I, then the Riemann function of this elastic mixed problem is given by the following:

$$G(x,y) = \begin{cases} E^I(x-y) - F^I(x,y) & \text{for} \quad x_3 < 0, \\ F^{II}(x,y) & \text{for} \quad x_3 > 0, \end{cases}$$

where $E^I(x)$ is the fundamental solution in Medium I describing an incident wave, is defined by

$$E^I(x) = (2\pi)^{-4} \int_{\mathbf{R}^4_\xi} e^{ix\cdot(\xi+i\eta)} P^I(\xi+i\eta)^{-1} d\xi, \qquad \eta \in -s\vartheta - \Gamma,$$

with a positive real s large enough. Here ϑ and Γ are defiend below. Taking partial Fourier-Laplace transform with respect to x' for the mixed problem, we obtain a interface value problem for ordinary differential equation with parameters. Then taking partial inverse Fourier-Laplace transform for the solution, we obtain expicit expressions of the reflected and refracted Riemann functions $F^I(x,y)$ and $F^{II}(x,y)$ which describe reflected and refracted waves, respectively.

In this paper, we give an inner estimate of the location of singularities of the reflected and refracted Riemann functions $F^I(x,y)$ and $F^{II}(x,y)$ by making use of the localization method. This method is first studied by M. F. Atiyah, R. Bott, L. Gårding [A-B-G] for initial value problem, then studied by M. Matsumura [Ma 1], M. Tsuji [Ts], and S. Wakabayashi [Wa 1], [Wa 2] for half-space mixed problem. Matsumura studied the singularities of the ordinary wave propagation problems in the stratified media by applying above methods [Ma 2], [Ma 3]. They are useful references to our study.

We define a localization of polynomials according to Atiyah-Bott-Gårding (cf. [A-B-G]):

Definition 1. Let $P(\xi)$ be a polynomial of degree $m \geq 0$ and develop $\nu^m P(\nu^{-1}\xi + \eta)$ in ascendeing power of ν:

$$\nu^m P\left(\nu^{-1}\xi + \eta\right) = \nu^p P_\xi(\eta) + O(\nu^{p+1}) \qquad \text{as} \quad \nu \to 0, \tag{1.4}$$

where $P_\xi(\eta)$ is the first coefficient that does not vanish identically in η. The polynomial $P_\xi(\eta)$ is the localization of P at ξ, the number p is the multiplicity of ξ relative to P.

Moreover we introduce the following:

Definition 2. $\Gamma = \Gamma(P,\vartheta)$ is the component of $\mathbf{R}^n_\eta \setminus \{\eta \in \mathbf{R}^n_\eta,\ P(\eta) = 0\}$ which contains $\vartheta = (1, 0, \cdots, 0) \in \mathbf{R}^n$. Moreover $\Gamma' = \Gamma'(P,\vartheta) = \{x \in \mathbf{R}^n |\ x\cdot\eta \geq 0,\ \eta \in \Gamma\}$ is the dual cone of Γ and is called the propagation cone.

We obtain the following main theorem. This main theorem means singular supports of the reflected and refracted Riemann functions $F^I(x,y)$ and $F^{II}(x,y)$ are estimated innerly by localizations $F^I_{\xi^0}(x,y)$ and $F^{II}_{\xi^0}(x,y)$ of $F^I(x,y)$ and $F^{II}(x,y)$ at ξ^0, respectively.

Main Theorem. *For $\xi^0 \in \mathbf{R}^4_\xi$ satisfying $(\det P^I_j)(\xi^0) = 0$ $(j \in \{p_1, s_1\})$, that is*

$$(\det P^I_{p_1})(\xi^0) = \xi_0^{0\,2} - c^2_{p_1}|\xi^{0''}|^2 = 0,$$

or

$$(\det P^I_{s_1})(\xi^0) = \xi_0^{0\,2} - c^2_{s_1}|\xi^{0''}|^2 = 0,$$

we have the following:
(1) For the reflected Riemann function $F^I(x,y)$, if $\xi^{0'}$ are not zeros of inner radical sign of $\tau^{\pm}_m(\zeta')$ $(m \in \{p_1, p_2, s_2\})$, then we have

$$\lim_{\nu\to\infty} \nu e^{-i\nu\{(x'-y')\cdot\xi^{0'} + x_3\tau^-_k(\xi^{0'}) - y_3\xi^0_3\}} F^I(x,y) = F^I_{j\xi^0 k}(x,y), \tag{1.5}$$
$$j \in \{p_1, s_1\}, \quad k \in \{p_1, s_1\},$$

if $\xi^{0'}$ are zeros of inner radical sign of $\tau^{\pm}_m(\zeta')$, that is, $\xi^{0'}$ satisfy $|\xi^{0'''}| = \frac{\xi^0_0}{c_m}$ $(m \in \{p_1, p_2, s_2\})$, then we have

(1.6)
$$\lim_{\nu\to\infty} \left\{\nu^{\frac{3}{2}} e^{-i\nu\{(x'-y')\cdot\xi^{0'} + x_3\tau^-_k(\xi^{0'}) - y_3\xi^0_3\}} F^I(x,y) - \nu^{\frac{1}{2}} F^I_{j\xi^0 k}(x,y)\right\}$$
$$= F^I_{j\xi^0 km}(x,y), \qquad j \in \{p_1, s_1\}, \quad k \in \{p_1, s_1\}, \quad m \in \{p_1, p_2, s_2\},$$

in the distribution sense with respect to $(x,y) \in \mathbf{R}^4_- \times \mathbf{R}^4_-$.
Moreover we have

$$\bigcup_{\xi^0 \neq 0} \left(\operatorname{supp} F^I_{j\xi^0 k}(x,y) \cup \operatorname{supp} F^I_{j\xi^0 km}(x,y)\right) \subset \operatorname{sing\,supp} F^I(x,y), \tag{1.7}$$

and

(1.8)
$$\operatorname{supp} F^I_{j\xi^0 k}(x,y) = (\Gamma_{j\xi^0})^I_k = \Big\{(x,y) \in \mathbf{R}^4_- \times \mathbf{R}^4_- :$$
$$((x'-y') + x_3 \operatorname{grad}_\xi \tau^-_k(\xi^{0'})) \cdot \eta' - y_3\,\eta_3 \geq 0,\ \eta \in \Gamma_{j\xi^0}\Big\},$$
$$j \in \{p_1, s_1\}, \quad k \in \{p_1, s_1\},$$

(1.9)
$$\operatorname{supp} F^I_{j\xi^0 km}(x,y) = (\Gamma_{j\xi^0 m})^I_k = \Big\{(x,y) \in \mathbf{R}^4_- \times \mathbf{R}^4_- :$$
$$((x'-y') + x_3 \operatorname{grad}_\xi \tau^-_k(\xi^{0'})) \cdot \eta' - y_3\,\eta_3 \geq 0,\ \eta \in \Gamma_{j\xi^0 m}\Big\},$$
$$j \in \{p_1, s_1\}, \quad k \in \{p_1, s_1\}, \quad m \in \{p_1, p_2, s_2\}.$$

(2) For the refracted Riemann function $F^{II}(x,y)$, if $\xi^{0'}$ are not zeros of inner radical sign of $\tau^{\pm}_m(\zeta')$ $(m \in \{p_2\})$, then we have

(1.10)
$$\lim_{\nu\to\infty} \nu e^{-i\nu\{(x'-y')\cdot\xi^{0'} + x_3\tau^+_k(\xi^{0'}) - y_3\xi^0_3\}} F^{II}(x,y) = F^{II}_{j\xi^0 k}(x,y),$$
$$j \in \{p_1, s_1\}, \quad k \in \{p_2, s_2\},$$

if $\xi^{0'}$ are zeros of inner radical sign of $\tau_m^+(\zeta')$ $(m \in \{p_2\})$, then we have

(1.11)

$$\lim_{\nu \to \infty} \left\{ \nu^{\frac{3}{2}} e^{-i\nu\{(x'-y')\cdot\xi^{0'} + x_3\tau_k^+(\xi^{0'}) - y_3\xi_3^0\}} F^{II}(x,y) - \nu^{\frac{1}{2}} F^{II}_{j\xi^0 k}(x,y) \right\}$$
$$= F^{II}_{j\xi^0 km}(x,y), \qquad j \in \{p_1, s_1\}, \quad k \in \{p_1, s_1\}, \quad m \in \{p_2\},$$

in the distribution sense with respect to $(x,y) \in \mathbf{R}^4_+ \times \mathbf{R}^4_-$.

Moreover we have

(1.12)
$$\bigcup_{\xi^0 \neq 0} (\operatorname{supp} F^{II}_{j\xi^0 k}(x,y) \cup \operatorname{supp} F^{II}_{j\xi^0 km}(x,y)) \subset \operatorname{sing\,supp} F^{II}(x,y),$$

and

(1.13)

$$\operatorname{supp} F^{II}_{j\xi^0 k}(x,y) = (\Gamma_{j\xi^0})^{II}_k = \Big\{ (x,y) \in \mathbf{R}^4_+ \times \mathbf{R}^4_- :$$
$$((x'-y') + x_3 \operatorname{grad}_\xi \tau_k^+(\xi^{0'})) \cdot \eta' - y_3\,\eta_3 \geq 0,\ \eta \in \Gamma_{j\xi^0} \Big\},$$
$$j \in \{p_1, s_1\}, \quad k \in \{p_2, s_2\}$$

(1.14)

$$\operatorname{supp} F^{II}_{j\xi^0 km}(x,y) = (\Gamma_{j\xi^0 m})^{II}_k = \Big\{ (x,y) \in \mathbf{R}^4_- \times \mathbf{R}^4_- :$$
$$((x'-y') + x_3 \operatorname{grad}_\xi \tau_k^+(\xi^{0'})) \cdot \eta' - y_3\,\eta_3 \geq 0,\ \eta \in \Gamma_{j\xi^0 m} \Big\},$$
$$j \in \{p_1, s_1\}, \quad k \in \{p_2, s_2\}, \quad m \in \{p_2\}.$$

where

(1.15)
$$\Gamma_{j\xi^0} = \Gamma((\det P^I_j)_{\xi^0}(\eta), \vartheta), \qquad \vartheta = (1,0,0,0), \quad j \in \{p_1, s_1\},$$

(1.16)
$$\Gamma_{j\xi^0 m} = \Gamma((\det P^I_j)_{\xi^0}(\eta), \vartheta) \cap \left\{ \Gamma\left(\frac{\xi_0^0}{c_m^2}\eta_0 - \xi_1^0\eta_1 - \xi_2^0\eta_2, \vartheta' \right) \times \mathbf{R}_\eta \right\},$$
$$\vartheta' = (1,0,0),\ j \in \{p_1, s_1\}, \quad m \in \{p_2\},$$

$$\tau^\pm_{p_1}(\xi') = \operatorname{sgn}(\mp\xi_0)\sqrt{\frac{\xi_0^2}{c_{p_1}^2} - (\xi_1^2 + \xi_2^2)}, \quad \text{if} \quad \frac{\xi_0^2}{c_{p_1}^2} - (\xi_1^2 + \xi_2^2) \geq 0,$$

and $\tau^\pm_{p_1}(\xi')$ is a branch of $\sqrt{\frac{\xi_0^2}{c_{p_1}^2} - (\xi_1^2 + \xi_2^2)}$ such that $\pm\mathrm{Im}\tau^\pm_{p_1}(\xi') > 0$ if $\frac{\xi_0^2}{c_{p_1}^2} - (\xi_1^2 + \xi_2^2) < 0$. $\tau^\pm_{s_1}(\xi')$, $\tau^\pm_{p_2}(\xi')$, and $\tau^\pm_{s_2}(\xi')$ are defined as the same as $\tau^\pm_{p_1}(\xi')$ substituting c_{p_1} for c_{s_1}, c_{p_2}, and c_{s_2}, respectively.

Remark.1. The $(\Gamma_{j\xi^0})^I_k$ $(j \in \{p_1, s_1\},\ k \in \{p_1, s_1\})$ represent k reflected wave for j incident wave. The $(\Gamma_{j\xi^0 m})^I_k$ $(j \in \{p_1, s_1\},\ k \in \{p_1, s_1\}\ m\{p_1, p_2, s_2\})$ represent m lateral wave of k reflected wave for j incident wave. The $(\Gamma_{j\xi^0})_k^{II}$ $(j \in \{p_1, s_1\}$, $k \in \{p_2, s_2\})$ represent k refracted wave for j incident wave. The $(\Gamma_{j\xi^0 m})_k^{II}$ $(j \in \{p_1, s_1\}$, $k \in \{p_2, s_2\}$ $m\{p_2\})$ represent m lateral wave of k refracted wave for j incident wave.

Remark.2. The $\tau_{p_1}^{\pm}(\xi')$, $\tau_{s_1}^{\pm}(\xi')$, $\tau_{p_2}^{\pm}(\xi')$, and $\tau_{s_2}^{\pm}(\xi')$ arise from

(1.17)
$$\begin{aligned}
\det P^I(\xi) &= \det P_1^I(\xi) \times \det P_2^I(\xi) \\
&= \{(-\xi_0^2 + c_{p_1}^2|\xi''|^2)(-\xi_0^2 + c_{s_1}^2|\xi''|^2)\} \times (-\xi_0^2 + c_{s_1}^2|\xi''|^2) \\
&= \det P_{p_1}^I(\xi) \times \{\det P_{s_1}^I(\xi)\}^2 \\
&= \{c_{p_1}^2 c_{s_1}^2 (\xi_3 - \tau_{p_1}^+(\xi'))(\xi_3 - \tau_{p_1}^-(\xi'))(\xi_3 - \tau_{s_1}^+(\xi'))(\xi_3 - \tau_{s_1}^-(\xi'))\} \\
&\quad \times \{c_{s_1}^2(\xi_3 - \tau_{s_1}^+(\xi'))(\xi_3 - \tau_{s_1}^-(\xi'))\},
\end{aligned}$$

and the factor of $\det P^{II}(\xi)$ given with replaced p_1, s_1 by p_2, s_2, respectively.

Remark.3. If $(\det P^I)_j(\xi^0) \neq 0$, $(j \in \{p_1, s_1\})$ then $(\det P^I)_{j\xi^0}(\eta)=(\det P^I)_j(\xi^0)$ and is constant. So $\Gamma_{j\xi^0} = \Gamma_{j\xi^0 m} = \mathbf{R}^4$ and thus $(\Gamma_{j\xi^0})_k^I = (\Gamma_{j\xi^0 m})_k^I = \{0\} \subset \mathbf{R}_-^4 \times \mathbf{R}_-^4$ $(j \in \{p_1, s_1\}$, $k \in \{p_1, s_1\}$ $m \in \{p_1, p_2, s_2\})$ and $(\Gamma_{j\xi^0})_k^{II} = (\Gamma_{j\xi^0 m})_k^{II} = \{0\} \subset \mathbf{R}_+^4 \times \mathbf{R}_-^4$ $(j \in \{p_1, s_1\}$, $k \in \{p_1, s_1\}$ $m \in \{p_2\})$.

Remark.4. By the assumption (1.2), there are not any real ξ that are roots of $\zeta_0^2 - c_{p_1}^2|\zeta''|^2 = 0$ and branch points of $\tau_{s_1}^+(\xi')$.

Remark.5. In (1.6) and (1.11), $F_{j\xi^0 k}^I(x,y)$ and $F_{j\xi^0 k}^I(x,y)$ are defined as (1.5) and (1.10) for ξ^0 such that $\xi^{0'}$ satisfying zeros of inner radical sign .

Lateral waves, in other words, glancing wave, arise from the presence of branch points of $\tau_{p_1}^{\pm}(\xi')$, $\tau_{s_1}^{\pm}(\xi')$, $\tau_{p_2}^{\pm}(\xi')$, and $\tau_{s_2}^{\pm}(\xi')$. In our problem, many lateral waves are appeared. More precise results are given in Section 4 below.

2. The reflected and refracted Riemann functions

In this section, we solve the mixed problem and show the explicit expressions of the reflected and refracted Riemann functions $F^I(x,y)$ and $F^{II}(x,y)$, respectively.
Note that if we put $\xi + i\eta = \zeta$, then

$$P^I(\zeta', D_3) = \mathrm{U}(\zeta''')\mathrm{C}\begin{pmatrix} P_1^I(\zeta', D_3) & 0_{2\times 1} \\ 0_{1\times 2} & P_2^I(\zeta', D_3) \end{pmatrix}(\mathrm{U}(\zeta''')\mathrm{C})^{-1},$$

where $P_1^I(\zeta', D_3)$ and $P_2^I(\zeta', D_3)$ are 2×2 and 1×1 ordinary differential operators with parameters, respectively, defined by

$$\begin{aligned}
P_1^I(\zeta', D_3) &= \begin{pmatrix} -\zeta_0^2 + \{c_{s_1}^2 D_3^2 + c_{p_1}^2(\zeta_1^2 + \zeta_2^2)\} & (c_{p_1}^2 - c_{s_1}^2)\sqrt{\zeta_1^2 + \zeta_2^2}D_3 \\ (c_{p_1}^2 - c_{s_1}^2)\sqrt{\zeta_1^2 + \zeta_2^2}D_3 & -\zeta_0^2 + \{c_{p_1}^2 D_3^2 + c_{s_1}^2(\zeta_1^2 + \zeta_2^2)\} \end{pmatrix}, \\
P_2^I(\zeta', D_3) &= \zeta_0^2 - c_{s_1}^2\{D_3^2 + (\zeta_1^2 + \zeta_2^2)\},
\end{aligned}$$

and

$$\mathrm{U}(\zeta''') = \frac{1}{\sqrt{\zeta_1^2+\zeta_2^2}} \begin{pmatrix} \zeta_1 & -\zeta_2 & 0 \\ \zeta_2 & \zeta_1 & 0 \\ 0 & 0 & \sqrt{\zeta_1^2+\zeta_2^2} \end{pmatrix}, \quad \mathrm{C} = \begin{pmatrix} 1 & 0 & 0 \\ 0 & 0 & 1 \\ 0 & 1 & 0 \end{pmatrix}.$$

Moreover

$$B^I(\zeta', D_3) = \mathrm{U}(\zeta''')\mathrm{C} \begin{pmatrix} B_1^I(\zeta', D_3) & 0_{2\times 1} \\ 0_{1\times 2} & B_2^I(\zeta', D_3) \end{pmatrix} (\mathrm{U}(\zeta''')\mathrm{C})^{-1},$$

$$B_1^I(\zeta', D_3) = i\rho_1 \begin{pmatrix} c_{s_1}^2 D_3 & c_{s_1}^2 \sqrt{\zeta_1^2+\zeta_2^2} \\ (c_{p_1}^2 - 2c_{s_1}^2)\sqrt{\zeta_1^2+\zeta_2^2} & c_{p_1}^2 D_3 \end{pmatrix}, \quad B_2^I(\zeta', D_3) = i\rho_1 c_{s_1}^2 D_3.$$

As shown in the preceding section, if we put unit impulse the Dirac delta $\delta(x-y)$ at position $y = (0,\ y'')$ with $y_3 < 0$, that is, put in Medium I, then the Riemann function $G(x,y)$ of this elastic mixed problem is given by the following:

$$G(x,y) = \begin{cases} E^I(x-y) - F^I(x,y) & \text{for } \ x_3 < 0, \\ F^{II}(x,y) & \text{for } \ x_3 > 0, \end{cases}$$

where $E^I(x)$ is the fundamental solution in Medium I describing an incident wave, is defined by

$$E^I(x) = (2\pi)^{-4} \int_{\mathbf{R}^4_\xi} e^{ix\cdot(\xi+i\eta)} P^I(\xi+i\eta)^{-1} d\xi, \qquad \eta \in -s\vartheta - \Gamma,$$

with a positive real s large enough. Taking partial Fourier-Laplace transform with respect to x' for the interface value problem of the mixed problem (1.3), we obtain the interface value problem for ordinary differential equations with parameters. Then taking partial inverse Fourier-Laplace transform for the solution, we obtain the following expressions of the reflected and refracted Riemann functions $F^I(x,y)$ and $F^{II}(x.y)$.

(2.1)

$$F^I(x,y) = (2\pi)^{-4} \int_{\mathbf{R}^3} e^{i(x'-y')(\xi'+i\eta')} \int_{\mathbf{R}} e^{-iy_3(\xi_3+i\eta_3)} \mathrm{U}(\xi'''+i\eta''')\mathrm{C}\times$$

$$\left(\begin{matrix} \dfrac{\begin{vmatrix} \{P_1^I(\xi+i\eta)^{-1}\}_1 & \cdots \\ & \cdots \\ B_1^I(\xi+i\eta)\{P_1^I(\xi+i\eta)^{-1}\}_1 & \cdots \\ & \cdots \end{vmatrix}}{R_1(\xi'+i\eta')} \begin{pmatrix} |\xi'''+i\eta'''| \\ -\tau_{p_1}^+(\xi'+i\eta') \end{pmatrix} e^{-i\tau_{p_1}^+(\xi'+i\eta')x_3} \\ + \dfrac{\begin{vmatrix} \cdot & \{P_1^I(\xi+i\eta)^{-1}\}_1 & \cdot\,\cdot \\ \cdot & & \cdot\,\cdot \\ \cdot & B_1^I(\xi+i\eta)\{P_1^I(\xi+i\eta)^{-1}\}_1 & \cdot\,\cdot \\ \cdot & & \cdot\,\cdot \end{vmatrix}}{R_1(\xi'+i\eta')} \begin{pmatrix} \tau_{s_1}^+(\xi'+i\eta') \\ |\xi'''+i\eta'''| \end{pmatrix} e^{-i\tau_{s_1}^+(\xi'+i\eta')x_3} \\ 0 \end{matrix} \right.$$

$$
\frac{\begin{vmatrix} & \{P_1^I(\xi+i\eta)^{-1}\}_2 & \cdot\ \cdot\ \cdot \\ & & \cdot\ \cdot\ \cdot \\ & B_1^I(\xi+i\eta)\{P_1^I(\xi+i\eta)^{-1}\}_2 & \cdot\ \cdot\ \cdot \\ & & \cdot\ \cdot\ \cdot \end{vmatrix}}{R_1(\xi'+i\eta')} \begin{pmatrix} |\xi'''+i\eta'''| \\ -\tau_{p_1}^+(\xi'+i\eta') \end{pmatrix} e^{-i\tau_{p_1}^+(\xi'+i\eta')x_3}
$$

$$
+\frac{\begin{vmatrix} \cdot & \{P_1^I(\xi+i\eta)^{-1}\}_2 & \cdot & \cdot \\ \cdot & & \cdot & \cdot \\ \cdot & B_1^I(\xi+i\eta)\{P_1^I(\xi+i\eta)^{-1}\}_2 & \cdot & \cdot \\ \cdot & & \cdot & \cdot \end{vmatrix}}{R_1(\xi'+i\eta')} \begin{pmatrix} \tau_{s_1}^+(\xi'+i\eta') \\ |\xi'''+i\eta'''| \end{pmatrix} e^{-i\tau_{s_1}^+(\xi'+i\eta')x_3}
$$

$$
0
$$

$$
\left.\begin{matrix} 0 \\ 0 \\ \dfrac{\begin{vmatrix} P_2^I(\xi+i\eta)^{-1} & \cdot \\ B_2^I(\xi+i\eta)P_2^I(\xi+i\eta)^{-1} & \cdot \end{vmatrix}}{R_2(\xi'+i\eta')} e^{-i\tau_{s_1}^+(\xi'+i\eta')x_3} \end{matrix}\right) \times (\mathrm{U}(\xi'''+i\eta''')\mathrm{C})^{-1} d\xi_3 d\xi'
$$

for $\quad x_3 < 0.$

(2.2)

$$
F^{II}(x,y) = (2\pi)^{-4} \int_{\mathbf{R}^3} e^{i(x'-y')(\xi'+i\eta')} \int_{\mathbf{R}} e^{-iy_3(\xi_3+i\eta_3)} \mathrm{U}(\xi'''+i\eta''')\mathrm{C}\times
$$

$$
\left(\begin{matrix} \dfrac{\begin{vmatrix} \cdot & \cdot & \{P_1^I(\xi+i\eta)^{-1}\}_1 & \cdot \\ \cdot & \cdot & & \cdot \\ \cdot & \cdot & B_1^I(\xi+i\eta)\{P_1^I(\xi+i\eta)^{-1}\}_1 & \cdot \\ \cdot & \cdot & & \cdot \end{vmatrix}}{R_1(\xi'+i\eta')} \begin{pmatrix} |\xi'''+i\eta'''| \\ \tau_{p_2}^+(\xi'+i\eta') \end{pmatrix} e^{i\tau_{p_2}^+(\xi'+i\eta')x_3} \\ + \dfrac{\begin{vmatrix} \cdot & \cdot & \cdot & \{P_1^I(\xi+i\eta)^{-1}\}_1 \\ \cdot & \cdot & \cdot & \\ \cdot & \cdot & \cdot & B_1^I(\xi+i\eta)\{P_1^I(\xi+i\eta)^{-1}\}_1 \\ \cdot & \cdot & \cdot & \end{vmatrix}}{R_1(\xi'+i\eta')} \begin{pmatrix} -\tau_{s_2}^+(\xi'+i\eta') \\ |\xi'''+i\eta'''| \end{pmatrix} e^{i\tau_{s_2}^+(\xi'+i\eta')x_3} \\ 0 \end{matrix}\right.
$$

$$
\frac{\begin{vmatrix} \cdot & \cdot & \{P_1^I(\xi+i\eta)^{-1}\}_2 & \cdot \\ \cdot & \cdot & & \cdot \\ \cdot & \cdot & B_1^I(\xi+i\eta)\{P_1^I(\xi+i\eta)^{-1}\}_2 & \cdot \\ \cdot & \cdot & & \cdot \end{vmatrix}}{R_1(\xi'+i\eta')} \begin{pmatrix} |\xi'''+i\eta'''| \\ \tau_{p_2}^+(\xi'+i\eta') \end{pmatrix} e^{i\tau_{p_2}^+(\xi'+i\eta')x_3}
$$

$$
+\frac{\begin{vmatrix} \cdot & \cdot & \cdot & \{P_1^I(\xi+i\eta)^{-1}\}_2 \\ \cdot & \cdot & \cdot & \\ \cdot & \cdot & \cdot & B_1^I(\xi+i\eta)\{P_1^I(\xi+i\eta)^{-1}\}_2 \\ \cdot & \cdot & \cdot & \end{vmatrix}}{R_1(\xi'+i\eta')} \begin{pmatrix} -\tau_{s_2}^+(\xi'+i\eta') \\ |\xi'''+i\eta'''| \end{pmatrix} e^{i\tau_{s_2}^+(\xi'+i\eta')x_3}
$$

$$
0
$$

$$\left.\frac{\begin{vmatrix} \cdot & \begin{matrix} 0 \\ 0 \\ P_2^I(\xi+i\eta)^{-1}_1 \end{matrix} \\ \cdot & B_2^I(\xi+i\eta)P_2^I(\xi+i\eta)^{-1} \end{vmatrix}}{R_2(\xi'+i\eta')} e^{i\tau_{s_2}^+(\xi'+i\eta')x_3}\right) \times (\mathrm{U}(\xi'''+i\eta''')\mathrm{C})^{-1} d\xi_3 d\xi'$$

for $x_3 < 0$.

Here $\cdot$ means the same component of the Lopatinski matrices $\mathcal{R}_1(\zeta')$ ($\zeta' = \xi' + i\eta'$) and $\mathcal{R}_2(\zeta')$ given below, $\{P_1^I(\zeta)^{-1}\}_1$ and $\{P_1^I(\zeta)^{-1}\}_2$ are the 1 and 2 columns, respectively, of the inverse matrix of $P_1^I(\zeta)$ given by

(2.3)

$$P_1^I(\zeta)^{-1} = (\{P_1^I(\zeta)^{-1}\}_1, \{P_1^I(\zeta)^{-1}\}_2) = \frac{\mathrm{cof}P_1^I(\zeta)}{\det P_1^I(\zeta)} = \frac{1}{(\zeta_0^2 - c_{p_1}^2|\zeta''|^2)(\zeta_0^2 - c_{s_1}^2|\zeta''|^2)}$$
$$\times \begin{pmatrix} -\zeta_0^2 + \{c_{p_1}^2\zeta_3^2 + c_{s_1}^2(\zeta_1^2+\zeta_2^2)\} & -(c_{p_1}^2 - c_{s_1}^2)\sqrt{\zeta_1^2+\zeta_2^2}\,\zeta_3 \\ -(c_{p_1}^2 - c_{s_1}^2)\sqrt{\zeta_1^2+\zeta_2^2}\,\zeta_3 & -\zeta_0^2 + \{c_{s_1}^2\zeta_3^2 + c_{p_1}^2(\zeta_1^2+\zeta_2^2)\} \end{pmatrix},$$

$R_1(\zeta')$ and $R_2(\zeta')$ are the Lopatinski determinants of the systems $\{P_1^I(\zeta', D_3)$, $P_1^{II}(\zeta', D_3)$, $B_1^I(\zeta', D_3)$, $B_1^{II}(\zeta', D_3)\}$ and $\{P_2^I(\zeta', D_3)$, $P_2^{II}(\zeta', D_3)$, $B_2^I(\zeta', D_3)$, $B_2^{II}(\zeta', D_3)\}$, respectively, given by

(2.4)
$$R_1(\zeta') = \det \mathcal{R}_1(\zeta'),$$

(2.5)

$$\mathcal{R}_1(\zeta') = \begin{pmatrix} |\zeta'''| & \tau_{s_1}^+(\zeta') & |\zeta'''| & -\tau_{s_2}^+(\zeta') \\ -\tau_{p_1}^+(\zeta') & |\zeta'''| & \tau_{p_2}^+(\zeta') & |\zeta'''| \\ -2\rho_1 c_{s_1}^2 \tau_{p_1}^+(\zeta')|\zeta'''| & -\rho_1 c_{s_1}^2(\tau_{s_1}^+(\zeta')^2 - |\zeta'''|^2) & 2\rho_2 c_{s_2}^2 \tau_{p_2}^+(\zeta')|\zeta'''| & -\rho_2 c_{s_2}^2(\tau_{s_2}^+(\zeta')^2 - |\zeta'''|^2) \\ \rho_1 c_{s_1}^2(\tau_{s_1}^+(\zeta')^2 - |\zeta'''|^2) & -2\rho_1 c_{s_1}^2 \tau_{s_1}^+(\zeta')|\zeta'''| & \rho_2 c_{s_2}^2(\tau_{s_2}^+(\zeta')^2 - |\zeta'''|^2) & 2\rho_2 c_{s_2}^2 \tau_{s_2}^+(\zeta')|\zeta'''| \end{pmatrix},$$

(2.6)
$$R_2(\zeta') = \det \mathcal{R}_2(\zeta'),$$

(2.7)
$$\mathcal{R}_2(\zeta') = \begin{pmatrix} 1 & 1 \\ \rho_1 c_{s_1}^2 \tau_{s_1}^+(\zeta') & -\rho_2 c_{s_2}^2 \tau_{s_2}^+(\zeta') \end{pmatrix}.$$

3. Proof of Main Theorem

In this section, we give a proof of Main Theorem. We prove for the reflected Riemann function $F^I(x,y)$. A similar proof is given for the refracted Riemann function $F^{II}(x,y)$.

We could obtain

$$\lim_{\nu\to\infty} < \nu e^{-i\nu\{(x'-y')\cdot\xi^{0\prime}+x_3\tau_k^-(\xi^{0\prime})-y_3\xi_3^0\}} F^I(x,y), \phi(x,y) >_{x,y}$$
$$=< F_{s_1\xi^0 p_1}(x,y), \phi(x,y) >_{x,y} \quad \text{for} \quad \phi(x,y) \in C_0^\infty(\mathbf{R}_-^4 \times \mathbf{R}_-^4),$$

by localization of the explicit expression of the Riemann function (2.1). Here

$$F^I_{s_1\xi^0 p_1}(x,y)$$
$$= (2\pi)^{-4} \int_{\mathbf{R}^3} e^{i(x'-y'-\mathrm{grad}\,\tau^+_{p_1}(\xi^{0\prime})x_3)\cdot(\kappa'+i\eta')} \int_{\mathbf{R}} e^{-iy_3(\kappa_3+i\eta_3)}\, \mathrm{U}(\xi^{0\prime\prime\prime})\mathrm{C}$$
$$\times \frac{1}{(\det P^I_{s_1})_{\xi^0}(\kappa+i\eta)} \begin{pmatrix} \frac{Q_1(\xi^0)}{R_1(\xi^{0\prime})}\begin{pmatrix} |\xi^{0\prime\prime\prime}| \\ -\tau^+_{p_1}(\xi^{0\prime}) \\ 0 \end{pmatrix} & \frac{Q_2(\xi^0)}{R_1(\xi^{0\prime})}\begin{pmatrix} |\xi^{0\prime\prime\prime}| \\ -\tau^+_{p_1}(\xi^{0\prime}) \\ 0 \end{pmatrix} & \begin{matrix} 0 \\ 0 \\ 0 \end{matrix} \end{pmatrix}$$
$$\left(\mathrm{U}(\xi^{0\prime\prime\prime})\mathrm{C}\right)^{-1} d\kappa_3 d\kappa',$$

where

$$(\det P^I_{s_1})_{\xi^0}(\kappa+i\eta) = 2\left(\xi_0^{0\,2} - c^2_{p_1}|\xi^{0\prime\prime}|^2\right)$$
$$\times \left\{\xi_0^0(\kappa_0+i\eta_0) - c^2_{s_1}(\xi_1^0(\kappa_1+i\eta_1)+\xi_2^0(\kappa_2+i\eta_2)+\xi_3^0(\kappa_3+i\eta_3))\right\},$$

and $Q_1(\xi^0)$ and $Q_2(\xi^0)$ are given by

$$Q_1(\xi^0) = (\det P^I_{p_1})_{\xi^0}(\xi^0) \begin{vmatrix} \{P_1^I(\xi^0)^{-1}\}_1 & \cdot & \cdot & \cdot \\ & \cdot & \cdot & \cdot \\ B_1^I(\xi^0)\{P_1^I(\xi^0)^{-1}\}_1 & \cdot & \cdot & \cdot \\ & \cdot & \cdot & \cdot \end{vmatrix},$$

$$Q_2(\xi^0) = (\det P^I_{p_1})_{\xi^0}(\xi^0) \begin{vmatrix} \{P_1^I(\xi^0)^{-1}\}_2 & \cdot & \cdot & \cdot \\ & \cdot & \cdot & \cdot \\ B_1^I(\xi^0)\{P_1^I(\xi^0)^{-1}\}_2 & \cdot & \cdot & \cdot \\ & \cdot & \cdot & \cdot \end{vmatrix},$$

where $\cdot$ means the same component of the Lopatinski matrix (2.5). Thus we prove the equation (1.5). The equation (1.6) is proved similarly, remarking that

$$\nu^{-\frac{1}{2}}\tau^+_m(\nu\xi^{0\prime}+\kappa'+i\eta')$$
$$\longrightarrow \sqrt{2\left\{\frac{\xi_0^0}{c_m^2}(\kappa_0+i\eta_0) - \xi^{0\prime\prime\prime}\cdot(\kappa'''+i\eta''')\right\}} \quad \text{as} \quad \nu\to\infty, \quad m\in\{p_1,p_2,s_2\},$$

where $\sqrt{\cdot}$ satisfies $\mathrm{Im}\sqrt{\cdot} > 0$.

Next we prove the inclusion relation (1.7). Let $V(\subset \mathbf{R}^4_+ \times \mathbf{R}^4_-)$ be the complement of sing supp$F^I(x,y)$ and

$$V \cap \text{sing supp } F^I(x,y) = \varnothing$$

for $g(x,y) \in C_0^\infty(V)$. For the points $\xi^{0\prime}$ that are not zeros of inner radical sign of $\tau_m^+(\zeta')$ $(m \in \{p_1, p_2, s_2\})$, by the Riemann-Lebesgue theorem

$$(3.1)\quad \int \nu e^{-i\nu\{(x'-y')\cdot\xi^{0\prime} - x_3\tau_k^+(\xi^{0\prime}) + y_3\xi_3^0\}} F^I(x,y)g(x,y)dxdy \longrightarrow 0 \quad \text{as} \quad \nu \to \infty.$$

On the other hand, by the localization method

$$\nu e^{-i\nu\{(x'-y')\cdot\xi^{0\prime} - x_3\tau_k^+(\xi^{0\prime}) + y_3\xi_3^0\}} F^I(x,y) \longrightarrow F^I_{\xi^0}(x,y) \quad \text{as} \quad \nu \to \infty,$$

so we have

$$V \cap \text{supp } F^I_{\xi^0}(x,y) = \varnothing.$$

For the points $\xi^{0\prime}$ that are zeros of inner radical sign of $\tau_m^+(\zeta')$ $(m \in \{p_1, p_2, s_2\})$, we have

$$\int \left\{\nu^{\frac{3}{2}} e^{-i\nu\{(x'-y')\cdot\xi^{0\prime} - x_3\tau_k^+(\xi^{0\prime}) + y_3\xi_3^0\}} F^I(x,y) - \nu^{\frac{1}{2}} F^I_{\xi^0}(x,y)\right\} g(x,y)dxdy \longrightarrow 0$$
$$\text{as} \quad \nu \to \infty,$$

because of the the Riemann-Lebesgue theorem and the limit relation (3.1). On the other hand, by the localization method

$$\nu^{\frac{3}{2}} e^{-i\nu\{(x'-y')\cdot\xi^{0\prime} - x_3\tau_k^+(\xi^{0\prime}) + y_3\xi_3^0\}} F^I(x,y) - \nu^{\frac{1}{2}} F^I_{\xi^0}(x,y) \longrightarrow F^I_{\xi^0 m}(x,y)$$
$$\text{as} \quad \nu \to \infty,$$

so we have

$$V \cap \text{supp } F^I_{\xi^0 m}(x,y) = \varnothing.$$

Thus we obtain the inclusion relation (1.7).

Finally we prove the last part. We could put

$$F^I_{s_1\xi^0 p_1}(x,y) = \text{Const.}(2\pi)^{-4} \times \int_{\mathbf{R}^4} \frac{e^{i(x'-y'-\text{grad}\,\tau_{p_1}^+(\xi^{0\prime})x_3)\cdot(\kappa'+i\eta') - y_3(\kappa_3+i\eta_3)}}{(\kappa_0+i\eta_0) - \frac{c_{s_1}^2}{\xi_0^0}\xi^{0\prime\prime}\cdot(\kappa''+i\eta'')} d\kappa,$$

and would like to obtain supp$F^I_{s_1\xi^0 p_1}(x,y)$. If we put

$$G(x) = (2\pi)^{-4} \int_{\mathbf{R}^4} \frac{e^{ix\cdot(\kappa+i\eta)}}{(\kappa_0+i\eta_0) - \frac{c_{s_1}^2}{\xi_0^0}\xi^{0\prime\prime}\cdot(\kappa''+i\eta'')} d\kappa,$$

then
$$F^I_{s_1\xi^0 p_1}(x,y) = G(x'-y'-\mathrm{grad}\tau^+_{p_1}(\xi^{0'})x_3, -y_3).$$

So it is sufficient that we consider supp$G$. From the Paley-Wiener-Schwartz theorem,
$$ch[\mathrm{supp}G] = \Gamma' = \{x \in \mathbf{R}^4 |\ x\cdot\eta \geq 0 \quad \text{for} \quad \forall\eta \in \Gamma\},$$

where ch denotes a convex hull. Thus
$$\mathrm{supp}G = \left\{ x \in \mathbf{R}^4 | \quad x = \lambda\left(1, -\frac{c_{s_1}^2}{\xi_0^0}\xi^{0''}\right), \quad \lambda \geq 0 \right\},$$

since Γ' is half-line and G is a homogeneous distribution.

This completes the proof of Main Theorem.

4. Location of Singularities

By using Main Theorem, we find an inner estimate of the location of singularities of the reflected and refracted Riemann functions $F^I(x,y)$ and $F^{II}(x,y)$.

In the expressions (2.1) and (2.2), the parts put between $U(\xi''' + i\eta''')C$ and $(U(\xi''' + i\eta''')C)^{-1}$ are decomposed into 2×2 and 1×1 matrices valued Riemann functions $F^I_{2\times 2}(x,y)$ and $F^I_{1\times 1}(x,y)$ for $F^I(x,y)$, and $F^{II}_{2\times 2}(x,y)$ and $F^{II}_{1\times 1}(x,y)$ for $F^{II}(x,y)$. The displacement vector of $F^\iota_{2\times 2}(x,y)$ ($\iota = \{I, II\}$) lies in $(x_1-y_1)(x_3-y_3)$-plane and that of $F^\iota_{1\times 1}(x,y)$ ($\iota = \{I, II\}$) lies in (x_2-y_2)-axis. Thus we can treat $F^\iota_{2\times 2}(x,y)$ and $F^\iota_{1\times 1}(x,y)$ ($\iota = \{I, II\}$) independently.

First we consider $F^I_{2\times 2}(x,y)$ and $F^{II}_{2\times 2}(x,y)$. For $F^I_{2\times 2}(x,y)$, we have the following 4 sets of ξ^0 that are roots of $\det P^I_1(\xi^0) = 0$ and are not zeros of inner radical sign of $\tau^+_m(\xi^0)$ ($m \in \{p_1, p_2, s_2\}$); roots of $\det P^I_{s_1}(\xi^0) = {\xi_0^0}^2 - c^2_{s_1}|\xi^{0''}|^2 = 0$ are

(4.1)
$$\xi^0 = (1, \xi^0_1, \xi^0_2, \tau^+_{s_1}(\xi^{0'})) \text{ for } \tau^-_{s_1}(\xi^{0'}) \text{ in (1.5) with } |\xi^{0'''}| < \frac{1}{c_{s_1}}, \neq \frac{1}{c_{p_1}}, \frac{1}{c_{s_2}}, \frac{1}{c_{p_2}},$$

(4.2)
$$\xi^0 = (1, \xi^0_1, \xi^0_2, \tau^+_{s_1}(\xi^{0'})) \text{ for } \tau^-_{p_1}(\xi^{0'}) \text{ in (1.5) with } |\xi^{0'''}| < \frac{1}{c_{p_1}}, \neq \frac{1}{c_{s_2}}, \frac{1}{c_{p_2}},$$

and roots of $\det P^I_{p_1}(\xi^0) = {\xi_0^0}^2 - c^2_{p_1}|\xi^{0''}|^2 = 0$ are

(4.3)
$$\xi^0 = (1, \xi^0_1, \xi^0_2, \tau^+_{p_1}(\xi^{0'})) \text{ for } \tau^-_{s_1}(\xi^{0'}) \text{ in (1.5) with } |\xi^{0'''}| < \frac{1}{c_{p_1}}, \neq \frac{1}{c_{s_2}}, \frac{1}{c_{p_2}},$$

(4.4)
$$\xi^0 = (1, \xi^0_1, \xi^0_2, \tau^+_{p_1}(\xi^{0'})) \text{ for } \tau^-_{p_1}(\xi^{0'}) \text{ in (1.5) with } |\xi^{0'''}| < \frac{1}{c_{p_1}}, \neq \frac{1}{c_{s_2}}, \frac{1}{c_{p_2}}.$$

(4.1) and (4.2) (resp. (4.3) and (4.4)) correspond to P_1 and S_1 reflected waves for S_1 (resp. P_1) incident wave, respectively.

We have the following 9 sets of ξ^0 that are roots of $\det P_1^I(\xi^0) = 0$ and are not zeros of inner radical sign of $\tau_m^+(\xi^0)$ $(m \in \{p_1, s_2, p_2\})$; (4.5)-(4.9) are roots of $\det P_{s_1}^I(\xi^0) = 0$ and

$$\xi^0 = (1, \xi_1^0, \xi_2^0, \tau_{s_1}^+(\xi^{0\prime})) \text{ for } \tau_{s_1}^-(\xi^{0\prime}) \text{ in (1.6) with } |\xi^{0'''}| = \frac{1}{c_{p_1}} \tag{4.5}$$

are zeros of inner radical sign of $\tau_{p_1}^+(\xi^{0\prime})$,

$$\xi^0 = (1, \xi_1^0, \xi_2^0, \tau_{s_1}^+(\xi^{0\prime})) \text{ for } \tau_{s_1}^-(\xi^{0\prime}) \text{ in (1.6) with } |\xi^{0'''}| = \frac{1}{c_{s_2}}, \tag{4.6}$$

$$\xi^0 = (1, \xi_1^0, \xi_2^0, \tau_{s_1}^+(\xi^{0\prime})) \text{ for } \tau_{p_1}^-(\xi^{0\prime}) \text{ in (1.6) with } |\xi^{0'''}| = \frac{1}{c_{s_2}} \tag{4.7}$$

are zeros of inner radical sign of $\tau_{s_2}^+(\xi^{0\prime})$,

$$\xi^0 = (1, \xi_1^0, \xi_2^0, \tau_{s_1}^+(\xi^{0\prime})) \text{ for } \tau_{s_1}^-(\xi^{0\prime}) \text{ in (1.6) with } |\xi^{0'''}| = \frac{1}{c_{p_2}}, \tag{4.8}$$

$$\xi^0 = (1, \xi_1^0, \xi_2^0, \tau_{s_1}^+(\xi^{0\prime})) \text{ for } \tau_{p_1}^-(\xi^{0\prime}) \text{ in (1.6) with } |\xi^{0'''}| = \frac{1}{c_{p_2}} \tag{4.9}$$

are zeros of inner radical sign of $\tau_{p_2}^+(\xi^{0\prime})$, moreover (4.10)-(4.13) are roots of $\det P_{s_1}^I(\xi^0) = 0$ and

$$\xi^0 = (1, \xi_1^0, \xi_2^0, \tau_{p_1}^+(\xi^{0\prime})) \text{ for } \tau_{s_1}^-(\xi^{0\prime}) \text{ in (1.6) with } |\xi^{0'''}| = \frac{1}{c_{s_2}}, \tag{4.10}$$

$$\xi^0 = (1, \xi_1^0, \xi_2^0, \tau_{p_1}^+(\xi^{0\prime})) \text{ for } \tau_{p_1}^-(\xi^{0\prime}) \text{ in (1.6) with } |\xi^{0'''}| = \frac{1}{c_{s_2}} \tag{4.11}$$

are zeros of inner radical sign of $\tau_{s_2}^+(\xi^{0\prime})$,

$$\xi^0 = (1, \xi_1^0, \xi_2^0, \tau_{p_1}^+(\xi^{0\prime})) \text{ for } \tau_{s_1}^-(\xi^{0\prime}) \text{ in (1.6) with } |\xi^{0'''}| = \frac{1}{c_{p_2}}, \tag{4.12}$$

$$\xi^0 = (1, \xi_1^0, \xi_2^0, \tau_{p_1}^+(\xi^{0\prime})) \text{ for } \tau_{p_1}^-(\xi^{0\prime}) \text{ in (1.6) with } |\xi^{0'''}| = \frac{1}{c_{p_2}} \tag{4.13}$$

are zeros of inner radical sign of $\tau_{p_2}^+(\xi^{0\prime})$. (4.5) corresponds to S_1 lateral or glancing wave for S_1 incident wave with P_1 influence. (4.6) and (4.7) (resp. (4.8) and (4.9)) correspond to P_1 and S_1 lateral waves for S_1 incident wave with S_2 (resp. P_2) influence, respectively. (4.10) and (4.11) (resp. (4.12) and (4.13)) correspond to P_1 and S_1 lateral waves for P_1 incident wave with S_2 (resp. P_2) influence, respectively.

For $F^{II}_{2\times 2}(x,y)$, we have the following the 4 sets of ξ^0 that are roots of $\det P^I_1(\xi^0) = 0$ and are not zeros of inner radical sign of $\tau^+_{p_2}(\xi^0)$; roots of $\det P^I_{s_1}(\xi^0) = 0$ are

(4.14)
$$\xi^0 = (1, \xi^0_1, \xi^0_2, \tau^+_{s_1}(\xi^{0\prime})) \text{ for } \tau^+_{s_2}(\xi^{0\prime}) \text{ in (1.10) with } |\xi^{0\prime\prime\prime}| < \frac{1}{c_{s_2}},\ \neq \frac{1}{c_{p_2}},$$

(4.15)
$$\xi^0 = (1, \xi^0_1, \xi^0_2, \tau^+_{s_1}(\xi^{0\prime})) \text{ for } \tau^+_{p_2}(\xi^{0\prime}) \text{ in (1.10) with } |\xi^{0\prime\prime\prime}| < \frac{1}{c_{p_2}},$$

and roots of $\det P^I_{p_1}(\xi^0) = 0$ are

(4.16)
$$\xi^0 = (1, \xi^0_1, \xi^0_2, \tau^+_{p_1}(\xi^{0\prime})) \text{ for } \tau^+_{s_2}(\xi^{0\prime}) \text{ in (1.10) with } |\xi^{0\prime\prime\prime}| < \frac{1}{c_{s_2}},\ \neq \frac{1}{c_{p_2}},$$

(4.17)
$$\xi^0 = (1, \xi^0_1, \xi^0_2, \tau^+_{p_1}(\xi^{0\prime})) \text{ for } \tau^+_{p_2}(\xi^{0\prime}) \text{ in (1.10) with } |\xi^{0\prime\prime\prime}| < \frac{1}{c_{p_2}}.$$

(4.14) and (4.15) (resp. (4.16) and (4.17)) correspond to S_2 and P_2 reflected waves for S_1 (resp. P_1) incident wave, respectively.

We have the following 2 sets of ξ^0 that are roots of det $P^I_1(\xi^0) = 0$ and are not zeros of inner radical sign of $\tau^+_{p_2}(\xi^0)$;

$$\text{(4.18)} \qquad \xi^0 = (1, \xi^0_1, \xi^0_2, \tau^+_{s_1}(\xi^{0\prime})) \text{ for } \tau^+_{s_2}(\xi^{0\prime}) \text{ in (1.11) with } |\xi^{0\prime\prime\prime}| = \frac{1}{c_{p_2}}$$

are roots of $\det P^I_{s_1}(\xi^0) = 0$ and zeros of inner radical sign of $\tau^+_{p_2}(\xi^0)$,

$$\text{(4.19)} \qquad \xi^0 = (1, \xi^0_1, \xi^0_2, \tau^+_{p_1}(\xi^{0\prime})) \text{ for } \tau^+_{s_2}(\xi^{0\prime}) \text{ in (1.11) with } |\xi^{0\prime\prime\prime}| = \frac{1}{c_{p_2}}$$

are roots of $\det P^I_{p_1}(\xi^0) = 0$ and zeros of inner radical sign of $\tau^+_{p_2}(\xi^0)$. (4.18) (resp. (4.19)) corresponds to S_2 lateral wave for S_1 (resp. P_1) incident wave with P_2 influence.

Remark. It is sufficient to consider only the case $\xi^0_0 = 1$ since $(\Gamma_{j\xi^0})^I_k = (\Gamma_{j(t\xi^0)})^I_k$, $(\Gamma_{j\xi^0 m})^I_k = (\Gamma_{j(t\xi^0)m})^I_k$ ($j = \{p_1, s_1\}$, $k = \{p_1, s_1\}$, $m = \{p_1, p_2, s_2\}$), and $(\Gamma_{j\xi^0})^{II}_k = (\Gamma_{j(t\xi^0)})^{II}_k$, $(\Gamma_{j\xi^0 m})^{II}_k = (\Gamma_{j(t\xi^0)m})^{II}_k$ ($j = \{p_1, s_1\}$, $k = \{p_1, s_1\}$, $m = \{p_2,\}$) for $t \in \mathbf{R} \setminus \{0\}$.

We calculate the singularity caused by the point (4.4) as a reflected wave. Localization of $\det P^I_1(\eta)$ at the point $\xi^0 = (1, \xi^0_1, \xi^0_2, \tau^+_{p_1}(\xi^{0\prime}))$ is given by

(4.20)
$$(\det P^I_{p_1})_{\xi^0}(\eta) = 2\{c^2_{s_1}({\xi^0_1}^2 + {\xi^0_2}^2 + \tau^+_{p_1}(\xi^{0\prime})^2) - 1\}\{-\eta_0 + c^2_{p_1}(\xi^0_1\eta_1 + \xi^0_2\eta_2 + \xi^0_3\eta_3)\},$$

where we note that $2\{c_{s_1}^2({\xi_1^0}^2+{\xi_2^0}^2+\tau_{p_1}^+(\xi^{0'})^2)-1\}$ is a non-zero constant. The $\Gamma_{p_1\xi^0}$ is given by

$$\Gamma_{p_1\xi^0}=\left\{\eta\in\mathbf{R}^4:\ \eta_0-c_{p_1}^2(\xi_1^0\eta_1+\xi_2^0\eta_2+\tau_{p_1}^+(\xi^{0'}))\eta_3>0\right\}.$$

The $(\Gamma_{p_1\xi^0})_{p_1}^I$ is calculated as follows.

$$\begin{aligned}
(\Gamma_{p_1\xi^0})_{p_1}^I &= \Big\{(x,y)\in\mathbf{R}_-^4\times\mathbf{R}_-^4:\\
&\quad\left((x'-y')+\frac{1}{\tau_{p_1}^-(\xi^{0'})}\left(\frac{1}{c_{p_1}^2},-\xi_1^0,-\xi_2^0\right)x_3\right)\cdot\eta'-y_3\eta_3\geq 0,\ \eta\in\Gamma_{\xi^0}\Big\}\\
&=\Big\{(x,y)\in\mathbf{R}_-^4\times\mathbf{R}_-^4:\\
&\quad\left((x_0-y_0)+\frac{1}{\tau_{p_1}^-(\xi^{0'})}\frac{x_3}{c_{p_1}^2},\ (x_1-y_1)-\frac{\xi_1^0x_3}{\tau_{p_1}^-(\xi^{0'})},\ (x_2-y_2)-\frac{\xi_2^0x_3}{\tau_{p_1}^-(\xi^{0'})},\ -y_3\right)\\
&\quad=u\left(1,\ -c_{p_1}^2\xi_1^0,\ -c_{p_1}^2\xi_2^0,\ -c_{p_1}^2\tau_{p_1}^+(\xi^{0'})\right),\quad u\geq 0\Big\}
\end{aligned}$$

(4.21)

$$\begin{aligned}
&=\Big\{(x,y)\in\mathbf{R}_-^4\times\mathbf{R}_-^4:(x_0-y_0)+\frac{1}{\tau_{p_1}^-(\xi^{0'})}\frac{x_3}{c_{p_1}^2}\geq 0,\\
&\qquad x_1-y_1=-c_{p_1}^2\xi_1^0(x_0-y_0),\ x_2-y_2=-c_{p_1}^2\xi_2^0(x_0-y_0),\\
&\qquad x_3+y_3=c_{p_1}^2\tau_{p_1}^+(\xi^{0'})(x_0-y_0)\Big\}.
\end{aligned}$$

From (4.21) and $\tau_{p_1}^+(\xi^{0'})^2+{\xi_1^0}^2+{\xi_2^0}^2=\frac{1}{c_{p_1}^2}$, we have

(4.22)

$$\begin{aligned}
\bigcup_{|\xi^{0'''}|<\frac{1}{c_{p_1}}}(\Gamma_{p_1\xi^0})_{p_1}^I&=\Big\{(x,y)\in\mathbf{R}_-^4\times\mathbf{R}_-^4:\frac{x_3}{-c_{p_1}^2\tau_{p_1}^-(\xi^{0'})}<x_0-y_0,\\
&\qquad(x_0-y_0)^2=\frac{1}{c_{p_1}^2}\{(x_1-y_1)^2+(x_2-y_2)^2+(x_3+y_3)^2\}\Big\}.
\end{aligned}$$

Secondly we calculate the singularity caused by the point (4.7) as a lateral wave. By (1.16), we have

$$\begin{aligned}
&\Gamma_{s_1\xi^0s_2}=\Gamma\left((\det P_{s_1}^I)_{\xi^0}(\eta),\vartheta\right)\cap\left\{\Gamma\left(\frac{\xi_0^0}{c_{s_2}^2}\eta_0-\xi_1^0\eta_1-\xi_2^0\eta_2,\vartheta'\right)\times\mathbf{R}_\eta\right\},\\
&\Gamma\left((\det P_{s_1}^I)_{\xi^0}(\eta),\vartheta\right)=\left\{\eta\in\mathbf{R}^4:\ \eta_0-c_{s_1}^2(\xi_1^0\eta_1+\xi_2^0\eta_2+\tau_{s_1}^+(\xi^{0'})\eta_3)>0\right\},\\
&\Gamma\left(\frac{\xi_0^0}{c_{s_2}^2}\eta_0-\xi_1^0\eta_1-\xi_2^0\eta_2,\vartheta'\right)\times\mathbf{R}_\eta=\{\eta\in\mathbf{R}^4:\eta_0-c_{s_2}^2(\xi_1^0\eta_1+\xi_2^0\eta_2)>0\}.
\end{aligned}$$

The $(\Gamma_{s_1\xi^0 s_2})^I_{p_1}$ is calculated as follows.

$$(\Gamma_{s_1\xi^0 s_2})^I_{p_1} = \Big\{(x,y) \in \mathbf{R}^4_- \times \mathbf{R}^4_- :$$

$$\left((x'-y') + \frac{1}{\tau^-_{p_1}(\xi^0)}\left(\frac{1}{c^2_{p_1}}, -\xi^0_1, -\xi^0_2\right)x_3\right)\cdot\eta' - y_3\eta_3 \geq 0, \quad \eta \in \Gamma_{s_1\xi^0 s_2}\Big\}$$

$$= \Big\{(x,y) \in \mathbf{R}^4_- \times \mathbf{R}^4_- :$$

$$\left((x_0-y_0) + \frac{1}{\tau^-_{p_1}(\xi^{0'})}\frac{x_3}{c^2_{p_1}},\ (x_1-y_1) - \frac{\xi^0_1 x_3}{\tau^-_{p_1}(\xi^{0'})},\ (x_2-y_2) - \frac{\xi^0_2 x_3}{\tau^-_{p_1}(\xi^{0'})},\ -y_3\right)$$

$$= u\left(1,\ -c^2_{s_1}\xi^0_1,\ -c^2_{s_1}\xi^0_2,\ -c^2_{s_1}\tau^+_{s_1}(\xi^{0'})\right)$$

$$+ v\left(1,\ -c^2_{s_2}\xi^0_1,\ -c^2_{s_2}\xi^0_2,\ 0\right),\quad u,\ v \geq 0\Big\}$$

$$= \Big\{(x,y) \in \mathbf{R}^4_- \times \mathbf{R}^4_- :$$

$$(x_0-y_0) + \frac{c_{s_2}}{c_{p_1}\sqrt{c^2_{s_2}-c^2_{p_1}}}x_3 + \frac{c_{s_2}}{c_{s_1}\sqrt{c^2_{s_2}-c^2_{s_1}}}y_3 \geq 0,$$

$$x_1 - y_1 = -c^2_{s_2}\xi^0_1(x_0-y_0) - \frac{c_{s_2}}{c_{p_1}}\sqrt{c^2_{s_2}-c^2_{p_1}}\,\xi^0_1 x_3 - \frac{c_{s_2}}{c_{s_1}}\sqrt{c^2_{s_2}-c^2_{s_1}}\,\xi^0_1 y_3,$$

$$x_2 - y_2 = -c^2_{s_2}\xi^0_2(x_0-y_0) - \frac{c_{s_2}}{c_{p_1}}\sqrt{c^2_{s_2}-c^2_{p_1}}\,\xi^0_2 x_3 - \frac{c_{s_2}}{c_{s_1}}\sqrt{c^2_{s_2}-c^2_{s_1}}\,\xi^0_2 y_3\Big\}.$$

Thus we have

(4.24)

$$\bigcup_{|\xi^{0'''}|=\frac{1}{c_{s_2}}} (\Gamma_{s_1\xi^0 s_2})^I_{p_1} = \Big\{(x,y) \in \mathbf{R}^4_- \times \mathbf{R}^4_- :$$

$$(x_0-y_0) + \frac{c_{s_2}}{c_{p_1}\sqrt{c^2_{s_2}-c^2_{p_1}}}x_3 + \frac{c_{s_2}}{c_{s_1}\sqrt{c^2_{s_2}-c^2_{s_1}}}y_3 \geq 0,$$

$$(x_1-y_1)^2 + (x_2-y_2)^2 = \left(c_{s_2}(x_0-y_0) + \frac{\sqrt{c^2_{s_2}-c^2_{p_1}}}{c_{p_1}}x_3 + \frac{\sqrt{c^2_{s_2}-c^2_{s_1}}}{c_{s_1}}y_3\right)^2\Big\}$$

If we cut at $x_1 = 0$ and $y' = 0$, then the sectional face is

(4.25)

$$\bigcup_{|\xi^{0'''}|=\frac{1}{c_{s_2}}} (\Gamma_{s_1\xi^0 s_2})^I_{p_1} \cap \{x_1 = 0\} \cap \{y' = 0\} = \Big\{(x_0, x_2, x_3, y_3) \in \mathbf{R}^3_- \times \mathbf{R}_- :$$

$$x_0 + \frac{c_{s_2}}{c_{p_1}\sqrt{c^2_{s_2}-c^2_{p_1}}}x_3 + \frac{c_{s_2}}{c_{s_1}\sqrt{c^2_{s_2}-c^2_{s_1}}}y_3 \geq 0,$$

$$x_2 = \pm\left(c_{s_2}x_0 + \frac{\sqrt{c_{s_2}^2 - c_{p_1}^2}}{c_{p_1}}x_3 + \frac{\sqrt{c_{s_2}^2 - c_{s_1}^2}}{c_{s_1}}y_3\right)\Bigg\}.$$

By similar calculus of other sets ξ^0, we obtain the following inner estimate of the location of singularities of the reflected and refracted Riemann functions $F^I_{2\times2}(x,y)$ and $F^{II}_{2\times2}(x,y)$ with pass of time as in Figure 2.

Remark 1. If $c_{p_1} = c_{s_2}$ in assumption (1.2), then the singularities corresponding to lateral waves caused by the points (4.10) and (4.11) do not appear.

Remark 2. If we consider the half-space problem, then the singularities corresponding to reflected waves caused by the points (4.1)-(4.4) and only the singularity corresponding to a lateral wave caused by the point (4.5) appear. Other singularities do not appear.

Remark 3. If we consider the stratified media problem of usual wave operator, then the singularity corresponding to a reflected (resp. refracted) wave caused by the point (4.1) (resp. (4.14)) and only the singularity corresponding to a lateral wave caused by the point (4.6) appear. Other singularities do not appear.

Remark 4. $F^I_{1\times1}(x,y)$ and $F^{II}_{1\times1}(x,y)$ are treated as the same as $p_1 = p_2 = 0$ in the case of $F^I_{2\times2}(x,y)$ and $F^{II}_{2\times2}(x,y)$.

$$x_0 < -\frac{y_3}{c_{p_1}}$$

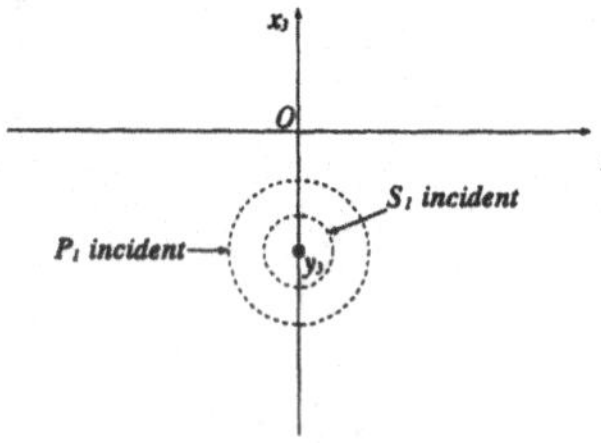

$$-\frac{y_3}{c_{p_1}} \le x_0 < \frac{c_{p_2}(-y_3)}{c_{p_1}\sqrt{c_{p_2}^2 - c_{p_1}^2}}$$

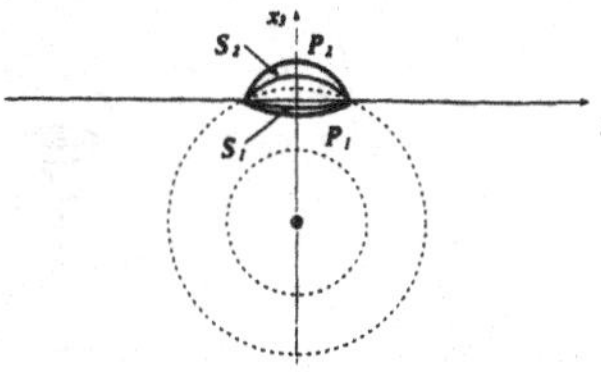

$$\frac{c_{p_2}(-y_3)}{c_{p_1}\sqrt{c_{p_2}^2 - c_{p_1}^2}} \leq x_0 < \frac{c_{s_2}(-y_3)}{c_{p_1}\sqrt{c_{s_2}^2 - c_{p_1}^2}}$$

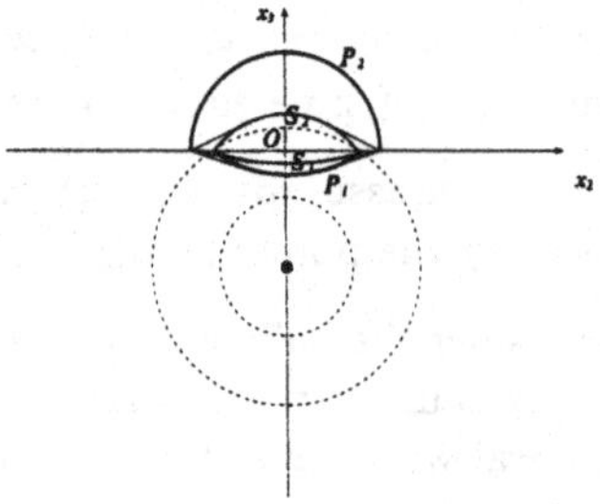

$$\frac{c_{s_2}(-y_3)}{c_{p_1}\sqrt{c_{s_2}^2 - c_{p_1}^2}} \leq x_0 < \frac{c_{p_2}(-y_3)}{c_{s_1}\sqrt{c_{p_2}^2 - c_{s_1}^2}}$$

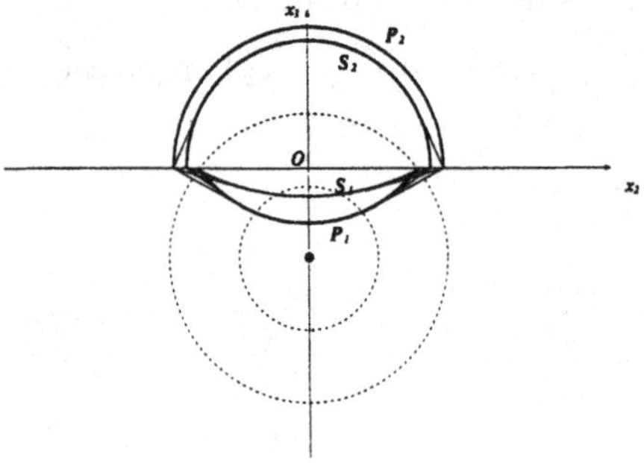

$$\frac{c_{p_2}(-y_3)}{c_{s_1}\sqrt{c_{p_2}^2 - c_{s_1}^2}} \leq x_0 < \frac{c_{s_2}(-y_3)}{c_{s_1}\sqrt{c_{s_2}^2 - c_{s_1}^2}}$$

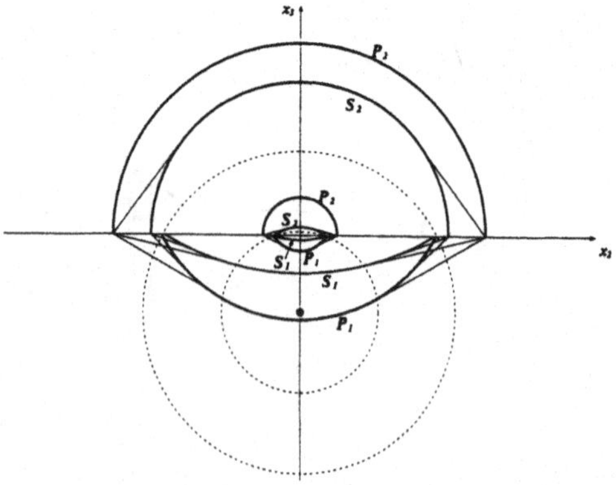

$$\frac{c_{s_2}(-y_3)}{c_{s_1}\sqrt{c_{s_2}^2 - c_{s_1}^2}} \leq x_0 < \frac{c_{p_1}(-y_3)}{c_{s_1}\sqrt{c_{p_1}^2 - c_{s_1}^2}}$$

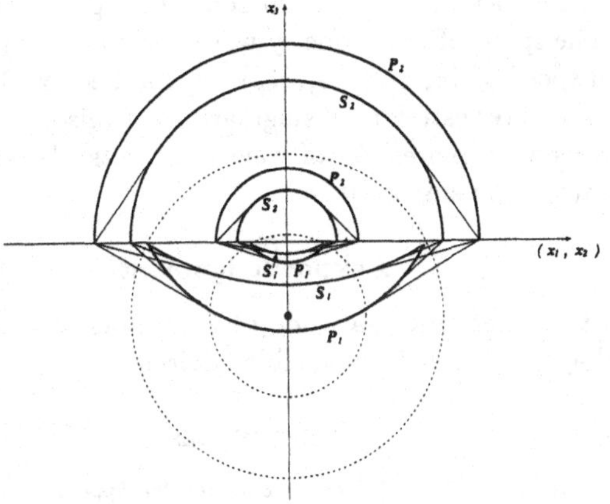

$$\frac{c_{p_1}(-y_3)}{c_{s_1}\sqrt{c_{p_1}^2 - c_{s_1}^2}} \leq x_0$$

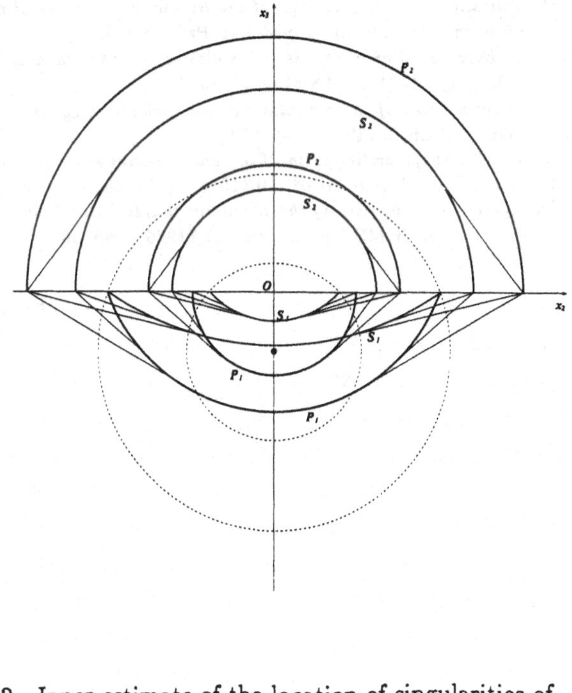

Figure 2 Inner estimate of the location of singularities of $F^{I}_{2\times 2}(x, y)$ and $F^{II}_{2\times 2}(x, y)$

Finally we mention about open problems. The one is the interface wave. If we put unit impulse δ on the interface $x_3 = 0$, a singularity appears which corresponds to the interface wave of which name is the Stoneley wave, in the case that the Lopationski determinant has one real zero. Concerning the Lopatinski determinant, there are two cases: One is the case that Lopatinski determinant has one real zero, and the other is the case that it has no zero. It depends on the Lamé constants and densities. The speed of the Stoneley wave c_{St} is less than or equal to c_{s_1} which is the minimum speed of $\{c_{s_1}, c_{p_1}, c_{s_2}, c_{p_2}\}$ (cf. [Sh Section 3]).

The other is the outer estimate of singularities of this problem. We might obtain outer estimates using a expression of wave front sets. It might be correspond to the inner estimate of the singularities.

Acknowledgement

I would like to express my gratitude to Professors Mutsuhide Matsumura and Seiichiro Wakabayashi for their invaluable advices.

References

[A-B-G] M. F. Atiyah, R. Bott, L. Gårding, *Lacunas for hyperbolic differential operators with constant coefficients I*, Acta Math. **124** (1970), 109–189.

[Ma 1] M. Matsumura, *Localization theorem in hyperbolic mixed problems*, Proc. Japan. Acad. **47** (1971), 115–119.

[Ma 2] M. Matsumura, *On the singularities of the Riemann functions of mixed problems for the wave equation in plane-stratified media I*, Proc. Japan. Acad. **52** (1976), 289–292.

[Ma 3] M. Matsumura, *On the singularities of the Riemann functions of mixed problems for the wave equation in plane-stratified media II*, Proc. Japan. Acad. **52** (1976), 293–295.

[Sh] S. Shimizu, *Eigenfunction expansions for elastic wave propagation problems in stratified media R^3*, Tsukuba J. Math. **18** (1994), 283–350.

[Ts] M. Tsuji, *Propagation of the singularities for hyperbolic equations with constant coefficients*, Japan J. Math. **2** (1976), 369–373.

[Wa 1] S. Wakabayashi, *Singularities of the Riemann functions of hyperbolic mixed problems in a quater-space*, Proc. Japan. Acad. **50** (1974), 821–825.

[Wa 2] S. Wakabayashi, *Singularities of the Riemann functions of hyperbolic mixed problems in a quater-space*, Publ. RIMS Kyoto Univ. **11** (1976), 785–807.

Faculty of Engineering
Shizuoka University
Hamamatsu 432
JAPAN

Exact boundary controllability of a first order, non-linear hyperbolic equation with non-local integral terms arising in epidemic modeling[1]

Irena Lasiecka and Roberto Triggiani
Applied Mathematics, Thornton Hall
University of Virginia
Charlottesville, VA 22901

Abstract

We consider a first order, non-linear hyperbolic equation with non-local integral terms, such as it arises in an age-structured epidemic model [B-I-T.1], and subject to the action of a boundary control. We study (local well-posedness as well as) local, possibly global, exact boundary controllability of the non-linear model. To this end we provide Carleman estimates and (global) exact controllability of the linearized model.

[1]Research partially supported by the National Science Foundation under Grant DMS-904822, and by the Army Research Office under Grant DAAH04-96-1-0059. This research was initiated while the authors were visiting the Mathematics Department, University of Trento, Italy, where a preliminary version was presented in a seminar series. The authors wish to thank Professor Mimmo Iannelli for his hospitality and useful conversations.

R.P. Gilbert et al.(eds.), Direct and Inverse Problems of Mathematical Physics, 363–398.

1. Introduction, model, statement of main results

Model. In this paper were consider the following nonlinear model in the scalar unknown $u(t,x)$, where x runs over the finite interval $I = (0,\ell)$:

$$u_t + u_x + \alpha(x)u + k(x)u^2 + u(Vu) - Vu \equiv 0, \quad \text{in } (0,T] \times (0,\ell) \tag{1.1a}$$

$$u(0,x) = u_0(x) \quad \text{in } (0,\ell); \quad u|_{x=0} = g(t) \quad \text{at } x=0 \tag{1.1b}$$

subject to the boundary control g, chosen in some function space over $(0,T)$; e.g., $g \in L_2(0,T)$, or else $g \in H_0^1(0,T)$. The above model, in its homogeneous form with $g \equiv 0$, is an age-structured epidemic model taken from [B-I-T.1, p.1068]. Here we assume

$$\alpha(\cdot) \in L_\infty(0,\ell); \qquad k(\cdot) \in L_\infty(0,\ell) \tag{1.2}$$

$$(Vf)(x) = \int_0^\ell K(x,\xi)\, p_\infty(\xi)\, f(\xi) d\xi : \text{ continuous } L_2(0,\ell) \to L_2(0,\ell) \tag{1.3}$$

In (1.3a) we have written the kernel as it arises in the age-structured epidemic model [B-I-T.1. p1068]. Verifiable sufficient conditions (Schur's Theorem) on the kernel to ensure the boundedness property (1.3b) are well known, e.g., [W.1, Corollary, p 155]:

$$\int_0^\ell |K(x,\xi)\, p_\infty(\xi)|\, d\xi \le C_1^2 \quad \text{a.e. in } x \in [0,\ell]; \quad \int_0^\ell |K(x,\xi)\, p_\infty(\xi)|\, dx \le C_2^2 \quad \text{a.e. in } \xi \in [0,\ell] \tag{1.4}$$

and then the boundedness property (1.3) for V follows with $\|V\| \le C_1 C_2$. We note that the non-linear model (1.1), as well as its linearized version (2.1a) below, contain the *non-local* integral term (Vu). Terms such as $\alpha(\cdot)u$ and (Vu), while benign (bounded perturbations) with respect to the *(trace regularity)* inequality (2.26) below, are notoriously a source of serious technical difficulties with respect to the *reverse* (*continuous observability*) inequality (3.3.7) or (3.4.4) below [unless $\sup_x |\alpha|$ and $\|V\|$ are 'very small', the non-interesting case]. Their influence on (3.3.7) or (3.4.4)—hence on the main goal of the present paper—will be pointed out below. Indeed, the main *goal* of the paper is to establish a local exact controllability result (in a neighborhood of the origin) for the non-linear model (1.1): see Theorem 6.1. [This local result becomes then, as usual, a *global* one, when zero is a stable equilibrium of the uncontrolled dynamics ($g \equiv 0$), see below.] To this end, it is critical to study first the (global) exact controllability of the linearized model (2.1) below. Accordingly, the paper is divided in two parts.

Part I provides—in Section 3.4—two exact controllability results for the linearized model (2.1): one on the state space $L_2(0,\ell)$ within the class of $L_2(0,T)$-controls g, with $T > 2\ell$ (Theorem 3.4.2); and one, on the domain $\mathcal{D}(A)$ of the generator A of the linearized model (2.1), as defined in (2.4) below, in particular, in the state space $H_0^1(0,\ell)$ within the class of $H_0^1(0,T)$-controls g with the same T (Theorem 3.4.5).

Both results are obtained, equivalently, by establishing the (same) continuous observability inequality (3.3.7), or (3.4.4) for the dual ψ-problem (2.27) in Theorem 3.3.2. Such key a-priori estimate (3.3.7) is definitely a non-trivial result [despite the perhaps deceiving appearance of the one-dimensionality of the (dual) linear problem], as we hinted above, and as we now explain. The dual linear equation (2.27a) contains two terms—$\alpha(x)\psi$ and $(V^*\psi)$—one with variable coefficient, the other nonlocal, which are both at the *energy level*, which for the problem in question is $L_2(0, \ell)$. Both these features are notorious sources of difficulties in achieving a (continuous observability) inequality such as (3.3.7). To overcome these, our first strategy is to establish Carleman estimates for the linearized equation (2.1a) for u, or (2.27a) for ψ. This is done in Theorem 3.1.1 and Theorem 3.1.5 of Section 3.1, respectively, by using the more general and more flexible multiplier in (3.1.12) than classical ones, see Remark 3.1.1. For the general topic of Carleman estimates for PDE solutions with compact support we refer to treatises such as [Ho.1]. The multiplier (3.1.12) is tuned to first order equations: being in differential form, it allows to carry out the consequent energy method entirely at the differential level. [For other dynamics and corresponding differential Carleman multipliers see [F-I.1] and [L-T.3] e.g.]. Pseudo-differential multipliers and related pseudo-differential energy methods were previously proposed in [Ta.1] for general evolution equations. Once the boundary conditions of the whole dual linearized ψ-problem (2.27a-b-c) are factored in, this yields the Carleman estimates of Theorem 3.1.1, Eq (3.3.3). This result takes care of the energy level term $\alpha(x)\psi$ with variable coefficient, but *not* of the term $(V^*\psi)$. In fact, Eq (3.3.3) comes corrupted by the $L_2(0, T; L_2(I))$—norm of $(V^*\psi)$. This is *not* a lower order term, and, moreover, is non-local. Thus, to absorb this term, the established compactness-uniqueness argument runs into difficulty for compactness (in time and space) and uniqueness. To overcome these obstacles, we need additional hypotheses on V, beyond the continuity (boundedness) assumed in (1.3). Assuming, in Theorem 3.3.2, that the kernel of V(or V^*) be Hilbert-Schmidt, i.e. in $L_2((0, \ell) \times (0, \ell))$, so that V and V^* are compact in space permits to absorb $\|V^*\psi\|_{L_2(0,T;L_2(I))}$ via the Lebesgue dominated theorem, provided that uniqueness of the corresponding overdetermined linear problem holds true. Uniqueness in the presence of nonlocal operators is notoriously an untractable problem in general. Here we provide some preliminary uniqueness results under the assumption that the kernel is separated (Theorem 3.2.1 for the original u-problem (3.2.1), (3.2.2); Theorem 3.2.3 for the dual ψ-problem (3.2.36), (3.2.37); or, *generically*, for any degenerate kernel (see Appendix A). In between, we manage to insert a uniqueness result (Theorem 3.2.2) for a very general kernel, with respect, however, to non-negative initial conditions. Possible extension of uniqueness to any L_2 (Hilbert-Schimdt)-kernel is discussed in the last paragraph of the Appendix as an issue of future research.

Part II first establishes a local well-posedness result (Theorem 4.1); and next establishes the following local exact controllability result: *if the initial condition* u_0 *is suitably small, in particular in* $H_0^1(0, \ell)$, *in the norm of* $\mathcal{D}(A)$ *and* $T > 2\ell$ *is given, there exists a boundary control* $g \in H_0^1(0, \ell)$ with *suitably small norm such that the solution* $u(\cdot)$ *of problem* (1.1) can be *brought to rest at time* $t = T$: i.e., $u(T) = 0$. A formal statement is given in Theorem 6.1

The above local exact controllability result may be boosted to a *global one* under the condition that zero is a stable equilibrium of the uncontrolled process ($g \equiv 0$), for which [B-T-T.1, Theorem 5.1, p 1075] provides sufficient conditions: in such case, given *any* initial condition u_0 in the state space, we first set $g \equiv 0$ over some $[0, T_1]$ so that the solution of the free dynamics $u(T_1)$ at $t = T_1$ hits the suitably small exact controllability neighborhood of the origin. Next, we use the above local exact controllability result to steer $u(T_1)$ to rest at time $t = T_2 > T_1 : u(T_2) = 0$, by means of a control $g \in H_0^1(T_1, T_2)$.

For all well-posedness results of Section 4 and related Frechet differentiability properties of Section 5, the basic assumption of boundedness for V as in (1.3) is adequate.

Remark 1.1. The results of the present paper can be extended to cover the B.C.

$$u\,|_{x=0} = \int_0^\ell \beta(\xi)u(t,\xi)d\xi + g_1(t) \tag{1.5}$$

instead of $u\,|_{x=0} = g(t)$ as in (1.1b). In (1.5) we have that β is non-negative and in $L_\infty(0,\ell)$. This version of an age-dependent model has been proposed for a linear equation (2.1) with $V = 0$ in [D-I.1] [see also BDDM.1, p247], as in this paper where the quadratic optimal control problem is investigated, rather than the exact controllability problem

To see the extension for the exact controllability of the linear equation (2.1a-b-c) (or the non-linear equation (1.1a-b), let $\bar{g}$ be the control steering the initial condition u_0 to rest at $t = T$, along the corresponding solution $\bar{u}$, so that $\bar{u}(T) = 0$. Then define

$$g_1(t) \equiv \bar{g}(t) - \int_0^\ell \beta(\xi)u(t,\xi)d\xi \tag{1.6}$$

and $g_1(t)$ steers u_0 to rest at $t = T$, along the corresponding solution of (1.1) (or (2.1)) with B.C. (1.5).□

2. Part I: Linearization. Abstract Model

In the present Part I, we analyze the linearized mixed problem of the nonlinear model (1.1) around $u = 0$. If I is the interval $I = (0,\ell)$, such *linearized model* is

$$u_t + u_x + \alpha u - Vu = 0 \quad \text{in } (0,T] \times (0,\ell) = Q \tag{2.1a}$$

$$u(0,\cdot) = u_0(x) \quad \text{in } (0,\ell); \quad u\,|_{x=0} = g \quad \text{at } x = 0 \tag{2.1b-c}$$

where the function $\alpha(x)$ and the operator V *satisfy properties* (1.2) *and* (1.3).

Abstract model. We shall show that the mixed problem (2.1) admits the abstract model

$$u_t = Au + Bg \quad \text{in } [D(A^*)]' \tag{2.2}$$

where the operators A and B are identified below in (2.3), (2.4), (2.11).

Dynamic operator A. On the space $L_2(I)$ we define the operator

$$(Af)(x) = -f'(x) - \alpha(x)f(x) + (Vf)(x) \tag{2.3}$$

$$\mathcal{D}(A) = \{f \in L_2(I) : f \quad \text{is absolutely continuous; } f(0) = 0\} \tag{2.4}$$

For $\lambda_0 > 0$ large enough, in fact $\lambda_0 > \|\alpha\|_{L_\infty(I)} + \|V\|_{\mathcal{L}(L_2(I))}$, the operator $(A - \lambda_0 I)$ is maximal dissipative on $L_2(I)$. Thus, A is *the generator of a s.c. semigroup* e^{At} *on* $L_2(I)$, via the Lumer-Phillips theorem [P.1, p15] applied to $(A - \lambda_0 I)$.

The boundary operator B. In order to introduce the boundary operator B, we begin by noting the following *Claim*: that the static problem corresponding to (2.1), i.e.,

$$\left[\frac{\partial}{\partial x} + \alpha(\cdot) - V + \lambda_0\right] f \equiv 0 \quad \text{in } I = (0,\ell); \;\; f\,|_{x=0} = v \in \mathcal{R} \tag{2.5a-b}$$

has a unique solution $f \in H^1(I)$ for any constant $v \in \mathcal{R}$, provided that $\lambda_0 > 0$ is sufficiently large. Indeed, multiplying Eq. (2.5a) by the integrating factor $\rho(x) = \exp\left[\int_0^x (\alpha(\xi) + \lambda_0)d\xi\right]$, and using the B.C. (2.5b) in the ensuing integration, we obtain the integral equation

$$f(x) = ve^{-\int_0^x(\alpha(\xi)+\lambda_0)d\xi} + \frac{1}{\rho(x)}\int_0^x (Vf)(\xi)\rho(\xi)d\xi \tag{2.6}$$

This equation has a unique fixed point solution, initially say in $L_2(I)$, as guaranteed by the contraction mapping principle with λ_0 real large enough, hence with $\max_I [\rho^{-1}(x)]$ small enough. Such solution may then be boosted to be in $H^1(I)$, by using the right-hand side of (2.6). Thus, the above Claim is established.

Next, with $\lambda_0 > 0$ sufficiently large as required by the above Claim, we define the (Dirichlet) map D_{λ_0} by

$$f = D_{\lambda_0}v \Leftrightarrow f \text{ satisfies (2.5a-b)} \Leftrightarrow \begin{cases} \left[\dfrac{\partial}{\partial x} + \alpha - V + \lambda_0\right](D_{\lambda_0}v) \equiv 0 \text{ in } I \\ (D_{\lambda_0}v)_{x=0} = v \end{cases} \tag{2.7}$$

By the Claim:

$$D_{\lambda_0}: \text{continuous } \mathcal{R} \to H^1(0,\ell) \tag{2.8}$$

Next, we return to problem (2.1) and obtain via (2.5a-b), or (2.7):

$$u_t = -\left[\frac{\partial}{\partial x} + \alpha(\cdot) - V + \lambda_0\right](u - D_{\lambda_0}g) + \lambda_0 u \qquad \text{in } (0,T] \times I \tag{2.9a}$$

$$\left[u - D_{\lambda_0} g\right]_{t=0} = u_0 - D_{\lambda_0} g(0) \qquad \text{in } I \tag{2.9b}$$

$$\left[u - D_{\lambda_0} g\right]_{x=0} = 0 \qquad \text{at } x = 0 \tag{2.9c}$$

Thus, recalling (2.3), (2.4), we may rewrite problem (2.9) abstractly as

$$u_t = (A - \lambda_0)(u - D_{\lambda_0} g) + \lambda_0 u \quad \text{in } L_2(I); \quad \text{or:} \quad u_t = Au + Bg \text{ in } [\mathcal{D}(A^*)]' \tag{2.10}$$

$$Bg = -(A - \lambda_0) D_{\lambda_0} g = -A D_{\lambda_0} g + \lambda_0 D_{\lambda_0} g \tag{2.11}$$

where, in going from the left-hand side equation to the right-hand side equation in (2.10), with duality with respect to the pivot space $L_2(I)$, we have extended the original operator A given by (2.3), (2.4) to the operator A: continuous $L_2(I) \to [\mathcal{D}(A^*)]'$ by isomorphism. The solution to the abstract equation (2.10) (right) is then

$$u(t) = e^{At} u_0 + (Lg)(t) \tag{2.12}$$

$$(Lg)(t) = \lambda_0 \int_0^t e^{A(t-\tau)} D_{\lambda_0} g(\tau) d\tau - A \int_0^t e^{A(t-\tau)} D_{\lambda_0} g(\tau) d\tau . \tag{2.13}$$

Since the first integral term in (2.13) is definitely smoother than the second one, we shall henceforth take without loss of generality

$$\lambda_0 = 0; \; D_{\lambda_0} = D, \text{ hence } B = -AD\text{: continuous } \mathcal{R} \to [\mathcal{D}(A^*)]' \tag{2.14}$$

and simply write

$$(Lg)(t) = -A \int_0^t e^{A(t-\tau)} Dg(\tau) d\tau \tag{2.15}$$

The regularity of L will be dealt with in Theorem 2.2, at the end of this section.

Adjoint operator A^*. The $L_2(I)$-adjoint A^* of the operator A defined by (2.3), (2.4) is

$$A^* h = h'(x) - \alpha(x) h(x) + (V^* h)(x); \quad (V^* h)(x) = p_\infty(x) \int_0^\ell K(\xi, x) h(\xi) d\xi \tag{2.16}$$

$$\mathcal{D}(A^*) = \left\{ h \in H^1(0, \ell) : h(\ell) = 0 \right\} \tag{2.17}$$

where V^* is the $L_2(I)$-adjoint of V in (1.3).

The Operator $D^*_{\lambda_0} A^*_{\lambda_0}$. With reference to the operator D_{λ_0} defined by (2.7), let $D^*_{\lambda_0} : L_2(I) \to \mathcal{R}$ be its L_2-adjoint. Moreover, let $A_{\lambda_0} = (A - \lambda_0 I)$, λ_0 real. We then have:

Lemma 2.1. Under (1.2), (1.3), the following holds true

$$D^*_{\lambda_0} A^*_{\lambda_0} f = -f\,|_{x=0} \qquad f \in \mathcal{D}(A^*_{\lambda_0}) = \mathcal{D}(A^*) \tag{2.18}$$

Proof. With $f \in \mathcal{D}(A^*)$, hence $f(\ell) = 0$ by (2.17), and $v \in \mathcal{R}$, we compute by integrating by parts using (2.16)

$$(D^*_{\lambda_0} A^*_{\lambda_0} f, v)_{\mathcal{R}} = ((A^* - \lambda_0 I)f,\, D_{\lambda_0} v)_{L_2(I)} = \left(\left[\frac{d}{dx} - \alpha(\cdot) + V^* - \lambda_0\right] f, D_{\lambda_0} v\right)_{L_2(I)} \tag{2.19}$$

$$= f(\ell)(D_{\lambda_0} v)(\ell) - f(0)(D_{\lambda_0} v)(0)$$

$$+ \left(f, \left[-\frac{d}{dx} - \alpha + V - \lambda_0\right](D_{\lambda_0} v)\right)_{L_2(I)} = -f(0)v \tag{2.20}$$

since $f(\ell) = 0$ by (2.17); while $(D_{\lambda_0} v)(0) = v$ and $\left[-\frac{d}{dx} - \alpha + V - \lambda_0\right](D_{\lambda_0} v) = 0$ by the definition of D_{λ_0}, see (2.7). Thus, (2.20) yields (2.18), as desired. □

We can now state an optimal regularity result for the operator L in (2.15) (i.e., for problem (2.1) with $u_0 = 0$), and its adjoint, where Lemma 2.1 will be needed.

Theorem 2.2. Assume (1.2), (1.3). With reference to the operator L in (2.15), we have

$$\text{(i)} \qquad L\text{: continuous } L_2(0,T) \to C([0,T];\ L_2(I)) \tag{2.21}$$

where $(Lg)(t) = u(t)$ solves problem (2.1) for $u_0 = 0$; in particular

$$L_T g \equiv (Lg)(T) = -A \int_0^T e^{A(T-t)} Dg(t)dt \tag{2.22}$$

$$: \text{ continuous } L_2(0,T) \longrightarrow L_2(I) \tag{2.23}$$

(ii) equivalently [L-T.1], the L_2-adjoint L^*_T of L_T satisfies

$$(L^*_T z)(t) = -D^* A^* e^{A^*(T-t)} z : \text{continuous } L_2(I) \to L_2(0,T)\ ; \tag{2.24}$$

$$|L^*_T z|^2_{L_2(0,T)} = \int_0^T \left|D^* A^* e^{A^* t} z\right|^2 dt \le c_T \|z\|^2_{L_2(I)} \tag{2.25}$$

(iii) in P.D.E. terms, the regularity inequality (2.25) means, via (2.18) (with $\lambda_0 = 0$ w.l.o.g.), that $\psi(t;\psi_0) \equiv e^{A^*(T-t)}\psi_0$ satisfies

$$\int_0^T |\psi(t;\psi_0)|_{x=0}|^2\, dt \le c_T \int_0^T |\psi_0(x)|^2\, dx \tag{2.26}$$

where ψ satisfies (recalling (2.16), (2.17)) the dual problem

$$\begin{cases} \psi_t = A^*\psi \\ \psi(T) = \psi_0 \end{cases} \iff \begin{cases} \psi_t - \psi_x + \alpha\psi - V^*\psi \equiv 0 & \text{in } Q = (0,T) \times I \\ \psi(T,\cdot) = \psi_0 & \text{in } I = (0,\ell) \\ \psi\,|_{x=\ell} = 0 & \text{in } (0,T) \times \{\ell\} \end{cases} \qquad \text{(2.27a–2.27c)}$$

Proof. One may either readily prove (2.21) directly using a solution formula as in [D-I.1, p95], or—preferably—one may readily prove the dual version (2.26) using a solution formula, as in [B-D-D-M.1, p249]. At the outset one may take $\alpha \equiv 0, V = 0$, since their presence contributes bounded perturbations at the energy level (which is $L_2(0,\ell)$).□

3. Part I: Linearization. Carleman estimates. Uniqueness. Continuous observability. Exact controllability

The goal of the present section is to establish the following result: that the linearized model (2.1) is exactly controllable over any finite time $T > 2\ell, I = [0,\ell]$, both on the state space $L_2(I)$ within the class of $L_2(0,T)$-controls g; as well as on the state space $H^1(I)$ within the class of $H_0^1(0,T)$-controls g. To this end, we shall provide Carleman estimates in Section 3.1, which yield then the desired continuous observability estimates of Section 3.1, albeit corrupted by an additional (nonlower order) term. This will then be absorbed by a compactness/uniqueness argument in Section 3.3 after establishing a uniqueness result in Section 3.2. Finally, exact controllability is achieved in Section 3.4.1 on $L_2(I)$ and in Section 3.4.2 on $H^1(I)$, by duality on the preceding continuous observability estimates of Section 3.3.

3.1. Carleman estimates for original Eq. (2.1a) and its adjoint (2.27a).

Eq. (2.1a). In this subsection we return to Eq. (2.1a) (with no boundary conditions), here re-written for convenience

$$u_t + u_x + \alpha u - Vu = 0 \qquad \text{in } (0,T] \times [0,\ell] = Q \qquad (3.1.1)$$

where the function α and the operator V obey assumptions (1.2), (1.3) and consider solutions of (3.1.1) satisfying the following regularity properties

$$u \in L_2\left(0,T; L_2(I)\right); \quad u\,|_{x=0}\,, u\,|_{x=\ell} \in L_2(0,T) \qquad (3.1.2)$$

For such solutions we shall establish a-priori estimates. To this end, we introduce the (pseudo-convex) function

$$\phi(x,t) = |x - x_0|^2 - c\left|t - \tfrac{T}{2}\right|^2, \quad 0 < c < 1 \qquad (3.1.3)$$

where, for sharp results, we select the parameters so that

$$cT > 2\max_{x\in I}|x-x_0|\,; \qquad \min_{x\in I}|x-x_0| = r > 0 \tag{3.1.4a}$$

We may select $x_0 = \ell + r$, hence $\max|x-x_0| = \ell + r$, with $r > 0$ sufficiently small so that $cT > 2(\ell + r)$ and T can be taken just any number greater than twice the diameter, i.e. $T > 2\ell$, with $c < 1$ but close to 1. Then, both requirements in (3.1.4a) are satisfied with optimal choice of T (henceforth kept fixed) and the following consequences hold true:

$$\phi(x,t) = |x-x_0|^2 - c\left|t-\tfrac{T}{2}\right|^2 \geq r^2 - c\left|t-\tfrac{T}{2}\right|^2 \geq p > 0, \text{ uniformly in } x \in I \tag{3.1.4b}$$

for a suitably small interval $[t_0, t_1] \subset [0, T]$ centered at $t = \frac{T}{2}$;

$$(x-x_0) + \frac{cT}{2} > 0 \text{ and } (x-x_0) - \frac{cT}{2} < 0, \text{ uniformly in } x \in I. \tag{3.1.4c}$$

The first requirement in (3.1.4c) holds true, since $\frac{cT}{2} > \max(x_0 - x) = \ell + r$, as seen above; the second requirement in (3.1.4c) holds true since $(x - x_0) < 0$ with our choice of $x_0 = \ell + r, r > 0$. In the sequel, we shall invoke both consequences (3.1.4b) and (3.1.4c) with the optimal choice of $T > 2\ell$. Finally, we introduce the 'energy' of Eq. (3.1.1), i.e.,

$$E(t) = \int_0^\ell |u(t,x)|^2\, dx \tag{3.1.5}$$

Theorem 3.1.1. (Carleman estimates) Assume (1.2), (1.3) for α and V. Let u be a solution of Eq. (3.1.1) in the class (3.1.2). Then, for $\tau > 0$ sufficiently large, the following one-parameter family of estimates holds true for any $\varepsilon > 0$ fixed

$$(BT)\,|_\Sigma + \frac{1}{\varepsilon}\int_0^T\int_0^\ell e^{\tau\phi}\,|Vu|^2\, dx dt \;\geq\; (1-c-\varepsilon)\int_0^T\int_0^\ell e^{\tau\phi}\, u^2 dx dt \tag{3.1.6}$$

$$\geq\; (1-c-\varepsilon)\, e^{\tau p}\int_{t_0}^{t_1} E(t)dt \tag{3.1.7}$$

where the function $\phi(x,t)$, hence the constants c, p and the time interval $[t_0, t_1]$, are defined in (3.1.3), (3.1.4). Moreover, the boundary terms $(BT)\,|_\Sigma$ are defined by

$$(BT)\,|_\Sigma \;\equiv\; \int_0^T \left\{ e^{\tau\phi(x,t)}u^2(t,x)\left[(x-x_0) - c\left(t - \frac{T}{2}\right)\right]\right\}_{x=0}^{x=\ell} dt \tag{3.1.8}$$

$$=\; \int_0^T \left\{ e^{\tau\phi(x,t)}u^2(t,x)\frac{1}{2}\left[\phi_x(x,t) + \phi_t(x,t)\right]\right\}_{x=0}^{x=\ell} dt \tag{3.1.9}$$

$$= \quad \text{difference of value at } x = \ell \text{ and of value at } x = 0.$$

Estimate (3.1.7) may be further pushed forward, by analyzing the behavior of the energy $E(t)$, as in Lemma 3.1.4, Eq. (3.1.24) below, to obtain the following result.

Theorem 3.1.2. Assume (1.2), (1.3) for α and V. Let u be a solution of Eq (3.1.1) in the class (3.1.2). Then, for $\tau > 0$ sufficiently large, the following one-parameter family of estimates holds true for any $\varepsilon > 0$ fixed:

$$(\overline{BT})\,|_{\Sigma} + \frac{1}{\varepsilon}\int_0^T\int_0^\ell e^{\tau\phi}\,|Vu|^2\,dxdt \geq (1-c-\varepsilon)\,e^{\tau p}k_{t_0,t_1}E(0) \tag{3.1.10a}$$

with k_{t_0,t_1} defined below (3.1.31), where we have now defined the boundary term $(\overline{BT})\,|_{\Sigma}$ by

$$(\overline{BT})\,|_{\Sigma} \equiv (BT)\,|_{\Sigma} + (1-c-\varepsilon)\,e^{\tau p}(t_1-t_0)\int_0^T [u^2(t,\ell)+u^2(t,0)]dt \tag{3.1.11}$$

with $(BT)\,|_{\Sigma}$ defined in (3.1.8). Alternatively, we also obtain

$$(\overline{BT})\,|_{\Sigma} + \wedge(T) + \frac{1}{\varepsilon}\int_0^T\int_0^\ell e^{\tau\phi}\,|Vu|^2\,dxdt \geq (1-c-\varepsilon)\,e^{\tau p}k_{t_0,t_1}e^{-C_{\alpha,V}T}E(T) \tag{3.1.10b}$$

where the boundary term $\wedge(T)$ is defined in (3.1.25) below

Proof of Theorem 3.1.1. **Step** 1 We shall multiply Eq (3.1.1) by the Carleman multiplier

$$e^{\tau\phi(x,t)}\left[(x-x_0)-c\left(t-\frac{T}{2}\right)\right] = \frac{1}{2}e^{\tau\phi(x,t)}\,[\phi_x(x,t)+\phi_t(x,t)] \tag{3.1.12}$$

with $\phi(x,t)$ defined by (3.1.3), and integrate by parts. We obtain:

Lemma 3.1.3. Assume (1.2), (1.3) for α and V. With reference to a solution u of Eq (3.1.1) in the class (3.1.2), the following identity holds true

$$\begin{aligned}(BT)\,|_{\Sigma} &= (1-c)\int_0^T\int_0^\ell e^{\tau\phi}\,u^2dxdt + \int_0^T\int_0^\ell \tau e^{\tau\phi}u^2\,4\left[(x-x_0)-c\left(t-\frac{T}{2}\right)\right]^2 dxdt \\ &\quad -2\int_0^T\int_0^\ell e^{\tau\phi}\alpha u^2\left[(x-x_0)-c\left(t-\frac{T}{2}\right)\right]dxdt \\ &\quad +2\int_0^T\int_0^\ell e^{\tau\phi}u(Vu)\left[(x-x_0)-c\left(t-\frac{T}{2}\right)\right]dxdt + \beta_{0,T}\end{aligned} \tag{3.1.13}$$

where $(BT)\,|_{\Sigma}$ is defined by (3.1.8), while $\beta_{0,T}$ is given by

$$\beta_{0,T} = \int_0^\ell e^{\tau\phi(x,0)}\,u^2(0,x)\left[(x-x_0)+\frac{cT}{2}\right]dx - \int_0^\ell e^{\tau\phi(x,T)}\,u^2(T,x)\left[(x-x_0)-\frac{cT}{2}\right]dx \tag{3.1.14}$$

Proof. Regarding the first term of (3.1.1), we use $u_t u = \frac{1}{2}\frac{\partial(u^2)}{\partial t}$, and integrating by parts in t, we obtain, using (3.1.2)

$$\int_0^\ell \int_0^T (u_t u) e^{\tau\phi} \left[(x-x_0) - c\left(t-\frac{T}{2}\right)\right] dt dx = \frac{1}{2}\int_0^\ell \left\{ e^{\tau\phi} u^2 \left[(x-x_0) - c\left(t-\frac{T}{2}\right)\right]\right\}_{t=0}^{t=T} dx$$

$$-\frac{1}{2}\int_0^T \int_0^\ell \tau e^{\tau\phi} u^2 \phi_t \left[(x-x_0) - c\left(t-\frac{T}{2}\right)\right] dt dx + \frac{1}{2} c \int_0^T \int_0^\ell e^{\tau\phi} u^2 dx dt \tag{3.1.15}$$

Regarding the second term of (3.1.1), we use $u_x u = \frac{1}{2}\frac{\partial(u^2)}{\partial x}$, and integrating by parts in x, we obtain, using (3.1.2):

$$\int_0^\ell \int_0^\ell (u_x u) e^{\tau\phi} \left[(x-x_0) - c\left(t-\frac{T}{2}\right)\right] dx = \frac{1}{2}\int_0^T \left\{ e^{\tau\phi} u^2 \left[(x-x_0) - c\left(t-\frac{T}{2}\right)\right]\right\}_{x=0}^{x=\ell} dt$$

$$-\frac{1}{2}\int_0^T \int_0^\ell \tau e^{\tau\phi} u^2 \phi_x \left[(x-x_0) - c\left(t-\frac{T}{2}\right)\right] dx dt - \frac{1}{2}\int_0^T \int_0^\ell e^{\tau\phi} u^2 dx dt \tag{3.1.16}$$

Thus, multiplying Eq (3.1.1) by the Carleman multiplier in (3.1.12), and using (3.1.15) and (3.1.16), yields identity (3.1.13) via (3.1.14), as desired, by virtue also of the relationships

$$\begin{aligned}[\phi_t + \phi_x]\left[(x-x_0) - c\left(t-\frac{T}{2}\right)\right] &= [\phi_t + \phi_x]\frac{[\phi_x + \phi_t]}{2} = \frac{1}{2}[\phi_x + \phi_t]^2 \\ &= 2\left[(x-x_0) - c\left(t-\frac{T}{2}\right)\right]^2 \end{aligned} \tag{3.1.17}$$

Step 2. Since $\alpha \in L_\infty(I)$ by (1.2), we obtain for any $\varepsilon > 0$

$$-2\int_0^T \int_0^\ell e^{\tau\phi} \alpha\, u\, u \left[(x-x_0) - c\left(t-\frac{T}{2}\right)\right] dx dt$$

$$\geq -\varepsilon \int_0^T \int_0^\ell e^{\tau\phi} u^2 dx dt - \frac{C_\alpha}{\varepsilon}\int_0^T \int_0^\ell e^{\tau\phi} u^2 \left[(x-x_0) - c\left(t-\frac{T}{2}\right)\right] dx dt \tag{3.1.18}$$

Similarly, using (1.3) and (3.1.2) for u:

$$2\int_0^T \int_0^\ell e^{\tau\phi} (Vu) u \left[(x-x_0) - c\left(t-\frac{T}{2}\right)\right] dx dt$$

$$\geq -\varepsilon' \int_0^T \int_0^\ell e^{\tau\phi} u^2 \left[(x-x_0) - c\left(t-\frac{T}{2}\right)\right]^2 dx dt - \frac{1}{\varepsilon'}\int_0^T \int_0^\ell e^{\tau\phi} |Vu|^2 dx dt \tag{3.1.19}$$

Thus, using (3.1.18) and (3.1.19) in identity (3.1.13), we obtain the following inequality

$$(BT)|_{\Sigma} + \frac{1}{\varepsilon'}\int_0^T\int_0^\ell e^{\tau\phi}|\nabla u|^2\,dxdt \geq (1-c-\varepsilon)\int_0^T\int_0^\ell e^{\tau\phi}u^2dxdt$$

$$+\left(4\tau - \frac{C_\alpha}{\varepsilon} - \varepsilon'\right)\int_0^T\int_0^\ell e^{\tau\phi}u^2\left[(x-x_0) - c\left(t-\frac{T}{2}\right)\right]^2 dxdt + \beta_{0,T} \quad (3.1.20)$$

Step 3. We return to $\beta_{0,T}$ defined by (3.1.14): by selecting c, x_0 and T as described in (3.1.4), we have with $I = [0,\ell]$

$$(x-x_0) + \frac{cT}{2} \geq \left[\inf_{x\in I}(x-x_0)\right] + \frac{cT}{2} = -\ell - r + \frac{cT}{2} > 0 \quad \text{for all } x \in I \quad (3.1.21)$$

$$(x-x_0) - \frac{cT}{2} < 0 \quad \text{for all } x \in I \quad (3.1.22)$$

as observed in (3.1.4c). Thus, by (3.1.21), (3.1.22) used in (3.1.14), and moreover by selecting τ sufficiently large (with $\varepsilon > 0$ and $\varepsilon' > 0$ fixed) so that $(4\tau - \frac{C_\alpha}{\varepsilon} - \varepsilon') > 0$, we obtain

$$\beta_{0,T} > 0; \quad \left(4\tau - \frac{C_\alpha}{\varepsilon} - \varepsilon'\right)\int_0^T\int_0^\ell e^{\tau\phi}u^2\left[(x-x_0) - c\left(t-\frac{T}{2}\right)\right]^2 dxdt > 0 \quad (3.1.23)$$

respectively. Finally, dropping the positive terms in (3.1.23) on the right-hand side of (3.1.20) with $\varepsilon' = \varepsilon$, yields estimate (3.1.6), as desired. Then, estimate (3.1.7) follows from (3.1.6) via property (3.1.4b), with $p > 0$. The proof of Theorem 3.1.1 is complete. □

Remark 3.1.1. The virtue of the presence of the free parameter τ in the Carleman multiplier (3.1.12) is seen precisely in the choice of τ sufficiently large as to obtain the positivity condition (3.1.23): dropping an integral term at the energy level which could not be absorbed by the other energy level integral term with coefficient $(1-c-\varepsilon)$ in (3.1.20).

Proof of Theorem 3.1.2. **Step** 1. **Lemma** 3.1.4 Let α, V satisfy (1.2), (1.3). Let u be a solution of Eq (3.1.1) in the class (3.1.2). Then the following inequalities held true for $T \geq t > 0$:

$$e^{-C_{\alpha,V}t}E(0) - \wedge(T) \leq E(t) \leq [E(0) + \wedge(T)]\,e^{C_{\alpha,V}t} \quad (3.1.24)$$

where we have set $\wedge(T)$ to be the boundary term

$$\wedge(T) = \int_0^T \left[u^2(t,0) + u^2(t,1)\right] dt \quad (3.1.25)$$

Proof. First, we multiply Eq (3.1.1) by u and integrate over $(s,t] \times I$ by parts in t, thus obtaining the following energy identity for $E(t)$ defined by (3.1.5)

$$E(t) = E(s) + \int_0^t \left[u^2(\sigma,0) - u^2(\sigma,1)\right] d\sigma - 2\int_s^t \int_0^\ell \alpha\, u^2\, dx d\sigma + 2\int_s^t \int_0^\ell (Vu)u\, dx d\sigma \quad (3.1.26)$$

Next, under present assumptions (1.2) for α and (1.3) for V, we have

$$\left| -2\int_s^t \int_0^\ell \alpha\, u^2\, dx d\sigma + 2\int_s^t \int_0^\ell (Vu)u\, dx d\sigma \right| = C_{\alpha,V} \int_s^t \left(\int_0^\ell u^2 dx \right) d\sigma$$

$$= C_{\alpha,V} \int_s^t E(\sigma) d\sigma \quad (3.1.27)$$

since $\alpha \in L_\infty(I)$ and $\|Vu\|^2 = \int_0^\ell |Vu|^2\, dx \leq \|V\|^2 \int_0^\ell u^2 dx$. Then, using (3.1.27) in (3.1.26) yields for $t \geq s \geq 0$, recalling (3.1.25)

$$E(t) \leq [E(s) + \wedge(T)] + C_{\alpha,V} \int_s^t E(\sigma) d\sigma \quad (3.1.28)$$

$$E(s) \leq [E(t) + \wedge(T)] + C_{\alpha,V} \int_s^t E(\sigma) d\sigma \quad (3.1.29)$$

We apply the classical argument of the Gronwall's inequality [B.1, p139] to (3.1.28), (3.1.29) where we note that the terms into the square brackets are independent of t in (3.1.28), and independent of s in (3.1.29). We thus obtain for $t \geq s \geq 0$:

$$E(t) \leq [E(s) + \wedge(T)]\, e^{C_{\alpha,V}(t-s)}; \; E(s) \leq [E(t) + \wedge(T)]\, e^{C_{\alpha,V}(t-s)} \quad (3.1.30)$$

Setting $s = 0$ and hence $t > 0$ in (3.1.30) yields (3.1.24).

Step 2. Using the left-hand side inequality in (3.1.24), we compute recalling (3.1.25)

$$\int_{t_0}^{t_1} E(t) dt \geq \int_{t_0}^{t_1} \left[e^{-C_{\alpha,V}t} E(0) - \wedge(T)\right] dt = k_{t_0,t_1} E(0) - (t_1 - t_0) \int_0^T \left[u^2(t,0) + u^2(t,1)\right] dt \quad (3.1.31)$$

$k_{t_0,t_1} = \int_{t_0}^{t_1} e^{-C_{\alpha,V}t} dt$. Next we multiply inequality (3.1.31) by $(1-c-\varepsilon)e^{\tau p}$ with $(1-c-\varepsilon) > 0$ and use the result on the right-hand side of inequality (3.1.7). This way, (3.1.7) yields (3.1.10a), as desired, recalling (3.1.11). Inserting (3.1.24) for $E(0)$ in (3.1.10a) yields (3.1.10b). □

Dual Eq. (2.27a). As a matter of fact, below we shall need not Theorem 3.1.2 for the original equation (3.1.1) directly, but its counterpart version for the dual equation (2.27a), here re-written for convenience:

$$\psi_t - \psi_x + \alpha\psi - V^*\psi \equiv 0 \quad \text{in} \quad Q = (0,T] \times (0,\ell) \quad (3.1.32)$$

with V^* defined in (2.16). The proof is, of course, identical.

Theorem 3.1.5. Assume (1.2) and (1.3) for α and V. Let ψ be a solution of the dual Eq. (2.27a) = (3.1.32) in the class (counterpart of (3.1.2)).

$$\psi \in L_2(0,T;\ L_2(I));\quad \psi|_{x=0},\quad \psi|_{x=\ell} \in L_2(0,T) \tag{3.1.33}$$

Then, for $\tau > 0$ sufficiently large, the following one-parameter family of estimates holds true for any $\varepsilon > 0$ fixed:

$$(\overline{BT}_\psi)|_\Sigma + \frac{1}{\varepsilon}\int_0^T\int_0^\ell e^{\tau\phi}|V^*\psi|^2\,dxdt \geq (1-c-\varepsilon)\,e^{\tau p}k_{t_0,t_1}E_\psi(0) \tag{3.1.34a}$$

where $(\overline{BT}_\psi)|_\Sigma$ and $E_\psi(\cdot)$ are given, respectively, by (3.1.11) and (3.1.5), with u solution of (3.1.1) replaced by ψ solution of (2.27a) = (3.1.32). Alternatively, via (3.1.25) for ψ:

$$(\overline{BT}_\psi)|_\Sigma + \Lambda_\psi(T) + \frac{1}{\varepsilon}\int_0^T\int_0^\ell e^{\tau\phi}|V^*\psi|^2\,dxdt \geq (1.-c-\varepsilon)\,e^{\tau p}k_{t_0,t_1}e^{C_{\alpha,V}T}E_\psi(T) \tag{3.1.34b}$$

3.2 **Uniqueness of original and adjoint problem.**

Original problem. With reference to (1.3a), we begin this section by considering the following homogeneous overdetermined problem:

$$u_t + u_x + \alpha u - \int_0^\ell K(x,\xi)p_\infty(\xi)\,u(t,\xi)d\xi \equiv 0 \quad \text{in} \quad (0,T_1]\times I \tag{3.2.1}$$

$$u|_{x\equiv 0} \equiv 0;\quad u|_{x\equiv\ell} \equiv 0 \qquad 0 \leq t \leq T_1 = \ell \tag{3.2.2}$$

under the standing assumptions (1.2) for $\alpha : \alpha \in L_\infty(I)$. Moreover, we shall at first assume the following hypothesis (a.1) on the kernel:

(a.1) The kernel is separated

$$K(x,\xi)\,p_\infty(\xi) = k_1(x)k_2(\xi);\quad k_1 \in L_2(0,2\ell),\ k_2 \in L_2(0,\ell) \tag{3.2.3}$$

and the following non-vanishing condition holds true

$$\int_0^\ell k_1(\sigma)\,Q(\sigma)d\sigma \neq 0;\quad q(s) \equiv \int_0^\ell k_1(s+\tau)\frac{k_2(\tau)d\tau}{\rho(\tau)}; \tag{3.2.4}$$

$$\rho(\tau) = e^{\int^\tau \alpha(\xi)d\xi};\quad Q(s) \equiv e^{\int_0^s q(\sigma)d\sigma},\ 0 \leq s \leq \ell\,. \tag{3.2.5}$$

Thus, in particular, the non-vanishing condition (3.2.4) holds true, if k_1 does not change sign on $I = (0,\ell)$, as is the case in the age-structured epidemic model in [B-I-T.1], where the kernel is non-negative.

Theorem 3.2.1. Let $u(t,x)$ be a solution of the over-determined problem (3.2.1), (3.2.2), subject to assumption (a.1): (3.2.3)–(3.2.5), due to $u_0 \in L_2(I)$. Then, in fact

$$u_0 = 0, \quad \text{and hence} \quad u(t,x) \equiv 0,\ t \geq 0,\ 0 \leq x \leq \ell \tag{3.2.6}$$

Proof. Case 1. It will be convenient to first assume that $\alpha \equiv 0$.

Step 1. We solve Eq (3.2.1) by the method of characteristics. The initial curve at $s = 0$ is given parametrically as follows

$$\text{at} \quad s = 0 : t \equiv 0,\ x = \tau,\ u = u_0(\tau),\ 0 \leq \tau \leq \ell \tag{3.2.7}$$

The characteristic equations are

$$\frac{dt(s,\tau)}{ds} \equiv 1; \ \frac{dx(s,\tau)}{ds} \equiv 1; \quad \text{hence} \quad t = s;\ x = s + \tau \quad \text{by (3.2.7)} \tag{3.2.8}$$

$$\frac{du(s,\tau)}{ds} = \int_0^\ell K(x,\xi)p_\infty(\xi)\, u(t,\xi)d\xi = k_1(x) \int_0^\ell k_2(\xi)\, u(s,\xi)d\xi \tag{3.2.9}$$

$$= k_1(s+\tau) \int_0^\ell k_2(\xi)\, u(s,\xi)d\xi,\ 0 \leq s \leq \ell - \tau \tag{3.2.10}$$

after using assumption (3.2.3), and $t = s,\ x = s + \tau \leq \ell$.

Figure 1

Next, multiplying Eq (3.2.9) by $k_2(\tau)$ and integrating in τ over $[0,\ell]$ we obtain, after recalling $q(s)$ (with $\alpha \equiv 0$, hence $\rho \equiv 1$) from (3.2.5)

$$\frac{d}{ds}\left(u(s,\cdot),\ k_2\right)_I = q(s)\left(u(s,\cdot),\ k_2\right)_I \tag{3.2.11}$$

where $(\cdot,\cdot)_I$ denotes the $L_2(0,\ell)$-inner product, $I = (0,\ell)$. Recalling $Q(s)$ from (3.2.5) (with $\alpha \equiv 0$ hence $\rho \equiv 1$), we have that the solution of the linear equation (3.2.11) is

$$\left(u(s,\cdot),k_2\right)_I = (u_0,k_2)_I\, Q(s), \quad u_0 = u(0,\cdot) \quad \text{at} \quad s = 0 \tag{3.2.12}$$

Inserting (3.2.12) into (3.2.10), we obtain

$$\frac{du(s,\tau)}{ds} = k_1(s+\tau)(u_0,k_2)_I\, Q(s),\ 0 \leq s \leq \ell - \tau;\ 0 \leq \tau \leq \ell \tag{3.2.13}$$

Integrating (3.2.13) in s from $s = 0$, along the characteristic base segment τ =const, we obtain an explicit formula for the solution (recall $u(0,\tau) = u_0(\tau)$ at $s = 0$):

$$u(s,\tau) - u_0(\tau) = (u_0, k_2)_I \int_0^s k_1(\sigma+\tau)\, Q(\sigma) d\sigma; \; 0 \le s \le \ell - \tau; \; 0 \le \tau \le \ell \tag{3.2.14}$$

We now evaluate (3.2.14) along the characteristic segment $\tau = 0$, with $s = \ell$ (end of this segment; see Fig. 1)

$$u(\ell,0) - u_0(0) = (u_0, k_2)_I \int_0^\ell k_1(\sigma)\, Q(\sigma) d\sigma \tag{3.2.15}$$

Step 2. Expression (3.2.14) has been obtained by using only Eq. (3.2.1), but not the B.C.s (3.2.2). By using (parts of) the latter, we shall now show that, for $u_0 \in L_2(I)$, we have

$$(u_0, k_2)_I = 0 \tag{3.2.16a}$$

In fact, let first $u_0 \in \mathcal{D}(A)$, see (2.4), so that u_0 satisfies the B.C. at the left-hand point, and $u_0(0) = 0$. Moreover, the B.C. (3.2.2) at the right-hand point implies (see Fig. 1)

$$u(\ell - \tau, \tau) \equiv 0, \; 0 \le \tau \le \ell; \text{ hence } u(\ell, 0) = 0 \tag{3.2.16b}$$

Thus, by the above, the left-hand side of Eq. (3.2.15) vanishes, and the non-vanishing assumption (3.2.4) then yields (3.2.16a), at least for any $u_0 \in \mathcal{D}(A)$. By density, we extend the validity of (3.2.16a) to all $u_0 \in L_2(I)$, as desired.

Step 3. Having established (3.2.16a), we return to Eq. (3.2.14) and obtain

$$u(s,\tau) \equiv u_0(\tau), \; 0 \le s \le \ell - \tau, \; 0 \le \tau \le \ell; \tag{3.2.17}$$

in particular, for $s = \ell - \tau$, recalling (3.2.16b)

$$0 \equiv u(\ell - \tau, \tau) \equiv u_0(\tau), \; 0 \le \tau \le \ell \tag{3.2.18}$$

and the proof of Theorem 3.2.1 is complete, at least for Case 1, under the assumption $\alpha \equiv 0$.

Case 2. The general case with $\alpha \in L_\infty(I)$ will be reduced to Case 1. In fact, multiplying Eq (3.2.1) with a general α, by the integrating factor

$$p(x) = e^{\int^x \alpha(\xi) d\xi}; \quad \text{with} \quad p_x = \alpha p \tag{3.2.19}$$

yields the equation

$$\mu_t + \mu_x - \int_0^\ell \left(\frac{K(x,\xi)\, p_\infty(\xi)}{p(\xi)} \right) \mu(t,\xi) d\xi \equiv 0 \tag{3.2.20}$$

along with the B.C.

$$\mu\,|_{x=0} = (\rho u)|_{x=0}\,; \quad \text{and } \mu\,|_{x=\ell} = (\rho u)|_{x=\ell} \equiv 0,\ 0 \le t \le T_1 = \ell \tag{3.2.21}$$

in the new variable $\mu(t,x) = \rho(x)\, u(t,x)$. Then problem (3.2.20), (3.2.21) is of the type considered under Case 1 in the variable $\mu(t,x)$ just with $p_\infty(\xi)$ replaced by $p_\infty(\xi)/\rho(\xi)$ and with initial condition $\mu_0(x) = \rho(x)\, u_0(x)$ on $I = [0,\ell]$. Then, the proof under Case 1 yields

$$\mu(t,x) \equiv \rho(x)\, u(t,x) \equiv 0, \quad \text{hence} \quad u(t,x) \equiv 0,\ t \ge 0,\ 0 \le x \le \ell \ \text{ as } \ \rho \ne 0 \text{ by (3.2.19)} \tag{3.2.22}$$

as desired. The proof of Theorem 3.2.1 is complete. □

We next present a uniqueness result for the over-determined u-problem (3.2.1), (3.2.2), under a broader class of kernels, as required in [B-I-T.1]; i.e. satisfying the following (a.2) assumption:

(a.2) the kernel satisfies, for some $\varepsilon > 0$:

$$\varepsilon\, k_1(x)\, k_2(\xi) \le K(x,\xi)\, p_\infty(\xi) \le k_1(x)\, k_2(\xi) \tag{3.2.23}$$

$$k_1(x) > 0 \quad \text{for} \quad 0 \le x \le \ell; \quad k_2(\xi) > 0 \quad \text{for} \quad 0 \le \xi \le \ell; \tag{3.2.24}$$

Under (a.2), we shall then show that the over-determined problem (3.2.1), (3.2.2) implies uniqueness within the class of solutions $u(t,x)$ originating from non-negative initial conditions $u_0(x)$ in $L_2(I)$.

Theorem 3.2.2. Let $u(t,x)$ be a solution of the over-determined problem (3.2.1), (3.2.2), subject to assumption (a.2): (3.2.23), (3.2.24), due to the initial condition $u_0(x) \ge 0$, a.e. $0 \le x \le \ell, u_0 \in L_2(I)$. Then, in fact

$$u(t,x) \equiv 0 \quad t \ge 0,\ 0 \le x \le \ell \tag{3.2.25}$$

Proof. We first observe that [B-I-T, 1. Theorem 3.2, p 1070]

$$u_0(x) \ge 0,\ 0 \le x \le \ell, \quad \text{implies} \quad u(t,x) \ge 0,\ t > 0,\ 0 \le x \le \ell \tag{3.2.26}$$

Case 1. It will be convenient to first assume that $\alpha \equiv 0$.

Step 1. We solve Eq (3.2.1) by the method of the characteristics, as in (3.2.7)–(3.2.9) We obtain $t = s, x = s + \tau$, and then (3.2.9) implies now

$$\varepsilon\, k_1(x) \int_0^\ell k_2(\xi)\, u(s,\xi) d\xi \le \frac{du(s,\tau)}{ds} \le k_1(x) \int_0^\ell k_2(\xi)\, u(s,\xi) d\xi \tag{3.2.27}$$

by use of the positivity property (3.2.26) and assumption (3.2.23). We proceed as in the proof of Theorem 3.2.1: multiplying Eq (3.2.27) by the positive $k_2(\tau)$, see assumption (3.2.24), and integrating over $[0,\ell]$, we obtain after recalling $q(s)$ (with $\alpha \equiv 0$, hence $\rho \equiv 1$) from (3.2.5)

$$\varepsilon\, q(s)\, (u(s,\cdot),\ k_2)_I \le \frac{d}{ds}(u(s,\cdot),\ k_2)_I \le q(s)\, (u(s,\cdot),\ k_2) \tag{3.2.28}$$

whose solution is, recalling $Q(s)$ (with $\alpha \equiv 0$) from (3.2.5, $\rho \equiv 1$)

$$v_0\, Q^{\varepsilon}(s) \le v(s) \equiv (u(s,\cdot),\ k_2)_I \le v_0\, Q(s),\ v_0 = (u_0, k_2)_I \tag{3.2.29}$$

Using (3.2.29) on both the left- and the right-hand side of inequality (3.2.27), we obtain since $x = s + \tau$:

$$\varepsilon\, k_1(s+\tau) v_0\, Q^{\varepsilon}(s) \le \frac{du(s,\tau)}{ds} \le k_1(s+\tau) v_0\, Q(s),\ \ 0 \le s \le \ell - \tau;\ 0 \le \tau \le \ell \tag{3.2.30}$$

Integrating (3.2.30) in s from $s = 0$, we obtain, since $u(0,\tau) = u_0(\tau)$ by (3.2.7)

$$\varepsilon\, v_0 \int_0^s k_1(\sigma+\tau)\, Q^{\varepsilon}(\sigma) d\sigma \le u(s,\tau) - u_0(\tau) \le v_0 \int_0^s k_1(\sigma+\tau)\, Q(\sigma) d\sigma,\ \ 0 \le s \le \ell - \tau;\ 0 \le \tau \le \ell \tag{3.2.31}$$

We now evaluate (3.2.31) along the characteristic segment $\tau = 0$, with $s = \ell$ at the end of this segment (see Fig. 1).

$$\varepsilon\, v_0 \int_0^{\ell} k_1(\sigma)\, Q^{\varepsilon}(\sigma) d\sigma \le u(\ell, 0) - u_0(0) \le v_0 \int_0^{\ell} k_1(\sigma)\, Q(\sigma) d\sigma \tag{3.2.32}$$

Step 2. Equations (3.2.31) and (3.2.32) are the counterpart of Eqs. (3.2.14) and (3.2.15) in the proof of Theorem 3.2.1. In particular, both (3.2.31) and (3.2.32) use only Eq. (3.2.1), but not the B.C.s (3.2.2). By using now (part of) these B.C.s, we shall show that, for $u_0 \in L_2(I)$ $[u_0(\tau) \ge 0,\ 0 \le \tau \le \ell]$ we have (recall (3.2.29)):

$$v_0 = (u_0, k_2)_I = 0 \tag{3.2.33}$$

In fact, let first $u_0 \in \mathcal{D}(A)$, thus $u_0(0) = 0$ by (2.4). Moreover, the B.C. (3.2.2) at $x = \ell$ yields

$$u(\ell - \tau, \tau) \equiv 0,\ \ 0 \le \tau \le \ell;\quad \text{hence}\quad u(\ell, 0) = 0 \tag{3.2.34}$$

Thus, with $u(\ell, 0) = u_0(0) = 0$, we obtain from both sides of (3.2.32) that $v_0 \le 0$, as well as $v_0 \ge 0$, since the integral terms are positive. We conclude that $v_0 = 0$, and (3.2.33) holds true at least for all $u_0 \in \mathcal{D}(A)$. By density, we extend the validity of (3.2.33) to all $u_0 \in L_2(I)$ [with $u_0(\tau) \ge 0,\ 0 \le \tau \le \ell$], as desired.

Step 3. Having established (3.2.33), we return to (3.2.31) and readily obtain

$$u(s,\tau) \equiv u_0(\tau),\ 0 \le s \le \ell - \tau;\ \ 0 \le \tau \le \ell;\quad \text{hence}\quad 0 \equiv u(\ell - \tau, \tau) = u_0(\tau),\ 0 \le \tau \le \ell \tag{3.2.35}$$

recalling (3.2.34), and the proof of Theorem 3.2.2 is complete, at least for Case 1, under the assumption $\alpha \equiv 0$.

Case 2. The general case with $\alpha \in L_\infty(I)$ is reduced to Case 1, as in the proof of Theorem 3.2.1, by the integrating factor ρ in (3.2.19) yielding problem (3.2.20), (3.2.21) with I.C. $\mu_0(x) = \rho(x)\, u_0(x) \geq 0$ on I, since $u_0(x) \geq 0$ on I. Then, (3.2.22) follows, as desired. □

Dual problem. With reference now to the dual (adjoint) overdetermined problem (see (2.27a-b-c)), or (3.1.32); and (2.16) for V^*)

$$\psi_t - \psi_x + \alpha\psi - p_\infty(x) \int_0^\ell K(\xi, x) u(t, \xi) d\xi \equiv 0 \quad \text{in } (0, T_1) \times I \tag{3.2.36}$$

$$\psi\,|_{x=\ell} \equiv 0; \quad \psi|_{x=0} \equiv 0, \;\; 0 \leq t \leq T = \ell \tag{3.2.37}$$

where we recall that $p_\infty(x) \geq 0$ [B-I-T.1]. As in the original problem, we shall at first assume the following hypothesis (b.1) on the kernel

(b.1) the kernel is separated

$$K(\xi, x)\, p_\infty(x) = k_1(\xi)\, k_2(x); \quad k_1 \in L_2(0, \ell), \; k_2 \in L_2(-\ell, \ell) \tag{3.2.38}$$

and the following non-vanishing condition holds true

$$\int_0^\ell k_2(-\sigma + \ell)\, \overline{Q}(\sigma) d\sigma \neq 0; \quad \overline{q}(s) = \int_0^\ell k_2(-s + \tau) \frac{k_1(\tau)}{\overline{\rho}(\tau)} d\tau \tag{3.2.39}$$

$$\overline{\rho}(x) = e^{-\int^x \alpha(\xi) d\xi}; \quad \overline{Q}(s) = \int_0^s e^{\overline{q}(\sigma)} d\sigma \tag{3.2.40}$$

Thus, in particular, the non-vanishing condition (3.2.39) holds true, if k_2 does not change sign on $I = (0, \ell)$, as is the case in the age-structured epidemic model in [B-I-T.1], where the kernel is non-negative.

Theorem 3.2.3. Let $\psi(t, x)$ be a solution of the over-determined problem (3.2.36), (3.2.37), subject to assumption (b.1): (3.2.38)–(3.2.40), due to the I.C. $\overline{\psi}_0 \in L_2(I)$ at $t = 0$. Then, in fact,

$$\overline{\psi}_0 = 0, \quad \text{hence} \quad \psi(t, x) \equiv 0 \quad t \geq 0, \quad 0 \leq x \leq \ell, \tag{3.2.41}$$

Proof. It is similar to that of Theorem 3.2.1. First for $\alpha \equiv 0$, the method of characteristics with initial curve at $s = 0$ as in (3.2.7) yields for $0 \leq s \leq \tau, 0 \leq \tau \leq \ell : dt(s, \tau)/ds \equiv 1, \; dx(s, \tau)ds \equiv -1$, hence $t = s, x = -s + \tau$.

$$\frac{d\psi(s, \tau)}{ds} = p_\infty(x) \int_0^\ell K(\xi, x)\, \psi(t, \xi) d\xi = k_2(-s + \tau) \int_0^\ell k_1(\xi)\, \psi(s, \xi) d\xi \tag{3.2.42}$$

counterpart of (3.2.10). We now multiply (3.2.42) by $k_1(\tau)$ and integrate in τ over $[0, \ell]$, thus obtaining

$$\frac{d(\psi(s, \cdot), k_1)_I}{ds} = \overline{q}(s)\; (\psi(s, \cdot), k_1)_I; \;\; (\psi(s, \cdot), k_1)_I = (\overline{\psi}_0, k_1)_I\, \overline{Q}(s), \;\; \psi(0, \tau) \equiv \overline{\psi}_0(\tau) \tag{3.2.43}$$

after recalling $\overline{q}(s)$ and $\overline{Q}(s)$ (for $\alpha \equiv 0$) from (3.2.39), (3.2.40). Inserting (3.2.43) (right) into (3.2.42) and integrating in s yields an explicit solution formula

$$\psi(s,\tau) - \overline{\psi}_0(\tau) = (\overline{\psi}_0, k_1)_I \int_0^s k_2(-\sigma+\tau)\,\overline{Q}(\sigma)d\sigma,\ 0 \le s \le \tau, 0 \le \tau \le \ell \qquad (3.2.44)$$

counterpart of (3.2.14)

Figure 2

Now we use (part of) the B.C.s (3.2.37). We evaluate (3.2.44) along the characteristic segment $\tau = \ell$, and $s = \ell$ to get

$$\psi(\ell,\ell) - \overline{\psi}_0(\ell) = (\overline{\psi}_0, k_1)_I \int_0^\ell k_2(-\sigma+\ell)\,\overline{Q}(\sigma)d\sigma \qquad (3.2.45)$$

counterpart of (3.2.15), where $\psi(\ell,\ell) = 0$ by the B.C. at $x = 0$ (see Fig. 2), while $\overline{\psi}_0(\ell) = 0$, at least for $\overline{\psi}_0 \in \mathcal{D}(A^*)$ (recall (2.17)). Thus, the left-hand side of (3.2.45) vanishes, and the non-vanishing assumption (3.2.39) then yields $(\overline{\psi}_0, k_1)_I = 0$, at first for $\overline{\psi}_0 \in \mathcal{D}(A^*)$, next for all $\overline{\psi}_0 \in L_2(I)$ by density. Then (3.2.44) yields

$$\psi(s,\tau) = \overline{\psi}_0(\tau),\ 0 \le s \le \tau, 0 \le \tau \le \ell;\ \text{hence } 0 \equiv \psi(\tau,\tau) = \overline{\psi}_0(\tau),\ 0 \le \tau \le \ell \qquad (3.2.46)$$

this time exploiting in full the B.C. at $x = 0$, see Fig. 2. Thus Theorem 3.2.3 is proved, at least of $\alpha \equiv 0$.

The case of a general $\alpha \in L_\infty(I)$ is reduced to the preceding case, as in the proof of Theorem 3.2.1, this time in the

variable $\overline{\mu}(t,x) = \overline{p}(x)\,\psi(t,x)$, with $\overline{p}(x)$ as in (3.2.40). Details are referred to the proof of Theorem 3.2.1. □

A uniqueness result counterpart of Theorem 3.2.2 holds true also for the dual problem (3.2.36) with I.C. $\overline{\psi}_0(\tau) \ge 0$ mutatis mutandis, but we shall omit it for lack of space.

3.3 Absorption of V-integral term from estimate (3.1.10) [resp. of V^*-integral term from estimate (3.1.34)] of original u-problem [resp. dual ψ-problem]. Final Continuous Observability estimate.

We return to the original u-equation (3.1.1) supplemented this time by the B.C at $x = 0$, as well as to the dual ψ-problem (2.27a-b-c):

$$\begin{cases} u_t + u_x + \alpha u - Vu \equiv 0 & \text{in } Q \\ u(0,\cdot) = u_0(\cdot) & \text{in } I \\ u|_{x=0} \equiv 0 & \text{in } (0,T)\times\{0\} \end{cases} \qquad \begin{cases} \psi_t - \psi_x + \alpha\psi - V^*\psi \equiv 0 & \text{in } Q \\ \psi(T,\cdot) = \psi_0 & \text{in } I \\ \psi|_{x=\ell} \equiv 0 & \text{in } (0,T)\times\{\ell\} \end{cases}$$

(3.3.1a-b-c)

We first specialize Theorem 3.1.2, Eq. (3.1.10a), as well as Theorem 3.1.5, Eq. (3.1.34), to the u- and ψ-problems (3.3.1a-b-c) by using the B.C. (3.3.1c).

Theorem 3.3.1. Assume (1.2), (1.3) for α, V.

(i) With reference to the u-problem (3.3.1), Eq (3.1.10a) of Theorem 3.1.2 specializes to

$$\bar{C}_{\varepsilon,\phi,T}\int_0^T u^2(t,\ell)dt + C_{\varepsilon,\phi,T}\int_0^T \|Vu\|^2_{L_2(I)}\,dt \geq (1-c-\varepsilon)\,e^{\tau p}k_{t_0,t_1}\int_0^\ell u^2(0,x)dx \tag{3.3.2}$$

(ii) Similarly, with reference to the ψ-problem (3.3.1), Eq (3.1.34b) of Theorem 3.1.5 specializes to

$$\bar{C}_{\varepsilon,\phi,T}\int_0^T \psi^2(t,0)dt + C_{\varepsilon,\phi,T}\int_0^T \|V^*\psi\|^2_{L_2(I)}\,dt \geq (1-c-\varepsilon)\,e^{\tau p}k_{t_0,t_1}\int_0^\ell \psi^2(T,x)dx \tag{3.3.3}$$

Proof. We use the B.C. $u\,|_{x=0}\equiv 0$, respectively, $\psi|_{x=\ell}\equiv 0$ in (3.3.1c) in the definition of the boundary terms $(\overline{BT})\,|_\Sigma$ and $\wedge(T)$ for u and ψ, in (3.1.11), (3.1.25). □

We next eliminate the V- and V^*-integral terms from estimates (3.3.2) and (3.3.3), respectively. They are *not* lower order terms.

Theorem 3.3.2. Assume (1.2), (1.3) for α, V. Moreover, assume that the kernel $K(x,\xi)p_\infty(\xi)\in L_2((0,\ell)\times(0,\ell))$ (so that the operator V is Hilbert-Schmidt, hence compact, $L_2(I)\rightarrow L_2(I)$).

(i) Assume, moreover, that the over-determined u-problem (3.2.1), (3.2.2) admits the unique trivial solution $u\equiv 0$ in $t\geq 0, 0\leq x\leq \ell$. [A sufficient condition for this uniqueness property to hold true is given by Theorem 3.2.1.] Then, with reference to (3.3.2), there exists a constant $C_T>0$, such that

(i1) $$\|Vu\|^2_{L_2(0,T;\,L_2(I))}\leq C_T\int_0^T u^2(t,\ell)dt = C_T\,\|u(\cdot,\ell)\|^2_{L_2(0,T)} \tag{3.3.4}$$

and hence estimate (3.3.2) for the u-problem (3.3.1) simplifies to

(i2) $$\int_0^T u^2(t,\ell)dt \geq \text{const}_{\phi,T,\varepsilon}\int_0^\ell u^2(0,x)dx \tag{3.3.5}$$

(ii) Assume, moreover, that the over-determined ψ-problem (3.2.36), (3.2.37) admits the unique trivial solution $\psi\equiv 0$ in $t\geq 0,\ 0\leq x\leq\ell$. [A sufficient condition for this uniqueness property to hold true is given by Theorem 3.2.3.] Then, with reference to (3.3.3), there exists a constant $C_T>0$, such that

(ii1) $$\|V^*\psi\|^2_{L_2(0,T;\,L_2(I))}\leq C_T\int_0^T \psi^2(t,0)dt = c_T\,\|\psi(\cdot,0)\|^2_{L_2(0,T)} \tag{3.3.6}$$

and hence estimates (3.3.3) for the ψ-problem (3.3.1) simplifies to

$$\int_0^T \psi^2(t,0)dt \geq \text{const}_{\phi,T,\varepsilon}\int_0^\ell \psi^2(T,x)dx \tag{3.3.7}$$

[In Section 3.4 we shall see that estimate (3.3.7) is a *continuous observability* inequality.]

Proof. (i1) By contradiction, let there be a sequence $\{u_n\}$ of solutions of the u-problem (3.3.1) such that

$$\|Vu_n\|^2_{L_2(0,T;\ L_2(I))} \equiv 1; \quad \text{and} \quad \|u_n(\cdot,\ell)\|^2_{L_2(0,T)} \to 0 \tag{3.3.8}$$

Then, since each u_n satisfies estimate (3.3.2), we then obtain from (3.3.8) that

$$\int_0^\ell u_n^2(0,x)dx = \|u_n(0,\cdot)\|^2_{L_2(I)} \leq \text{const}, \quad \forall n\,. \tag{3.3.9}$$

By semigroup well-posedness of the u-problem (3.3.1) [recall the s.c. semigroup e^{At} below (2.4)], (3.3.9) implies

$$\|u_n(\cdot)\|_{C([0,T];\ L_2(I))} = \left\|e^{A\cdot}u_n(0,\cdot)\right\|_{C([0,T];\ L_2(I))} \leq C_T\,\|u_n(0,\cdot)\|_{L_2(I)} \leq \text{const}_T \quad \forall n \tag{3.3.10}$$

Thus, by (3.3.10), there is a subsequence, still called u_n, such that

$$u_n \to \text{ some } \overline{u}, \text{ weakly in } L_2(0,T;\ L_2(I)); \text{ and weak}^* \text{ in } L_\infty(0,T;\ L_2(I)) \tag{3.3.11}$$

By a limit argument on (3.3.1), we then obtain that $\overline{u}$ satisfies problem (3.3.1), i.e.,

$$\overline{u}_t + \overline{u}_x + \alpha\overline{u} - V\overline{u} \equiv 0 \text{ in } Q \tag{3.3.12a}$$

$$\overline{u}\,|_{x=0} \text{ in } (0,T)\times\{0\}; \text{ as well as } \overline{u}|_{x=\ell} = 0, \quad 0 \leq t \leq T \tag{3.3.12b}$$

recalling the limit to zero in (3.3.8). Since $T > 2\ell$ (recall the statement below (3.1.4a), we conclude by the assumed uniqueness property of the over-determined problem (3.2.1), (3.2.2)—a fortiori of (3.3.12)—that, in fact, $\overline{u} = 0, t \geq 0$.

On the other hand, by the $L_2(I)$-weak convergence of u_n to $\bar{u} = 0$ at each $t \in (0,T]$ noted in (3.3.11), we have since $K(x,\cdot)p_\infty(\cdot) \in L_2(0,\ell)$ a.e. in x:

$$\left|\int_0^\ell K(x,\xi)\,p_\infty(\xi)\,u_n(t,\xi)d\xi\right| \to 0 \quad \text{a.e. in } x, \text{ and for all } t \in (0,T] \tag{3.3.13}$$

Moreover, under present assumption that the kernel is in $L_2(0,\ell)\times(0,\ell)$, we have by the Schwarz inequality

$$\left|\int_0^\ell K(x,\xi)\,p_\infty(\xi)\,u_n(t,\xi)d\xi\right| \leq G(x)\int_0^\ell u_n^2(t,\xi)d\xi \leq C_T G(x) \tag{3.3.14}$$

by recalling (3.3.10) on the last step, where $G(x) \in L_1(0,\ell)$

$$G(x) = \int_0^\ell K^2(x,\xi)\,p_\infty^2(\xi)d\xi; \ \int_0^\ell G(x)dx = \int_0^\ell\int_0^\ell K^2(x,\xi)\,p_\infty^2(\xi)d\xi\,dx < \infty \tag{3.3.15}$$

Then the Lebesgue Dominated Convergence Theorem applies and yields from (3.3.8), (1.3):

$$1 \equiv \|Vu_n\|^2_{L_2(0,T;\, L_2(I))} = \int_0^T \int_0^\ell \left| \int_0^\ell K(x,\xi)\, p_\infty(\xi)\, u_n(t,\xi) d\xi \right|^2 dx dt \to 0 \tag{3.3.16}$$

by virtue of (3.3.13), a contradiction. Thus, part (i1), Eq (3.3.4) is proved.

(ii1) The proof is identical for (3.3.6). □

3.4. **Exact controllability of the linear problem** (2.1) **on** $L_2(I)$ **and on** $\mathcal{D}(A)$.

In this section we return to the continuous operator L_T in (2.22) and investigate its surjectivity in two settings.

3.4.1. **Exact controllability of the linear problem** (2.1) **on** $L_2(I)$, **using** $L_2(0,T)$**-controls.**

The inequalities (3.4.3), (3.4.4) below are the reverse inequalities of (2.25), (2.26), respectively.

Proposition 3.4.1. Assume (1.2), (1.3). The following conditions are equivalent:

(i) (exact controllability) the (continuous) operator L_T in (2.22) satisfies

$$L_T : L_2(0,T) \quad \text{onto } L_2(I)\,; \tag{3.4.1}$$

(ii) (continuous observability) There is a constant $c_T > 0$ such that

$$|L_T^* z|_{L_2(0,T)} \geq c_T \|z\|_{L_2(I)} \qquad \forall z \in L_2(I) \tag{3.4.2}$$

where L_T^* is the L_2-adjoint of L_T defined by (2.24); explicitly, by (2.24)

$$\int_0^T \left| D^* A^* e^{A^* t} z \right|^2 dt \geq c_T \|z\|^2_{L_2(I)} \tag{3.4.3}$$

(iii) in P.D.E. terms, if $\psi(t;\psi_0) = e^{A^*(T-t)}\psi_0$ is the solution of problem (2.27), then

$$\int_0^T |\psi(t;\psi_0)|_{x=0}|^2\, dt \geq c_T \int_0^\ell |\psi_0(x)|^2\, dx \tag{3.4.4}$$

(iv) given $T > 0, u_0 \in L_2(I)$ and $u_T \in L_2(T)$, there exists a control $g \in L_2(0,T)$ such that the corresponding solution of the linear problem (2.1) satisfies: $u(T) = u_T$. In particular, one can use the $L_2(0,T)$-minimal norm control. [L-T.2, Appendix].

Theorem 3.4.2. (Exact controllability on $L_2(I)$) Assume the hypotheses of Theorem 3.3.2(ii). Then, all equivalent conditions of Proposition 3.4.1 hold true.

Proof. Theorem 3.3.2(ii), Eq. (3.3.7) proves the continuous observability estimate (3.4.4). □

3.4.2. **Exact controllability of the linear problem** (2.1) **on** $\mathcal{D}(A)$, **in particular on** $H_0^1(I)$, **using** $H_0^1(0,T)$**-controls.** We recall from (2.4) that $H_0^1(I) \subset \mathcal{D}(A)$, since $f \in H_0^1(I)$ implies $f|_{x=0} = f|_{x=\ell} = 0$.

Claim: let $g \in H_0^1(0,T)$, so that $g(0) = g(T) = 0, \dot{g} \in L_2(0,T)$; then, recalling (2.22)

$$L_T g = -A \int_0^T e^{A(T-t)} Dg(t)dt = \int_0^T \frac{de^{A(T-t)}}{dt} Dg(t)dt = -\int_0^T e^{A(T-t)} \dot{g}(t)dt \in \mathcal{D}(A) \tag{3.4.5}$$

where the indicated regularity follows by (2.23): i.e.

$$AL_T g = L_T \dot{g}, \; g \in H_0^1(0,T); \; L_T : \text{continuous } H_0^1(0,T) \to \mathcal{D}(A) \tag{3.4.6}$$

We now investigate the surjectivity of the map (3.4.6). To this end, we need the corresponding adjoint operator $L_T^\# : \mathcal{D}(A) \to H_0^1(0,T)$; for $g \in H_0^1(0,T), \; z \in \mathcal{D}(A)$, $L_T^\#$ is defined by

$$(L_T g, z)_{\mathcal{D}(A)} = (g, L_T^\# z)_{H_0^1(0,T)} = \left(\dot{g}, \frac{d}{dt}(L_T^\# z) \right)_{L_2(0,T)} \tag{3.4.7}$$

where we take the gradient norm on $H_0^1(0,T)$ throughout.

Lemma 3.4.3. With reference to (3.4.7), we have

$$\frac{d}{dt}(L_T^\# z)(t) = (L_T^* Az)(t) \in L_2(0,T); \quad (L_T^\# z)(t) = \int^t (L_T^* Az)(\tau) d\tau \tag{3.4.8}$$

where L_T^* is the L_2-adjoint given in (2.24).

Proof. We compute with $g \in H_0^1(0,T)$ and $z \in \mathcal{D}(A)$, recalling (3.4.6) and (3.4.7)

$$\begin{aligned} (L_T g, z)_{\mathcal{D}(A)} &= (AL_T g, Az)_{L_2(I)} = (L_T \dot{g}, Az)_{L_2(I)} = \\ \text{(by (3.4.7))} &= (\dot{g}, L_T^* Az)_{L_2(0,T)} = \left(\dot{g}, \frac{d}{dt}(L_T^\# z) \right)_{L_2(0,T)} \end{aligned} \tag{3.4.9}$$

and (3.4.8) follows from (3.4.9). □

The desired surjectivity characterization of the map L_T in (3.4.6) is given next.

Proposition 3.4.4. The following conditions are equivalent:

(i) (exact controllability) the continuous operator L_T in (3.4.6) satisfies

$$L_T : H_0^1(0,T) \text{ onto } \mathcal{D}(A) ; \tag{3.4.10}$$

(ii) (continuous observability) there is a constant $c_T > 0$ such that

$$\left| L_T^\# z \right|_{H_0^1(0,T)} \geq c_T \|z\|_{\mathcal{D}(A)} \tag{3.4.11}$$

(iii) By (3.4.8), we may re-write (3.4.11) as

$$\left| L_T^\# z \right|_{H_0^1(0,T)} = \left| \frac{d}{dt}(L_T^\# z) \right|_{L_2(0,T)} = |L_T^* Az|_{L_2(0,T)} \geq c_T \|z\|_{\mathcal{D}(A)} = c_T \|Az\|_{L_2(I)} \tag{3.4.12}$$

Since A may be taken to be an isomorphism $\mathcal{D}(A) \to L_2(I)$ (without loss of generality we are taking $\lambda_0 = 0$, see (2.14)), then (3.4.12) says that: inequality (3.4.11) is *equivalent* to inequality (3.4.2) hence to inequality (3.4.4); or surjectivity (3.4.10) is *equivalent* to surjectivity (3.4.1).

(iv) Exact controllability (from the origin; to the origin; from point to point) on $\mathcal{D}(A)$ within the class of $H_0^1(0,T)$-controls is *equivalent* to exact controllability (from the origin; to the origin; from point to point) on $L_2(I)$ within the class of $L_2(0,T)$-controls. □

Theorem 3.4.5. (Exact controllability on $\mathcal{D}(A)$). Assume the hypotheses of Theorem 3.3.2(ii).

Then, all equivalent conditions of Proposition 3.4.4 hold true.

Proof. Use Theorem 3.4.2 and Proposition 3.4.4. □

4 Part II: Nonlinear problem (1.1). Local solution in $H^1(I)$.

We introduce the Banach space

$$X_T \equiv C\left([0,T];\ H^1(I)\right) \cap H^1(0,T;L_2(I)) \subset H^{1,1}(Q),\ Q = (0,T] \times I, \tag{4.1a}$$

$$\|f\|^2_{X_T} \equiv \|f\|^2_{C([0,T];\ H^1(I))} + \|f_t\|^2_{L_2(0,T;\ L_2(I))} \tag{4.1b}$$

which will provide the setting for the local solvability of the non-linear problem (1.1).

Theorem 4.1. Assume (1.2), (1.3). With reference to the non-linear problem (1.1), let $u_0 \in \mathcal{D}(A)$, $g \in H_0^1(0,T)$, with $T > 0$ preassigned. There exists a constant $r_T > 0$ depending on $T > 0$, and explicitly given in (4.36) below, such that, if

$$\|u_0\|_{\mathcal{D}(A)} \leq r_T\,; \qquad \|g\|_{H_0^1(0,T)} \leq r_T \tag{4.2}$$

then the non-linear problem (1.1) admits a unique solution $u \in X_T$, see (4.1), in a ball $\mathcal{B}(0,R) \subset X_T$, with radius $R < 1$. This solution satisfies the integral equation

$$u(t) = e^{At}u_0 + (Lg)(t) - \int_0^t e^{A(t-s)}\left[ku^2(s) + u(s)\,(Vu)\,(s)\right] ds \tag{4.3}$$

Proof. Orientation. The proof is by fixed point (contraction mapping principle) on the space X_T in (4.1). With $u_0 \in \mathcal{D}(A)$, $g \in H_0^1(0,T)$ given, construct the map $v \to u \equiv \mathcal{T}_{u_0,g}v$, for v running over X_T, defined as the unique solution of the linear mixed problem

$$u_t + u_x + \alpha(x)u - Vu = -k(x)v^2 - v(Vv) \quad \text{in } (0,T] \times I = Q \tag{4.4a}$$

$$u\,|_{t=0} = u_0 \quad \text{in } I;\ u|_{x=0} = g \quad \text{in } (0,T] \times \{0\} \tag{4.4b}$$

Thus, such map $u \equiv \mathcal{T}_{u_0,g}v$ is given explicitly via (2.3), (2.4) by

$$u(t) = (\mathcal{T}_{u_0,g}v)(t) = e^{At}u_0 + (Lg)(t) - \int_0^t e^{A(t-s)}\left[kv^2(s) + v(s)(Vv)(s)\right] ds \tag{4.5}$$

recalling L in (2.15). Let $\mathcal{B}(0, R)$ be the closed ball of the space X_T, centered at 0 and of radius $R > 0$. We shall prove that, if u_0 and g satisfy the local condition (4.2) for some suitable r_T, depending on T, then:

$$\mathcal{T}_{u_0,g} \text{ as a map from the ball } \mathcal{B}(0,R) \subset X_T \text{ into } \mathcal{B}(0,R) \subset X_T \text{ admits a unique} \tag{4.6}$$

fixed point for a suitable value of $R_T > 0$, depending on T.

We shall achieve objective (4.6) with R_T given in (4.36) in the next two Propositions, each dealing with a component space of X_T in (4.1).

Step 1. Proposition 4.2. Assume the hypotheses of Theorem 4.1.

With reference to problem (4.5), or the map $\mathcal{T}$ in (4.6), let $u_0 \in \mathcal{D}(A)$, $g \in H_0^1(0,T)$, and let $v \in X_T$. Then

$$\left\|\mathcal{T}_{u_0,g}v\right\|_{C([0,T];\, H^1(I))} \leq C_T\left\{\|v\|^2_{X_T} + \|u_0\|_{\mathcal{D}(A)} + \|g\|_{H_0^1(0,T)}\right\} \tag{4.7}$$

Proof. We shall deal with each of the three terms defining $\mathcal{T}_{u_0,g}$ in (4.5) separately.

(i) We begin with the last integral term: we shall show that if $v \in X_T$, then

$$\left\|\int_0^t e^{A(t-s)}\left[kv^2(s) + v(s)(Vv)(s)\right] ds\right\|_{C([0,T];H^1(I))} \leq C_T\,\|v\|^2_{X_T} \tag{4.8}$$

To this end, let $f \in X_T$ given by (4.1). Then, integrating by parts

$$\int_0^t e^{A(t-s)}f(s)ds = -A^{-1}\int_0^t \frac{d\left(e^{A(t-s)}\right)}{ds}f(s)ds = -A^{-1}f(t) + A^{-1}e^{At}f(0) \tag{4.9}$$

$$+ A^{-1}\int_0^t e^{A(t-s)}f'(s)ds \in C\left([0,T];\mathcal{D}(A)\right) \tag{4.10}$$

Specializing (4.9) to $f(s) = [kv^2(s) + v(s)(Vv)(s)]$, with $v \in X_T$, we obtain

$$\begin{aligned} A\int_0^t e^{A(t-s)}\left[kv^2(s) + v(s)(Vv)(s)\right] ds \;=\; & -\left[kv^2(t) + v(t)(Vv)(t)\right] \\ & + e^{At}\left[kv^2(0) + v(0)(Vv)(0)\right] \end{aligned}$$

$$+ \int_0^t e^{A(t-s)}\left[2kv(s)v_t(s) + v_t(s)(Vv)(s) + v(s)(Vv_t)(s)\right] ds \in C\left([0,T];L_2(I)\right) \tag{4.11}$$

Next, from (2.4), we see that if $h \in \mathcal{D}(A)$, then the $H^1(I)$-norm of h and the $\mathcal{D}(A)$-norm of h are equivalent:

$$h \in \mathcal{D}(A) \Rightarrow \|h\|_{H^1(I)} \text{ equivalent to } \|h\|_{\mathcal{D}(A)} \tag{4.12}$$

Thus, by (4.12) and (4.11), we obtain (4.8), as desired, via (1.2), (1.3)

$$\left\| \int_0^t e^{A(t-s)} \left[kv^2(s) + v(s)(Vv)(s)\right] ds \right\|_{C([0,T];H^1(I))}$$

$$\leq \left\| A \int_0^t e^{A(t-s)} \left[kv^2(s) + v(s)(Vv)(s)\right] ds \right\|_{C([0,T];L_2(I))} \tag{4.13}$$

$$\text{(by (4.11))} \quad \leq C_T \left\{ \|v\|_{C([0,T];H^1(I))} \left[\|v\|_{C([0,T];L_2(I))} + \|v_t\|_{L_2(0,T;L_2(I))} \right] \right\} \tag{4.14}$$

$$\text{(by (4.1))} \quad \leq C_T \|v\|_{X_T} \|v\|_{X_T} . \tag{4.15}$$

(ii) Let $u_0 \in \mathcal{D}(A)$ so that $e^{At}u_0 \in \mathcal{D}(A)$ and (4.12) yields

$$\|e^{A\cdot}u_0\|_{C([0,T];H^1(I))} \leq C_T \|Ae^{A\cdot}u_0\|_{C([0,T];L_2(I))} \leq C_T \|Au_0\|_{L_2(I)} = C_T \|u_0\|_{\mathcal{D}(A)} \tag{4.16}$$

(iii) Finally, let $g \in H_0^1(0,T)$, then with reference to (2.15), we have

$$\|Lg\|_{C([0,T];H^1(I))} \leq C_T \|g\|_{H_0^1(0,T)} \tag{4.17}$$

In fact, since $g(0) = 0$, integrating by parts (as in (3.4.5)) we obtain from (2.15), we have

$$(Lg)(t) = \int_0^t \frac{d\left(e^{A(t-s)}\right)}{ds} Dg(s)ds = -Dg(t) + \int_0^t e^{A(t-s)} D\dot{g}(s)ds \tag{4.18}$$

$$= -Dg(t) - A^{-1}(L\dot{g})(t) . \tag{4.19}$$

By the regularity of D in (2.8), we have

$$\|Dg\|_{C([0,T];H^1(I))} \leq C \|g\|_{C[0,T]} \tag{4.20}$$

Moreover, by the regularity of L in (2.21) and by (4.12), we have

$$\|A^{-1}(L\dot{g})\|_{C([0,T];H^1(I))} \equiv C \|A^{-1}(L\dot{g})\|_{C([0,T];\mathcal{D}(A))} = C \|L\dot{g}\|_{C([0,T];L_2(I))} \tag{4.21}$$

$$\text{(by (2.21))} \quad \leq C_T \|\dot{g}\|_{L_2(0,T)} \tag{4.22}$$

Using (4.20) and (4.22) in (4.19) yields (4.17), as desired.

(iv) Finally, the regularities in (4.8), (4.16), and (4.17) used in the definition (4.5) of $\mathcal{T}_{u_0,g}$, yield estimate (4.7), as desired. □

Step 2. Proposition 4.3. Assume the hypotheses of Theorem 4.1.

With reference to problem (4.5), or the map $\mathcal{T}$ in (4.6), let $u_0 \in \mathcal{D}(A)$, $g \in H_0^1(0,T)$, and let $v \in X_T$. Then

$$\|\mathcal{T}_{u_0,g}v\|_{H^1(0,T);L_2(I))} \leq C_T\left\{\|v\|_{X_T}^2 + \|u_0\|_{\mathcal{D}(A)} + \|g\|_{H_0^1(0,T)}\right\} \tag{4.23}$$

Proof. We shall deal with each of the three terms defining $\mathcal{T}_{u_0,g}$ in (4.5) separately.

(a) Let $f \in H_0^1(0,T;L_2(I))$. Then, setting $t - s = \sigma$:

$$\frac{d}{dt}\int_0^t e^{A(t-s)}f(s)ds = \frac{d}{dt}\int_0^t e^{A\sigma}f(t-\sigma)d\sigma = e^{At}f(0) + \int_0^t e^{A(t-s)}f'(t-\sigma)d\sigma \tag{4.24}$$

$$= e^{At}f(0) + \int_0^t e^{A(t-s)}f'(s)ds \tag{4.25}$$

Specializing (4.25) to $f(s) = [kv^2(s) + v(s)(Vv)(s)]$ for $v \in X_T$ yields

$$\frac{d}{dt}\int_0^t e^{A(t-s)}\left[kv^2(s) + v(s)(Vv)(s)\right]ds = e^{At}\left[kv^2(0) + v(0)(Vv)(0)\right]$$

$$+ \int_0^t e^{A(t-s)}\left[2kv(s)v_t(s) + v_t(s)(Vv)(s) + v(s)(Vv_t)(s)\right]ds \tag{4.26}$$

Thus, (as in going from (4.13) to (4.14) using (4.11)), we obtain from (4.26)

$$\left\|\int_0^t e^{A(t-s)}\left[kv^2(s) + v(s)(Vv)(s)\right]ds\right\|_{H^1(0,T);L_2(I))}$$

$$\leq C_T\left\{\|v\|_{C([0,T];H^1(I))}\left[\|v\|_{C([0,T];L_2(I))} + \|v_t\|_{L_2(0,T;L_2(I))}\right]\right\} \tag{4.27}$$

$$\leq C_T\|v\|_{X_T}\,\|v\|_{X_T} \tag{4.28}$$

(b) Let $g \in H_0^1(0,T)$, so that $g(0) = 0$. Then, we obtain from (2.15)

$$\frac{d}{dt}(Lg)(t) = \frac{d}{dt}\left(-A\int_0^t e^{A(t-s)}Dg(s)ds\right) = \frac{d}{dt}\left(-A\int_0^t e^{A\sigma}Dg(t-\sigma)d\sigma\right) \tag{4.29}$$

$$= 0 - A\int_0^t e^{A\sigma}D\dot{g}(t-\sigma)d\sigma = -A\int_0^t e^{A(t-s)}D\dot{g}(s)ds = (L\dot{g})(t) \tag{4.30}$$

Thus, recalling the regularity of L in (2.21), we obtain from (4.30)

$$\|Lg\|_{H^1(0,T;L_2(I))} \le \|L\dot{g}\|_{L_2(0,T;L_2(I))} \le C_T \|g\|_{H_0^1(0,T)} \tag{4.31}$$

(c) Moreover, if $u_0 \in \mathcal{D}(A)$

$$\|e^{At}u_0\|_{H^1(0,T;L_2(I))} \le C_T \|Au_0\|_{L_2(I)} = C_T \|u_0\|_{\mathcal{D}(A)} \tag{4.32}$$

(d) Finally, using the regularities (4.28), (4.31), and (4.32) in the definition (4.5) of $\mathcal{T}_{u_0,g}$ yields estimate (4.23), as desired. □

Step 3. Corollary 4.4. Assume the hypotheses of Theorem 4.1.

The map $\mathcal{T}$ in (4.5) satisfies the following regularity, for $u_0 \in \mathcal{D}(A)$, $g \in H_0^1(0,T)$, and $v \in X_T$:

$$\|\mathcal{T}_{u_0,g}v\|_{X_T} \le C_T \left\{ \|v\|_{X_T}^2 + \|u_0\|_{\mathcal{D}(A)} + \|g\|_{H_0^1(0,T)} \right\} \tag{4.33}$$

Proof. Combine estimate (4.7) of Proposition 4.2 with estimate (4.23) of Proposition 4.3, and recall X_T in (4.1a-b). □

Step 4. Let now

$$v \in \mathcal{B}(0,R) \subset X_T\,;\ \|u_0\|_{\mathcal{D}(A)} \le r,\ \|g\|_{H_0^1(0,T)} \le r \tag{4.34}$$

for $R > 0$ and $r > 0$ arbitrary. Then Corollary 4.4, Eq (4.34) implies

$$\|\mathcal{T}_{u_0,g}v\|_{X_T} \le C_T \{R^2 + 2r\} \tag{4.35}$$

We seek to obtain: $C_T \{R^2 + 2r\} = C_T R^2 + C_T 2r < R$, and this can be achieved by taking

$$C_T R^2 < \frac{R}{2}, \text{ hence } R < R_T \equiv \frac{1}{2C_T};\ C_T 2r < \frac{R}{2}, \text{ hence } r < \frac{R}{4C_T} < r_T \equiv \frac{1}{8C_T^2} \tag{4.36}$$

Corollary 4.5. Assume the hypotheses of Theorem 4.1.

(i) With C_T the constant in (4.33), define R_T and r_T as in (4.36). Then for v, u_0, g satisfying (4.34) with $R < R_T$ and $r < r_T$, we obtain

$$\|\mathcal{T}_{u_0,g}v\|_{X_T} \le R, \text{ i.e. } u = \mathcal{T}_{u_0,g}v \in \mathcal{B}(0,R) \subset X_T \tag{4.37}$$

and $\mathcal{T}_{u_0,g}$ maps $\mathcal{B}(0,R) \subset X_T$ into $\mathcal{B}(0,R) \subset X_T$.

(ii) In particular, if we select $R < 1$, then $\mathcal{T}_{u_0,g}$ is a contraction mapping and admits a unique fixed point in $\mathcal{B}(0,R)$. Such a fixed point is the unique solution of the original non-linear problem (1.1) (for data u_0 and g as in (4.2)). Theorem 4.1 is proved. □

5. Auxiliary results on the Frechet differentiability of the maps $\{u_0, g\} \to u$, and $\{u_0, g\} \to u(T)$.

According to Theorem 4.1, if $0 < T < \infty$ is given, then there exists a constant $0 < r_T < \infty$ such that : if u_0 and g satisfy (4.2); i.e. if

$$u_0 \in S_1(0, r_T) \equiv \left\{ z \in \mathcal{D}(A) : \|z\|_{\mathcal{D}(A)} \le r_T \right\}, \; g \in S_2(0, r_T) \equiv \left\{ f \in H_0^1(0,T) : \|f\|_{H_0^1(0,T)} \le r_T \right\} \tag{5.1}$$

then there exists a unique solution $u \in \mathcal{B}(0,R) \subset X_T$, $R < 1$, in the ball of radius R in the space X_T defined by (4.1), of the non-linear problem (1.1): such u is also the unique solution of the integral equation (4.3). Accordingly, under these circumstances, we may define two maps:

(i)
$$\mathcal{H}(u_0, g) \equiv u_{u_0,g}(\cdot) \equiv e^{A\cdot} u_0 + Lg - \int_0^{\cdot} e^{A(\cdot - s)} \left[k u_{u_0,g}^2(s) + u_{u_0,g}(s) \left(V u_{u_0,g} \right)(s) \right] ds \tag{5.2a}$$

$$: \text{continuous } S_1(0, r_T) \times S_2(0, r_T) \subset \mathcal{D}(A) \times H_0^1(0,T) \to \mathcal{B}(0,R) \subset X_T \tag{5.2b}$$

(ii)
$$H(u_0, g; T) \equiv u_{u_0,g}(T) \equiv e^{AT} u_0 + L_T g - \int_0^T e^{A(T-s)} \left[k u_{u_0,g}^2(s) + u_{u_0,g}(s) \left(V u_{u_0,g} \right)(s) \right] ds \tag{5.3a}$$

$$: \text{continuous } S_1(0, r_T) \times S_2(0, r_T) \subset \mathcal{D}(A) \times H_0^1(0,T) \to \mathcal{D}(A) \tag{5.3b}$$

We next show Frechet differentiability properties of the map $\mathcal{H}(u_0, g)$.

Proposition 5.1. Assume the hypotheses of Theorem 4.1. Let $\{u_0, g\}$ satisfy (5.1), so that $\mathcal{H}(u_0, g) \in X_T$ is well defined by (5.2).

(a) Let $u_0 \in S_1(0, r_T)$ be fixed. Then, the map $g \to \mathcal{H}(u_0, g)$ from $S_2(0, r_T)$ to $\mathcal{B}(0,R) \subset X_T, R < 1$, is Frechet differentiable, with Frechet derivative $\mathcal{H}'_g(u_0, g)$ which satisfies

$$\mathcal{H}'_g(u_0, g)\Delta g \equiv w \in X_T; \quad \Delta g \in H_0^1(0,T) \tag{5.4}$$

where w in (5.4) is the unique solution of the integral equation

$$w \equiv \mathcal{M}_T w \equiv L(\Delta g) - \int_0^{\cdot} e^{A(\cdot - s)} \left[2k u_{u_0,g}(s) w(s) + w(s) \left(V u_{u_0,g} \right)(s) + u_{u_0,g}(s) (Vw)(s) \right] ds \tag{5.5}$$

Moreover:
$$\mathcal{H}'_g(u_0, g) : \text{continuous } H_0^1(0,T) \to X_T$$

$$\left\| \mathcal{H}'_g(u_0, g)\Delta g \right\|_{X_T} \le c_T \left(\|u_{u_0,g}\|_{X_T} \right) \|\Delta g\|_{H_0^1(0,T)} \tag{5.6b}$$

$$\text{(by Theorem 4.1)} \qquad \le C_T \left(\|u_0\|_{\mathcal{D}(A)}, \|g\|_{H_0^1(0,T)} \right) \|\Delta g\|_{H_0^1(0,T)} \tag{5.6c}$$

(b) Let $g \in S_2(0, r_T)$ be fixed. Then, the map $u_0 \to \mathcal{H}(u_0, g)$ from $S_1(0, r_T)$ to $\mathcal{B}(0, R) \subset X_T, R < 1$, is Frechet differentiable, with Frechet derivative $\mathcal{H}'_{u_0}(u_0, g)$ which satisfies

$$\mathcal{H}'_{u_0}(u_0, g)\Delta u_0 = q, \quad \Delta u_0 \in \mathcal{D}(A) \tag{5.7}$$

where q in (5.7) is the unique solution of the integral equation

$$q \equiv \mathcal{N}_T q \equiv e^{A\cdot}(\Delta u_0) - \int_0^{\cdot} e^{A(\cdot - s)}[2ku_{u_0,g}(s)q(s) + q(s)(Vu_{u_0,g})(s) + u_{u_0,g}(s)(Vq)(s)]\,ds \tag{5.8}$$

Moreover:

$$\mathcal{H}'_{u_0}(u_0, g) : \text{continuous } \mathcal{D}(A) \to X_T \tag{5.9a}$$

$$\left\|\mathcal{H}'_{u_0}(u_0, g)\Delta u_0\right\|_{X_T} \leq c_T\left(\|u_{u_0,g}\|_{X_T}\right)\|\Delta u_0\|_{\mathcal{D}(A)} \tag{5.9b}$$

$$\text{(by Theorem 4.1)} \leq C_T\left(\|u_0\|_{\mathcal{D}(A)}, \|g\|_{H_0^1(0,T)}\right)\|\Delta u_0\|_{\mathcal{D}(A)} \tag{5.9c}$$

Proof. **(i) Step 1.** One first shows—as in Section 4—that the map $\mathcal{M}_T$ defined by (5.5) is contraction on the space X_T (defined by (4.1)) for $0 < T$ sufficiently small. Thus, Eq. (5.5) has a unique solution in the space X_T, for $0 < T$ sufficiently small. Since Eq. (5.6) is linear, such unique solution can be extended to an arbitrary finite interval $[0, T]$.

Step 2. The result of Step 1 means that the map $\mathcal{H}(u_0, g)$ defined by (5.2) has, for fixed $u_0 \in S_1(0, r_T)$, a Frechet derivative $\mathcal{H}'_g(u_0, g)$, which then satisfies (5.4)–(5.6c).

(ii) The proof of part (ii) is similar □

By the use of Proposition 5.1, one next obtains Frechet differentiability properties of the map $H(u_0, g; T)$ defined by (5.3).

Corollary 5.2. Assume the hypotheses of Theorem 4.1.

Let $\{u_0, g\}$ satisfy (5.1), so that $H(u_0, g; T) \in \mathcal{D}(A)$ is well-defined by (5.3)

(a) Let $u_0 \in S_1(0, r_T)$ be fixed. Then, the map $g \to H(u_0, g; T)$ from $S_2(0, r_T)$ to $\mathcal{D}(A)$ is Frechet differentiable with Frechet derivative $H'_g(u_0, g; T)$ which satisfies

$$H'_g(u_0, g; T)\Delta g = w(T) \in \mathcal{D}(A); \quad \Delta g \in H_0^1(0, T) \tag{5.10}$$

where w is the function defined in (5.4), (5.5) of Proposition 5.1, so that recalling L_T from (2.22), we have

$$w(T) = L_T(\Delta g) - \int_0^T e^{A(T-s)}[2ku_{u_0,g}(s)w(s) + w(s)(Vu_{u_0,g})(s) + u_{u_0,g}(s)(Vw)(s)]\,ds \tag{5.11}$$

Moreover:

$$H'_g(u_0, g; T) : \text{continuous } H_0^1(0,T) \to \mathcal{D}(A) \tag{5.12a}$$

$$\left\|H'_g(u_0, g; T)\Delta g\right\|_{\mathcal{D}(A)} \le c_T\left(\|u_{u_0,g}\|_{X_T}\right) \|\Delta g\|_{H_0^1(0,T)} \tag{5.12b}$$

(by Theorem 4.1)

$$\le C_T\left(\|u_0\|_{\mathcal{D}(A)}, \|g\|_{H_0^1(0,T)}\right) \|\Delta g\|_{H_0^1(0,T)} \tag{5.12c}$$

(b) Let $g \in S_2(0, r_T)$. Then, the map $u_0 \to H(u_0, g; T)$ from $S_1(0, r_T)$ to $\mathcal{D}(A)$ is Frechet differentiable with Frechet derivative $H'_{u_0}(u_0, g; T)$ which satisfies

$$H'_{u_0}(u_0, g; T)\Delta u_0 = q(T) \in \mathcal{D}(A), \quad \Delta u_0 \in \mathcal{D}(A) \tag{5.13}$$

where q is the function defined by (5.7) or (5.8) of Proposition 5.1, so that

$$q(T) = e^{AT}(\Delta u_0) - \int_0^T e^{A(T-s)} \left[2ku_{u_0,g}(s)q(s) + q(s)\left(Vu_{u_0,g}\right)(s) + u_{u_0,g}(s)(Vq)(s)\right] ds \tag{5.14}$$

Moreover:

$$H'_{u_0}(u_0, g; T) : \text{conotinuous } \mathcal{D}(A) \to \mathcal{D}(A) \tag{5.15a}$$

$$\left\|H'_{u_0}(u_0, g; T)\Delta u_0\right\|_{\mathcal{D}(A)} \le c_T\left(\|u_{u_0,g}\|_{X_T}\right) \|\Delta u_0\|_{\mathcal{D}(A)} \tag{5.15b}$$

(by Theorem 4.1)

$$\le C_T\left(\|u_0\|_{\mathcal{D}(A)}, \|g\|_{H_0^1(0,T)}\right) \|\Delta u_0\|_{\mathcal{D}(A)}$$

6. Local exact controllability of (1.1)

Theorem 6.1. Let $T > r\ell$. Assume the hypotheses of Theorem 3.3.2 (ii), as well as that $\|u_0\|_{\mathcal{D}(A)} \le r_T$, see (5.1). Then, there exists $g \in S_2(0, r_T)$, see (5.1), such that the unique solution of problem (1.1) [guaranteed by Theorem 4.1] satisfies $u(T) = 0$.

Proof. Let $\{u_0, g\} \in S_1(0, r_T) \times S_2(0, r_T)$ as in (5.1), so that, by Theorem 4.1, the map $H(u_0, g;\ T) = u_{u_0,g}(T)$ in (5.3) is well defined. With u_0 fixed and given, we further seek $g \in S_2(0, r_T)$ such that:

$$H(u_0, g;\ T) = u_{0,g}(T) = 0\,. \tag{6.1}$$

Since $H(0, 0;\ T) = 0$, we then have that—via the implicit Function Theorem (IFT)—the desired relation (6.1) is indeed satisfied, provided that we justify the application of the IFT. To this end, we must ascertain that the following assumptions of the IFT [Lu. p 266] are satisfied:

(i) $H(\cdot, \cdot; T)$ is Frechet differentiable in a neighborhood of $(0, 0)$: this is guaranteed by Corollary 5.2.

(ii) The following surjectivity assumption attains:

$$H'_{u_0}(\cdot,\cdot;T)\Delta u_0 + H'_g(\cdot,\cdot;T)\Delta g: \text{ from a neighborhood of the origin of } \mathcal{D}(A)\times H_0^1(0,T) \text{ onto } \mathcal{D}(A) \tag{6.2}$$

But by the surjectivity in (6.2) is equivalent to the surjectivity property

$$e^{AT}(\Delta u_0) + L_T(\Delta g): \text{ from a neighborhood of the origin of } \mathcal{D}(A)\times H_0^1(0,T) \text{ onto } \mathcal{D}(A) \tag{6.3}$$

However, (6.3) does hold true by virtue of the exact controllability result in Theorem 3.4.5. Thus the IFT is applicable, and Theorem 6.1 is proved. □

Appendix

Assume the following form for the kernel

$$K(x,\xi)\, p_\infty(\xi) = k_1(x)k_2(\xi) + h_1(x)h_2(\xi) \tag{A.1}$$

with k_1, h_1 and k_2, h_2 linearly independent. We shall extend Theorem 3.2.1 *generically.*

Step 1. The counterpart of Eq (3.2.109) is then

$$\frac{du(s,\tau)}{ds} = k_1(s+\tau)(k_2, u(s,\cdot))_I + h_1(s+\tau)(h_2, u(s,\cdot))_I \tag{A.2}$$

Next, multiply (A.2) first by $k_2(\tau)$, next by $h_2(\tau)$, and integrate each time in τ over $[0,\ell] = I$, to obtain the system

$$\frac{d}{ds}\begin{bmatrix} (k_2, u(s,\cdot))_I \\ (h_2, u(s,\cdot))_I \end{bmatrix} = A(s)\begin{bmatrix} (k_2, u(s,\cdot))_I \\ (h_2, u(s,\cdot))_I \end{bmatrix}, \; A(s) = [a_{ij}(s)] \tag{A.3}$$

$$a_{11}(s) = \int_0^\ell k_1(s+\tau)k_2(\tau)d\tau; \quad a_{12}(s) = \int_0^\ell h_1(s+\tau)k_2(\tau)d\tau \tag{A.4}$$

$$a_{21}(s) = \int_0^\ell k_1(s+\tau)h_2(\tau)d\tau; \quad a_{22}(s) = \int_0^\ell h_1(s+\tau)h_2(\tau)d\tau \tag{A.5}$$

counterpart of Eq (3.2.11).

If $\Phi(s,0)$ is the 2×2 fundamental matrix corresponding to $A(s)$ and explicitly given by the Peano-Bekker series, then the solution of the linear Eq (A.3) is

$$\begin{bmatrix} (k_2, u(s,\cdot))_I \\ (h_2, u(s,\cdot))_I \end{bmatrix} = \Phi(s,0)\begin{bmatrix} (k_2, u_0)_I \\ (h_2, u_0)_I \end{bmatrix}, \quad u(0,\cdot) = u_0(\cdot) \tag{A.6}$$

which is the counterpart of Eq (3.2.11). Inserting (A.6) into (A.2) yields

$$\frac{du(s,\tau)}{ds} = [k_1(s+\tau), h_1(s+\tau)]\Phi(s,0) \begin{bmatrix} (k_2, u_0)_I \\ (h_2, u_0)_I \end{bmatrix} \tag{A.7}$$

counterpart of (3.2.13). Integrating (A.7) in s from $s = 0$ (so that $u(0,\tau) = u_0(\tau)$) along the characteristic $\tau =$ const, yields for $0 \le s \le \ell - \tau$; $0 \le \tau \le \ell$:

$$u(s,\tau) - u_0(\tau) = \left\{ \int_0^s [k_1(\sigma+\tau), h_1(\sigma+\tau)]\, \Phi(\sigma,0) d\sigma \right\} \begin{bmatrix} (k_2, u_0)_I \\ (h_2, u_0)_I \end{bmatrix} \tag{A.8}$$

which is the counterpart of Eq (3.2.14). Specializing (A.8) with $s = \ell, \tau = 0$, and $u_0 \in \mathcal{D}(A)$, so that $u_0(0) = 0$, we obtain by virtue of the B.C. $u(\ell, 0) = 0$ as in (3.2.16b)

$$\left\{ \int_0^\ell [k_1(\sigma), h_1(\sigma)]\, \Phi(\sigma,0) d\sigma \right\} \begin{bmatrix} (k_2, u_0)_I \\ (h_2, u_0)_I \end{bmatrix} = 0 \tag{A.9}$$

Step 2. To obtain a second equation in the unknown $[(k_2, u_0)_I, (h_2, u_0)_I$ we specialize this time (A.8) with $s = \ell - \tau$, so that $u(\ell - \tau, \tau) \equiv 0$ by the B.C. as in (3.2.16b), and obtain

$$-u_0(\tau) = \left\{ \int_0^{\ell-\tau} [k_1(\sigma+\tau), h_1(\sigma+\tau)]\, \Phi(\sigma,0) d\sigma \right\} \begin{bmatrix} (k_2, u_0)_I \\ (h_2, u_0)_I \end{bmatrix} \tag{A.10}$$

Multiplying (A.10) by $k_2(\tau)$ [or else by $h_2(\tau)$] and integrating in τ over $[0, \ell]$ yields

$$-(k_2, u_0)_I = \left\{ \int_0^\ell \int_0^{\ell-\tau} [k_1(\sigma+\tau), h_1(\sigma+\tau)]\, \Phi(\sigma,0) d\sigma d\tau \right\} \begin{bmatrix} (k_2, u_0)_I \\ (h_2, u_0)_I \end{bmatrix} \tag{A.11}$$

which is the sought-after second equation. The linear homogeneous system in the unknown $[(k_2, u_0)_I, (h_2, u_0)_I]$ consisting of Eq (A.9) and Eq (A.11) has a coefficient matrix which depends smoothly in the data k_1, k_2, h_1, h_2. Thus, such coefficient matrix is *generically* (at least) non-singular, so that we then obtain

$$(k_2, u_0)_I = 0, \quad (h_2, u_0)_I = 0 \tag{A.12}$$

which is the counterpart of Eq (3.2.16a). From here on, the proof can be concluded as before (Step 3 of proof of Theorem 3.2.1): inserting (A.12) in (A.8) yields

$$u(s,\tau) = u_0(\tau),\ 0 \le s \le \ell - \tau,\ 0 \le \tau \le \ell, \quad \text{hence} \quad 0 \equiv u(\ell - \tau, \tau) \equiv u_0(\tau),\ 0 \le \tau \le \ell \tag{A.13}$$

as desired.

We conclude. The uniqueness Theorem 3.2.1 can be *generically* extended to kernels of the form (A.1); indeed, by the procedure of the present Appendix, *genericall* y to so called *degenerate* kernels of the form

$$K(x,\xi)\, p_\infty(\xi) = \sum_{i=1}^n k_i(x)\, q_i(\xi) \tag{A.14}$$

for any finite n, $\{k_i\}$ and $\{q_i\}$ being both linearly independent sets. It is known [T-L.1, p204] that an integral operator such as V in (1.3) can be arbitrarily approximated in the uniform norm by an integral operator with a *degenerate* kernel as in (A.14). It may be possible to obtain the *continuous observability* estimate (3.4.4) [i.e. (3.3.7) of Theorem 3.3.2] of the dual linearized ψ-problem for any operator V with an L_2-kernel, thus extending to such V the ultimate local exact controllability result in Theorem 6.1 for the non-linear model (1.1). This issue has not been investigated yet.

References

[B.1] L.D. Berkovitz, *Optimal Control Theory*, Springer-Verlag 1974.

[B-D-D-M.1] A. Bensoussan, M. Delfour, G.Da Prato, S. Mitter, *Representation and Control of Infinite Dimensional Systems*, Birkhauser 1993, Vol II.

[B-I-Th.1] S.N. Busenberg, M. Iannelli and H.R. Thieme, Global behavior of an age-structured epidemic model, *SIAM J. Mathem. Analysis*, Vol. 22 (91991), 1065–1080.

[D-I.1] G. DaPrato and M. Iannelli, Boundary control problem for age-dependent equations, Evolution equations, control theory, and biomathematics, Lectures Notes in Pure & Applied Mathematics, Vol. 155.

[F-I.1] A.V. Fursikov and O. Yu. Imanuvilov, *Controllability of Evolution Equations*, Lecture Notes Series N. 34. Research Institute of Mathematics, Seoul National University, Seoul 151–742, Korea.

[Ho.1] L. Hormander, *The analysis of linear partial differential operators*, vol I-IV, Springer-Vergla, 1983.

[L-T.1] I. Lasiecka and R. Triggiani. Differential and Algebraic Riccati Equations with applications to boundary point control problems. In *Lecture Notes in Control and Information Sciences*, volume 164. Springer-Verlag, 1991.

[L-T.2] I. Lasiecka and R. Triggiani. Exact Boundary controllability for the Wave Equation with Neumann Boundary Control, *Appl Math and Optimiz*, 19 (1989).

[L-T.3] I. Lasiecka and R. Triggiani, Carleman estimates and exact boundary controllability for a system of coupled non-conservative second-order hyperbolic equations, *Marcel Dekker Lectures Notes in Pure and Applied Mathematics, Partial Differential equation methods in control and shape analysis*, vol 188, G. Da Prato and J.P. Zolesio, Editors 1997.

[Lu.1] D. Luenberger, *Optimization by vector space methods*, John Wiley 1969.

[P.1] A. Pazy, *Semigroups of linear operators and applications to partial differential equations*, Springer-Verlag 1983.

[T-L.1] A Taylor and D. Lay, *Introductioni to Functional Analysis*, John Wiley 2nd Edit 1980.

[W.1] J. Weidmann, *Linear operators in Hilbert spaces*, Springer-Verlag 1980.

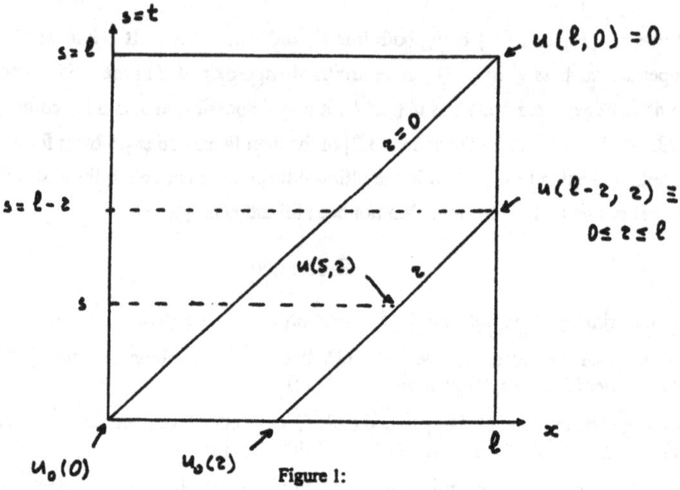

Figure 1:

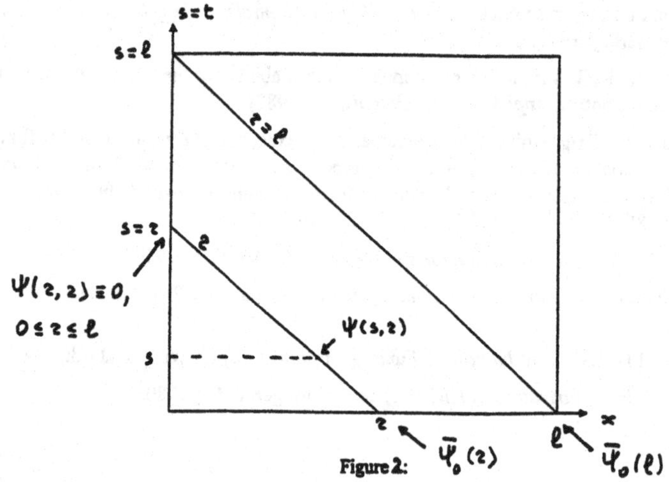

Figure 2:

Singularities of solutions for nonlinear hyperbolic equations of second order

Mikio TSUJI *
Dept. of Math., Kyoto Sangyo University
Kamigamo, Kita-ku, kyoto 603-8555, Japan

1 Introduction

We consider the Cauchy problem for nonlinear hyperbolic partial differential equations of second order. Then the Cauchy problem does not generally admit a classical solution in the large, that is to say, singularities generally appear in finite time. The typical example of singularity is "shock wave". Our problem is to extend the solution beyond the singularities. Though there are many papers concerning this subject, the existence and uniqueness of weak solution is even now unsolved. In this talk we consider the above problem from geometrical point of view.

Let us explain our method briefly. First we lift the solution surface into higher dimensional space so that the lifted surface does not contain the singularities. To do so, we need to solve certain hyperbolic system of first order. Though it is not easy to solve the system, we can sometimes get the smooth solution in the large. As the singularites do not appear in the solution of the system, the extension of the solution is unique in higher dimensional space. The above subject will be discussed in §2 and §3.

---

*1991 *Mathematics Subject Classification.* Primary 35L67, 35L70, 57R45; Secondary 35L65, 57R55, 58C27; *Key words and phrases.* Nonlinear equations, contact structure, weak solutions, singularities of smooth mappings. The author's research was partially supprted by Grant-in-Aid for Scientific Research (C) (No.10640219), Ministry of Education, Science and Culture (Japan)

R.P. Gilbert et al.(eds.), Direct and Inverse Problems of Mathematical Physics, 399–413.

Next we project the lifted solution surface to the base space. In a neighbourhood of a point where this projection is regular, we can get a smooth solution. This is just the characteristic method for partial differential equations of second order. In a neighbourhood of a point where the mapping is singular, we use the singularity theory of smooth mappings and we can construct a solution with singularities.

Finally we apply our method to certain nonlinear wave equation in §4 and §5, and also to certain system of conservation law, in §6, which is called as "p-system". Then we see that one can not construct a weak solution in the sense of Lax by the characteristic method. This may suggest us that the definition of weak solution may not be appropriate.

If the equations are of first order, then we could construct reasonable weak solutions by the above method. For example, refer to J. Guckenheimer [6], M. Tsuji [18, 19], S. Nakane [16,17], S. Izumiya [8], S. Izumiya and G. T. Kossioris [9,10], etc. etc.

2 Integration of Monge-Ampère equations I

In this section and the following one, we will discuss the method of integration of second order nonlinear partial differential equations. As the equations of second order are too general, we will treat the equations of Monge-Ampère type as follows:

$$F(x, y, z, p, q, r, s, t) = Ar + Bs + Ct + D(rt - s^2) - E = 0 \qquad (2.1)$$

where $p = \partial z/\partial x, q = \partial z/\partial y, r = \partial^2 z/\partial x^2, s = \partial^2 z/\partial x \partial y$, and $t = \partial^2 z/\partial y^2$. Here we assume that A, B, C, D and E are real smooth functions of (x, y, z, p, q). Partial differential equations of second order which appear in physics and geometry are generally written by the above form. The aim of this talk is to consider the structure of singularities of solutions of (2.1) in the case where equation (2.1) is hyperbolic. The definition of "hyperbolicity" will be given later. For our aim, we have to represent the solutions explicitely. In this section, we recall the characteristic method developed principally by D. Darboux and E. Goursat ([2], [4], [5]). As it seems to us that the method is not familiar today, we will explain it in this section "from our point of view". The principal idea of Darboux and Goursat is how to reduce the solvability of (2.1) to the integration of first order partial differential equations, though their theory is local. But, as their method is constructive, it is very useful for our purpose. Now we recall the notion of "strip". Let

$$\Gamma : (x, y, z, p, q) = (x(\alpha), y(\alpha), z(\alpha), p(\alpha), q(\alpha)), \quad \alpha \in \mathbf{R}^1,$$

be a smooth curve in $\mathbf{R}^5 = \{(x, y, z, p, q)\}$, and suppose that it satisfies the following "strip condition"

$$\frac{dz}{d\alpha}(\alpha) = p(\alpha)\frac{dx}{d\alpha}(\alpha) + q(\alpha)\frac{dy}{d\alpha}(\alpha) . \tag{2.2}$$

This reflects that p and q are corresponding to the first order derivatives of $z = z(x, y)$. If we may state the above by "geometrical terminology", the condition (2.2) means that we introduce the contact structure $dz = pdx + qdy$ into $\mathbf{R}^5$.

Let Γ be any strip in $\mathbf{R}^5$, and consider equation (2.1) in its open neighbourhood. As a "characteristic" strip means that one can not determine the values of the second order derivatives of solution along the strip, we have the following

Definition 2.1 *A curve Γ in $\mathbf{R}^5 = \{(x, y, z, p, q)\}$ is a "characteristic strip" if it satisfies* (2.2) *and*

$$\det \begin{bmatrix} F_r & F_s & F_t \\ \dot{x} & \dot{y} & 0 \\ 0 & \dot{x} & \dot{y} \end{bmatrix} = F_t\dot{x}^2 - F_s\dot{x}\dot{y} + F_r\dot{y}^2 = 0 \tag{2.3}$$

where $F_t = \partial F/\partial t, F_s = \partial F/\partial s, F_r = \partial F/\partial r, \dot{x} = dx/d\alpha$ and $\dot{y} = dy/d\alpha$.

Denote the discriminant of (2.3) by Δ, then

$$\Delta = F_s^2 - 4F_rF_t = B^2 - 4(AC + DE) .$$

If $\Delta < 0$, equation (2.1) is called to be elliptic. If $\Delta > 0$, equation (2.1) is hyperbolic. In this note, we will treat the equations of hyperbolic type. More precisely, we assume $\Delta \geq 0$ and also $D \neq 0$. Let λ_1 and λ_2 be the solutions of $\lambda^2 + B\lambda + (AC + DE) = 0$. Then the characteristic strip satisfies the following equations:

$$dz - pdx - qdy = 0, \quad Ddp + Cdx + \lambda_1 dy = 0, \quad Ddq + \lambda_2 dx + Ady = 0, \tag{2.4}$$

or

$$dz-pdx-qdy=0,\quad Ddp+Cdx+\lambda_2dy=0,\quad Ddq+\lambda_1dx+Ady=0. \quad (2.5)$$

Let us denote $\omega_0 = dz - pdx - qdy$, $\omega_1 = Ddp + Cdx + \lambda_1 dy$ and $\omega_2 = Ddq+\lambda_2dx+Ady$. Take an exterior product of ω_1 and ω_2, and substitute into their product the contact relations $\omega_0 = 0$, $dp = rdx+sdy$ and $dq = sdx+rdy$. Then we get

$$\omega_1 \wedge \omega_2 = D\Big\{Ar + Bs + Ct + D(rt - s^2) - E\Big\}\, dx \wedge dy\ . \qquad (2.6)$$

In a space whose dimension is greater than two, the decomposition as above is not possible in general. Here we introduce the notion of "first integral".

Definition 2.2 *A function $V = V(x, y, z, p, q)$ is called "first integral" of $\{\omega_0, \omega_1, \omega_2\}$ if $dV \equiv 0 \bmod\{\omega_0, \omega_1, \omega_2\}$.*

Remark We can easily see that a function $V = V(x, y, z, p, q)$ is the "first integral" of (2.4) (or (2.5)) if it is constant on any solution of (2.4) (or of (2.5) respectively).

G. Darboux [2] and E. Goursat [4,5], especially in [5], had considered equation (2.1) under the assumption that (2.4), or (2.5), has at least two independent first integrals. We denote them by u and v. Then we get the following

Proposition 2.3 *Assume that $\lambda_1 \neq \lambda_2$, and that (2.4), or (2.5), has two independent first integrals $\{u, v\}$. Then there exists a function $k = k(x, y, z, p, q) \neq 0$ satisfying*

$$du \wedge dv = k\, \omega_1 \wedge \omega_2 = kD\{Ar + Bs + Ct + D(rt - s^2) - E\}dx \wedge dy\ . \quad (2.7)$$

If equation (2.1) is written as (2.7), it would be obvious that (2.4), or (2.5), has two independent first integrals $\{u, v\}$. Therefore the representation (2.7) gives the characterization of "Monge-Ampère equations of Darboux-Goursat type". Let $\{u, v\}$ be two independent first integrals of (2.4). For any function g of two variables whose gradient does not vanish, $g(u, v) = 0$ is called an "intermediate integral" of (2.1). Let C_0 be an initial strip defined in $\mathbf{R}^5 = \{(x, y, z, p, q)\}$. If the strip C_0 is not characteristic, we can find an "intermediate integral" $g(u,v)$ which vanishes on C_0. Here we put $g(u,v) = f(x,y,z,p,q)$. The Cauchy problem for (2.1) satisfying the initial condition C_0 is to look for a solution $z = z(x,y)$ of (2.1) which contains the strip C_0, i.e., the two dimensional surface $\{(x,y,z(x,y), \partial z/\partial x(x,y), \partial z/\partial y(x,y))\}$ in $\mathbf{R}^5$ contains the strip C_0. The representation (2.7) assures that, as $du \wedge dv = 0$ on a surface $g(u,v) = 0$, a smooth solution of $f(x,y,z,\partial z/\partial x, \partial z/\partial y) = 0$ satisfies equation (2.1). Therefore we get the following

Theorem 2.4 ([2], [4], [5]) *Assume that the initial strip C_0 is not characteristic. Then a function $z = z(x,y)$ is a solution of the Cauchy problem for* (2.1) *with the initial condition C_0 if and only if it is a solution of $f(x,y,z,\partial z/\partial x, \partial z/\partial y) = 0$ satisfying the same initial condition C_0.*

3 Integration of Monge-Ampère equations II

In this section we will consider the method of integration of Monge-Ampère equations (2.1) in the case where (2.4), and (2.5) also, has not two independent first integrals. We start from the point at which equation (2.1) is represented as a product of one forms as (2.6). We suppose $D \neq 0$ for simplicity, though it is not indispensable for our study. The essential condition for our following discussion is $\Delta \neq 0$. We will here take heuristic approach to get solutions of (2.1). As the preparation, we will give another representation of (2.1) which is similar to (2.6).

Exchanging λ_1 and λ_2 in ω_1 and ω_2, we define ϖ_1 and ϖ_2 by

$$\varpi_1 = Ddp + Cdx + \lambda_2 dy, \quad \varpi_2 = Ddq + \lambda_1 dx + Ady$$

Then we get the following identity:

$$\omega_1 \wedge \omega_2 = \varpi_1 \wedge \varpi_2 = D\, F(x,y,z,p,q,r,s,t)\, dx \wedge dy. \tag{3.1}$$

Let us suppose that a solution can be represented by two parameters, i.e.,

$$x = x(\alpha,\beta), y = y(\alpha,\beta), z = z(\alpha,\beta), p = p(\alpha,\beta), q = q(\alpha,\beta). \tag{3.2}$$

Then ω_i and ϖ_i $(i = 1,2)$ are written as $\omega_i = c_{i1}d\alpha + c_{i2}d\beta$, $\varpi_i = d_{i1}d\alpha + d_{i2}d\beta$ $(i = 1,2)$. Hence we have $\omega_1 \wedge \omega_2 = (c_{11}c_{22} - c_{12}c_{21})\, d\alpha \wedge d\beta$ and

$\varpi_1 \wedge \varpi_2 = (d_{11}d_{22} - d_{12}d_{21})\, d\alpha \wedge d\beta$. As $\omega_1 \wedge \omega_2 = \varpi_1 \wedge \varpi_2 = 0$ on $F = 0$, a sufficient condition so that (3.2) is a solution of (2.1) is

$$c_{11} = c_{21} = d_{12} = d_{22} = 0. \tag{3.3}$$

Adding the contact relation $dz = pdx + qdy$ to (3.3), we get a system of first order partial differential equation as follows:

$$\begin{cases} \dfrac{\partial z}{\partial \alpha} - p\dfrac{\partial x}{\partial \alpha} - q\dfrac{\partial y}{\partial \alpha} = 0 \\ D\dfrac{\partial p}{\partial \alpha} + C\dfrac{\partial x}{\partial \alpha} + \lambda_1\dfrac{\partial y}{\partial \alpha} = 0, \quad D\dfrac{\partial q}{\partial \alpha} + \lambda_2\dfrac{\partial x}{\partial \alpha} + A\dfrac{\partial y}{\partial \alpha} = 0 \\ D\dfrac{\partial p}{\partial \beta} + C\dfrac{\partial x}{\partial \beta} + \lambda_2\dfrac{\partial y}{\partial \beta} = 0, \quad D\dfrac{\partial q}{\partial \beta} + \lambda_1\dfrac{\partial x}{\partial \beta} + A\dfrac{\partial y}{\partial \beta} = 0 \end{cases} . \tag{3.4}$$

If $(x(\alpha,\beta), y(\alpha,\beta), z(\alpha,\beta), p(\alpha,\beta), q(\alpha,\beta))$ satisfies system (3.4), one can prove $\partial z/\partial\beta - p\partial x/\partial\beta - q\partial y/\partial\beta = 0$. Therefore we do not need to add this to (3.4). This means that (3.4) is just the "determined" system. The local solvability of (3.4) is already proved by H. Lewy [14] and J. Hadamard [7].

Let us denote the solution of the Cauchy problem for (3.4) by $(x(\alpha,\beta), y(\alpha,\beta), z(\alpha,\beta), p(\alpha,\beta), q(\alpha,\beta))$. We can prove that, if the initial strip is not characteristic, the Jacobian $D(x,y)/D(\alpha,\beta)$ does not vanish in a neighbourhood of the initial strip. Therefore we can uniquely solve the system of equations $x = x(\alpha,\beta)$, $y = y(\alpha,\beta)$ with respect to (α,β). Then the solution of (2.1) with the initial condition is given by $z(x,y) = z(\alpha(x,y), \beta(x,y))$.

If we may say the above characteristic method from the geometrical point of view, it is the method to construct a submanifold of the surface $\{(x,y,z,p,q,r,s,t) \in \mathbf{R}^8; f(x,y,z,p,q,r,s,t) = 0\}$ on which the contact structure of second order $dz = pdx + qdy$, $dp = rdx + sdy$, $dq = sdx + tdy$ is satisfied.

4 Nonlinear wave equations

Let us consider the Cauchy problem for nonlinear wave equations as follows:

$$\frac{\partial^2 z}{\partial x^2} - \frac{\partial}{\partial y} f\left(\frac{\partial z}{\partial y}\right) = 0 \quad \text{in } \{x > 0, y \in \mathbf{R}^1\} \equiv \mathbf{R}^2_+, \tag{4.1}$$

$$z(0,y) = z_0(y), \quad \frac{\partial z}{\partial x}(0,y) = z_1(y) \qquad \text{on} \quad \{x = 0, y \in \mathbf{R}^1\} \tag{4.2}$$

where $f(q)$ is in $C^\infty(\mathbf{R}^1)$ and $f'(q)$ is positive. Here $z = z(x,y)$ is an unknown function of $(x,y) \in \mathbf{R}^2$, and we assume that the initial functions $z_i(y)$ ($i = 0,1$) are sufficiently smooth. Equation (4.1) is also of Monge-Ampère type. For example, if we may put $A = 1, B = D = E = 0$, and $C = -f'(q)$ in (2.1), then we get (4.1).

It is well known that the Cauchy problem (4.1)-(4.2) does not have a classical solution in the large. For example, see N. F. Zabusky [23] for the case where $f'(q) = (1+\varepsilon q)^{2\alpha}$ and P. D. Lax [12] for 2×2 hyperboric systems of conservation law. After them, many people have considered the life-span of classical solutions. As the number of papers on this subject is too many, we do not mention here their contributions.

Our problem is how to extend the solution of (4.1) after the apperance of sigularities. In this section, as we will consider this problem from the physical point of view, we will look for single-valued solutions of (4.1). The first question is what kinds of singularities may appear. In [23], N. J. Zabusky showed that the first order derivatives remain bounded, and that the second order derivatives tend to infinity in finite time. Therefore, for equations which have similar properties, we are led to the following

Definition 4.1 *Let $z = z(x,y)$, $(\partial z/\partial x)(x,y)$ and $(\partial z/\partial y)(x,y)$ be bounded and measurable. The function $z = z(x,y)$ is a weak solution of* (4.1)-(4.2) *if it satisfies equation* (4.1) *in distribution sense, i.e.,*

$$\int_{\mathbf{R}^2_+} \{ \frac{\partial z}{\partial x}\frac{\partial \varphi}{\partial x} - f(\frac{\partial z}{\partial y})\frac{\partial \varphi}{\partial y} \} dxdy + \int_{\mathbf{R}^1} z_1(y)\varphi(0,y)\, dy = 0 \tag{4.3}$$

and $z(0,y) = z_0(y)$ for all $\varphi(x,y) \in C_0^\infty(\mathbf{R}^2)$.

Remark In Definition 4.1, $\partial z/\partial x$ and $\partial z/\partial y$ are the derivatives of $z = z(x,y)$ in classical sense, not in distribution sense. Therefore, even if $z = z(x,y)$ may have jump discontinuities, Dirac measure does not appear in $\partial z/\partial x$ and $\partial z/\partial y$.

Let $z = z(x,y)$ be a weak solution of (4.1) in the sense of Definition 4.1. If $(\partial z/\partial x)$ $(x,y) \equiv p(x,y)$ and $(\partial z/\partial y)(x,y) \equiv q(x,y)$ have jump discontinuities along a curve $y = \gamma(x)$, we get by (4.3) the jump condition of

Rankine-Hugoniot type as follows:

$$[p]\dot{\gamma} + [f(q)] = 0. \tag{4.4}$$

where [ ] means the quantity of difference, i.e., $[p] = p(x, \gamma(x)+0) - p(x, \gamma(x)-0)$. Moreover suppose that $z = z(x, y)$ is continuous along the curve $y = \gamma(x)$, though $p(x, y)$ and $q(x, y)$ have jump discontinuities along it. Then, differentiating $z(x, \gamma(x)+0) = z(x, \gamma(x)-0)$ with respect to x, we get

$$[q]\dot{\gamma} + [p] = 0. \tag{4.5}$$

This means that the curve $y = \gamma(x)$ must satisfy two different differential equations (4.4) and (4.5). Therefore it seems to us that one can not generally prove the existence of continuous weak solutions of (4.1). But we can construct discontinuous weak solutions of (4.1) in the sense of Definition 4.1 by the characteristic method.

We think that many people would accept the above Definition 4.1. As we will write in §6, equation (4.1) can be transformed into a system of conservation law (6.1). For systems of conservation law, P. D. Lax [13] introduced the notion of weak solutions. Then we can show that weak solutions of (4.1) can not be transformed to weak solutions of systems of conservation law (6.1). Moreover, if we might accept the above definition, we can construct piecewise smooth weak solution of (4.1) by the characteristic method just as in the case of first order partial differential equations. But we can not do so for systems of conservation law. As our solutions are very natural from our point of view, we have some question on the definition of weak solution of systems of conservation law. We think that the best method to explain our considerations is to present some concrete example. Therefore, in the following section, we will solve an example exactly without stating our results in general form.

5 Example

In this section we will construct an exact solution of the Cauchy problem as follows:

$$\frac{\partial^2 z}{\partial x^2} - (1+\epsilon q)^{-4} \frac{\partial^2 z}{\partial y^2} = 0 \qquad \text{in} \quad \{x > 0, y \in \mathbf{R}^1\}, \tag{5.1}$$

$$z(0,y)=\sin y,\ (\partial z/\partial x)(0,y)=\epsilon^{-1}(1+\epsilon\cos y)^{-1}\quad \text{on } \{x=0, y\in \mathbf{R}^1\}. \tag{5.2}$$

where $0<\epsilon=$ constant $\leq 1/2$. We can get the solution of (5.1)-(5.2) by solving the following Cauchy problem:

$$\frac{\partial z}{\partial x}-\epsilon^{-1}(1+\epsilon\frac{\partial z}{\partial y})^{-1}=0 \quad \text{in } \{x>0, y\in \mathbf{R}^1\}, \tag{5.3}$$

$$z(0,y)=\sin y \quad \text{on } \{x=0, y\in \mathbf{R}^1\}. \tag{5.4}$$

Concerning the geometric meaning of (5.3), see G. Darboux [2], E. Goursat [4, 5] and M. Tsuji [20, 21]. The characteristic equations for (5.3)-(5.4) are written by

$$\frac{dx}{d\beta}=1,\quad \frac{dy}{d\beta}=(1+\epsilon q)^{-2},\quad \frac{dz}{d\beta}=p+q(1+\epsilon q)^{-2},\quad \frac{dp}{d\beta}=\frac{dq}{d\beta}=0, \tag{5.5}$$

$$x(0)=0,\ \ y(0)=\alpha,\ \ z(0)=\sin\alpha,\ \ p(0)=\epsilon^{-1}(1+\epsilon\cos\alpha)^{-1},\ \ q(0)=\cos\alpha. \tag{5.6}$$

The solutions of (5.5)-(5.6) are obtained as

$$y=\alpha+(1+\epsilon\cos\alpha)^{-2}x \tag{5.7}$$

$$z=\sin\alpha+\epsilon^{-1}(1+2\epsilon\cos\alpha)(1+\epsilon\cos\alpha)^{-2}x \tag{5.8}$$

$$p=\epsilon^{-1}(1+\epsilon\cos\alpha)^{-1},\quad q=\cos\alpha. \tag{5.9}$$

When $(\partial y/\partial\alpha)(x,\alpha)\neq 0$, we can uniquely solve the equation $y=y(x,\alpha)$ with respect to α and denote it by $\alpha=\alpha(x,y)$. Substituting $\alpha=\alpha(x,y)$ into $z=z(x,\alpha), p=p(\alpha)$ and $q=q(\alpha)$, we define $z(x,y)=z(x,\alpha(x,y))$, $p(x,y)=p(\alpha(x,y))$ and $q(x,y)=q(\alpha(x,y))$. Then it holds $(\partial/\partial x)z(x,y)=p(x,y)$ and $(\partial/\partial y)z(x,y)=q(x,y)$. Moreover we see that $z=z(x,y)$ satisfies (5.1))-(5.2) in a neighbourhood of $x=0$. As (5.9) means $p(x,y)=(\partial z/\partial x)(x,y)=\epsilon^{-1}\{1+\epsilon(\partial z/\partial y)\}^{-1}=\epsilon^{-1}(1+\epsilon\ q)^{-1}$, we get

$$\frac{\partial p}{\partial x}+\frac{\epsilon^2}{3}\frac{\partial}{\partial y}p^3=0 \quad \text{in } \{x>0, y\in \mathbf{R}^1\}. \tag{5.10}$$

$$p(0,y)=\epsilon^{-1}(1+\epsilon\cos y)^{-1} \quad \text{on } \{x=0, y\in \mathbf{R}^1\}. \tag{5.11}$$

As $0<\epsilon\leq 1/2$, (5.10)-(5.11) is a single convex conservation law. The characteristic strip of (5.10)-(5.11) is same to the one of (5.3)-(5.4), i.e., it is written by (5.7), (5.8) and (5.9). Calculating $(\partial y/\partial \alpha)(x,\alpha)$, we can easily find a point (x^0,α^0) where $(\partial y/\partial \alpha)(x,\alpha)$ vanishes for the first time, i.e., $(\partial y/\partial \alpha)(x,\alpha) \neq 0$ for $x < x^0$ and $\alpha \in \mathbf{R}^1$, and $(\partial y/\partial \alpha)(x^0,\alpha^0) = 0$. By elementary calculation, we see that, for $x > x^0$, $\alpha = \alpha(x,y)$ takes three values in a neighbourhood of (x^0,y^0) where $y^0 = y(x^0,\alpha^0)$. Denote them by $\alpha = \alpha_1(x,y) < \alpha_2(x,y) < \alpha_3(x,y)$, and define $z_i(x,y) \equiv z(x,\alpha_i(x,y))$, $p_i(x,y) \equiv p(\alpha_i(x,y))$, and $q_i(x,y) \equiv q(\alpha_i(x,y))$ $(i = 1,2,3)$. Then the discontinuity curve $y = \gamma(x)$ of $p = p(x,y)$ is obtained as a solution of the following Cauchy problem

$$\frac{d\gamma}{dx} = \frac{1}{3}\epsilon^2 \frac{p_3(x,\gamma)^3 - p_1(x,\gamma)^3}{p_3(x,\gamma) - p_1(x,\dot{\gamma})} \quad , \quad x > x^0, \tag{5.12}$$

$$\gamma(x^0) = y^0. \tag{5.13}$$

The Cauchy problem (5.12)-(5.13) has a unique solution $\gamma = \gamma(x)$, though the right Handside Is not Lipschitz continous in a neighbourhood of (x^0,y^0). See M. Tsuji [19]. Now we define the solution $z = z(x,y)$ of (5.1)-(5.2) by

$$z(x,y) = \begin{cases} z_1(x,y) & (y < \gamma(x)) \ , \\ z_3(x,y) & (y > \gamma(x)) \ . \end{cases}$$

Then we can prove

$$z(x,\gamma(x)-0) - z(x,\gamma(x)+0) = z_1(x,\gamma(x)) - z_3(x,\gamma(x)) \neq 0, \quad x > x^0. \tag{5.14}$$

In fact, if we may put $w(x) \equiv z_1(x,\gamma(x)) - z_3(x,\gamma(x))$, $w(x)$ satisfies

$$\frac{dw(x)}{dx} = -\frac{\{p_1(x,\gamma(x)) - p_3(x,\gamma(x))\}^3}{3p_1(x,\gamma(x))p_3(x,\gamma(x))} \neq 0, \quad x > x^0, \tag{5.15}$$

$$w(x^0) = 0. \tag{5.16}$$

Hence $w(x) \neq 0$ for $x > x^0$, i.e., we get (5.14). As $z = z(x,y)$ has jump discontinuity along the curve $y = \gamma(x)$, we take the derivatives of $z = z(x,y)$ in classical sense, i.e.,

$$\frac{\partial z}{\partial x}(x,y) = \begin{cases} p_1(x,y) & (y < \gamma(x)) \ , \\ p_3(x,y) & (y > \gamma(x)) \ . \end{cases} \tag{5.17}$$

Since $y = \gamma(x)$ is the solution of (5.12)-(5.13), $z = z(x, y)$ satisfies

$$[p]\dot{\gamma} - [3^{-1}\epsilon^2 p^3] = [p]\dot{\gamma} + [f(q)] = 0$$

where $f(q) = -3^{-1}\epsilon^{-1}(1+\epsilon q)^{-3}$. Hence $z = z(x, y)$ is a weak solution of (5.1)-(5.2) in the sense of Definition 4.1. Moreover we can prove that $y = \gamma(x)$ exists in the large, and that another jump discontinuity curve $y = \gamma_n(x)$ starts at a point $(x^0, y^0 + 2n\pi)$ for any integer n, because all data are the periodic functions of period 2π. Then it holds

$$\gamma_n(x) - \gamma(x) = 2n\pi \ , \quad x > x^0 \ .$$

Theorem 5.1 *The Cauchy probrem* (5.1)-(5.2) *admits a weak solution in the large in the sense of* Definition 4.1. *Moreover the weak solution is analytic except on the curves $y = \gamma_n(x)$ and it does not satisfy $[q]\dot{\gamma} + [p] = 0$ along $y = \gamma_n(x)$ for any integer n.*

At today's point, we do not have any criterion on the uniqueness of weak solutions of (4.1). Therefore we do not insist that our solution is reasonable. But it has several nice properties. The first one is that the jump discontinuity of $p = p(x, y)$ along the curve $y = \gamma(x)$ satisfies the entropy condition for the single convex conservation law (5.10)-(5.11). The second one is that $z = z(x, y)$ is analytic except on the curve $y = \gamma_n(x)$. We think that the analytic extension is acceptable in physical sciences.

Remark 1 Equation (5.3) is convex Hamilton-Jacobi equation. Therefore we can unique-ly find a curve $y = \eta(x)$ on which it holds $z_1(x, \eta(x)) = z_3(x, \eta(x))$. See M. Tsuji [18]. If we may define a solution $z = z(x, y)$ by

$$z(x, y) = \begin{cases} z_1(x, y) & (y < \eta(x)) \ , \\ z_3(x, y) & (y > \eta(x)) \ , \end{cases}$$

then $z = z(x, y)$ satisfies $[q]\dot{\eta} + [p] = 0$. But it does not satisfy (5.1) in distribution sense, though it does (5.1) except on the curve $y = \eta(x)$ in the classical sense.

Remark 2 Similar discussions are possible for $\partial^2 z/\partial x^2 - (1+\epsilon q)^{2\alpha}\partial^2 z/\partial y^2 = 0$ $(\alpha \neq -1)$, and also for $\partial^2 z/\partial x^2 + A(\partial/\partial y)(\partial z/\partial y)^{-\alpha} = 0$ ($A =$ constant $> 0, \ \alpha > 1$).

6 Systems of conservation law

Let us recall a well-known relation between equation (4.1) and certain first order system of conservation law. We write $p = \partial z/\partial x$ and $q = \partial z/\partial y$, and put $U(x,y) =^t (p,q)$, $F(U) =^t (f(q),p)$ and $U_0(y) =^t (z_1(y), z_0'(y))$. Then we get

$$\frac{\partial}{\partial x}U - \frac{\partial}{\partial y}F(U) = 0 \quad \text{in} \quad \{x > 0, y \in \mathbf{R}^1\}, \tag{6.1}$$

$$U(0,y) = U_0(y) \quad \text{on} \quad \{x = 0, y \in \mathbf{R}^1\}. \tag{6.2}$$

P. D. Lax [13] introduced the notion of weak solutions of (8.1)-(8.2) as follows;

Definition 6.1 *A bounded and measurabule 2-vecter function U=U(x,y) is a weak solution of* (6.1)-(6.2) *if it satisfies* (6.1)-(6.2) *in distribution sense, i.e.,*

$$\int_{\mathbf{R}^2_+} \{U(x,y)\frac{\partial \varphi}{\partial x}(x,y) - F\,(U)\frac{\partial \varphi}{\partial y}(x,y)\}dxdy + \int_{\mathbf{R}^1} U_0(y)\varphi(0,y)dy = 0 \tag{6.3}$$

for any two-vector function $\varphi(x,y) \in C_0^\infty(\mathbf{R}^2)$.

If $U = U(x,y)$ is a weak solution of (6.1) which has jump discontinuity along a curve $y = \gamma(x)$, we get the jump condition of Rankine-Hugoniot as follows:

$$[p]\dot{\gamma} + [f(q)] = 0 \tag{6.4}$$

$$[q]\dot{\gamma} + [p] = 0. \tag{6.5}$$

Equations (6.4) and (6.5) are same to (4.4) and (6.5). Let us transform the example given in §5 to a 2×2 system of conservation law. Put $U(x,y) =^t (p,q)$, $F(U) =^t (f(q),p)$ and $U_0 = (\epsilon^{-1}(1+\epsilon\cos y)^{-1}, \cos y)$ where $f(q) = -3^{-1}\epsilon^{-1}(1+\epsilon q)^{-3}$. Then the Cauchy problem (5.1)-(5.2) is expressed by the following form:

$$\frac{\partial U}{\partial x} - \frac{\partial}{\partial y}F(U) = 0 \quad \text{in} \quad \{x > 0, y \in \mathbf{R}^1\}, \tag{6.6}$$

$$U(0,y) = U_0(y) \quad \text{on} \quad \{x = 0, y \in \mathbf{R}^1\}. \tag{6.7}$$

The aim of this section is to get an exact solution of (6.6)-(6.7). We see by the characteristic method that a solution of (6.6)-(6.7) is written by the same characteristic strips as (5.7) and (5.8), i.e.,

$$y = \alpha + (1 + \epsilon \cos \alpha)^{-2} x \tag{6.8}$$

$$p = \epsilon^{-1}(1 + \epsilon \cos \alpha)^{-1}, \quad q = \cos \alpha. \tag{6.9}$$

We repeat here the same discussions as in §5, using the same notations introduced in §5. A point (x^0, α^0) is a point where where $(\partial y/\partial \alpha)(x, \alpha)$ vanishes for the first time, i.e., $(\partial y/\partial \alpha)(x, \alpha) \neq 0$ for $x < x^0$ and $\alpha \in \mathbf{R}^1$, and $(\partial y/\partial \alpha)(x^0, \alpha^0) = 0$. We solve the equation $y = y(x, \alpha)$ with respect to α for $x > x^0$. As $\alpha = \alpha(x, y)$ takes three values in a neighbourhood of (x^0, y^0) where $y^0 = y(x^0, \alpha^0)$, we denote them by $\alpha = \alpha_1(x, y) < \alpha_2(x, y) < \alpha_3(x, y)$. Here we define $p_i(x, y) \equiv p(\alpha_i(x, y))$ and $q_i(x, y) \equiv q(\alpha_i(x, y))$ $(i = 1, 2, 3)$, then $U = U(x, y)$ takes also three values $\{U_i(x, y) \equiv (p_i(x, y), q_i(x, y))\ ; i = 1, 2, 3\}$ in a neighbourhood of (x^0, y^0). Our problem is how to pick up only one value from $\{U_i(x, y)\ ; i = 1, 2, 3\}$ so that $U = U(x, y)$ is a weak solution of (6.6)-(6.7) in the sense of Definition 6.1. For our aim, we would like to find a curve $y = \rho(x)$ so that, if we may define a solution $U = U(x, y)$ by

$$U(x, y) = \begin{cases} U_1(x, y) & (y < \rho(x)) \\ U_3(x, y) & (y > \rho(x)) \end{cases},$$

$U = U(x, y)$ is a weak solution of (6.6)-(6.7). Then the jump conditions (6.4) and (6.5) mean that $y = \rho(x)$ must satisfy the following differential equations:

$$\frac{d\rho}{dx} = -\frac{f(q_3(x, \rho(x)) - f(q_1(x, \rho(x))}{p_3(x, \rho(x)) - p_1(x, \rho(x))}, \quad x > x^0, \tag{6.10}$$

$$\frac{d\rho}{dx} = -\frac{p_3(x, \rho(x) - p_1(x, \rho(x))}{q_3(x, \rho(x) - q_1(x, \rho(x))}, \quad x > x^0, \tag{6.11}$$

$$\rho(x^0) = y^0. \tag{6.12}$$

As the Cauchy problem (6.10)-(6.12) is identical to (5.12)-(5.13) , a solution of (6.10)-(6.12) is equal to the solution, $\gamma = \gamma(x)$, of (5.12)-(5.13). Similarly a solution of (6.11)-(6.12) is the same as the solution $y = \eta(x)$ appeared in Remark 1 of §5. Moreover we can prove $\gamma(x) \neq \eta(x)$. Therefore, we could not succeed to get a weak solution of (6.6)-(6.7) in the sense of Definition 6.1 by the characteristic method. This might suggest us that we had better to introduce another definition of weak solutions of (6.6)-(6.7), because it seems to us that the solution constructed in the above would be acceptable.

References

[1] R. Courant and D. Hilbert, *Method of Mathematical Physics*, vol.2, Interscience, New York, 1962.

[2] G. Darboux, *Leçon sur la théorie générale des surfaces*, tome 3, Gauthier-Villars, Paris, 1894.

[3] B. Gaveau, *Evolution of a shock for a single conservation law in 2+1 dimensions*, Bull. Sci. Math., **113** (1989), 407-442.

[4] E. Goursat, *Leçons sur l'intégration des équations aux dérivées partielles du second ordre*, tome 1, Hermann, Paris, 1896.

[5] E. Goursat, *Cours d'analyse mathématique*, tome 3, Gauthier-Villars, Paris, 1927.

[6] J. Guckenheimer, *Solving a single conservation law*, Lecture Notes in Math. (Springer-Verlag) **468** (1975), 108-134.

[7] J. Hadamard, *Le problème de Cauchy et les équations aux dérivées partielles linéaires hyperboliques*, Hermann, Paris, 1932.

[8] S. Izumiya, *Geometric singularities for Hamilton-Jacobi equations*, Adv. Studies in Math. **22** (1993), 89-100.

[9] S. Izumiya and G. T. Kossioris, *Semi-local classification of geometric singuarities for Hamilton-Jacobi equations*, J. Diff. Eq. **118** (1995), 166-193.

[10] S. Izumiya and G. T. Kossioris, *Formation of singularities for viscosity solutions of Hamilton-Jacobi equations*, Banach Center Publications **33** (1996), 127-148.

[11] G. Jennings, *Piecewise smooth solutions of single conservation law exist*, Adv. in Math. **33** (1979), 192-205.

[12] P. D. Lax, *Development of singularities of solutions of nonlinear hyperbolic partial differential equations*, J. Math. Physics **5** (1964), 611-613.

[13] P. D. Lax, *Hyperbolic systems of conservation laws II*, Comm. Pure Appl. Math. **10** (1957), 537-566.

[14] H. Lewy, *Über das Anfangswertproblem einer hyperbolischen nichtlinearen partiellen Differentialgleichung zweiter Ordnung mit zwei unabhängen Veränderlichen*, Math. Ann. **98** (1928), 179-191.

[15] V. V. Lychagin, *Contact geometry and non-linear second order differential equations*, Russian Math. Surveys **34** (1979), 149-180.

[16] S. Nakane, *Formation of shocks for a single conservation law*, SIAM J. Math. Anal. **19** (1988), 1391-1408.

[17] S. Nakane, *Formation of singularities for Hamilton-Jacobi equations in several space variables*, J. Mat. Soc. Japan **43** (1991), 89-100.

[18] M. Tsuji, *Formation of singularities for Hamilton-Jacobi equation II*, J. Math. Kyoto Univ. **26** (1986), 299-308.

[19] M. Tsuji, *Prolongation of classical solutions and singularities of generalized solutions*, Ann. Inst. H. Poincarè - Analyse nonlnéaire **7** (1990), 505-523.

[20] M. Tsuji, *Formation of singularities for Monge-Ampère equations*, Bull. Sci. math. **119** (1995), 433-457.

[21] M. Tsuji, *Monge-Ampère equations and surfaces with negative Gaussian curvature*, Publications of Banach Center **39** (1997), 161-170.

[22] H. Whitney, *On singularities of mappings of Euclidean spaces I*, Ann. Math. **62** (1955), 374-410.

[23] N. J. Zabusky, *Exact solution for vibrations of a nonlinear continuous model string*, J. Math. Physics **3** (1962), 1028-1039.

POSITIVE SOLUTIONS OF SEMILINEAR ELLIPTIC BOUNDARY VALUE PROBLEMS IN CHEMICAL REACTOR THEORY

Kenichiro Umezu

Maebashi Institute of Technology
Maebashi 371, Japan

Kazuaki Taira

Department of Mathematics
Hiroshima University,
Higashi-Hiroshima 739, Japan

R.P. Gilbert et al.(eds.), Direct and Inverse Problems of Mathematical Physics, 415–422.

Abstract: This paper is devoted to the study of semilinear elliptic boundary value problems arising in chemical reactor theory which obey the simple Arrhenius rate law and Newtonian cooling. We prove that ignition and extinction phenomena occur in the stable steady temperature profile at some critical values of a dimensionless heat evolution rate. Moreover the asymptotic behavior of the stable steady temperature is also studied.

1.1 INTRODUCTION AND RESULTS

Let D be a bounded domain of Euclidean space $\boldsymbol{R}^N$, $N \geq 2$, with smooth boundary ∂D; its closure $\overline{D} = D \cup \partial D$ is an N-dimensional, compact smooth manifold with boundary. We let

$$Au(x) = -\sum_{i=1}^{N} \frac{\partial}{\partial x_i} \left(\sum_{j=1}^{N} a^{ij}(x) \frac{\partial u}{\partial x_j}(x) \right) + c(x)u(x)$$

be a second-order, *elliptic* differential operator with real smooth coefficients on $\overline{D}$ such that:

(1) $a^{ij}(x) = a^{ji}(x)$, $1 \leq i,j \leq N$, and there exists a constant $a_0 > 0$ such that

$$\sum_{i,j=1}^{N} a^{ij}(x)\xi_i\xi_j \geq a_0|\xi|^2, \quad x \in \overline{D},\ \xi \in \boldsymbol{R}^N.$$

(2) $c(x) > 0$ in D.

In this paper we consider the following semilinear elliptic boundary value problem stimulated by a problem of chemical reactor theory (cf. [BGW]).

$$(*)_\lambda \qquad \begin{cases} Au = \lambda \exp\left[\dfrac{u}{1+\varepsilon u}\right] & \text{in } D, \\ Bu = a\dfrac{\partial u}{\partial \nu} + (1-a)u = 0 & \text{on } \partial D. \end{cases}$$

Here:

(1) λ and ε are positive parameters.

(2) $a \in C^\infty(\partial D)$ and $0 \leq a(x) \leq 1$ on ∂D.

(3) $\partial/\partial\nu$ is the conormal derivative associated with the operator A:

$$\frac{\partial}{\partial \nu} = \sum_{i,j=1}^{N} a^{ij} n_j \frac{\partial}{\partial x_i},$$

where $\boldsymbol{n} = (n_1, n_2, \ldots, n_N)$ is the unit exterior normal to the boundary ∂D.

The nonlinear term

$$f(t) = \exp\left[\frac{t}{1+\varepsilon t}\right]$$

describes the temperature dependence of reaction rate for exothermic reactions obeying the simple *Arrhenius rate law* in circumstances in which heat flow is purely conductive. In this context the parameter ε is a dimensionless ambient temperature and the parameter λ is a dimensionless heat evolution rate. The equation

$$Au = \lambda f(u) = \lambda \exp\left[\frac{u}{1+\varepsilon u}\right]$$

represents heat balance with reactant consumption ignored, where u is a dimensionless temperature excess.

On the other hand, the boundary condition

$$Bu = a\frac{\partial u}{\partial \nu} + (1-a)u = 0$$

represents the exchange of heat at the surface of the reactant by *Newtonian cooling*. Moreover the boundary condition $Bu = 0$ is called the isothermal condition (or Dirichlet condition) if $a \equiv 0$ on ∂D, and is called the adiabatic condition (or Neumann condition) if $a \equiv 1$ on ∂D. We remark that problem $(*)_\lambda$ becomes a degenerate boundary value problem from an analytical point of view. This is due to the fact that the well-known Shapiro–Lopatinskii complementary condition is violated at the points where $a(x) = 0$. In the non-degenerate case or one-dimensional case, problem $(*)_\lambda$ has been studied by many authors (see [CL], [Co], [Pa], [LW2], [BIS]).

A function $u \in C^2(\overline{D})$ is called a *solution* of problem $(*)_\lambda$ if it satisfies the equation $Au - \lambda f(u) = 0$ and the boundary condition $Bu = 0$. A solution u is said to be *positive* if it is positive everywhere in D.

Our starting point is the following existence theorem for problem $(*)_\lambda$ (see [TU3, Theorem 1 and Corollary 2]):

Theorem 1.1.1 *For each $\lambda > 0$ and $\varepsilon > 0$, problem $(*)_\lambda$ has at least* one *positive solution. Furthermore, problem $(*)_\lambda$ has a* unique *positive solution for every $\lambda > 0$ if $\varepsilon \geq 1/4$.*

Theorem 0 says that if the activation energy is so low that the parameter ε exceeds the value $1/4$, then only a *smooth* progression of reaction rate with imposed ambient temperature can occur; such a reaction may be very rapid but it is only accelerating and lacks the discontinuous change associated with criticality and ignition.

Next we study the case where $0 < \varepsilon < 1/4$. In order to state our multiplicity theorem for problem $(*)_\lambda$, we define a function

$$\nu(t) = \frac{t}{f(t)} = \frac{t}{\exp\left[t/(1+\varepsilon t)\right]}, \quad t \geq 0.$$

It is easy to see that if $0 < \varepsilon < 1/4$, then the function $\nu(t)$ has a unique local maximum at $t = t_1(\varepsilon)$:

$$t_1(\varepsilon) = \frac{1 - 2\varepsilon - \sqrt{1 - 4\varepsilon}}{2\varepsilon^2},$$

and has a unique local minimum at $t = t_2(\varepsilon)$:

$$t_2(\varepsilon) = \frac{1 - 2\varepsilon + \sqrt{1 - 4\varepsilon}}{2\varepsilon^2}.$$

On the other hand, we let $\phi \in C^\infty(\overline{D})$ be the unique positive solution of the linear boundary value problem

$$\begin{cases} Au = 1 & \text{in } D, \\ Bu = 0 & \text{on } \partial D, \end{cases} \tag{1.1}$$

and let

$$\|\phi\|_\infty = \max_{x \in \overline{D}} \phi(x).$$

Now we can state our multiplicity theorem for problem $(*)_\lambda$:

Theorem 1.1.2 *We can find a constant $\beta > 0$, independent of ε, such that if $0 < \varepsilon < 1/4$ is so small that*

$$\frac{\nu(t_2(\varepsilon))}{\beta} < \frac{\nu(t_1(\varepsilon))}{\|\phi\|_\infty}, \tag{1.2}$$

then there exist at least three *distinct positive solutions of problem $(*)_\lambda$ for all λ satisfying the condition*

$$\frac{\nu(t_2(\varepsilon))}{\beta} < \lambda < \frac{\nu(t_1(\varepsilon))}{\|\phi\|_\infty}.$$

Theorem 1 is a generalization of [Wi, Theorem 4.3] to the degenerate case (see also [Pa], [LW2], [BIS]). We remark that, as $\varepsilon \downarrow 0$,

$$\frac{\nu(t_2(\varepsilon))}{\beta} \sim \frac{1}{\varepsilon^2} \exp\left[\frac{-1}{\varepsilon + \varepsilon^2}\right], \tag{1.3}$$

$$\frac{\nu(t_1(\varepsilon))}{\|\phi\|_\infty} \sim \exp\left[\frac{-1}{1 + \varepsilon}\right], \tag{1.4}$$

so that condition (1.2) makes sense.

Secondly we state two existence and uniqueness theorems for problem $(*)_\lambda$. Let λ_1 be the first eigenvalue of the linear eigenvalue problem

$$\begin{cases} Au = \lambda u & \text{in } D, \\ Bu = 0 & \text{on } \partial D. \end{cases}$$

The next two theorems assert that problem $(*)_\lambda$ is uniquely solvable for sufficiently small and for sufficiently large λ if $0 < \varepsilon < 1/4$:

Theorem 1.1.3 *Let $0 < \varepsilon < 1/4$. If the parameter λ is so small that*

$$0 < \lambda < \frac{\lambda_1 \exp\left[\frac{2\varepsilon - 1}{\varepsilon}\right]}{4\varepsilon^2},$$

then problem $(*)_\lambda$ has a unique *positive solution.*

Theorem 1.1.4 *Let $0 < \varepsilon < 1/4$. One can find a constant $\Lambda > 0$, independent of ε, such that if the parameter λ is so large that $\lambda > \Lambda$, then problem $(*)_\lambda$ has a* unique *positive solution.*

Theorems 2 and 3 are generalizations of [Wi, Theorems 2.9 and 2.6], to the degenerate case, respectively. Here it is worth while to point out (see conditions (1.3) and (1.4)) that we have, as $\varepsilon \downarrow 0$,

$$\frac{\nu(t_2(\varepsilon))}{\beta} \sim \frac{\lambda_1 \exp\left[\frac{2\varepsilon - 1}{\varepsilon}\right]}{4\varepsilon^2},$$
$$\frac{\nu(t_1(\varepsilon))}{\|\phi\|_\infty} \sim \Lambda.$$

By virtue of Theorems 1, 2 and 3, we can define two positive numbers μ_I and μ_E by the formulas

$$\mu_I = \inf\{\mu > 0 : \text{problem } (*)_\lambda \text{ is uniquely solvable for each } \mu < \lambda\},$$
$$\mu_E = \sup\{\mu > 0 : \text{problem } (*)_\lambda \text{ is uniquely solvable for each } 0 < \lambda < \mu\}.$$

Then it is easy to see that an *ignition* phenomenon occurs at $\lambda = \mu_I$ and an *extinction* phenomenon occurs at $\lambda = \mu_E$, respectively. In other words, a small increase in λ causes a large jump in the stable steady temperature profile at $\lambda = \mu_I$ and $\lambda = \mu_E$. More precisely the minimal positive solution $\underline{u}(\lambda)$ is continuous in $\lambda > \mu_I$ but is not continuous at $\lambda = \mu_I$, while the maximal positive solution $\overline{u}(\lambda)$ is continuous in $0 < \lambda < \mu_E$ but is not continuous at $\lambda = \mu_E$.

In this paper, proofs of Theorems 1, 2 and 3 are omitted. For the proofs we refer to [TU4].

Finally we study the asymptotic behavior of positive solutions of problem $(*)_\lambda$ as $\lambda \downarrow 0$ and as $\lambda \uparrow \infty$ for any $\varepsilon > 0$, which is the main purpose of this paper. The following theorem is a generalization for our nonlinearity f of [Da, Theorem 1] to the degenerate case.

Theorem 1.1.5 *Let $\varepsilon > 0$, and let u_λ the unique positive solution of problem $(*)_\lambda$ for every small or large λ (see Theorems 2 and 3). Then we have*

$$u_\lambda \sim \lambda\phi, \quad \lambda \downarrow 0, \tag{1.5}$$

$$u_\lambda \sim \lambda e^{1/\varepsilon}\phi, \quad \lambda \uparrow \infty. \tag{1.6}$$

More precisely, we have

$$\frac{u_\lambda}{\lambda} \longrightarrow \phi \quad \text{in } C^1(\overline{D}), \quad \lambda \downarrow 0,$$

$$\frac{u_\lambda}{\lambda} \longrightarrow e^{1/\varepsilon}\phi \quad \textit{in } C^1(\overline{D}), \quad \lambda \uparrow \infty.$$

The rest of this paper is organized as follows. Section 2 is devoted to the proof of Theorem 4 which is based on a method inspired by Dancer [Da, Theorem 1].

1.2 PROOF OF THEOREM 4

In this section we prove Theorem 4. First we prove assertion (1.5). Let u_λ be the unique positive solution of problem $(*)_\lambda$ for small λ where the uniqueness of positive solutions for any small λ has been established in Theorem 2. For the positive solution ϕ of problem (1.1) we see that $\lambda e^{1/\varepsilon}\phi$ is a supersolution and $\lambda\phi$ is a subsolution of problem $(*)_\lambda$ for all $\lambda > 0$. Indeed, noting that

$$1 < f(t) < e^{1/\varepsilon}, \quad t > 0,$$

we have

$$A\left(\lambda e^{1/\varepsilon}\phi\right) - \lambda f\left(\lambda e^{1/\varepsilon}\phi\right) = \lambda\left\{e^{1/\varepsilon} - f\left(\lambda e^{1/\varepsilon}\phi\right)\right\} > 0 \quad \text{in } D,$$

$$A(\lambda\phi) - \lambda f(\lambda\phi) = \lambda\{1 - f(\lambda\phi)\} < 0 \quad \text{in } D.$$

By the method of super-subsolutions ([TU2, Theorem 1.1]), there exists a positive solution $v \in C^2(\overline{D})$ of problem $(*)_\lambda$ such that

$$\lambda\phi \leq v \leq \lambda e^{1/\varepsilon}\phi \quad \text{in } \overline{D}.$$

¿From the uniqueness of positive solutions it follows that $v \equiv u_\lambda$, and hence that

$$\lambda\phi \leq u_\lambda \leq \lambda e^{1/\varepsilon}\phi \quad \text{in } \overline{D}.$$

It follows that for $x \in \overline{D}$

$$u_\lambda(x) \longrightarrow 0, \quad \lambda \downarrow 0.$$

Since $f(t)$ is uniformly bounded in $t \in [0, \infty)$, so is $f(u_\lambda(x))$ in $x \in \overline{D}$ and $\lambda > 0$. From this assertion, the condition that $f(0) = 1$, and the Lebesgue convergence theorem it follows that for $1 < p < \infty$

$$f(u_\lambda) \longrightarrow 1 \quad \text{in } L^p(D), \quad \lambda \downarrow 0. \tag{2.7}$$

For our purpose we need the following proposition on the existence and uniqueness of solutions in the framework of L^p spaces ($1 < p < \infty$) for the linear degenerate elliptic boundary value problem:

$$\begin{cases} Au = g & \text{in } D, \\ Bu = 0 & \text{on } \partial D. \end{cases}$$

Proposition 1.2.1 ([Um, Theorem 1]) *Let* $1 < p < \infty$. *Then the mapping:*

$$\begin{aligned} \mathcal{A}\colon W^{2,p}_B(D) &\longrightarrow L^p(D) \\ u &\longmapsto Au, \end{aligned}$$

is isomorphic where

$$W_B^{2,p}(D) = \{u \in W^{2,p}(D) : Bu = 0 \text{ on } \partial D\}.$$

For the positive solution ϕ of problem (1.1) we have from assertion (2.1),

$$A\left(\frac{u_\lambda}{\lambda} - \phi\right) = f(u_\lambda) - 1 \longrightarrow 0 \quad \text{in } L^p(D), \quad \lambda \downarrow 0.$$

It follows from Proposition 2.1 that

$$\frac{u_\lambda}{\lambda} - \phi \longrightarrow 0 \quad \text{in } W^{2,p}(D), \quad \lambda \downarrow 0.$$

Taking $p > N$ and using Sobolev's imbedding theorem, we get

$$\frac{u_\lambda}{\lambda} \longrightarrow \phi \quad \text{in } C^1(\overline{D}), \quad \lambda \downarrow 0.$$

Next we prove assertion (1.6). The proof is accomplished in the same way as that of assertion (1.5). Let u_λ be the unique positive solution for large λ where the uniqueness has been established in Theorem 3. By the same argument we observe that

$$\lambda\phi \le u_\lambda \le \lambda e^{1/\varepsilon}\phi \quad \text{in } \overline{D}.$$

It follows that for $x \in D$

$$u_\lambda(x) \longrightarrow \infty, \quad \lambda \uparrow \infty.$$

Hence we have assertion (1.6) since $f(t) \to e^{1/\varepsilon}$ as $t \to \infty$.

The proof of Theorem 4 is now complete. □

References

[BGW] Boddington, T., P. Gray and G. C. Wake. (1977). *Criteria for thermal explosions with and without reactant consumption*, Proc. R. Soc. London A., Vol. 357, (pages 403–422).

[BIS] Brown, K. J., M. M. A. Ibrahim and R. Shivaji. (1981). *S-shaped bifurcation curves problems*, Nonlinear Analysis, TMA, Vol. 5, (pages 475–486).

[Co] Cohen, D. S. (1971). *Multiple stable solutions of nonlinear boundary value problems arising in chemical reactor theory*, SIAM J. Appl. Math., Vol. 20, (pages 1–13).

[CL] Cohen, D. S. and T. W. Laetsch. (1970). *Nonlinear boundary value problems suggested by chemical reactor theory*, J. Differential Equations, Vol. 7, (pages 217–226).

[Da] Dancer, E. N. (1986). *On the number of positive solutions of weakly nonlinear elliptic equations when a parameter is large*, Proc. London Math. Soc., Vol. 53, (pages 429–452).

[LW2] Legget, R. W. and L. R. Williams. (1979). *Multiple fixed point theorems for problems in chemical reactor theory*, J. Math. Anal. Appl., Vol. 69, (pages 180–193).

[Pa] Parter, S. V. (1974). *Solutions of a differential equation in chemical reactor processes*, SIAM J. Appl. Math., Vol. 26, (pages 687–715).

[TU2] Taira, K. and K. Umezu. (1996). *Bifurcation for nonlinear elliptic boundary value problems III*, Adv. Differential Equations, Vol. 1, (pages 709–727).

[TU3] Taira, K. and K. Umezu. (1997). *Positive solutions of sublinear elliptic boundary value problems*, Nonlinear Analysis, TMA, Vol. 29, (pages 761–771).

[TU4] Taira, K. and K. Umezu. (to appear). *Semilinear elliptic boundary value problems in chemical reactor theory*, J. Differential Equations.

[Um] Umezu,, K. (1994). *L^p-approach to mixed boundary value problems for second- order elliptic operators*, Tokyo J. Math., Vol. 17, (pages 101–123).

[Wi] Wiebers, H. (1985). *S-shaped bifurcation curves of nonlinear elliptic boundary value problems*, Math. Ann., Vol. 270, (pages 555–570).

FAST SOLVERS OF THE LIPPMANN–SCHWINGER EQUATION

Gennadi Vainikko
Helsinki University of Technology
Institute of Mathematics
P.O. Box 1100
FIN–02015 HUT, Finland
e-mail: Gennadi.Vainikko@hut.fi

Abstract The electromagnetic and acoustic scattering problems for the Helmholtz equation in two and three dimensions are equivalent to the Lippmann–Schwinger equation which is a weakly singular volume integral equation on the support of the scatterer. We propose for the Lippmann–Schwinger equation two discretizations of the optimal accuracy order, accompanied by fast solvers of corresponding systems of linear equations. The first method is of the second order and based on simplest cubatures; the scatterer is allowed to be only piecewise smooth. The second method is of arbitrary order and is based on a fully discrete version of the collocation method with trigonometric test functions; the scatterer is assumed to be smooth on whole space $\mathbb{R}^n$ and of compact support.

1. INTRODUCTION

In this paper we deal with the integral equation formulation on the scattering problem for the Helmholtz equation in the inhomogeneous media. We assume that the inhomogeneity is smooth or piecewise smooth and of compact support containing the origin, with possibly complex-valued smooth or piecewise smooth refractive index $b : \mathbb{R}^n \to \mathbb{C}$, $n = 2$ or 3, $b(x) = 1$ outside the inhomogeneity. The formulation of the problem reads as follows: find $u : \mathbb{R}^n \to \mathbb{C}$ ($n = 2$ or 3) such that

$$\Delta u(x) + \kappa^2 b(x) u(x) = 0\,, \qquad x \in \mathbb{R}^n, \tag{1.1}$$

R.P. Gilbert et al.(eds.), Direct and Inverse Problems of Mathematical Physics, 423–440.

The formulation of the problem reads as follows: find $u : \mathbb{R}^n \to \mathbb{C}$ ($n = 2$ or 3) such that

$$\Delta u(x) + \kappa^2 b(x) u(x) = 0\,, \qquad x \in \mathbb{R}^n, \tag{1.1}$$

$$u = u^i + u^s, \tag{1.2}$$

$$\lim_{r=|x|\to\infty} r^{(n-1)/2}\Big(\frac{\partial u^s}{\partial r} - i\kappa u^s\Big) = 0 \quad \text{uniformly for } \frac{x}{|x|} \in S(0,1) \tag{1.3}$$

where u^i (the incident field) is a given entire solution of the Helmholtz equation $\Delta u + \kappa^2 u = 0$, $x \in \mathbb{R}^n$ (usually u^i is given as a plain wave: $u^i(x) = \exp(i\kappa d \cdot x)$, $d \in \mathbb{R}^n$, $|d| = 1$), u^s is the scattered field, $\kappa > 0$ is the wave number; (3) means that u^s must satisfy the Sommerfeld radiation condition. We refer to [2] for more details concerning problem (1)–(3).

Problem (1)–(3) is equivalent to the Lippmann–Schwinger integral equation (see [2])

$$u(x) = u^i(x) - \kappa^n \int l_{\mathbb{R}^n} \Phi(\kappa|x-y|) a(y) u(y) dy \tag{1.4}$$

where $a = 1 - b$ is smooth or piecewise smooth and of compact support,

$$\Phi(r) = \left\{ \begin{array}{ll} \S\frac{i}{4} H_0^{(1)}(r), & n = 2 \\ \S\frac{1}{4\pi}\frac{e^{ir}}{r} \quad, & n = 3 \end{array} \right\}, \quad r > 0\,, \tag{1.5}$$

$H_0^{(1)}$ is the Hankel function of the first kind of order zero (see [1], formula 9.1.3). For $r \to 0$, $H_0^{(1)}(r) \sim -\frac{1}{2\pi} \ln r$. Thus integral equation (4) is weakly singular both in cases $n = 2$ and $n = 3$. The integration over $\mathbb{R}^n$ can be replaced by the integration over sup a.

Problem (1)–(3) and integral equation (4) are uniquely solvable if and only in the homogeneous integral equation corresponding to (4) has only the trivial solution or, equivalently, the homogeneous problem corresponding to, (1)–(3), i.e. the problem with $u^i = 0$, has only the trivial solution.

The unboundedness of the domain $\mathbb{R}^n$ in problem (1)–(3) causes some numerical difficulties. A simplest idea is to use grid methods in a large ball $B(0,R)$, with boundary condition $\frac{u^s}{r} - i\kappa u^s = 0$ for $|x| = R$. This method produces very large discrete problems. Another, more popular idea elaborated in [5] is to use coupled finite and boundary methods: in a ball $B(0,\rho)$ containing the support of $a = 1 - b$, the problem is treated by finite elements; a boundary integral equation and the Nystrom method are used to treat the problem in the domain $|x| > \rho$; finally, a special equation is derived to produce appropriate boundary values of $u + i\kappa\frac{u}{r}$ for $|x| = \rho$. This approach is complicated in it essence. As mentioned in [2], the volume potential approach, i.e. the solution of Lippmann–Schwinger equation (4) instead of (1)–(3), has the advantage that the problem in an unbounded domain is handled in a simple and natural way; a disadvantage is that one has to approximate multidimensional weakly singular integrals and that the discrete problem derived from (4) has a non-sparse matrix. In the present paper we try to show that actually these disadvantages are not serious. First, the optimal convergence order

$$\|v_N - au\|_\lambda \le cN^{\lambda-\mu}\|au\|_\mu \qquad (0 \le \lambda \le \mu)$$

in the scale of Sobolev norms with any $\mu > n/2$ can be achieved by trigonometric collocation method (cf. [6]) applied to a periodized version of (4) if a and u^i are sufficiently smooth ($a \in W^{\mu,2}(\mathrm{IR}^n)$ and $u^i \in W^{\mu,2}_{loc}(\mathrm{IR}^n)$). Secondly, the N^n parameters of v_N can be computed in $\emptyset(N^n \ln N)$ arithmetical operations. Finally, the algorithm needs to store $\emptyset(N^n)$ quantities. The method is treated in Section 3. In Section 2 we discuss a method of the second accuracy order:

$$\max_j |u_{j,h} - u(jh)| \le ch^2(1 + |\ln h|), \qquad h = 1/N;$$

here a may be only piecewise smooth. This method is a modification of a simplest cubature formula method examined in [7] for more general weakly singular integral equations. The purpose of the modification is to obtain a convolution system as the discrete counterpart of (4) maintaining the second order of the approximation. The convolution system can be solved in $\emptyset(N^n \log N)$ arithmetical operations using FFT and two grid iterations.

1.2 THE CASE OF PIECEWISE SMOOTH A

First we somewhat simplify the form of the integral equation (4). The change of variables

$$\tilde{x} = \kappa x, \quad \tilde{y} = \kappa y, \quad \tilde{u}(\tilde{x}) = u(x), \quad \tilde{a}(\tilde{y}) = a(y), \quad \tilde{u}^i(\tilde{x}) = u^i(x)$$

transforms (4) into

$$\tilde{u}(\tilde{x}) = \tilde{u}^i(\tilde{x}) - \int l_{\mathrm{IR}^n}\Phi(|\tilde{x} - \tilde{y}|)\tilde{u}(\tilde{y})d\tilde{y}$$

which is (4) with $\kappa = 1$. Thus, without a loss of generality, we put $\kappa = 1$ in (4). To a great wave number κ now there corresponds a large support of $\tilde{a}$, namely, $\sup \tilde{a} = \kappa \sup a$. Further, instead of u^i, an entire solution to Helmholtz equation, we consider an arbitrary sufficiently smooth function $f : \mathrm{IR}^n \to \mathbb{C}$. Thus our problem reads as follows: given a piecewise smooth function $a : \mathrm{IR}^n \to \mathbb{C}$ with support in an open bounded set $G \subset \mathrm{IR}^n$, and a smooth function $f : \mathrm{IR}^n \to \mathbb{C}$, find $u : G \to \mathrm{IR}^n$ satisfying the integral equation

$$u(x) = f(x) - \int_G \Phi(|x - y|)a(y)u(y)dy \qquad (x \in G). \tag{2.6}$$

Recall that Φ is given by formula (5). Solving (6) we obtain $u(x)$ for $x \in G$; for $x \in \mathrm{IR}^n \backslash G$, $u(x)$ can be obtained after that by simple integration. Notice that due to the Fredholm alternative, equation (6) remains to be uniquely solvable if it is uniquely solvable for $f(x) = u^i(x)$, i.e. if problem (1)–(3) is uniquely given a piecewise smooth function $a: \mathbb{R}^n \to \mathbb{C}$ with support in an open bounded set $G \subset \mathbb{R}^n$, and a smooth function $f: \mathbb{R}^n \to \mathbb{C}$, find $u: G \to \mathbb{R}^n$ satisfying the integral equation

$$u(x) = f(x) - \int_G \Phi(|x-y|)a(y)u(y)dy \qquad (x \in G). \tag{2.6}$$

Recall that Φ is given by formula (1.5). Solving (2.6) we obtain $u(x)$ for $x \in G$; for $x \in \mathbb{R}^n \backslash G$, $u(x)$ can be obtained after that by simple integration. Notice that due to the Fredholm alternative, equation (2.6) remains to be uniquely solvable if it is uniquely solvable for $f(x) = u^i(x)$, i.e. if problem (1.1)–(1.3) is uniquely solvable. Now we make precise conditions on a used in this section. We assume that

$$a \in W^{2,\infty}(G \backslash \Gamma)$$

where Γ consists of a finite number of piecewise smooth compact surfaces Γ_i (for $n = 3$) or curves Γ_i (for $n = 2$) which may meet each other along manifolds of dimension $\leq n-2$. More precisely, every Γ_i satisfies the following condition (PS): there exist constants $c_0 > 0$ and $r_0 > 0$ such that, for $x \in \Gamma_i$, the piece $\Gamma_i \cap \overline{B}(x, r_0)$ of Γ_i is representable in the form $z_n = \varphi(z')$, $z' = (z_1, z_{n-1}) \in Z_{i,x}$ where $z_1,, z_n$ is a suitable orthogonal system of coordinates obtained from the original system $x_1,, x_n$ by the translation of the origin into the point x and a rotation of axes; $Z_{i,x} \subset \mathbb{R}^{n-1}$ is a bounded closed region and $\varphi = \varphi_{i,x}$ is a continuous function on $Z_{i,x}$ which is continuously differentiable with $|\text{grad}\varphi(z')| \leq c_0$ everywhere in $Z_{i,x}$ except possibly a manifold of dimension $\leq n-2$. The surfaces of a ball, cylinder, cone and cube are simplest examples of Γ_i satisfying (PS), together they may build more complicated configurations, e.g. two tangential balls (one in another).

Let $h > 0$ be a discretization step. For $j = (j_1, j_n) \in \mathbb{Z}^n$, denote

$$B_{j,h} = \{x = (x_1, \ldots, x_n) \in \mathbb{R}^n : \Big(j_k - \frac{1}{2}\Big)h < x_k < \Big(j_k + \frac{1}{2}\Big)h,\ k = 1, \ldots, n\}.$$

This is a rectangular cell with the center at jh. We define the grid approximation of a as follows. We put

$$a_{j,h} = a(jh) \ \text{ if } \ B_{j,h} \cap \Gamma = \emptyset;$$

in particular,

$$a_{j,h} = 0 \ \text{ if } \ B_{j,h} \cap \text{ supp } a = \emptyset.$$

For $B_{j,h} \cap \Gamma \neq \emptyset$ we put

$$a_{j,h} = h^{-n} \sum_{p=1}^{q_j} a(x_{j,h}^{(p)}) \text{ meas } B_{j,h}^{(p)} \ \text{ with some } x_{j,h}^{(p)} \in B_{j,h}^{(p)}$$

where $B_{j,h}^{(p)}$ $(p = 1, \ldots, q_j)$ are the connectivity components of the open set $B_{j,h} \backslash \Gamma$. The measure of $B_{j,h}^{(p)}$ may be computed approximately with an accuracy $\emptyset(h^{n+1})$.

Further, define

$$\Phi_{j,h} = \begin{cases} \Phi(|j|h) & 0 \neq j \in \mathbb{Z}^n, \\ 0 & j = 0\,. \end{cases}$$

Take a sufficiently large $N \in \mathbb{N}$ and an open bounded set $G \in \mathbb{R}^n$ (independent of h) such that

$$\operatorname{supp} a \subset \sum_{j \in \mathbb{Z}_N^n} \overline{B}_{j,h} \subset G \tag{2.7}$$

where

$$\mathbb{Z}_N^n = \{j \in \mathbb{Z}^n : -\frac{N}{2} < j_k \leq \frac{N}{2}\ k = 1, \ldots, n\}.$$

We approximate (6) by the discrete problem

$$u_{j,h} = f(jh) - h^n \sum_{k \in \mathbb{Z}_N^n} \Phi_{j-k,h} a_{k,h} u_{k,h} \qquad (j \in \mathbb{Z}_N^n)\,.$$

This is a modification of the cubature formula method examined in [7]. The modification is performed so that the discrete problem maintains the convolution structure. The following convergence result can be followed from the general result of

Theorem 2..1 . *Assume that* $\operatorname{supp} a$ *satisfies (7),* $f \in C^2(\overline{G})$ *and* $a \in W^{2,\infty}(G \backslash \Gamma)$ *with* Γ *satisfying condition (PS). Finally, assume that the homogeneous integral equation corresponding to (6) possesses only the trivial solution. Then system (8) is uniquely solvable for all sufficiently small* $h > 0$*, and*

$$\max_{j \in \mathbb{Z}_N^n} |u_{j,h} - u(jh)| \leq ch^2(1 + |\ln h|)$$

where $u \in C(\overline{G})$ *is the solution of (2.6) and* $u_{j,h}$ $(j \in \mathbb{Z}_N^n)$ *is the solution of system (8)*.

An application of the convolution multi-matrix of system (8) to the multi-vector $\underline{a}_h \underline{u}_h$ costs $\mathcal{O}(N^n \log N)$ arithmetical operations if FFT techniques is involved. This enables to solve system (8) with the accuracy $\mathcal{O}(h^2(1 + |\ln h|)$ in $\mathcal{O}(N^n \ln N)$ arithmetical operations using two-grid iterations. For instance, putting $M \sim N^{1/3}$ for the coarse grid, 5 iterations are sufficient; the M-multisystems can be solved e.g. by Gauss elimination. There are different other strategies for the two-grid methods. We quote to [7] for numerical schemes. In Section 3.7 we present more details in the case of smooth a and trigonometric collocation as the basis of the discretization.

3. THE CASE OF A SMOOTH A

3.1 PERIODIZATION OF THE PROBLEM

From now we assume that

$$\operatorname{supp} a \subset \overline{B}(0,\rho), \quad a \in W^{\mu,2}(\mathcal{R}^n), \quad f \in W^{\mu,2}_{\mathrm{loc}}(\mathcal{R}^n), \quad \mu > \frac{n}{2}. \tag{3.8}$$

Due to the Sobolev imbedding theorem, it follows from (9) that a and f are continuous on $\mathcal{R}^n$. Denote

$$G_R = \{\in \mathcal{R}^n : |x_k| < R,\ k = 1,\ldots,n\}$$

where $R \geq 2\rho$ is a parameter. Multiplying both sides of (6) by $a(x)$ we rewrite equation (2.6) with respect to $a(x)u(x)$:

$$a(x)u(x) = a(x)f(x) - a(x)\int_{G_R} \Phi(|x-y|)[a(y)u(y)]dy \qquad (x \in G_R).$$

We are interested in finding of $a(x)u(x)$ for $x \in \operatorname{supp} a \subseteq \overline{B}(0,\rho)$. For those x, only values from $\overline{B}(0,2\rho)$ of the function $\Phi(|x|)$ are involved; changing Φ outside this ball, the solution $a(x)u(x)$ does not change in $\overline{B}(0,\rho)$. We exploit this observation and define a new kernel $K(x)$ which coincides with $\Phi(|x|)$ for $x \in \overline{B}(0,2\rho)$. The simplest possibility is to cut $\Phi(|x|)$ off at $|x| = R$:

$$K(x) = \begin{cases} \Phi(|x|) & |x| \leq R \\ 0 & x \in G_R \backslash B(0,R)\,, \end{cases} \qquad R \geq 2\rho. \tag{3.9}$$

We also consider the possibility with smooth cutting:

$$K(x) = \Phi(|x|)\psi(|x|)\,, \qquad x \in G_R\,,\ R > 2\rho\,, \tag{3.10}$$

with $\psi :\ [0,\infty) \to \mathcal{R}$ satisfying the conditions

$$\psi \in C^\infty[0,\infty)\,, \quad \psi(r) = 1 \ \text{ for } \ 0 \leq r \leq 2\rho\,, \psi(r) = 0 \ \text{ for } \ r \geq R \tag{3.11}$$

After that we extend functions K, a and af from G_R to $\mathcal{R}^n$ as $2R$-periodic functions with respect to $x_1,\ldots,x_n$; for extensions we use the same designations. Thus we have a multiperiodic integral equation

$$v(x) = a(x)f(x) - a(x)\int_{G_R} K(x-y)v(y)dy. \tag{3.12}$$

It is easy to see that a unique solvability of (6) involves a unique solvability of (13). As already explained, the solutions are related by

$$v(x) = a(x)u(x) \quad \text{for} \quad x \in \overline{B}(0,\rho)\,,$$

moreover, v is the $2R$-periodization of au restricted to G_R. Further,

$$u(x) = f(x) - \int_{B(0,\rho)} \Phi(|x-y|)v(y)dy \quad \text{for } x \in \mathcal{R}^n \tag{3.13}$$

and in particular

$$u(x) = f(x) - \int_{G_R} K(x-y)v(y)dy \quad \text{for} \quad x \in \overline{B}(0,\rho)\,.$$

3.2 PERIODIC SOBOLEV SPACES H^λ

The trigonometric orthonormal basis of $L^2(G_R)$ is given by

$$\varphi_j(x) = (2R)^{-n/2}\exp\left(i\pi j\cdot\frac{x}{R}\right), \qquad j = (j_1,\ldots,j_n) \in \mathbb{Z}^n. \tag{3.14}$$

Introduce the Sobolev space $H^\lambda = H^\lambda(G_R)$, $\lambda \in \mathcal{R}$, which consists of $2R$-multiperiodic functions (distributions) u having the finite norm

$$\|u\|_\lambda = \Big(\sum_{j\in\mathbb{Z}^n} \underline{j}^{2\lambda}|\hat{u}(j)|^2\Big)^{\frac{1}{2}}.$$

Here

$$\hat{u}(j) = \int_{G_R} u(x)\overline{\varphi_j(x)}dx = \langle u, \varphi_{-j}\rangle\,, \qquad j \in \mathbb{Z}^n,$$

are the Fourier coefficients of u, and

$$\underline{j} = \begin{cases} |j| = (j_1^2 + \ldots + j_n^2)^{\frac{1}{2}} & 0 \neq j \in \mathbb{Z}^n \\ 1 & j = 0\,. \end{cases}$$

Notice that $H^\lambda \subset W^{\lambda,2}_{\text{loc}}(\mathcal{R}^n)$, thus a function $u \in H^\lambda$ with $\lambda > n/2$ is continuous. We will also use the relation

$$u, v \in H^\lambda, \quad \lambda > n/2 \;\Rightarrow\; uv \in H^\lambda, \quad \|uv\|_\lambda \leq c_\lambda \|u\|_\lambda \|v\|_\lambda\,;$$

a proof can be constructed as in [4] where somewhat different Sobolev spaces are used.

3.3 TRIGONOMETRIC COLLOCATION

Recall the designation

$$\mathcal{Z}_N^n = \{j \in \mathcal{Z}^n : -\frac{N}{2} < j_k \leq \frac{N}{2},\ k = 1, \ldots, n\}.$$

Let $\mathcal{T}_N$ be the N^n-dimensional linear space of trigonometric polynomials of the form $v_N = \sum_{j\in \mathcal{Z}_N^n} c_j \varphi_j, c_j \in \mathcal{C}$. The formula

$$P_N v = \sum_{j\in \mathbf{Z}_N^n} \hat{v}(j)\varphi_j$$

defines the orthogonal projection P_N in H^λ to $\mathcal{T}_N$. Clearly,

$$\|v - P_N v\|_\lambda \leq \left(\frac{N}{2}\right)^{\lambda-\mu} \|v\|_\mu \ \text{ for } \ \lambda \leq \mu,\ \lambda, \mu \in \mathcal{R}. \tag{3.15}$$

For $v \in H^\mu$, $\mu > \frac{n}{2}$, we define the interpolation projection $Q_N v$ claiming

$$Q_N v \in \mathcal{T}_N, \quad (Q_N v)(jh) = v(jh)\,, \quad j \in \mathbf{Z}_N^n\,, \ \text{ where } \ h = 2R/N\,.$$

The error of the trigonometric interpolation can be estimated by (cf. (16))

$$\|v - Q_N v\|_\lambda \leq c_{\lambda,\mu} N^{\lambda-\mu} \|v\|_\mu \ \text{ for } \ 0 \leq \lambda \leq \mu,\ \mu > \frac{n}{2};$$

$$\text{for } \ 0 \leq \lambda \leq \mu,\ \mu > \frac{n}{2}; \tag{3.16}$$

a proof with a characterization of the constant $c_{\lambda,\mu}$ can be constructed following [8].

We solve the equation (13) by trigonometric collocation method

$$v_N = Q_N(af) - Q_N(aKv_N) \tag{3.17}$$

where K is the integral operator from (13):

$$(Kv)(x) = \int_{G_R} K(x - y)v(y)dy\,.$$

Since $K(x)$ is $2R$-periodic, the eigenvalues and eigenfunctions of the convolution operator K are known to be $\hat{K}(j)$ and $\varphi_j(x)$, respectively:

$$K\varphi_j = \hat{K}(j)\varphi_j \qquad (j \in \mathbb{Z}^n)\,.$$

In the case of cutting (10), closed formulae for $\hat{K}(j)$ are presented in Section 3.8. According to those formulae,

$$|\hat{K}(j)| \leq c_R \begin{cases} |j|^{-3/2} & n = 2 \\ |j|^{-2} & n = 3 \end{cases} \qquad (j \neq 0)\,.$$

In the case of smooth cutting (11),

$$|\hat{K}(j)| \leq c_R |j|^{-2} \quad (j \neq 0) \text{ for } n=2 \text{ and } n=3.$$

We present in Section 3.8 a cheap algorithm to approximate $\hat{K}(j)$, $j \in \mathbb{Z}_N^n$, in this case.

We quote to [6] for the study of the trigonometric collocation method in a more complicated but one dimensional situation.

3.4 MATRIX FORM OF THE COLLOCATION METHOD

We have two representations of a trigonometric polynomial $v_N \in \mathcal{T}_N$: (i) through its Fourier coefficients by

$$v_N(x) = \sum_{k \in \mathbb{Z}_N^n} \hat{v}_N(k) \varphi_k(x)$$

with φ_k defined in (15); (ii) through its nodal values by

$$v_N(x) = \sum_{j \in \mathbb{Z}_N^n} v_N(jh) \varphi_{N,j}(x), \qquad h = \frac{2R}{N},$$

with $\varphi_{N,j} \in \mathcal{T}_N$ satisfying $\varphi_{N,j}(kh) = \delta_{j,k}$ (the Kronecker symbol), $j, k \in \mathbb{Z}_N^n$. An explicit formula for $\varphi_{N,j}$ is given by

$$\varphi_{N,j}(x) = N^{-n} \sum_{k \in \mathbb{Z}_N^n} \exp\left(i\pi k \cdot \left(\frac{x}{R} - \frac{2j}{N}\right)\right), \qquad j \in \mathbb{Z}_N^n.$$

It is easy to change the type of representation. Having the nodal values $\underline{v}_N$ of $v_N \in \mathcal{T}_N$, its Fourier coefficients $\hat{v}_N = \mathcal{F}_N \underline{v}_N$ are given by the discrete Fourier transformation $\mathcal{F}_N$:

$$\hat{v}_N(k) = \int_{G_R} v_N(x) \varphi_{-k}(x) dx = (2R)^{n/2} N^{-n} \sum_{j \in \mathbb{Z}_N^n} v_N(jh) \exp\left\{-i\pi k \cdot \frac{2j}{N}\right\}, \quad k \in \mathbb{Z}_N^n.$$

Conversely, if we have the Fourier coefficients $\hat{v}_N$ of $v_N \in \mathcal{T}_N$, then its nodal values $\underline{v}_N = \mathcal{F}_N^{-1} \hat{v}_N$ are given by the inverse discrete Fourier transformation $\mathcal{F}_N^{-1}$:

$$v_N(jh) = (2R)^{-n/2} \sum_{k \in \mathbb{Z}_N^n} \hat{v}_N(k) \exp\left\{i\pi k \cdot \frac{2j}{N}\right\}, \qquad j \in \mathbb{Z}_N^n.$$

Using the transforms $\mathcal{F}_N$ and $\mathcal{F}_N^{-1}$, the collocation method (18) can be represented as

$$\underline{v}_N = (\underline{af})_N - \underline{a}_N \mathcal{F}_N^{-1} \hat{K}_N \mathcal{F}_N \underline{v}_N$$

where $\underline{a}_N$ and $(\underline{af})_N$ present the nodal values at jh, $j \in \mathbb{Z}_N^n$, of a and af, respectively, and $\hat{K}_N$ multiplies the Fourier coefficients $\hat{v}_N(k)$ of $\hat{v}_N = \mathcal{F}_N \underline{v}_N$ by $\hat{K}(k)$, $k \in \mathbb{Z}_N^n$; the product of $\underline{a}_N$ and $\mathcal{F}_N^{-1}\hat{K}\underline{v}_N$ is to be taken pointwise at nodes. Thus, the matrix form of the collocation method (18) is given by

$$A_N \underline{v}_N = \underline{g}_N, \quad A_N = I_N + \underline{a}_N \mathcal{F}_N^{-1} \hat{K}_N \mathcal{F}_N, \quad \underline{g}_N = (\underline{af})_N. \quad (3.18)$$

An application of $\mathcal{F}_N$ or $\mathcal{F}_N^{-1}$ to a N^n-vector $\underline{v}_N$ or $\hat{v}_N$ in a usual way costs N^{2n} multiplications and additions but the FFT does this in $\emptyset(N^n \log N)$ arithmetical operations. The application of diagonal operations $\hat{K}_N$ and $\underline{a}_N$ cost N^n multiplications. Thus, the application of A_N is cheap and therefore it is appropriate to solve (19) by some iteration method. We return to this question in Sections 3.6–3.7.

With respect to Fourier coefficients, the matrix form of the collocation method (18) reads as follows:

$$\hat{A}_N \hat{v}_N = \hat{g}_N, \quad \hat{A}_N = I_N + \mathcal{F}_N \underline{a}_N \mathcal{F}_N^{-1} \hat{K}_N, \quad \hat{g}_N = \mathcal{F}_N(\underline{af})_N.$$

3.5 CONVERGENCE OF THE COLLOCATION METHOD

Lemma 3..1 *Assume that $a \in H^\mu$, $\mu > n/2$, and $|\hat{K}(j)| \le c|j|^{-2}$ ($j \neq 0$). Then*

$$\|aK - Q_N aK\|_{L(H^\lambda, H^\lambda)} \le c \begin{cases} N^{\lambda-\mu} & 0 \le \lambda \le \mu, \ \mu \le 2, \\ N^{-2} & 0 \le \lambda \le \mu - 2, \ \mu \ge 2, \\ N^{\lambda-\mu} & \mu - 2 \le \lambda \le \mu, \ \mu \ge 2. \end{cases} \quad (3.19)$$

Proof. First notice that $K \in L(H^\lambda, H^{\lambda+2})$ for any $\lambda \in \mathcal{R}$. Consider the case $\mu \ge 2$, $0 \le \lambda \le \mu - 2$. Due to (17),

$$\|aKv - Q_N(aKv)\|_\lambda \le cN^{-2}\|aKv\|_{\lambda+2} \le c'N^{-2}\|a\|_{\lambda+2}\|Kv\|_{\lambda+2} \le c''N^{-2}\|a\|_{\lambda+2}\|v\|_\lambda$$

resulting to $\|aK - Q_N aK\|_{L(H^\lambda,H^\lambda)} \le cN^{-2}$. The other cases can be analyzed in a similar way. □

In the case of cutting (10), $n = 2$, we have $|\hat{K}(j)| \le c|j|^{-3/2}$ ($j \neq 0$), and instead of (20) we obtain

$$\|aK - Q_N aK\|_{L(H^\lambda, H^\lambda)} \le c \begin{cases} N^{\lambda-\mu} & 0 \le \lambda \le \mu, \ \mu \le \frac{3}{2}, \\ N^{-3/2} & 0 \le \lambda \le \mu - \frac{3}{2}, \ \mu \ge \frac{3}{2}, \\ N^{\lambda-\mu} & \mu - \frac{3}{2} \le \lambda \le \mu, \ \mu \ge \frac{3}{2}. \end{cases} \quad (3.20)$$

Theorem 3..2 *Assume that the functions a and f satisfy (9), and the homogeneous problem corresponding to (1.1)-(1.3), with $\kappa = 1$, $u^i = 0$, has only the trivial solution. Then equation (13) has a unique solution $v \in H^\mu$, collocation equation (18) has a unique solution $v_N \in \mathcal{T}_N$ for $N \geq N_0$, and*

$$\|v_N - v\|_\lambda \leq c\|v - Q_N v\|_\lambda \leq c'\|v\|_\mu N^{\lambda-\mu}, \qquad 0 \leq \lambda \leq \mu. \quad (3.21)$$

Proof. The bounded inverse to $I + aK$ in $\mathrm{L}(H^\lambda, H^\lambda)$ exists since $aK \in \mathrm{L}(H^\lambda, H^\lambda)$ is compact and the homogeneous integral equation corresponding to (13) has only the trivial solution. Using (20) or (21) we obtain that the inverse to $I + Q_N aK$ in $\mathrm{L}(H^\lambda, H^\lambda)$ exists for all sufficiently great N, and

$$\|(I + Q_N aK)^{-1}\|_{\mathrm{L}(H^\lambda, H^\lambda)} \leq c \qquad (0 \leq \lambda \leq \mu,\ N \geq N_0). \quad (3.22)$$

Error estimate (22) follows form (21) and the equality

$$(I + Q_N aK)(v_N - v) = Q_N(af) - v - Q_N aKv = Q_N v - v. \qquad \square$$

Thus we have an approximation $v_N \in \mathcal{T}_N$ to the $2R$-periodic extension v of au. An approximation to u outside $B(0, \rho)$ can be defined by the discretization of (14):

$$u_N(x) = f(x) - h^n \sum_{j \in \mathbb{Z}_N^n} \Phi(|x - jh|) v_N(jh), \quad h = 2R/N, \quad |x| > \rho.$$

It can be deduced from (22) that

$$|u_N(x) - u(x)| \leq c|x|^{-(n-1)/2}\|v\|_\mu N^{-\mu}, \qquad |x| \geq 2\rho, \qquad (3.23)$$

where the constant c is independent of x and N (this constant has a bad behaviour as $|x| \to \rho$, therefore we restricted us to $|x| \geq 2\rho$).

The following well known asymptotic formula for the solution of (2.6) follows from (14)and the properties of $\Phi(r)$ as $r \to \infty$ (for the behaviour of $H_0^1(r)$, see formula 9.2.3 in [1]):

$$u(x) = f(x) - \frac{e^{i|x|}}{|x|^{(n-1)/2}} u_\infty(\hat{x}) + \emptyset(|x|^{-(n+1)/2}), \qquad |x| \to \infty,$$

where $\hat{x} = x/|x|$ and the far field pattern u_∞ is defined by

$$u_\infty(\hat{x}) = \gamma_n \int_{G_R} e^{-i\hat{x}\cdot y} v(y) dy, \qquad \gamma_n = \begin{cases} \S\frac{1+i}{4\sqrt{\pi}} & n = 2, \\ \S\frac{1}{4\pi} & n = 3. \end{cases}$$

A natural approximation to u_∞ is given by

$$u_{\infty,N}(\hat{x}) = \gamma_n h^n \sum_{j\in\mathbb{Z}_N^n} e^{-i\hat{x}\cdot jh} v_N(jh), \quad \hat{x}\in S(0,1).$$

Under conditions of Theorem 2,

$$\max_{\hat{x}\in S(0,1)} |u_{\infty,N}(\hat{x}) - u_\infty(\hat{x})| \le c\|v\|_\mu N^{-\mu}. \tag{3.24}$$

3.6 SOLUTION OF THE SYSTEM OF THE COLLOCATION METHOD

As already mentioned, iteration methods are most natural to solve the system(19)ı. Due to (23), for the condition number γ_N of the system, with respect to the spectral norm, we have

$$\gamma_N := \|A_N\|\,\|A_N^{-1}\| = \|I+Q_N aK\|_{\mathrm{L}(H^0,H^0)}\|(I+Q_N aK)^{-1}\|_{\mathrm{L}(H^0,H^0)} \le \gamma$$

where the constant γ is independent of N. If v_N^k denotes the kth iteration approximation by the conjugate gradient method applied to the symmetrized system

$$A_N^* A_N \underline{v}_N = A_N^* \underline{g}_N, \qquad A_N^* = I_N + \mathcal{F}_N^{-1}\hat{K}_N^*\mathcal{F}_N \underline{a}_n^*,$$

then (see e.g. [3])

$$\|v_N^k - v_N\|_0 \le cq^k\|v_N^0 - v_N\|_0, \qquad q = (\gamma-1)/(\gamma+1),$$

and the accuracy $\|v_N^k - v_N\|_0 \le cN^{-\mu}$ (cf. (22)) will be achieved in $\emptyset(\log N/|\log q|)$ iteration steps. Since every iteration step costs $\emptyset(N^n \log N)$ arithmetical operations, the whole cost of the method is $\emptyset(N^n \log^2 N)$ arithmetical operations. This amount of the work can be reduced to $\emptyset(N^n \log N)$ arithmetical operations with the help of two grid iteration schemes.

3.7 TWO GRID ITERATIONS

Denoting

$$g_N = Q_N(af) \in \mathcal{T}_N, \qquad T_N = Q_N aK \in \mathrm{L}(H^\lambda, H^\lambda),$$

the collocation equation(18)can be rewritten as $v_N + T_N v_N = g_N$. Take a $M \in \mathcal{N}$ of order $M \sim N^\Theta$, $0 < \Theta < 1$. The collocation equation is equivalent to $(I+T_M)^{-1}(I+T_N)v_N = (I+T_M)^{-1}g_N$, or

$$v_N = T_{M,N}v_N + g_{M,N}$$

with

$$T_{M,N} = (I+T_M)^{-1}(T_M - T_N), \qquad g_{M,N} = (I+T_M)^{-1} g_N .$$

Under conditions of Lemma 1 and Theorem 2 we have for $0 \le \lambda \le \mu - 2$, $\mu \ge 2$, the estimate (see (20) and (23))

$$\|T_{M,N}\|_{\mathrm{L}(H^\lambda, H^\lambda)} \le cM^{-2} \le c' N^{-2\Theta}.$$

Thus, the norm of the operator $T_{M,N}$ is small, and we may apply the iterations

$$v_N^k = T_{M,N} v_N^{k-1} + g_{M,N} \qquad (k = 1, 2, \ldots) \tag{3.25}$$

starting e.g. from $v_N^0 = v_M = (I+T_M)^{-1} g_M$. For the exact collocation solution v_N we have $v_N = T_{M,N} v_N + g_{M,N}$ and

$$\begin{aligned} v_N^k - v_N &= T_{M,N}(v_N^{k-1} - v_N) = \dot{=} T_{M,N}^k (v_N^0 - v_N), \\ \|v_N^k - v_N\|_\lambda &\le \|T_{M,N}\|_{\mathrm{L}(H^\lambda, H^\lambda)}^k (\|v_N - v\|_\lambda + \|v_M - v\|_\lambda) \\ &\le c' c^k N^{-2\Theta k + \Theta(\lambda - \mu)} \|v\|_\mu \le \varepsilon N^{\lambda - \mu} \|v\|_\mu \qquad (0 \le \lambda \le \mu) \end{aligned}$$

with a small $\varepsilon > 0$ (cf. (22)) provided that $(2k + \mu - \lambda)\Theta > \mu - \lambda$. This condition is most strong for $\lambda = 0$:

$$k > \frac{1-\Theta}{2\Theta}\mu \ \text{ for fixed } \ \Theta \in (0,1),$$

or equivalently

$$\Theta > \frac{\mu}{\mu + 2k} \quad \text{for fixed } \ k \in .$$

So only few iterations (26) are needed to achieve the accuracy (22) by v_N^k, and this number of iterations may be taken to be independent of N. On the other hand, if we put $\Theta > \frac{\mu}{\mu+2}$, then only one iteration (26) is sufficient, i.e. asymptotically already v_N^1 achieves the accuracy(22).

To present the matrix form of the two-grid iterations (26), notice that

$$(I+T_M)^{-1} = I - (I+T_M)^{-1} T_M .$$

Thus (26) can be written in the form where $(I - T_M)^{-1}$ is applied only to functions from $_M$:

$$v_N^k = [I - (I+T_M)^{-1} T_M][(T_M - T_N) v_N^{k-1} + g_N] .$$

With respect to the Fourier coefficients of v_N^k, the matrix form of the two-grid iterations (26) is as follows:

$$\begin{aligned} \hat{v}_N^k &= [I_N - \hat{P}_{N,M} \hat{A}_M^{-1} \mathcal{F}_M \breve{a}_M R_{M,N} \mathcal{F}_N^{-1} \hat{K}_N][\hat{g}_N \\ &\qquad + (\hat{P}_{N,M} \mathcal{F}_M \breve{a}_M R_{M,N} -_N \breve{a}_N)_N^{-1} \hat{K}_N \hat{v}_N^{k-1}] \end{aligned} \tag{3.26}$$

The designations $\hat{A}_N$, ${}_N$, ${}_N^{-1}$, $\hat{K}_N$, $\breve{a}_N$, $\hat{g}_N$ have been explained in Section 3.4; $R_{M,N}\breve{w}_N$ restricts $\breve{w}_N$ from the net $hZ_N^n = \{hj : j \in \mathbb{Z}_N^n\}$, $h = \frac{2R}{N}$, to the subnet $h'\mathbb{Z}_M^n$, $h' = \frac{2R}{M}$ (we assume that $h'/h = N/M$ is an integer); the prolongation operator $\hat{P}_{N,M}$ is defined by

$$(\hat{P}_{N,M}\hat{w}_M)(j) = \begin{cases} \hat{w}_M(j), & j \in \mathbb{Z}_M^n, \\ \quad 0 \quad, & j \in \mathbb{Z}_N^n \backslash \mathbb{Z}_M^n . \end{cases}$$

Of course, a vector $u_M = \hat{A}_M^{-1} w_M$ is computed solving the M^n-system $\hat{A}_M u_M = w_M$. This can be done e.g. by the conjugate gradient method in $\emptyset(M^n \log^2 M) = \emptyset(N^{\Theta n} \log^2 N)$ arithmetical operations as explained in Section 3.6. For $0 < \Theta \leq 1/3$, also a direct solution of the M^n-system e.g. by the Gauss elimination holds the amount of work in $\emptyset(N^n)$ arithmetical operations.

Most costful operations in (3.26) are ${}_N$ and ${}_N^{-1}$. During one iterations, they occur three times, plus once to compute $\hat{g}_N$. Asymptotically most cheap version of (3.26) is obtained putting $\Theta > \mu/(\mu+2)$. As explained, then only one iteration(26) is sufficient to achieve the accuracy(22); respectively, only once we have to solve M^n-system in iteration(3.26), and once it should be done to compute the initial guess $\hat{v}_N^0 = \hat{v}_M = \hat{A}_M^{-1}\hat{g}_M$. The whole amount of the computational work is $\emptyset(N^n \log N)$ arithmetical operations, and it is caused by 4 operations with ${}_N$ and ${}_N^{-1}$; all other operations cost $\emptyset(N^n)$ or less.

Recall that this analysis is based on (20) for $0 \leq \lambda \leq \mu - 2$, $\mu \geq 2$. It is easy to complete the analysis considering other cases in (20) and (21).

3.8 APPENDIX: FOURIER COEFFICIENTS OF $K(X)$

Clearly, $(\Delta + 1)\varphi_j = (1 - \pi^2|j|^2/R^2)\varphi_j$. For $\pi|j| \neq R$, denoting $\lambda_j = R^2/(\pi^2|j|^2 - R^2)$, with help of the Green formula we obtain

$$\begin{aligned}
\hat{K}(j) &= \int_{G_R} K(x)\varphi_{-j}(x)dx = -\lambda_j \int_{G_R} K(x)(\Delta+1)\varphi_{-j}(x)dx \\
&= -\lambda_j \lim_{\delta\to 0} \int_{B(0,R)\backslash B(0,\delta)} K(x)(\Delta+1)\varphi_{-1}(x)dx \\
&= -\lambda \lim_{\delta\to 0} \left\{ \left(\int_{S(0,R)} - \int_{S(0,\delta)} \right) \left(K\frac{\partial \varphi_{-j}}{\partial r} - \frac{\partial K}{\partial r}\varphi_{-j} \right) dS \right.
\end{aligned}$$

$$+ \int\limits_{B(0.R)\setminus B(0,\delta)} ((\Delta+1)K(x))\varphi_{-j}(x)dx \Bigg\}$$

where $\frac{\partial}{\partial r} = \sum_{k=1}^{n} \frac{x_k}{|x|}\frac{\partial}{\partial x_k}$. According to the construction (see (10) and (11)) $K(x) = \Phi(|x|)$ on the sphere ($n = 3$) or circle ($n = 2$) $S(0,\delta)$. Taking into account the asymptotics of $\Phi(r)$ and $\Phi'(r)$ as $r \to 0$, we obtain

$$\hat{K}(j) = \lambda_j \Bigg\{ \varphi_{-j}(0) - \int\limits_{S(0,R)} \left(K\frac{\partial\varphi_{-j}}{\partial r} - \frac{\partial K}{\partial r}\varphi_{-j} \right) dS$$

$$- \lim_{\delta\to 0} \int\limits_{B(0,R)\setminus B(0,\delta)} ((\Delta+1)K(x))\varphi_{-j}(x)dx \Bigg\} \qquad (3.27)$$

Cutting (10) , $n=3$. According to (10) , since $(\Delta+1)K(x) = (\Delta+1)\Phi(|x|) = 0$ for $0 \neq x \in B(0,R)$, (3.27) reduces to

$$\hat{K}(j) = \lambda \Bigg\{ \varphi_{-j}(0) - \Phi(R) \int\limits_{S(0,R)} \frac{\partial\varphi_{-j}}{\partial r} dS + \Phi'(R) \int\limits_{S(0,R)} \varphi_{-j} dS \Bigg\}.$$

We use the symmetry argument to evaluate

$$\int\limits_{S(0,R)} e^{i\pi j\cdot x/R} dS = R^2 \int\limits_{S(0,1)} e^{i\pi j\cdot x} dS = R^2 \int\limits_{S(0,1)} e^{i\pi|j|x_1} dS$$

$$= R^2 \int_{-1}^{1} e^{i\pi|j|x_1} 2\pi dx_1 = \frac{4R^2}{|j|}\sin(\pi|j|), \qquad j \neq 0.$$

Similarly

$$\int\limits_{S(0,R)} \frac{\partial}{\partial r} e^{i\pi j\cdot x/R} dS = \frac{1}{R} \int\limits_{S(0,R)} \left(i\pi j \cdot \frac{x}{R} \right) e^{i\pi j\cdot x/R} dS$$

$$= R \int\limits_{S(0,1)} (i\pi \cdot x) e^{i\pi j\cdot x} dS = R \int\limits_{S(0,1)} i\pi|j|x_1 e^{i\pi|j|x_1} dS$$

$$= Ri\pi|j| \int_{-1}^{1} x_1 e^{i\pi|j|x_1} 2\pi dx_1 = 4\pi R\cos(\pi|j|) - \frac{4R}{|j|}\sin(\pi|j|), \; j \neq 0.$$

Recalling that $\varphi_{-j}(x) = (2R)^{-3/2}\exp\left(-i\pi j\cdot \frac{x}{R}\right)$ for $n = 3$, this results to

$$\hat{K}(j) = \frac{R^2}{\pi^2|j|^2 - R^2}(2R)^{-3/2}\left[1 - e^{iR}\left(\cos(\pi|j|) - i\frac{R}{\pi|j|}\sin(\pi|j|)\right)\right], \ j \neq 0$$

$$\hat{K}(0) = -(2R)^{-3/2}[1 - e^{iR}(1 - Ri)]. \ .$$

For $\pi|j| = R$ we obtain

$$\hat{K}(j) = -i2^{-5/2}R^{-1/2}(1 - e^{iR}R^{-1}\sin R)$$

by the L'Hospital rule or directly noticing that

$$\Phi(|x|) = \frac{1}{8\pi i}(\Delta + 1)e^{i|x|} \quad \text{for} \quad n = 3.$$

Clearly $|\hat{K}(j)| \leq c|j|^{-2}$ $(j \neq 0)$ in the case of cutting (10), n = 3.

Cutting (10), n = 2 . Now

$$\int\limits_{S(0,R)} e^{i\pi j\cdot x/R}ds = R\int\limits_{S(0,1)} e^{i\pi j\cdot x}ds = R\int\limits_{S(0,1)} e^{i\pi|j|x_1}ds$$

$$= 2R\int_{-1}^{1} e^{i\pi|j|x_1}(1 - x_1^2)^{-1/2}ds_1 = 4R\int_0^1 \cos(\pi|j|x_1)(1 - x_1^2)^{-1/2}dx_1$$

$$= 2\pi R J_0(\pi|j|)$$

and similarly

$$\int\limits_{S(0,R)} \frac{\partial}{\partial r}e^{i\pi j\cdot x/R}ds = -2\pi^2|j|J_1(\pi|j|)$$

(see [1], formulae 9.1.20 and 9.1.28 for the Bessel functions J_0 and J_1). This results to

$$\hat{K}(j) = \frac{R^2}{\pi^2|j|^2 - R^2}(2R)^{-1}\left\{1 + \frac{1}{2}i\pi\left[\pi|j|J_1(\pi|j|)H_0^1(R) - RJ_0(\pi|j|)H_1^1(R)\right]\right.$$

$$\text{for} \quad \pi|j| \neq R, \qquad j \neq 0,$$

$$\hat{K}(0) = -(2R)^{-1} - \pi\Phi'(R) = -(2R)^{-1} + \frac{\pi i}{4}H_1^{(1)}(R),$$

$$\hat{K}(j) = \frac{1}{8}\pi Ri\left[J_0(R)H_0^{(1)}(R) + J_1(R)H_1^{(1)}(R)\right] \quad \text{for} \quad \pi|j| = R.$$

Since $J_\nu(r) \sim \sqrt{2/(\pi r)}\cos(r - \frac{1}{2}\nu\pi - \frac{1}{4}\pi)$ as $r \to \infty$ (see [1], formula 9.2.1), we have $|\hat{K}(j)| \leq c|j|^{-3/2}$ $(j \neq 0)$ in the case of cutting (10), $n = 2$.

Cutting (11), $n = 2$ *or* $n = 3$. Using the soft cutting (11) we obtain from (3.27)

$$\hat{K}(j) = \lambda_j \left\{ (2R)^{-n/2} - \int\limits_{G_R} (2\nabla\Phi\cdot\nabla\psi + \Phi\Delta\psi)\varphi_{-j}dx \right\} = \lambda_j\{(2R)^{-n/2} - \hat{\chi}(j)\}$$

where the function

$$\chi(x) := 2\nabla\Phi(|x|)\cdot\nabla\psi(|x|) + \Phi(|x|)\Delta\psi(|x|) = 2\Phi'(r)\psi'(r) + \Phi(r)[\psi''(r) + 2r^{-1}\psi'(r)]$$

is C^∞-smooth and supported on the annulus $2\rho \leq r \leq R$ (see (12). Therefore $|\hat{\chi}(j)| \leq c_p|j|^{-p}$ $(0 \neq j \in \mathbb{Z}^n)$ with any $p > 0$, and $|\hat{K}(j)| \leq c|\lambda_j| \leq c'|j|^{-2}$ $(j \neq 0)$. Approximating χ by $Q_M\chi$, $M \sim N^\Theta, 0 < \Theta \leq 1$, we have

$$\max_{x \in G_R} |\chi(x) - (Q_M\chi)(x)| \leq c\|\chi - Q_M\chi\|_2 \leq c_p M^{-p}\|\chi\|_{p+2}$$

with any $p > 0$. The computation of $(\widehat{Q_M\chi})(j)$, $j \in \mathbb{Z}^n_M$, ¿from grid values of χ by FFT costs $\emptyset(M^n \log M)$ arithmetical operations. Thus, we have a cheap way to compute the approximations

$$\hat{K}_N(j) = \frac{R^2}{\pi^2|j|^2 - R^2}\left\{(2R)^{-n/2} - (\widehat{Q_M\chi})(j)\right\} \text{ for } j \in \mathbb{Z}^n_M\,,$$

$$\hat{K}_N(j) = \frac{R^2}{\pi^2|j|^2 - R^2}(2R)^{-n/2} \text{ for } j \in \mathbb{Z}^n_N \backslash \mathbb{Z}^n_M$$

to $\hat{K}(j)$ of a high accuracy:

$$\max_{j \in \mathbb{Z}^n_N} |\hat{K}_N(j) - \hat{K}(j) \leq c_p N^{-\Theta p}\|\chi\|_{p+2} \text{ with any } p > 0\,. \tag{3.28}$$

There is also a possibility for an exact expression of $\hat{\chi}(j)$ through integrals on $(2\rho, R)$ from some smooth functions containing the multipliers $\psi'(r)$ and $\psi''(r)$.

Let us briefly discuss also the construction of cutting function ψ satisfying (12). Take a function $\varphi \in C^\infty[0,\infty)$ such that $\varphi^{(k)}(0) = 0$ for all $k = 0, 1, 2, , e.g. \varphi(s) = e^{-\delta/s}$, $\delta > 0$. Then

$$\psi(r) := \begin{cases} 1 & , \quad 0 \leq r \leq 2\rho \\ c_0^{-1}\int_r^R \varphi(s - 2\rho)\varphi(R - s)ds, & 2\rho \leq r \leq R \\ 0 & , \quad r \geq R \end{cases}$$

with

$$c_0 = \int_{2\rho}^R \varphi(s - 2\rho)\varphi(R - s)ds$$

satisfies (12). A fortune is that we need only derivatives $\psi'(r)$ and $\psi''(r)$, not $\psi(r)$ itself, and the derivatives of ψ are available from the formula. The only integral defining c_0 can be approximated with a high accuracy by the trapezoidal rule since the integrand and all its derivatives vanish at the end points 2ρ and R. If we put $\varphi(s) = s^m$ we have $c_0 = \frac{(m!)^2}{(2m)!}(R - 2\rho)^{2m-1}$ and $\psi \in C^m[0, \infty)$. With m sufficiently large, also this cutting function ψ is suitable for our purposes.

References

[1] Abramowitz, M. and I. A. Stegun. (1965). *Handbook of Mathematical Functions*, 4th Printing, United States Department of Commerce.

[2] Colton, D. and R. Kress. (1992). *Inverse Acoustic and Electromagnetic Scattering Theory*, Springer.

[3] Golub, G. H. and C. F. van Loan. (1989). *Matrix Computations*, John Hopkins Univ. Press, Baltimore, London.

[4] Kelle, O. and G. Vainikko. (1995). *A fully discrete Galerkin method for integral and pseudodifferential equations on closed curves*, J. for Anal. and its Appl., Vol. 14(3), (pages 593–622).

[5] Kirsch, A. and P. Monk. (1994). *An analysis of coupling of finite element and Nyström methods in acoustic scattering*, IMA J. Numer. Anal., Vol. 14, (pages 523–544).

[6] Saranen, J. and G. Vainikko. (1996). *Trigonometric collocation methods with product integration for boundary integral equations on closed curves*, SIAM J. Numer. Anal., Vol. 33(4), (pages 1577–1596).

[7] Vainikko, G. (1993). *Multidimensional Weakly Singular Integral Equations*, Lecture Notes in Math., Vol. 1549, Springer.

[8] Vainikko, G. (1996). *Periodic Integral and Pseudodifferential Equations*, Helsinki University of Technology, Report C13.

INVERSE SOURCE PROBLEM FOR THE STOKES SYSTEM

Oleg Yu Imanuvilov

Korean Institute for Advanced Study
207-43 Cheongryangri-dong, Dongdaemun-gu,
Seoul 130-012, Korea
e-mail : olegkias.kaist.ac.kr

Masahiro Yamamoto

Department of Mathematical Sciences
The University of Tokyo
3-8-1 Komaba, Meguro, Tokyo 153 Japan
e-mail : myamams.u-tokyo.ac.jp

R.P. Gilbert et al.(eds.), Direct and Inverse Problems of Mathematical Physics, 441-451.

Abstract: We consider a Stokes system with external force term:

$$\frac{\partial y}{\partial t} = \Delta y - \nabla p + r(t)f(x), \quad \nabla \cdot y = 0 \quad \text{in } (0,T) \times \Omega$$

and

$$y_{|(0,T)\times\partial\Omega} = 0,$$

where $\Omega \subset \mathbb{R}^n$, $n = 2,3$ is a bounded domain. We discuss an inverse source problem of determining $f = (f_1, ..., f_n)$ from $y_{|\omega\times(0,T)}$, $p_{|\omega\times(0,T)}$, $y(\theta,\cdot)$ and $p(\theta,\cdot)$, provided that $\omega \subset \Omega$ is an arbitrary domain, $\theta > 0$ and real-valued r are given. Our main result is the Lipschitz stability in the inverse problem and the proof is based on a Carleman estimate for the Stokes system.

1.1 INTRODUCTION

Let us consider a Stokes system:

$$\frac{\partial y}{\partial t}(t,x) = \Delta y(t,x) - \nabla p(t,x) + r(t)f(x) \quad \text{in}(0,T) \times \Omega$$

$$\nabla \cdots y = 0 \quad \text{in } (0,T) \times \Omega$$

$$y_{|(0,T)\times\partial\Omega} = 0,$$

where $\Omega \subset \mathbb{R}^n$, $n = 2,3$ is a bounded domain whose boundary $\partial\Omega$ is of the class C^2. Here $y(t,x) = (y_1(t,x), ..., y_n(t,x))$ denotes a velocity field of the imcompressible fluid flow, p a pressure, and a real-valued function $r = r(t)$ and a vector-valued $f(x) = (f_1(x), ..., f_n(x))$ are a time varying factor and a spatial density of a distributed external force causing the flow.

The purpose of this paper is to discuss

1.1.1 Inverse Source Problem

Let $r = r(t)$ be known, and let $\omega \subset \Omega$ be an arbitrary subdomain, and $T > 0$ and $\theta \geq 0$ be given arbitrarily. Then determine $f = f(x)$, $x \in \Omega$ from

$$y(t,x), \quad p(t,x), \quad 0 < t < T, x \in \omega$$

and

$$y(\theta,x), \quad p(\theta,x), \quad x \in \Omega.$$

The data (1.5) are initial values of y and p in the case of $\theta = 0$. This kind of problem is meaningful in the fluid dynamics, and the inverse source problems for hydrodynamics are considered in Kamynin and Vasin [14], Prilepko, Orlovskii and Vasin [18], Prilepko and Vasin [21], Vasin [22], [23] where r is mainly assumed to depend on both x and t, and different data from ours are used for the determination. As for inverse parabolic problems with final overdetermination, we can further refer to Choulli and Yamamoto [3], [4], Isakov [12], Prilepko and Kostin [17], Prilepko and Solov'ev [19], Prilepko and Tikhonov [20]. For determining f in the form of $r(t)f(x)$ in hyperbolic equations, we can refer to Yamamoto [24], [25].

Here we adopt overdetermining data (1.4) and (1.5), and discuss the stability for the inverse source problem for the Stokes system. The key is a Carleman estimate for the Stokes system.

The Carleman estimate has been found useful also for the inverse problem as well as for the uniqueness in Cauchy problems (e.g. Bukhgeim and Klibanov [1]), and for the uniqueness in multidimensional inverse problems by the Carleman estimate, we can refer to Isakov [11], [13], Klibanov [15], Kubo [16]. However, for inverse problems for equations of parabolic type, it is not easy to derive Lipschitz stability over the whole domain by means of the Carleman estimates used in the above-mentioned papers. On the other hand, for equations of parabolic type including the Stokes system, the first named author and A.V. Fursikov have proved a new Carleman estimate with a weight function which is globally given over the whole domain $(0,T) \times \Omega$ (e.g. Chae, Imanuvilov and Kim [2], Fursikov and Imanuvilov [5], [6], [7], Imanuvilov [8], [9], [10]).

Our main result is the Lipschitz stabilty in our inverse source problem for the Stokes system. The rest of this paper is composed of three sections: In Section 2, we present our main result and in Section 3, we show the Carleman estimate. The final Section 4 is devoted to the proof of the main result.

1.1.2 Main Result

Let us set

$$V = \{v = v(t,x) \in \{H^1(\Omega)\}^n | \nabla \cdot v = 0, \quad v_{|\partial\Omega} = 0\}$$

and

$$C^{1,2}((0,T) \times \overline{\Omega}) = \{g = g(t,x) | g, \frac{\partial g}{\partial t}, \frac{\partial g}{\partial x_i}, \tag{1.1}$$

$$\frac{\partial^2 g}{\partial x_i \partial x_j} \in C((0,T) \times \overline{\Omega}), \quad 1 \le i,j \le n\}. \tag{1.2}$$

Furthermore $H^2(\Omega)$, $H^1(\Omega)$, $L^2(\Omega)$ denote usual Sobolev spaces and the Lebesgue space, and we set

$$\|g\|_{H^1(0,T;L^2(\omega))} = \left(\int_0^T \int_\omega |g(t,x)|^2 + \left| \frac{\partial g}{\partial t}(t,x) \right|^2 dx dt \right)^{\frac{1}{2}}.$$

We can state our main result:

Theorem 1.1.1 *Let* $0 < \theta < T$ *and let us assume*

$$r \in C^2[0,T], \qquad r(\theta) \neq 0.$$

Moreover let y *and* p *satisfy (1.1) - (1.3) and* $y, \frac{\partial y}{\partial t} \in C^{1,2}((0,T) \times \overline{\Omega})$, $p \in C^1((0,T) \times \overline{\Omega})$. *Then there exists a constant* $C = C(\Omega, T, \omega, \theta, r) > 0$ *such that*

$$\|f\|_{L^2(\Omega)} \le C(\|y(\theta,\cdot)\|_{H^2(\Omega)} + \|\nabla p(\theta,\cdot)\|_{L^2(\Omega)} \tag{1.3}$$

$$+ \|p\|_{H^1(0,T;L^2(\omega))} + \|y\|_{H^1(0,T;L^2(\omega))}). \tag{1.4}$$

Here the constant $C > 0$ *is independent of* y *and* p.

1.1.3 Remark

For any small $\delta > 0$, we can replace (2.2) by

$$\|f\|_{L^2(\Omega)} \le \quad C(\|y(\theta,\cdot)\|_{H^2(\Omega)} + \|\nabla p(\theta,\cdot)\|_{L^2(\Omega)} \tag{1.5}$$

$$+ \quad \|p\|_{H^1(\theta-\delta,\theta+\delta;L^2(\omega))} + \|y\|_{H^1(\theta-\delta,\theta+\delta;L^2(\omega))}). \tag{1.6}$$

In other words, our theorem requires data only for a time interval including θ, not for the whole $(0,T)$.

Our theorem asserts: in order that our observation data $y_{|(0,T)\times\omega}$ and $p_{|(0,T)\times\omega}$ can estimate $\|f\|_{L^2(\Omega)}$, we have to extra specify $y(\theta,\cdot)$ and $p(\theta,\cdot)$. This formulation is actually overdetermining because $p(\theta,\cdot)$ is not necessary for the well-posedness in an initial/boundary value problem for the Stokes system (1.1) - (1.3). We can take "$\nabla\times$" of (1.1) and eliminate p for estimates without p. However this way requires more boundary conditions of y and we discuss it in a forthcoming paper. Moreover even for the uniqueness in a similar problem for a simple heat equation, it is open whether we can choose $\theta = 0$ (e.g. Isakov [11]).

1.1.4 Carleman Estimate for the Stokes System

First we show

Lemma 1.1.2 *Let ω_0 be an arbitrary subdomain such that $\overline{\omega_0} \subset \omega$. Then there exists a function $\psi \in C^2(\overline{\Omega})$ such that*

$$\psi > 0 \text{ in } \Omega, \qquad \psi_{|\partial\Omega} = 0, \qquad |\nabla\psi(x)| > 0, \quad x \in \overline{\Omega\setminus\omega_0}.$$

For the proof, see Chae, Imanuvilov and Kim [2], for example.

For $0 < \gamma_1 < \gamma_2 < T$ and a fixed large $\lambda > 0$, we set

$$\varphi(t,x) = \varphi_{\gamma_1,\gamma_2}(t,x) = \frac{e^{\lambda\psi(x)}}{(t-\gamma_1)^2(\gamma_2-t)^2}$$

$$\alpha(t,x) = \alpha_{\gamma_1,\gamma_2}(t,x) = \frac{e^{\lambda\psi(x)} - e^{2\lambda^2\|\psi\|_{C(\overline{\Omega})}}}{(t-\gamma_1)^2(\gamma_2-t)^2}.$$

We set

$$C^{1,2}([\gamma_1,\gamma_2]\times\overline{\Omega}) = \qquad \{g = g(t,x) | g, \tfrac{\partial g}{\partial t}, \tfrac{\partial g}{\partial x_i}, \tag{1.7}$$

$$\tfrac{\partial^2 g}{\partial x_i \partial x_j} \in C([\gamma_1,\gamma_2]\times\overline{\Omega}), \quad 1 \le i,j \le n\}. \tag{1.8}$$

We consider a Stokes system:

$$\frac{\partial z}{\partial t} = \Delta z - \nabla q + F \quad \text{in } (\gamma_1,\gamma_2)\times\Omega$$

$$\nabla\cdot z = 0 \quad \text{in } (\gamma_1,\gamma_2)\times\Omega$$

$$z_{|(\gamma_1,\gamma_2)\times\partial\Omega} = 0.$$

Then the following Carleman estimate is proved (Imanuvilov [10]);

Lemma 1.1.3 *Let* $(z,q) \in C^{1,2}([\gamma_1,\gamma_2] \times \overline{\Omega}) \times C^1([\gamma_1,\gamma_2] \times \overline{\Omega})$ *satisfy (3.4) - (3.6) with* $F \in L^2(\gamma_1,\gamma_2;V)$*. Then there exists* $s_0 > 0$ *such that*

$$\int_{(\gamma_1,\gamma_2)\times\Omega} \left(\frac{1}{s\varphi} \left(\left|\frac{\partial z}{\partial t}\right|^2 + \sum_{i,j=1}^n \left|\frac{\partial^2 z}{\partial x_i \partial x_j}\right|^2 \right) + s\varphi|\nabla z|^2 + s^3\varphi^3|z|^2 \right) e^{2s\alpha} dx dt \quad (1.9)$$

$$\leq C \int_{(\gamma_1,\gamma_2)\times\omega} (s^2\varphi^2 q^2 + s^3\varphi^3|z|^2) e^{2s\alpha} dx dt \quad (1.10)$$

$$+ C \int_{(\gamma_1,\gamma_2)\times\Omega} s\varphi|F|^2 e^{2s\alpha} dx dt \quad (1.11)$$

for all $s > s_0$*. Here the constant* $C > 0$ *is independent of* s *and* (z,q).

1.1.5 Proof of the Main Result

By the assumption (2.1), we can take a sufficiently small $\delta > 0$ such that

$$r(t) \neq 0, \qquad \theta - \delta < t < \theta + \delta.$$

Henceforth for simplicity, we set

$$Q_\delta = (\theta - \delta, \theta + \delta) \times \Omega, \quad Q_\omega = (0,T) \times \omega.$$

Let us consider (1.1) in Q_δ. We set

$$v = \frac{y}{r} \qquad \text{in } Q_\delta.$$

Then

$$\frac{\partial v}{\partial t} = \Delta v - \frac{r_t}{r} v - \nabla \left(\frac{p}{r}\right) + f, \quad \nabla \cdot v = 0 \quad \text{in } Q_\delta$$

and $v_{|(\theta-\delta,\theta+\delta)\times\partial\Omega} = 0$. We set

$$z = \frac{\partial v}{\partial t}.$$

Then

$$\frac{\partial z}{\partial t} = \Delta z - \frac{r_t}{r} z - \frac{\partial}{\partial t}\left(\frac{r_t}{r}\right) v - \nabla \left(\frac{\partial}{\partial t}\left(\frac{p}{r}\right)\right) \quad \text{in } Q_\delta,$$

$$\nabla \cdot z = 0 \quad \text{in } Q_\delta$$

and

$$z_{|(\theta-\delta,\theta+\delta)\times\partial\Omega} = 0.$$

Therefore we can apply Lemma 2 to (4.4) - (4.6) in Q_δ, so that we can obtain

$$\int_{Q_\delta} \frac{1}{s\varphi} \left|\frac{\partial z}{\partial t}\right|^2 e^{2s\alpha} dx dt + \int_{Q_\delta} s^3\varphi^3 |z|^2 e^{2s\alpha} dx dt \quad (1.12)$$

$$\leq \quad C \int_{Q_\omega} \left(s^2\varphi^2 \left|\frac{\partial}{\partial t}\left(\frac{p}{r}\right)\right|^2 + s^3\varphi^3|z|^2\right) e^{2s\alpha} dx dt \quad (1.13)$$

$$+ \quad C \int_{Q_\delta} s\varphi \left|\frac{r_t}{r} z + \frac{\partial}{\partial t}\left(\frac{r_t}{r}\right) v\right|^2 e^{2s\alpha} dx dt \quad (1.14)$$

for all large $s > 0$. Here and henceforth $C > 0$ denotes a generic constant which is independent of $s > 0$, p, y, but dependent on T, θ, r, Ω, ω. Therefore

$$\int_{Q_\delta} \frac{1}{s\varphi} \left|\frac{\partial z}{\partial t}\right|^2 e^{2s\alpha} dxdt + \int_{Q_\delta} s^3\varphi^3 |z|^2 e^{2s\alpha} dxdt \tag{1.15}$$
$$\leq \; C \int_{Q_\omega} (s^2\varphi^2(|p_t|^2 + |p|^2) + s^3\varphi^3(|y_t|^2 + |y|^2))e^{2s\alpha} dxdt \tag{1.16}$$
$$+ \quad C \int_{Q_\delta} s\varphi(|z|^2 + |v|^2)e^{2s\alpha} dxdt. \tag{1.17}$$

On the other hand, by (4.2), we have

$$v(t,x) = \int_\theta^t z(\eta, x)d\eta + \frac{y(\theta, x)}{r(\theta)}, \quad x \in \Omega.$$

We can prove

Lemma 1.1.4 *Let $g \in L^2(Q_\delta)$. Then*

$$\int_{Q_\delta} \varphi(t,x)e^{2s\alpha(t,x)} \left|\int_\theta^t g(\eta, x)d\eta\right|^2 dtdx \tag{1.18}$$
$$\leq \quad \delta^2 \int_{Q_\delta} \varphi(t,x)|g(t,x)|^2 e^{2s\alpha(t,x)} dtdx \tag{1.19}$$

and

$$\int_{Q_\delta} \varphi(t,x)^2 e^{2s\alpha(t,x)} \left|\int_\theta^t g(\eta, x)d\eta\right|^2 dtdx \tag{1.20}$$
$$\leq \quad \delta^2 \int_{Q_\delta} \varphi(t,x)^2 |g(t,x)|^2 e^{2s\alpha(t,x)} dtdx \tag{1.21}$$

for sufficiently large $s > 0$.

Similar lemmata are proved in Isakov [11], p.153 and Klibanov [15]. For convenience, in Appendix we will give the proof of Lemma 3.

In terms of (4.8), we apply (4.9) to the last term of the right hand side of (4.7), and we see

$$\int_{Q_\delta} s\varphi|v|^2 e^{2s\alpha} dxdt \leq Cs\|y(\theta, \cdot)\|^2_{L^2(\Omega)} + C \int_{Q_\delta} s\varphi|z|^2 e^{2s\alpha} dxdt.$$

Hence for large $s > 0$, the left hand side of (4.7) can absorb the last term of the right hand side of (4.7), so that

$$\int_{Q_\delta} \frac{1}{s\varphi} \left|\frac{\partial z}{\partial t}\right|^2 e^{2s\alpha} dxdt + \int_{Q_\delta} s^3\varphi^3 |z|^2 e^{2s\alpha} dxdt \tag{1.22}$$
$$\leq \; C \int_{Q_\omega} (s^2\varphi^2(|p_t|^2 + |p|^2) + s^3\varphi^3(|y_t|^2 + |y|^2))e^{2s\alpha} dxdt \tag{1.23}$$
$$+ \quad C\|y(\theta, \cdot)\|^2_{L^2(\Omega)}. \tag{1.24}$$

Noting

$$\frac{\partial y}{\partial t}(\theta - \delta, \cdot)\frac{e^{s\alpha(\theta-\delta,\cdot)}}{\sqrt{\varphi(\theta - \delta, \cdot)}} = 0$$

by the definitions (3.2) and (3.3), we have

$$\frac{\partial y}{\partial t}(\theta,x)\frac{e^{s\alpha(\theta,x)}}{\sqrt{\varphi(\theta,x)}} = \int_{\theta-\delta}^{\theta} \frac{\partial}{\partial t}\left(\frac{\partial y}{\partial t}\left(\frac{e^{s\alpha}}{\sqrt{\varphi}}\right)\right)(t,x)dt \tag{1.25}$$

$$= \int_{\theta-\delta}^{\theta}\left(\frac{e^{s\alpha}}{\sqrt{\varphi}}\frac{\partial^2 y}{\partial t^2} + \frac{\partial}{\partial t}\left(\frac{e^{s\alpha}}{\sqrt{\varphi}}\right)\frac{\partial y}{\partial t}\right)(t,x)dt. \tag{1.26}$$

Therefore

$$\left(\frac{e^{2s\alpha}}{s\varphi}\left(\frac{\partial y}{\partial t}\right)^2\right)(\theta,x) \le C\int_{\theta-\delta}^{\theta}\left(\frac{e^{2s\alpha}}{s\varphi}\left|\frac{\partial^2 y}{\partial t^2}\right|^2 + \frac{1}{s}\left|\frac{\partial}{\partial t}\left(\frac{e^{s\alpha}}{\sqrt{\varphi}}\right)\right|^2\left|\frac{\partial y}{\partial t}\right|^2\right)dt \tag{1.27}$$

$$= C\int_{\theta-\delta}^{\theta}\left(\frac{1}{s\varphi}\left|\frac{\partial^2 y}{\partial t^2}\right|^2 + \frac{(\varphi^{-\frac{1}{2}}\alpha_t s - \frac{1}{2}\varphi^{-\frac{3}{2}}\varphi_t)^2}{s}\left|\frac{\partial y}{\partial t}\right|^2\right)e^{2s\alpha}dt. \tag{1.28}$$

Here we directly see $|\alpha_t| \le C\varphi^{\frac{3}{2}}$, $|\varphi_t| \le C\varphi^{\frac{3}{2}}$ in Q_δ, and so

$$\frac{(\varphi^{-\frac{1}{2}}\alpha_t s - \frac{1}{2}\varphi^{-\frac{3}{2}}\varphi_t)^2}{s} \le Cs\varphi^2$$

in Q_δ for large $s > 0$. Therefore, since $\frac{\partial y}{\partial t} = rz + \frac{\partial r}{\partial t}v$ and $\frac{\partial^2 y}{\partial t^2} = r\frac{\partial z}{\partial t} + 2\frac{\partial r}{\partial t}z + \frac{\partial^2 r}{\partial t^2}v$ in Q_δ by (4.2) and (4.3), we obtain

$$\left(\frac{e^{2s\alpha}}{s\varphi}\left(\frac{\partial y}{\partial t}\right)^2\right)(\theta,x) \tag{1.29}$$

$$\le C\int_{\theta-\delta}^{\theta}\left(\frac{1}{s\varphi}\left|\frac{\partial^2 y}{\partial t^2}\right|^2 + s\varphi^2\left|\frac{\partial y}{\partial t}\right|^2\right)e^{2s\alpha}dt \tag{1.30}$$

$$\le C\int_{\theta-\delta}^{\theta}\left(\frac{1}{s\varphi}\left(\left|\frac{\partial z}{\partial t}\right|^2 + |z|^2 + |v|^2\right) + s\varphi^2(|z|^2 + |v|^2)\right)e^{2s\alpha}dt. \tag{1.31}$$

Hence

$$\int_\Omega \frac{e^{2s\alpha(\theta,x)}}{s\varphi(\theta,x)}\left|\frac{\partial y}{\partial t}(\theta,x)\right|^2 dx \tag{1.32}$$

$$\le C\int_\Omega\int_{\theta-\delta}^{\theta}\left(\frac{1}{s\varphi}\left(\left|\frac{\partial z}{\partial t}\right|^2 + |z|^2 + |v|^2\right) + s\varphi^2(|z|^2 + |v|^2)\right)e^{2s\alpha}dtdx \tag{1.33}$$

$$\le C\int_{Q_\delta}\left(\frac{1}{s\varphi}\left|\frac{\partial z}{\partial t}\right|^2 + s\varphi^2|z|^2\right)e^{2s\alpha}dxdt \tag{1.34}$$

$$+ C\int_{Q_\delta} s\varphi^2|v|^2e^{2s\alpha}dtdx. \tag{1.35}$$

Similarly to (4.11), application of (4.10) and (4.8) yields

$$\int_{Q_\delta} s\varphi^2|v|^2e^{2s\alpha}dtdx \le C\int_{Q_\delta} s\varphi^2|z|^2e^{2s\alpha}dtdx + Cs\|y(\theta,\cdot)\|^2_{L^2(\Omega)}.$$

Now apply (4.12) and (4.15) in (4.14), and we obtain

$$\int_\Omega \frac{e^{2s\alpha(\theta,x)}}{s\varphi(\theta,x)}\left|\frac{\partial y}{\partial t}(\theta,x)\right|^2 dx \tag{1.36}$$

$$\le C\int_{Q_\omega}(s^2\varphi^2(|p_t|^2 + |p|^2) + s^3\varphi^3(|y_t|^2 + |y|^2))e^{2s\alpha}dtdx + Cs\|y(\theta,\cdot)\|^2_{H^1(\Omega)} \tag{1.37}$$

Since

$$\frac{e^{2s\alpha(\theta,x)}}{s\varphi(\theta,x)} \geq Cs^{-1}\delta^4 e^{-\frac{C}{\delta^4}s} \equiv \delta_0(s) > 0, \quad x \in \overline{\Omega},$$

we see

$$\begin{aligned} & \left\| \frac{\partial y}{\partial t}(\theta,\cdot) \right\|^2_{L^2(\Omega)} && (1.38)\\ \leq & \ C\delta_0(s)^{-1} \int_{Q_\omega} \left(s^2\varphi^2(|p_t|^2+|p|^2) + s^3\varphi^3(|y_t|^2+|y|^2)\right) e^{2s\alpha} dtdx && (1.39)\\ + & \ Cs\delta_0(s)^{-1}\|y(\theta,\cdot)\|^2_{H^1(\Omega)}. && (1.40) \end{aligned}$$

By (1.1) and (2.1) we have

$$f = \frac{1}{r(\theta)}\left(\frac{\partial y}{\partial t}(\theta,\cdot) - \Delta y(\theta,\cdot) + \nabla p(\theta,\cdot)\right),$$

so that

$$\|f\|_{L^2(\Omega)} \leq C\left(\left\|\frac{\partial y}{\partial t}(\theta,\cdot)\right\|_{L^2(\Omega)} + \|\Delta y(\theta,\cdot)\|_{L^2(\Omega)} + \|\nabla p(\theta,\cdot)\|_{L^2(\Omega)}\right) \quad (1.41)$$

By noting that $|s^2\varphi^2 e^{2s\alpha}|$, $|s^3\varphi^3 e^{2s\alpha}| \leq C(s)$ in Q_δ with a constant dependent on $s > 0$, application of (4.16) completes the proof of Theorem.

1.1.6 Appendix. Proof of Lemma 3

Let $\mu \in C^1(\overline{Q_\delta})$ satisfy

$$\mu(t,x) > 0, \quad (t-\theta)\frac{\partial \mu}{\partial t}(t,x) \leq 0, \qquad \theta-\delta \leq t \leq \theta+\delta,\ x \in \overline{\Omega}.$$

We will show

$$\int_{Q_\delta} \mu(t,x)\left|\int_\theta^t g(\eta,x)d\eta\right|^2 dtdx \leq \delta^2 \int_{Q_\delta} \mu(t,x)|g(t,x)|^2 dtdx.$$

First, applying Schwarz's inequality and changing orders of integration, we have

$$\begin{aligned} & \int_{\theta-\delta}^\theta \mu(t,x)\left|\int_\theta^t g(\eta,x)d\eta\right|^2 dt \leq \int_{\theta-\delta}^\theta \mu(t,x)(\theta-t)\left(\int_t^\theta |g(\eta,x)|^2 d\eta\right)dt && (1.42)\\ \leq & \ \delta\int_{\theta-\delta}^\theta \left(\int_{\theta-\delta}^\eta \mu(t,x)dt\right)|g(\eta,x)|^2 d\eta && (1.43)\\ \leq & \ \delta^2\int_{\theta-\delta}^\theta \mu(\eta,x)|g(\eta,x)|^2 d\eta. && (1.44) \end{aligned}$$

At the last inequality, we use (1). Similarly we have

$$\int_\theta^{\theta+\delta} \mu(t,x)\left|\int_\theta^t g(\eta,x)d\eta\right|^2 dt \leq \delta^2 \int_\theta^{\theta+\delta} \mu(\eta,x)|g(\eta,x)|^2 d\eta.$$

Therefore (2) follows.

Thus for the completion of the proof of Lemma 3, it is sufficient to verify that $\mu = \varphi e^{2s\alpha}$ and $\mu = \varphi^2 e^{2s\alpha}$ satisfy (1). Direct calculations show $\frac{\partial}{\partial t}(\varphi e^{2s\alpha}) = e^{2s\alpha}(\varphi_t + 2\varphi s\alpha_t) = e^{2s\alpha}\varphi_t(1 - 2\varphi sl)$, where we set

$$l(x) = \frac{e^{2\lambda^2\|\psi\|_{C(\overline{\Omega})}} - e^{\lambda\psi(x)}}{e^{\lambda\psi(x)}}, \quad x \in \overline{\Omega}.$$

Noting that $\varphi > 0$ and $l > 0$ on $\overline{\Omega}$ for large $\lambda > 0$, we see that (1) is equivalent to $(t-\theta)\frac{\partial\varphi}{\partial t} \ge 0$ for large $s > 0$, which is directly seen. Hence we know that $\varphi e^{2s\alpha}$ satisfies (1). Finally, since we have already proved that $\varphi e^{s\alpha}$ is monotonously increasing in $t \in [\theta - \delta, \theta]$ and monotonously decreasing in $t \in [\theta, \theta + \delta]$ for large $s > 0$, the squared function $\varphi^2 e^{2s\alpha} = (\varphi e^{s\alpha})^2$ satisfies the same property. Thus the proof of Lemma 3 is complete.

1.1.7 Acknowledgements

The first named author is supported by KIAS under the grant number M97003. The second named author is partially supported by Sanwa Systems Development Co., Ltd. (Japan).

References

[1] Bukhgeim, A. L. and M. V. Klibanov. (1981). *Global uniqueness of a class of multidimensional inverse problems*, Soviet Math. Dokl., Vol. 24, (pages 244-247).

[2] Chae, D., O. Yu. Imanuvilov and S. M. Kim. (1996). *Exact controllability for semilinear parabolic equations with Neumann boundary conditions*, J. Dynamical and Control Systems, Vol. 2, (pages 449-483).

[3] Choulli, M. and M. Yamamoto. (1996). *Generic well-posedness of an inverse parabolic problem - the Hölder-space approach*, Inverse Problems, Vol. 12, (pages 195-205).

[4] Choulli, M. and M. Yamamoto. (1997). *An inverse parabolic problem with non-zero initial condition*, Inverse Problems, Vol. 13, (pages 19–27).

[5] Fursikov, A. V. and O. Yu. Imanuvilov. (1994). *On exact boundary zero-controllability of two-dimensional Navier-Stokes equations*, Acta Applicandae Mathematicae, Vol. 37, (pages 67–76).

[6] Fursikov, A. V. and O. Yu. Imanuvilov. (1996). *Exact controllability for 2-D Navier-Stokes system*, Preprint, RIM-GARC Preprint Series 96-35, Seoul National University.

[7] Fursikov, A. V. and O. Yu. Imanuvilov. (1996). *Controllability of Evolution Equations*, Lecture Notes Series No. 34, Seoul National University, Seoul, Korea.

[8] Imanuvilov, O. Yu. (1993). *Exact boundary controllability of the parabolic equation*, Russian Math. Surveys, Vol. 48, (pages 211–212).

[9] Imanuvilov, O. Yu. (1995). *Controllability of parabolic equations*, Sbornik Mathematics, Vol. 186, (pages 879–900).

[10] Imanuvilov, O. Yu. (Submitted). *On exact controllability for Navier-Stokes system.*

[11] Isakov, V. (1990). *Inverse Source Problems*, American Mathematical Society, Providence, Rhode Island.

[12] Isakov, V. (1991). *Inverse parabolic problems with the final overdetermination*, Commun. Pure Appl. Math., Vol. 54, (pages 185–209).

[13] Isakov, V. (1997). *Inverse Problems in Partial Differential Equations*, Springer-Verlag, Berlin.

[14] Kamynin, V. L. and I. A. Vasin. (1992). *Inverse problems for linearized Navier-Stokes equations with integral overdetermination. unique solvability and passage to the limit*, Ann. Univ. Ferrara Sez. VII-Sc. Mat., Vol. 38, (pages 229–247).

[15] Klibanov, M. V. (1992). *Inverse problems and Carleman estimates*, Inverse Problems, Vol. 8, (pages 575–596).

[16] Kubo, M. (1995). *Identification of the potential term of the wave equation*, Proceedings of the Japan Academy, Ser. A, Vol. 71, (pages 174–176).

[17] Prilepko, A. I. and A. B. Kostin. (1993). *On certain inverse problems for parabolic equations with final and integral observation*, Russian Acad. Sci. Sb. Math., Vol. 75, (pages 473–490).

[18] Prilepko, A. I., D. G. Orlovskii and I.A. Vasin. (1992). *Inverse problems in mathematical physics: "Ill-posed Problems in Natural Sciences*, VSP BV, AH Zeist, the Netherlands.

[19] Prilepko, A. I. and V. V. Solov'ev. (1987). *Solvability of the inverse boundary-value problem of finding a coefficient of a lower-order derivative in a parabolic equation*, Differential Equations, Vol. 23, (pages 101–107).

[20] Prilepko, A. I. and I. V. Tikhonov. *Recovery of the nonhomogeneous term in an abstract evolution equation*, Russian Acad. Sci. Izv. Math., Vol. 44, (pages 373–394).

[21] Prilepko, A. I. and I. A. Vasin. (1991). *On a non-linear non-stationary inverse problem of hydrodynamics*, Inverse Problems, Vol. 7, (pages L13–L16).

[22] Vasin, I. A. (1992). *Inverse boundary value problems in viscous fluid dynamics: "Ill-posed Problems in Natural Sciences"*, VSP BV, AH Zeist, the Netherlands, (pages 423–430).

[23] Vasin, I. A. (1995). *On a nonlinear inverse problem of simultaneous reconstruction of the evolution of two coefficients in Navier-Stokes equations*, Differential Equations, Vol 31, (pages 736–744).

[24] Yamamoto, M. (1994). *Well-posedness of some inverse hyperbolic problem by the Hilbert Uniqueness Method*, J. Inverse and Ill-posed Problems, Vol. 2, (pages 349–368).

[25] Yamamoto, M. (1995). *Stability, reconstruction formula and regularization for an inverse source hyperbolic problem by a control method*, Inverse Problems, Vol. 11, (pages 481–496).

International Society for Analysis, Applications and Computation

1. H. Florian et al. (eds.): *Generalized Analytic Functions*. Theory and Applications to Mechanics. 1998 ISBN 0-7923-5043-X
2. H.G.W. Begehr et al. (eds.): *Partial Differential and Integral Equations*. 1999 ISBN 0-7923-5482-6
3. S. Saitoh, D. Alpay, J.A. Ball and T. Ohsawa (eds.): *Reproducing Kernels and their Applications*. 1999 ISBN 0-7923-5618-7
4. R.P. Gilbert, J. Kajiwara and Y.S. Xu (eds.): *Recent Developments in Complex Analysis and Computer Algebra*. 1999 ISBN 0-7923-5999-2
5. R.P. Gilbert, J. Kajiwara and Y.S. Xu (eds.): *Direct and Inverse Problems of Mathematical Physics*. 2000 ISBN 0-7923-6005-2
6. H.G.W. Begehr, A. Okay Celebi and W. Tutschke (eds.): *Complex Methods for Partial Differential Equations*. 1999 ISBN 0-7923-6000-1

KLUWER ACADEMIC PUBLISHERS – DORDRECHT / BOSTON / LONDON